Protein Biosynthesis
in Eukaryotes

NATO ADVANCED STUDY INSTITUTES SERIES

A series of edited volumes comprising multifaceted studies of contemporary scientific issues by some of the best scientific minds in the world, assembled in cooperation with NATO Scientific Affairs Division.

Series A: Life Sciences

Recent Volumes in this Series

This series is published by an international board of publishers in conjunction with NATO Scientific Affairs Division

A Life Sciences	Plenum Publishing Corporation
B Physics	London and New York
C Mathematical and Physical Sciences	D. Reidel Publishing Company Dordrecht, Boston, and London
D Behavioral and Social Sciences	Sijthoff & Noordhoff International Publishers
E Applied Sciences	Alphen aan den Rijn, The Netherlands, and Germantown, U.S.A.

Protein Biosynthesis in Eukaryotes

Edited by

R. Pérez-Bercoff

The Institute of Virology
University of Rome
Rome, Italy

PLENUM PRESS • NEW YORK AND LONDON
Published in cooperation with NATO Scientific Affairs Division

Library of Congress Cataloging in Publication Data

NATO Advanced Study Institute on Protein Biosynthesis in Eukaryotes (1980: Maratea, Italy)
    Protein biosynthesis in eukaryotes.

    (NATO advanced study institute series. Series A, Life sciences; v. 41)
    "Proceedings of a NATO Advanced Study Institute on Protein Biosynthesis in Eukaryotes, held in Maratea, Italy, September 7 – 17, 1980" – T.p. verso.
    Bibliography: p.
    Includes index.
1. Protein biosynthesis – Congresses. 2. Eukaryotic cells – Congresses. I. Pérez-Bercoff, R. II. Title. III. Series.
QP551.N36   1980   574.19′296   81-22720
ISBN 978-1-4684-4126-0   ISBN 978-1-4684-4124-6   (eBook)   AACR2
DOI   10.1007/978-1-4684-4124-6

Proceedings of a NATO Advanced Study Institute on Protein Biosynthesis in Eukaryotes, held in Maratea, Italy, September 7-17, 1980.

© 1982 Plenum Press, New York
Softcover reprint of the hardcover 1st edition 1982

A Division of Plenum Publishing Corporation
233 Spring Street, New York, N.Y. 10013

PROTEIN BIOSYNTHESIS IN EUKARYOTES

(or "HOW DIVERGENT THEIR WAYS MAY BE OF *E. coli* AND ELEPHANTS...")

     Protein biosynthesis is the most complex biological process known
to occur at the cellular level: Even if we disregard the directly
related steps which *precede* the initiation of translation (e.g. transcription, polyadenylation, "capping", and splicing of messenger RNA,
among others), or those that usually take place *after* a protein has
been made (e.g. cleavage, modification, maturation), the synthesis
of the simplest peptide still constitutes the most formidable task a
cell is faced with... .

The word *protein*, coined one and a half century ago from the
προτειοσ ("proteios" = of primary importance), underlines the "primary
importance" ascribed to proteins from the time they were described as
biochemical entities. But the unmatched complexity of the process
involved in their biosynthesis was (understandably) overlooked.
Indeed, protein biosynthesis was supposed to be nothing more than the
reverse of protein degradation, and the same enzymes known to split
a protein into its constituent amino acids were thought to be able,
under adequate conditions, to reconstitute the peptide bond. This
oversimplified view persisted for more than 50 years: It was just
in 1940 that Borsook and Dubnoff examined the thermodynamical aspects
of the process, and concluded that protein synthesis could *not* be the
reverse of protein degradation, such an "uphill task being thermody-
namically impossible... ."

The next quarter of a century witnessed the unravelling of the
basic mechanisms of protein biosynthesis, a predictable aftermath of
the Copernican revolution in biology which followed such dramatic developments as the discovery of the nature of the genetic material,
the double helical structure of DNA, and the determination of the genetic code. Our present understanding of the sophisticated mechanisms of regulation and control is a relatively novel acquisition, and
recent studies have shed some light into the structure and organization of the eukaryotic gene.

First unravelled in bacterial systems, then in eukaryotes, the
basic mechanisms of protein biosynthesis proved to be surprisingly
similar in *E. coli* and *Proboscida elephantidae*[†], a circumstance
that led to the enunciation, by the mid-sixties, of a unifying theory on bacteria and elephants: *"...what holds true for E. coli..."*
read the versicle of the Holy Scripture of those days. Later observations, however, established that elephants find it difficult to
march over the footsteps of bacteria, and apparently one day, several
million years ago, they decided to make proteins their own way...
So great a success crowned their search for originality, that last
September we could devote a N.A.T.O. Advanced Study Institute to review and discuss our present understanding (and misunderstandings...)
of the mechanism of protein biosynthesis in eukaryotes ... . This
book is, indeed, the natural "fall out" of this N.A.T.O. Conference,
and features the edited version of the course.

For a course it was, and held at the highest possible level
thanks to the unlimited collaboration of our invited speakers: Their
response to my original request to contribute "a fresh, original and
interpretative *review* lecture" on topics I knew they had unchallenged
command of, was equalled only by the interest of all participants.
Needless to add, therefore, that all the merits of this book should be

---

[†]commonly named "elephants"...

accounted to their credit, while I take the blame for the imperfections that might be found in it.

Thanks are due to the Italian National Research Council (CNR), for its financial contribution, partly to defray the costs of preparing the edited manuscript for publication.

Finally, it is my pleasant duty to express, on behalf of all those who had the privilege of taking part at this N.A.T.O. Advanced Study Institute, my gratitude to the local authorities of Maratea, the Mayor Professor Sisini, and the Chairman of the Tourist Office, Mr. B. Vitolo, who so decisively contributed to make of our time in their blessed seaside resort an unforgettable experience.

R. Pérez-Bercoff
Rome, November 1980

CONTENTS

SECTION V: INHIBITION OF PROTEIN SYNTHESIS AT SELECTED LEVELS

SECTION I:

THE PROTEIN SYNTHESIZING MACHINERY OF EUKARYOTES

STRUCTURE AND FUNCTION OF tRNA AND

AMINOACYL tRNA SYNTHETASES IN EUKARYOTES

James Ofengand

Roche Institute of Molecular Biology

Nutley, N. J. 07110 (U.S.A.)

INTRODUCTION

In both prokaryotes and eukaryotes, transfer RNA (tRNA) occupies a central position in protein synthesis because it serves as the translator of the genetic message. It does this by acting as an adapter to convert the small structural differences among amino acids into a clearly distinct form capable of reading the specifying sequence of ribonucleotides in the mRNA. The need for such an adapter was recognized long ago (1; reviewed in 2), and it is instructive to note that even at that time the basic elements of tRNA and the aminoacylation process were recognized. Thus, tRNAs were postulated to be a set of specific oligonucleotides, 3-6 residues in length, which would be enzymatically attached to their amino acids, by a corresponding set of enzymes capable of selecting out both the given amino acid and its cognate oligonucleotide adapter from all others. The triplet codon would then be read by base-pairing with three of the adapter nucleotides (the anticodon). Crick (1) further suggested that the nature of the link between the amino acid and adapter would likely provide the energy for amino acid polymerization. Except for a failure to appreciate the ribosome binding function of tRNA which required a considerably larger molecule, this hypothesis proposed even before the existence of tRNA was known has proven to be surprisingly accurate in its description of the role of tRNA and aminoacyl-tRNA synthetases in protein synthesis.

The ribosome binding property of tRNA is a different kind of adapter function which provides a mechanism for fixing both the amino acid being added to the polypeptide chain and the growing peptide chain itself to the ribosomal surface where the enzymatic activity

for forming the peptide bond is located. It was probably easier in
an evolutionary sense to increase the size of the adapter tRNA to
include this common function than to design ribosomal binding sites
that could recognize and accommodate all varieties of small oligonucleotide adapters while at the same time keeping the anticodon
free for mRNA recognition.

As noted above, the process of translation proceeds in two distinct stages, both involving tRNA. The first is aminoacylation, the
process by which an amino acid is attached to the specific tRNA in-
tended for that amino acid. The second stage is that of codon dependent binding of aminoacyl tRNA (AA-tRNA)to the ribosomal A site, a
process by which tRNA anticodon-codon recognition is used to select
the proper AA-tRNA. Since only the tRNA portion of AA-tRNA is involved in the second stage, aminoacylation must be at least as specific as codon-anticodon pairing in order to insure the fidelity of
protein synthesis. The way this specific recognition between tRNA
and synthetases is accomplished will be discussed below as well as
the means by which the fidelity of codon reading is maintained and
modulated.

This review will be confined to a summary of the current state
of knowledge about tRNA and synthetases in eukaryotes. In general,
the properties of prokaryotic systems are similar, as might be expected given the similarity in their mechanism of protein synthesis.
In those cases where it seems particularly appropriate, I have compared eukaryotes and prokaryotes, but this has not been done rigorously throughout this chapter in view of the limitations on space.
More definitive and detailed discussion can be found in recent reviews on the general properties of tRNA and synthetases (2-4), and
in several recent comprehensive monographs (5-7). To conserve space,
references have been confined to recent work, and to reviews where
reference to the earlier literature can be found.

AMINOACYLATION

Stoichiometry and Energetics

The process of aminoacylation proceeds as shown in eq. $\underline{1}$ - $\underline{2}$[§]

$\underline{1}$ $E_i.tRNA_i + ATP + AA_i \rightleftharpoons AA_i\text{-}AMP.E_i.tRNA_i + PP$

[§] AA = amino acid; E = synthetase; AA-AMP = aminoacyl adenylate;
PP = pyrophosphate; EF1 = elongation factor 1; superscript i denotes a particular amino acid or a species specific for that amino
acid, ., denotes a non-covalent interaction.

2      $AA_i-AMP.E_i.tRNA_i \rightleftarrows AA_i-tRNA_i.E_i + AMP$

3      $AA_i-tRNA_i.E_i + EF1.GTP \rightleftarrows AA_i-tRNA_i.EF1.GTP + E_i$

In the first step, amino acids are activated by formation of an enzyme-bound aminoacyladenylate (Fig. 1). The amino acid is then transferred to tRNA with release of AMP. The stoichiometry of reactions 1 and 2 has been amply verified as has its reversibility (2). The AA-tRNA formed preserves the high group transfer potential of ATP, as shown by overall equilibrium constants for the sum of reactions 1 and 2, as written, of 0.32, 0.37, and 0.25 for valine, threonine, and arginine, respectively (2). This high group transfer potential finds expression in the facile reaction with nucleophiles such as the amino group of amino acids in peptide bond formation, as originally suggested by Crick. In practical terms, it also means that the covalent bond between AA and tRNA is labile, particularly in mild alkaline solution. This may cause experimental problems, especially if the tRNA is being traced by means of its attached radioactive amino acid. It is worth noting, therefore, that the lability varies widely (>30-fold) depending on the amino acid involved (8), so that a suitable choice of amino acid for labelling purposes can be experimentally quite important. N-acylation of the aminoacyl moiety also reduces its lability 5-20 fold (8,9). Further discussion of the stability of the amino acid-tRNA bond can be found in (2).

Structure of Aminoacyl-tRNA

The structure of AA-tRNA is shown in Fig. 2. The amino acid is attached via its carboxl group in an ester linkage to either the 2' or 3' hydroxyl group of the 3'-terminal adenosine of the tRNA. Although in solution the amino acid equilibrated rapidly (in milliseconds) between the two positions, it is now known that synthetases specifically select one hydroxyl or the other for acylation in a manner keyed to their amino acid specificity.

It became possible to determine this only when non-isomerizable but aminoacylatable analogs of tRNA became available. Three types

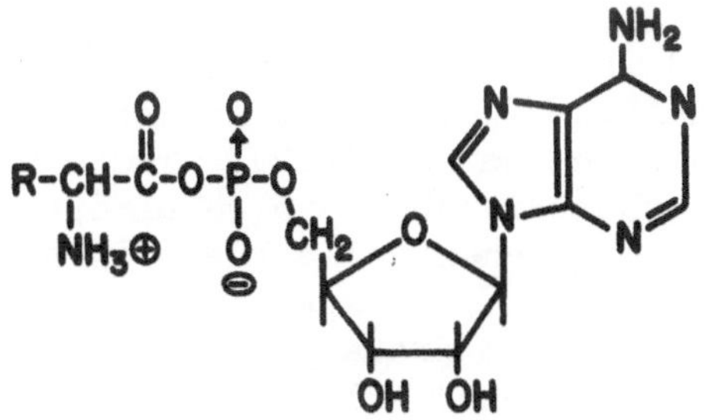

Fig. 1. Structure of aminoacyl adenylate. R, amino acid sidechain.

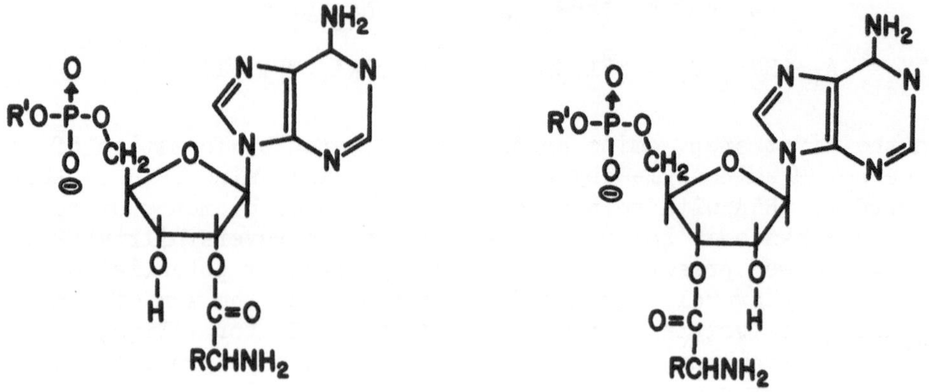

4   J. OFENGAND

Fig. 2. Structure of the two isomeric forms of AA-tRNA. R, amino
         acid side chain; R', the remainder of the tRNA molecule
         $(pN)_{70-91}pCpC$.

of such analogs have been used, namely tRNA with the terminal A replaced either by 2'- or 3'-deoxyadenosine, by 2'- or 3'-aminoadenosine
or by tRNA whose $C_{2'}-C_{3'}$ bond was first cleaved by periodate oxidation
followed by borohydride reduction. In the first case, only one
hydroxyl is available. In the second case, the amino group acts as
an irreversible trap since once the amide bond is made, it cannot be
broken. However, this latter analog can give ambiguous results since
while usually the hydroxyl *not* replaced by the amino group is
acylated, in some cases the amino group itself reacts in the absence
of the proper hydroxyl isomer. That is, 2'-hydroxyl, 3'-amino tRNA
given to a 3'-specific synthetase may be acylated on the 3'-amino
group rather than not be acylated at all. In such cases, the
erroneous conclusion is reached that the 2'-hydroxyl was reactive.
In the third type of analog, destruction of the vicinal nature of the
hydroxyl groups blocks amino acid migration. With these isomers,
however, the position of acylation must be determined by NMR
spectroscopy (10). This procedure requires large amounts of tRNA and
is technically difficult when a number of tRNAs in limited supply
must be analyzed.

     The results listed in Table 1 were obtained, except as noted,
with the deoxy tRNA analogs. With a few exceptions, the results
with the amino analogs were the same (12). There are several interesting points. First, there is about an equal distribution of 2'
vs. 3' specificity, although chemically there is a 2:1 bias in favor
of the 3'-isomer (17). Second, the specificity for a given amino
acid is preserved on going from prokaryotes to higher eukaryotes,
except for Trp (and possibly Gln). In a few cases, Asn, Asp, Cys,
Glu, and Tyr, there is no specificity, but this depends also on the
species examined.

Table 1. Position of Aminoacylation for Each Amino Acid on Prokaryotic and Eukaryotic tRNA (11-13)

Amino Acid	E. coli	Yeast	Calf Liver
Ala	3'	3'	3'
Arg	2'	2'	2'
Asn	2'(14);2',3'(15)	2',3'	2',3'
Asp	3'[a]	2',3'	3'
Cys	2',3'	2',3'	3'
Gln	2'[a]	3'	2'[b]
Glu	2'	-	2',3'[b]
Gly	3'	3'	3'
His	3'	3'	3'
Ile	2'	2'	2'
Leu	2'	2'	2'
Lys	3'	3'	3'
Met	2'	2'	-
Phe	2'	2'	2'
Pro	3'	3'	-
Ser	3'	3'	3'
Thr	3'	3'	3'
Trp	2'	3'	3'
Tyr	2',3'	2',3'[c]	2',3'
Val	2'	2'	2'

[a] with amino-tRNA (16)

[b] with wheat germ amino-tRNA (12)

[c] prefers 2'

The source of the specificity has been shown to reside in the synthetase, except in the case of Trp. When a 2'-specific synthetase like E^{Phe} is used to misacylate phenylalanine onto a 3'-specific tRNA like tRNA^{Lys} or tRNA^{Ala}, the amino acid acylates the 2'-hydroxy, 3'-H tRNA but not the other isomer (11,18). Similar results have been obtained with the pair E^{Val} and tRNA^{Thr} (19). The normal specificity for the cognate tRNA can be broken down by the use of abnormal (so-called mischarging) aminoacylation conditions so that both tRNA isomers are acylated. This has been shown for E^{Thr} and E^{Ser} (19) but is not always the case since E^{Phe} retained its specificity under mischarging conditions (10,19). In the case of tRNA^{Trp}, the specificity appears to reside in the tRNA. tRNA^{Trp} of *E. coli* was acylated at the 2'-hydroxyl with either *E. coli* or yeast synthetase while yeast tRNA^{Trp} was acylated at the 3'-hydroxyl by the same two enzymes (20). It is not difficult to understand how an enzyme could show specificity for one hydroxyl over the other, but it is completely unclear how this could be done by a tRNA molecule, especially when the 3' terminal CCA is the same.

It has been pointed out by Hecht (13,18) that all those amino acids whose second codon letter is U are 2', while all those with a second letter of C are 3', and that these amino acid groups also share certain chemical properties. However, other amino acids with purines as the second letter of the codon which also are 2' (or 3') do not share those chemical characteristics so that it is not obvious what relation, if any, the specificity for the site of aminoacylation has to any property of the amino acid. Since the site of aminoacylation is in most cases controlled by the synthetase, a correlation with some structural property of the synthetases might be envisaged. A comparison of Tables 1 and 5 shows that this also is not the case. Considering the approximately equal distribution between 2' and 3', it is possible that the specificity was randomly chosen and then became fixed at some point in evolution. The question of why there should be any specificity at all may be related to the mechanism for insuring the specific recognition of the cognate amino acid and will be discussed below. The existence of such a functional rationale could also explain why the specificity has not, with one exception, drifted during the course of evolution.

Mechanism of Aminoacylation

According to equations 1 and 2, enzyme-bound aminoacyladenylates (Fig.1) are obligatory intermediates in the synthesis of aminoacyl-tRNA. The original work leading to that conclusion is discussed in (2,3). Subsequently, this mechanism was called into question by Loftfield (21,22) who proposed a concerted mechanism. However, recent studies by Fersht and coworkers on yeast (23,24) and prokaryotic synthetases (25,26) as well as the work of others (27-31) have clearly established the role of aminoacyladenylates as obligate intermediates in a number of eukaryotic and prokaryotic systems.

Reaction 1 requires Mg^{++} (29,32,33) which can be sometimes replaced by Mn^{++} (34) but not by polyamines such as spermine (32).
Reaction 2 does not require divalent cations in the presence of sufficient monovalent cations to stabilize the structure of the tRNA
(29). The ability of polyamines to replace Mg^{++} in reaction 2
(32,33) and the stimulatory effects observed on reaction 2 (33) and
on the overall aminoacylation reaction (35) are likely to be due to
their known function in maintaining the tertiary structure of the
tRNA molecule (36,37,90). A more detailed description of the role
of cations and polyamines in reactions 1 and 2 with references to
earlier work can be found in (2,3,33). An additional role for
polyamines may be to stabilize the active structure of synthetases as
a very striking activation of dilution-inactivated yellow lupin E^{Val}
was recently reported (38).

The order of addition of reactants has been studied for a number of the synthetases. There does not appear to be any unified
order of substrate binding and product release among the various
synthetases studied. In some cases, addition is random and in
others, ATP or amino acid adds first (2,32,39-41).

The stereochemistry of the transfer of AMP from ATP to amino
acid is now known for E^{Met} and E^{Tyr} of *E. coli*. In both cases,
there is a net inversion of configuration at P_α of ATP. The simplest interpretation is that there is a direct "in-line" displacement (170). There is also preliminary evidence for covalent bond
formation between tryptophan and bovine E^{Trp} subsequent to AA-AMP
formation and before transfer to the tRNA. The bond is thought to
be a mixed anhydride between a carboxylic group on the enzyme and
the carboxyl group of the amino acid (171).

In equations 1 and 2, both the unacylated tRNA and AA-tRNA are
shown in complex with synthetase. Although this is not usually the
case *in vitro* where only catalytic quantities of synthetase are used,
it is very likely true in the cell. At least in prokaryotes, where
approximately equivalent μM quantities of each tRNA and synthetase
are to be found (42,43,43a), the association constants of ca 10^7
for both E·tRNA and E·AA-tRNA complexes insures that these is essentially no free tRNA in the bacterial cell. The situation with regard
to eukaryotic cells is less clear, as the necessary values for the
cellular concentrations of tRNA and synthetase are not generally
available. In yeast cells, the values are comparable to *E. coli*
(44). Eq. 3 describes the situation in *E. coli* where the EFTu concentration has been estimated at 0.2-0.26 mM (45,46), the total
tRNA concentration is ca 0.14 mM, and the affinity constant of
AA-tRNA for EFTu·GTP ranges from 5×10^6 to 5×10^7 (47,48). A
similar situation may exist in eukaryotes, since the cellular concentration of EF1 in the reticulocyte is 0.1-0.3 mM based on the
number of molecules per cell (48a) and estimates of intracellular
volume. Unfortunately, the other parameters are not known, and data
for other eukaryotes are not available. Thus equation 3 is still

largely speculative for eukaryotes.

STRUCTURE OF tRNA

Multiplicity and Location of tRNA in the Cell

In the cytoplasm of the eukaryotic cell, there are typically
40-60 tRNA species (49) with distinctly different sequences. Thus,
there must be several tRNAs for each amino acid. These multiple
tRNAs are termed isoacceptors. Isoacceptor species differ in the
sequence of their anticodon, since their function is to translate
the several different codons for a given amino acid. They are usually
distinctly different in their sequence, although in some cases isoacceptors from a given species show extensive sequence homology with
each other. *Pseudo* isoacceptor species are tRNAs with different
chromatographic or other separation properties that possess the same,
or nearly the same, sequence. They result usually from incomplete
modification of the numerous unusual nucleotides in tRNA (see Table 4).
These unusual bases are made by individual modifications of a normal
base after the polynucleotide chain is **synthesized**. Since some of
these reactions are considerably slower than others, apparent isoacceptors may actually correspond to one primary sequence in various
stages of completion (2,50-54). Under- or over-modication can also
explain most or all of the changes in tRNA fractionation patterns
observed during growth or differentiation (54a). Unsuspected chemical or enzymatic alteration, usually but not always to the unusual bases, is known to change the fractionation properties of
tRNA (54,55), and in still other cases, tRNA dimers and denatured
conformers have been shown to possess altered fractionation properties (56). Another class of *pseudo* isoacceptor species are native,
fully modified tRNAs which differ from each other by only a few
bases. These probably arose by gene duplication and subsequent
mutation. Numerous examples are known in both prokaryotes and eukaryotes, for example, chicken tRNA^{Pro} (Table 2), rabbit liver
tRNA^{Lys}, *B. mori* tRNA^{Ala}, yeast tRNA^{Ser}, and others (57). So far,
with the sole exception of initiator tRNA$_i^{Met}$ and tRNA$_m^{Met}$, no true
isoacceptor tRNA for the same species shares the same anticodon, in
keeping with the concept that the primary function of isoacceptors
is to utilize the degenerate nature of the genetic code. Mechanisms
for recognition of degenerate codons are discussed in more detail
below.

In addition to cytoplasmic tRNA, there are 20-30 tRNAs,
representing most or all amino acids, in organelles such as mitochondria and chloroplasts (58,59). In yeast mitochondria, 24 tRNAs
have been found corresponding to all amino acids (60), and corres-
pondingly all 20 aminoacyl-tRNAs hybridize to mitochondrial DNA
(61,62). No additional tRNAs are expected to be found (60). In
mammalian mitochondria, 23 tRNAs representing all 20 amino acids
have been found so far (63). Only the tRNA^{Arg} which recognizes AG$_G^A$

Table 3. Species Distribution of Sequenced Eukaryotic tRNAs Active in Protein Synthesis

Species	Ala	Arg	Asn	Asp	Cys	Gln	Glu	Gly	His	Ile	Leu	Lys	Met^f	Met^m	Phe	Pro	Ser	Thr	Trp	Tyr	Val
CYTOPLASMIC																					
Yeast (S. cerev.)	1	3		1	1		1	1			2	2	1	1	1		3	2	1	1	3
Yeast (S. pombe)							1				1						1			1	1
Yeast (T. utilis)	1								1	1	1	1							1		1
S. obliquus														1							
Euglena												1			1						
N. crassa											1				1						
Bombyx mori	2							2							1						
Drosophila		1					1					1	1		1						
Starfish												1									
Plants								1				2			1						
Birds																1			1		
Mammals				1				2			2	4	1	1	1		2		1		2
ORGANELLE																					
Yeast (S. cerev.) mitochondria									1			1					1	1	1	1	
N. crassa mitochondria	1				1						2	1					1	1		1	
Euglena chloroplast															1						
Bean chloroplast											3	1			1						

Data were summarized from (57) and Table 2. Mammalian mitochondrial tRNAs are not included.

is still undetected, and in fact may not exist. *Xenopus* mitochondria have 21-22 distinct tRNA sites, there are about 25 in locust
mitochondria, plants have 16-18, and Euglena have 26 tRNA sites (59).
The tRNAs found in organelles are used for the synthesis of organelle-
specific proteins (59). In yeast mitochondria no cytoplasmic tRNAs
are needed. The one cytoplasmic tRNA found in yeast mitochondria
(tRNA^{Lys}) is non-functional in the mitochondrion (64). There is no
evidence that mammalian mitochondria needs cytoplasmic tRNAs in order
to synthesize their proteins. However, the situation has not been
as well studied as in yeast mitochondria. If there really are only
23-24 tRNA species in mitochondria of eukaryotes, all 61 codons
cannot be translated unless other mechanisms for codon-anticodon
recognition are used in addition to that described by the wobble
hypothesis (65). This topic is discussed in more detail below as
part of the section on codon-anticodon recognition.

Primary and Secondary Structure

Primary Sequence. There are currently over 150 distint tRNA
sequences known from prokaryotes, lower and higher eukaryotes, and
from eukaryotic organelles. A very extensive list can be found in
Sprinzl *et al.* (57) and in Singhal and Fallis (66). Additional published sequences (up to Aug. 1980) not found in (57) are listed in
Table 2 (placed at the end of this chapter) using the format of
ref. 57. All amino acids are represented. The list includes numerous examples of isoacceptors from a single species, tRNAs specific
for the same amino acid from many different species, and numerous
tRNAs from the same species specific for different amino acids. Thus
there are many possibilities to search for sequence homologies related to species, for amino acid specificity homologies independent
of species, and for isoacceptor homologies. Although some relationships can be deduced, no clear principles have emerged. Approximately 90 of the sequenced tRNAs are from eukaryotes, with a species
and amino acid distribution as summarized in Table 3. Both lower
and higher eukaryotes are represented, as are a number of chloroplast
tRNAs and mitochondrial tRNAs from lower eukaryotes. Mammalian mitochondrial tRNAs are not included here because of their unique structural features. They will be considered separately below.

Secondary Structure. In all of the above cases, the tRNA sequence can be arranged into a two-dimensional cloverleaf form consisting of 4-5 double-stranded helical regions (a-e) called *stems*,
connected by 4 single-stranded regions (I-IV) called loops (Fig. 3).
A tRNA *arm* consists of its *loop* plus its *stem* and is denoted by its
loop number. The presence of stem c is variable and only occurs
when sufficient residues are present in arm III. The amino acid is
attached at the end of stem e to the A of the common CCA sequence,
and the anticodon triplet is located in the center of the 7-mem-
bered loop II at the end of stem b. The size of each element of
the cloverleaf is constant except at loops I, III, and stem e, where

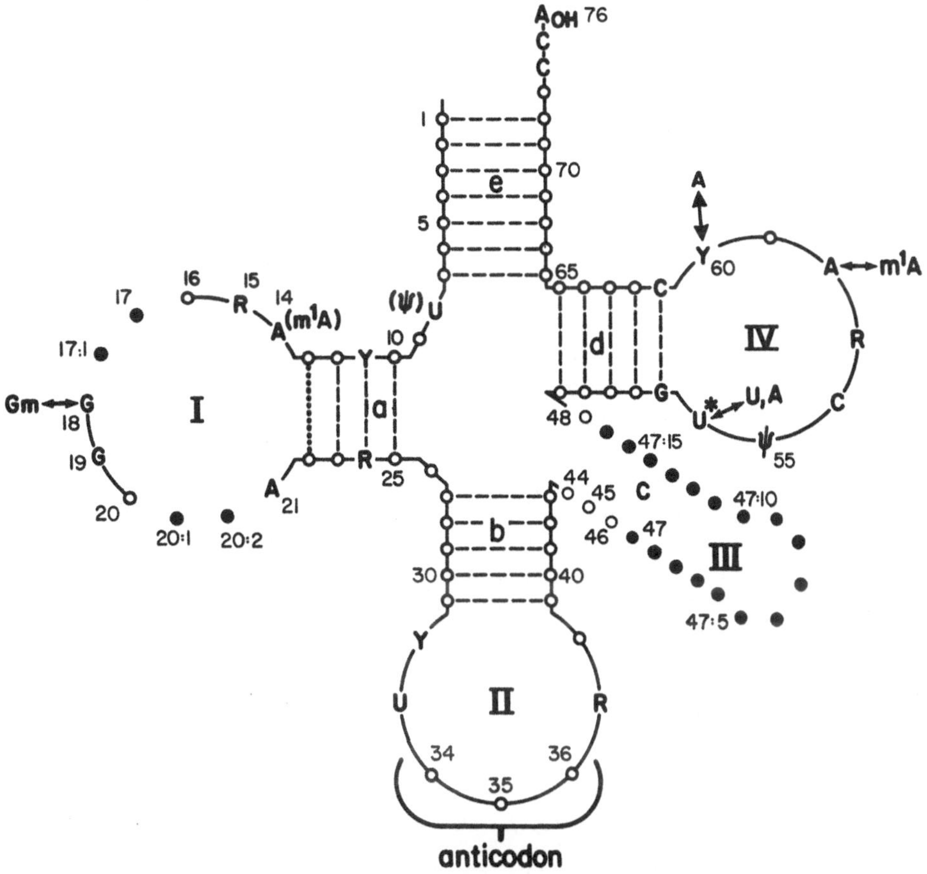

Fig. 3. Cloverleaf model of tRNA showing common features of cytoplasmic elongator tRNAs of eukaryotes according to the
sequences listed in (57) and Table 2. Nucleoside positions
are indicated by circles or by letters. Invariant or semi-
invariant bases are indicated by letters. Nucleoside positions always present (see text) are indicated by open
circles and those that may or may not be present by filled
circles. A solid line connecting the nucleosides represents the phosphodiester backbone and the base pairs are
indicated by dashed lines (the dotted line in stem a represents a base pair that is not always present). An arrow
between two nucleosides means the residue outside the loop
is found two times or more. A nucleoside in brackets is
the single exception to the commonly found nucleoside. The
5'-end of the molecule is position 1. Symbols for unusual
bases are given in Table 4. Y, uridine, cytidine, or their
modified forms; R, guanosine, adenosine, or their modified
forms; U*, modified uridine. The numbering system follows
that proposed in Appendix I of (6).

all of the variation in the length (73-94 bases of tRNA is accommodated. The location of the variably present bases, except for the extra base in stem e, is shown by the filled circles in Fig. 3. These are the two regions in loop I, positions 17, 17:1 and 20:1, 20:2, and positions 47-47:16 in arm III. Loop I ranges in size from 7-11 residues not counting residues 13 and 22 which are sometimes not base-paired (indicated by the dotted line).

The majority of tRNAs have a small arm III consisting of from 4-5 residues. Only tRNA^{Leu}, tRNA^{Ser}, and prokaryotic tRNA^{Tyr} have a large arm III which ranges from 13 to 21 residues. There are no examples yet of an intermediate arm length. Two examples exist of a 3 base pair arm III, namely baker's yeast tRNA^{Gly} and *T. utilis* tRNA^{Val}, but it has been argued that these must be incorrect since a minimum of 4 bases is required to conform to the three-dimensional crystal structure of yeast tRNA^{Phe} (67). None of the large number of tRNAs sequenced since these two were determined have less than 4 residues in arm III, and it has been recently shown that yeast tRNA^{Gly} has 4 bases in arm III (cited in reference 68). I have, therefore, assumed that a 3 residue arm III is unlikely and show 4 as the minimum number in Fig.3.

Exceptions to these rules occur in the case of organelle tRNAs. *N. crassa* mitochondrial tRNA^{Tyr} has the large arm III characteristic of prokaryotes rather than the short arm III that distinguishes eukaryotic cytoplasmic tRNA, and one of the three sequenced tRNA^{Leu} from bean chloroplasts has only 11 residues in arm III, although the other two are normal with 13 and 16 residues, respectively. Additional residues are found in mitochondrial tRNAs. In yeast tRNA^{Phe} and tRNA^{Lys} there is an extra base in stem d, in yeast tRNA^{Thr} there is an extra U at position 32, in *N. crassa* tRNA^{Tyr} UU replaces the constant A$_{14}$, and in yeast tRNA^{Ser} there is an extra base at position 26 and a deleted one at position 16.

Additional exceptions to the above rules about length and base-pairs are as follows. Prokaryotic, yeast mitochondrial, and Drosophila tRNA^{His} (the only three sequenced) have an additional base to the 5'-side of residue 1, which in the first two examples base-pairs with N$_{73}$. Thus these two tRNAs have 8 base-pairs in stem e. The base pair 31-39 at the loop II side of stem b is lacking in four cases, tRNA^{Leu} of *S. pombe* where it is A-C, tRNA$_m^{Met}$ in yeast and mammals, where it is Ψ-Ψ, and tRNA^{Thr} in *N. crassa* mitochondria where it is also Ψ-Ψ. The base pair 27-43 at the other end of stem b is lacking in five tRNAs, tRNA^{Cys} and tRNA^{His} of yeast mitochondria, tRNA^{Tyr} (*S. pombe*), and tRNA^{Val} and tRNA^{Trp} of *N. crassa* mitochondria. There are, in addition, occasional absences of a base-pair from one stem or the other which appear to be distributed at random, and in addition, numerous examples exist of the G-U "wobble" (65) base-pair in stems a-e. A detailed listing of the number and location of G-U pairs is given by Clark (68) who has noted that their likely function is to relieve backbone strain in the stems.

Invariant bases. In cytoplasmic tRNA of eukaryotes, there are 14 invariant bases if modified bases are counted equal to their parent. They are shown in Fig. 3. Since there are 40-44 residues in stems, only 19-40 truly variable positions remain to specify the uniqueness of individual tRNAs. Moreover, there are in addition a number of semi-invariant positions of several types. At positions 11, 32, and 60, only pyrimidines are found, except that in *B. mori* tRNA^{Ala} and mammalian tRNA^{Val}, Y_{60} is replaced by A. Position 14 is always A, or in one case, m^1A. Positions 15, 24, 37 and 57 are always purines, R_{37} being frequently modified (see Table 4). The variations shown at positions 18 and 58 occur frequently. U_{54} is usually (but not always) modified to T, Ψ or Tm, and is replaced by A in *B. mori* tRNA^{Ala}. Other semi-invariant positions not shown in Fig. 3 are the pairs 10-25 and 15-48. G or m^2G and Y are usually found at positions 10 and 25, respectively, but are replaced in two *S. pombe* tRNAs, Ler and Ser, by C_{10}-G_{25} and U_{10}-A_{25}. Position 48 is usually Y in order to pair with R_{15}, but is an A in two tRNA^{Ala} from *B. mori* and a G in *S. pombe* tRNA^{Glu}. In these three cases, and also in yeast t RNA$_3^{Glu}$, there is no correlation with position 15. In all other tRNAs, the alternation between A_{15} and G_{15} is related to a parallel U_{48} to C_{48} change. As with many other invariant positions, this feature is explained by the tertiary structural interactions which are described below. A detailed discussion of invariant bases can be found in (66).

There are many more exceptions to the rules of invariant and semi-invariant positions when elongator tRNAs of organelles are examined. The differences noted so far include Y_{11} to G_{11}, A_{14} to U or m^6A, G_{18} to A, A_{21} to G, R_{24} to C, Y_{32} to R_{32}, R_{37} to U, Ψ_{55} to G, Y_{60} to A_{60}, G_{10}-C_{25} to a variety of other base-pairs, and R_{15}-Y_{48} to non-pairing residues. A more detailed analysis can be obtained from Table 2 and referenced therein, the lists in (57), and discussions in (58,66,68,69). In general, it can be said that organelle tRNAs obey fexer of the rules that appear to constrain cytoplasmic tRNAs.

Mammalian mitochondrial tRNAs. Bovine and human mitochondrial tRNA show such extensive structural variation from the tRNAs described above that they merit discussion in a separate section. The sequences of twenty-three different tRNAs have now been determined (63) although only 6 have been published so far (70-72). Five are shown in Fig. 4. As these sequences were obtained from the DNA, the location of modified bases is not known, and therefore, the structural variations detected are limited to those of size, base-pairing and invariant bases. It is known, however, that a number of modified bases occur in bovine mitochondrial tRNAs, among them Ψ, m^5c, m^1G, m^2G, m^1A, and t^6A (Roe, B.A., personal communication). A more detailed tabulation of the minor base composition of mitochondrial tRNAs from a variety of species is given by Dirheimer (58).

It is immediately obvious from Fig. 4 that the usual rules governing arms I and IV no longer apply. Stem a can be as small as 2 base pairs, loop I as small as 3 residues, and the invariant $G_{18}G_{19}$ pair is absent in all but tRNA^{Leu}. Moreover, in a tRNA^{Ser} (not shown) which is only 62 residues in length, arm I has disappeared altogether and is replaced by 5 residues joining stems b and e (72). Similarly, stem d can be as small as 3 or as large as 6 base-pairs, loop IV varies from 5-9 residues, and most importantly, the common sequence GTTCA or GTTCG in loop IV is absent in all but tRNA^{Leu}. The absence of GTTCG from hamster mitochondrial tRNA has also been shown (72a). In tRNA^{Thr} (not shown) there are only 3 bases in loop IV (Barrell, B.G., personal communication). This phenomenon is not confined to the 7 tRNAs mentioned, since in at least two other cases, tRNA^{Met} and tRNA^{Pro} the size of loops I and IV are also abnormally small (Roe, B.A., personal communication).

The evidence that these tRNAs function in protein synthesis in the mitochondrion is so far only indirect, since it has not yet been possible to obtain an *in vitro* protein-synthesizing system from this organelle, nor even to aminoacylate the tRNAs *in vitro* with mitochondrial synthetases. Aminoacylation is possible in the intact mitochondrion (72). However, by several criteria, there do not appear to be any other tRNAs available in the mitochondrion to translate these functional codons (72), so one is forced to conclude that these aberrant structures *do* function.

It is worth noting that the two arms which vary most are those which interact in the tertiary structure (see below) and which are thought to be involved with binding to the ribosome. The amino acid carrying stem is constant at 7 base pairs, and the size, base-pairs, and invariant residues of arm II, the anticodon arm, are also normal. These facts suggest that there may be something unusual in the way mammalian mitochondrial tRNAs interact with their ribosomes. It is well known that mitochondrial ribosomes differ markedly from their cytoplasmic counterparts. It is conceivable that under the extreme requirement to be sparing with gene sequences which is implied by the small size of the mitochondrial DNA, 1×10^7 daltons, the tRNA sequences have shed all possible excess baggage, including part of their ribosome-recognition region. This might not be deleterious if the ribosome and/or ribosomal proteins were suitably redesigned. Since the ribosomal proteins are all (or nearly all) imported from the cytoplasm (73), there would be no additional genetic burden on the mitochondrial DNA. The resolution of this and many other important questions will have to wait until suitable *in vitro* protein synthesizing systems are available.

Minor bases. As indicated above and in Fig. 3, tRNAs contain a number of unusual bases, made by sometimes quite elaborate modifi-

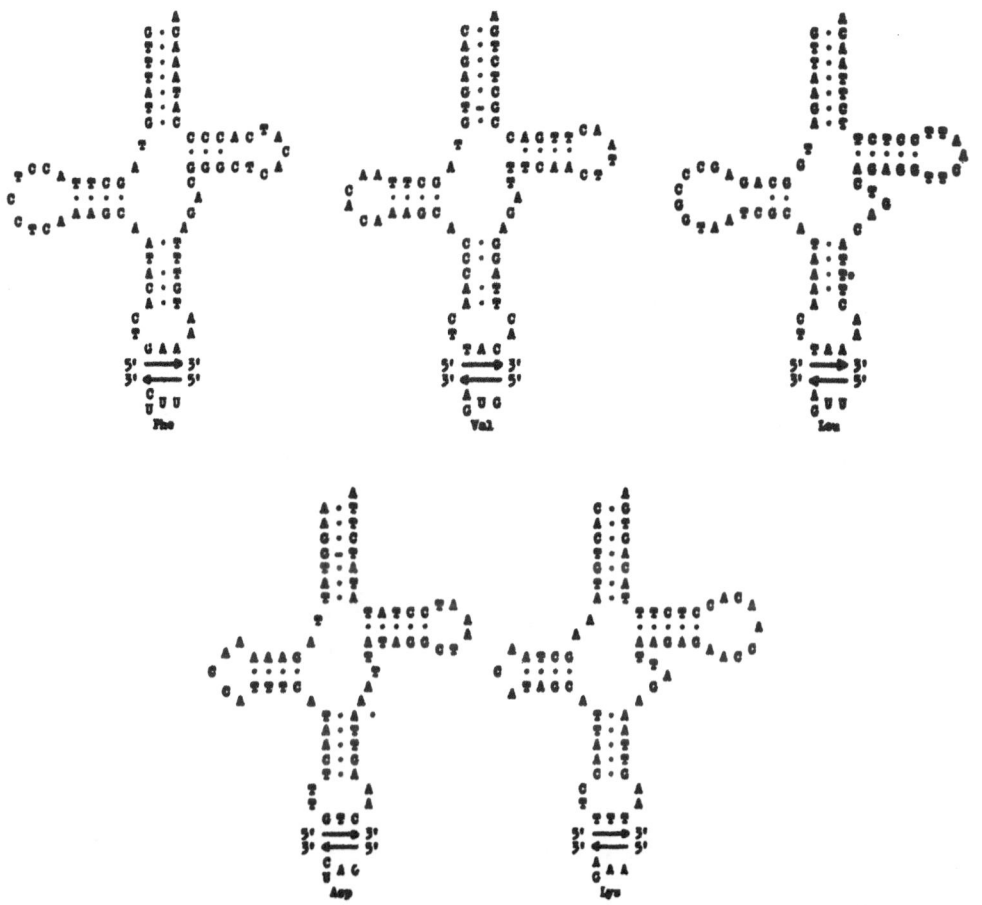

Fig. 4. The sequences of 5 human mitochondrial tRNAs obtained by
         DNA sequencing. Phe, Val, Leu are from Eperon *et al.* (70),
         and Asp and Lys are from Barrel *et al.* (71).

cations of one of the four normal nucleotides. More than 40 such
bases have been identified in sequenced tRNAs to date (53,54,58).
They are listed in Table 4. It is clear from the Table as well as
Fig. 3 that modified bases occur only at restricted sites in tRNA,
and in many cases, a particular base is found only at a single position. Particularly favored sites for modification, sometimes in a
most baroque manner, are positions 34 and 37, where they play a role
in codon-anticodon recognition (see below). Other favored sites are
loop I for D residues, and position 55 in loop IV for Ψ. In order
to conserve space in Table 4, the location and distribution for each
base were grouped together. However, in a number of cases, particular positions are unique to either prokaryotes or eukaryotes with
organelle tRNAs sometimes following one pattern and sometimes the
other.

Table 4. Unusual Nucleotides in tRNA

Name[a]	Symbol	Location[b]	Distribution[c]
Pseudouridine	Ψ	d	P,E,M,C
Dihydrouridine	D	16,17,17:1,20,20:1,20:2,21,47	P,E,M,C
4-Thiouridine	s^4U	8,9	P
Ribothymidine	T	54	P,E,M,C
2-Thiouridine	s^2U	34	E
5-Methyl-2-thiouridine	m^5s^2U	54	P
2'-O-methyl-ribothymidine	Tm	54	E
2'-O-methyluridine	Um	4,32,44	P,E
2'-O-methylpseudouridine	Ψm	32,39	E
5-Methoxyuridine	mo^5U	34	P
5-(Carboxymethoxy)uridine	cmo^5U	34	P
3-(3-amino-3-carboxypropyl)uridine	acp^3U	20:1,47	P,E,C
5-(Methoxycarbonylmethyl)uridine	mcm^5U	34	E
5-(Carboxyhydroxymethyl)uridine	chm^5U	34	E
5-(Methoxycarbonylmethyl)-2-thiouridine	mcm^5s^2U	34	E
5-(Methylaminomethyl)-2-thiouridine	mnm^5s^2U, mam^5s^2U	34	E
2'-O-methylguanosine	Gm	18,19,34,39,64	P,E,C
1-Methylguanosine	m^1G	9,37	P,E,M,C
N2-Methylguanosine	m^2G	6,7,9,10,26	E
N2,N2-Dimethylguanosine	m^2_2G	26	E,M
N7-Methylguanosine	m^7G	36,46	P,E,C
Queosine and derivatives	Q	34	P,E
Wybutosine and derivatives	yW	37	E

Name[a]	Symbol	Location[b]	Distribution[c]
Inosine	I	34	P,E
1-Methylinosine	m^1I	37	E
1-Methyladenosine	m^1A	14,22,58	P,E
2-Methyladenosine	m^2A	37	P
N6-Methyladenosine	m^6A	14,37	P,E
N6-isopentenyladenosine	i^6A	37	E,C
2-Methylthio-N6-isopentenyl adenosine	ms^2i^6A	37	P,C
N6-(threoninocarbonyl)adenosine	t^6A	37	P,E
N6-Methyl-N6-(threonino-carbonyl)adenosine	mt^6A	37	P,E
2-Methylthio-N6-(Threonino-carbonyl)adenosine	ms^2t^6A	37	P,E
5-Methylcytidine	m^5C	34,38,40,48,49,50,72	P,E,M
3-Methylcytidine	m^3C	32,47:3	E
2-thiocytidine	s^2C	32	P
2'-O-methylcytidine	Cm	4,32,34	P,E
N4-acetylcytidine	ac^4C	12,34	P,E

Data summarized from (57) and Table 2.

[a] A useful compilation of the structures associated with these names can be found in Appendix II of (6).

[b] Position number according to Fig. 3.

[c] P, prokaryotic; E, eukaryotic cytoplasm; M, mitochondria; C, chloroplast.

[d] occurs at positions 1, 8, 13, 20:1, 20:2, 25, 26, 27, 28, 31, 32, 35, 38, 39, 40, 45, 46, 47:1, 54, 55, 65, 67, 68, 72.

Three examples will suffice to illustrate this point (symbols as in Table 4). Ac p^3U is normally found at position 47 in prokaryotes and in chloroplasts. There is a single case where it is found in a sequenced eukaryote tRNA, rat liver $tRNA^{Asn}$, but in this case it is located at position 20:1. This may be a more frequent feature in eukaryotes, since there is evidence that ac p^3U may be present in tRNAs from tyrosine, threonine, isoleucine, and cysteine as well (74). m^7G frequently occurs in both prokaryotes and eukaryotes at position 46. In one case, bean chloroplast $tRNA^{Leu}$, m^7G is found in the anticodon at position 36. This is the first example of a modified base occurring at this position of the anticodon. Ψ is found at position 35 but only in $tRNA^{Tyr}$ from yeast species. The 5'-anticodon base, residue 34, is frequently modified (see above). ac^4C is found in both prokaryotes and eukaryotes, but only at position 34 in prokaryotes and only at position 12 in eukaryotes. Thus the modifying enzyme must have a different sequence specificity in the two cases. A more detailed discussion can be found in Dirheimer (58).

Another distinguishing feature of a number of minor bases is that they are unique, not only in location, but also to tRNA species specific for certain amino acids. Thus Q and its derivatives is found only in Asn, Asp, His and Tyr tRNAs of mammals as well as prokaryotes, but not in yeast. This is no doubt related to its location in the anitcodon wobble position, as the tRNAs involved read the codons NA_C^U. Similar explanations can be invoked for the other minor bases at positions 34 and 37, which are also tRNA-specific (53,54), but it should be noted that the same minor base is not always used in both prokaryotes and eukaryotes. For example, yW and its derivatives are specific to $tRNA^{Phe}$ of eukaryotes, whereas $tRNA^{Phe}$ of prokaryotes and organelles use m^1G_{37} or $ms^2i^6A_{37}$. However, as always, there are exceptions. The insect $tRNA^{Phe}$ species from *Drosophila* and *B. mori* have m^1G instead of yW, although *N. crassa* cytoplasm has yW. Another interesting case is that of initiator tRNA which in prokaryotes is always an unmodified A_{37}, but in eukaryotes is always t^6A_{37}. The three exceptions are organelle $tRNA_i^{Met}$ species. Yeast and *N. crassa* mitochondria have m^1G_{37}, while bean chloroplast has A_{37}, like the prokaryotes.

Initiator tRNA. The essential functional distinction between eukaryotic initiator $tRNA^{Met}$ and elongator $tRNA^{Met}$ is that one can participate in the formation of initiation complexes while the other cannot, and conversely, elongator tRNA is preferred over initiator tRNA in the formation of EFT complexes destined for binding to the ribosomal A site (2). Structurally, the main difference between initiator and elongator tRNA lies in loop IV, residues 54-60. In all 9 cytoplasmic initiator tRNAs examined to date, the sequence is $AUCGm^1AAA$, except in three cases where U_{55} is Ψ_{55} as in elongator tRNAs (57, and Table 2). This sequence is very different from the general elongator tRNA pattern (see Fig. 3) which is also shared by prokaryotes, including tRNA$_f^{Met}$. Recently, however, this distinction has become somewhat blurred since two species of, presumably, elongator tRNA^{Ala} from *Bombyx mori* have the almost identical sequence AΨCGm^{1}AUA. Nevertheless, it seems likely that this sequence is related to the ability of eIF-2 to specifically select out Met-tRNA$_i^{Met}$ from among all other AA-tRNAs since the prokaryotic initiator, Met-tRNA$_f^{Met}$, which has the sequence GTΨCA common to all elongator tRNAs, is also selected against by eIF-2 (75). The reciprocal case is not true, however. Both fMet-tRNA$_f^{Met}$ (prokaryote) and fMet-tRNA$_i^{Met}$ (eukaryote) are able to form complexes with IF2 (2). An additional feature of initiator tRNAs from higher eukaryotes is the replacement of U$_{33}$ by C.

There are only four organelle initiator tRNA sequences known, one each from yeast and *N. crassa* mitochondria, and two from chloroplasts (57, and Table 2). As noted above, organelle tRNAs tend to vary more than their cytoplasmic counterparts, and to sometimes mimic eukaryotic patterns of invariance and sometimes the prokaryotic pattern. This is also true of the initiator tRNAs. The base pair 1-72, which is characteristically absent from all prokaryotic initiator tRNAs, is also absent from yeast mitochondria and spinach or bean chloroplast tRNA$_f^{Met}$, but is present in *N. crassa* mitochondria, as in eukaryotic cytoplasm. In bean and spinach chloroplast tRNA, Y$_{11}$-R$_{24}$ is replaced by A$_{11}$-U$_{24}$, in yeast mitochondrial tRNA, Y$_{32}$ is replaced by G, and in *N. crassa*, the almost sacrosanct G$_{18}$ is replaced by A, and there is also a lack of 15-48 base-pairing. In three of the tRNAs, the prokaryotic loop IV sequence, TΨCAAAU, is found, but in *N. crassa* mitochondria T$_{54}\Psi_{55}$ is replaced by U$_{54}$G$_{55}$. This does not correspond to any previous sequence. In all other initiator tRNAs, residue 55 is either uridine or pseudouridine.

It might be suspected that the coordinate change of G$_{18}$-Ψ_{55}, to A$_{18}$-G$_{55}$ in the *N. crassa* tRNA$_i^{Met}$ might be related to its function as an initiator, since residues 18 and 55 are known to be involved with each other in a tertiary interaction (see below). However, mitochondrial *N. crassa* tRNA^{Thr} and yeast tRNA^{Asp} have the same A$_{18}$-G$_{55}$ change, and are elongator tRNAs. Thus, although all three tRNAs with G$_{55}$ have A$_{18}$, there is no distinction between initiator and elongator tRNAs. Results such as these tend to discourage a search for structure-function relationships in tRNA based on simple sequence comparisons. One supposes that the sequence variations seen, which are after all very distinctive, even if not all encompassing, reflect some distinctive tertiary structure which serves to delineate initiator from elongator tRNAs. One can only suppose that there is more than one way to reach the desired tertiary structure and that those apparently aberrant tRNAs, like *B. mori* tRNA^{Ala} and *N. crassa* mitochondrial tRNA$_i^{Met}$, have found such an alternate route. So far, however, the three-dimensional structural distinctions between initiator and elongator tRNAs have remained elusive.

CCA Sequence. All tRNAs terminate in this common sequence
where the amino acid is attached. Viral RNAs which can be aminoacylated (see below) also require this sequence. A special enzyme,
nucleotidyl transferase (76), exists in all cells including mitochondria to add this sequence to newly synthesized tRNA molecules
since many tRNA genes, especially in eukaryotes, lack it (see below).
The biological rationale for such a flexible, unpaired sequence
carrying the amino acid is probably to allow proper stereochemical
orientation of the petidyl- and aminoacyl-moieties on the ribosome
so that peptide bond formation can occur. An extensive review of
the properties of the CCA end can be found in (11).

Tertiary Structure

Crystal Structure. The crystal structures of four tRNAs are
now known. The structure of yeast tRNA^{Phe} in two crystal forms has
been extensively reviewed (2,3,36,37,68,69,77,78) since it was the
first tRNA structure to be solved by X-ray methods. Since then, the
structures of two initiator tRNAs, one from yeast (79) and one from
E. coli (80) have been reported, and that of a second elongator tRNA
from yeast, tRNA^{Asp}, has recently been described (81). All four
structures are similar, with only slight deviations of the latter
three from the yeast tRNA^{Phe} structure. This fact confirms earlier
conclusions that all tRNA structures would likely be similar, based
on a large body of physical and chemical studies on tRNA in solution
(2,3,68,78). It is disappointing, however, that neither the prokaryotic nor eukaryotic initiator tRNA shows obvious structural differences that could be correlated with their distinctive function.
We first review the features of yeast tRNA^{Phe}, and then discuss the
differences between it and the other X-ray structures.

The three-dimensional structure of yeast tRNA^{Phe} is shown in
Fig. 5, in two different schematic forms to facilitate discussion.
The molecule consists of two helical stems at approximately right
angles to each other which are held together by a variety of tertiary
structural interactions. The length of each arm is approximately
70A, the thickness is about 22A, and the distance between the two ends
is about 80A. It is a flat molecule. The amino acid binding end is
at the right extreme of the molecule, and the anticodon is at the
extreme bottom. The loop IV sequence suggested to be involved in
ribosome binding (reviewed in 72) is at the extreme top left, residues
54-56. Thus these three functionally important regions are maximally
separated from each other. This probably facilitates the proper
functioning of tRNA on the ribosomal surface.

The horizontal helix is made up of stems e and d in which the
3'-ended chain is continuous. The vertical more irregular helix
consists of stems a and b, plus additional bases. The central part
of the molecule, including the region where the two helices come

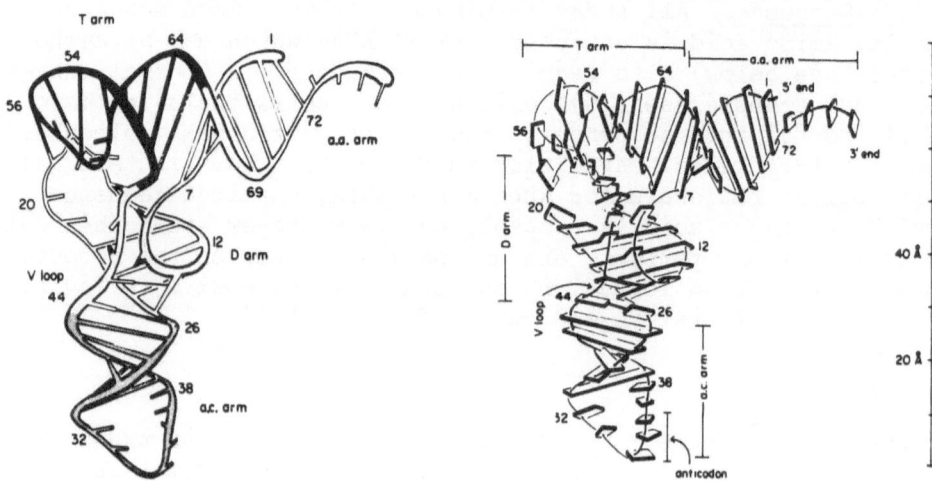

Fig. 5.   (Left) Schematic drawing of yeast tRNA^Phe showing the ribose-
phosphate backbone as a continuous tube, base pairs as long
bars, and single bases as short bars. Tertiary H-bonds
between the bases are indicated by solid dark rods. Stem e
(aa-arm), unshaded; residues 7 and 8, horizontal lines;
arm I (D-arm) unshaded; arm II (ac-arm), light stipple;
arm III (V-loop), unshaded; arm IV (T-arm), heavy stipple.
(Right) Drawing of yeast tRNA^Phe to illustrate the extensive
base stacking. Secondary and tertiary base pairs are shown
as bent or fused slabs or by connecting two slabs with dark
rods. Reproduced, with permission, from (83).

together, is stabilized by a variety of tertiary interactions. Loops
I and IV are held together by (1) a G_{19}-C_{56} base pair of the Watson-
Crick type, which is both twisted and bent, (2) a non-standard
G_{18}-Ψ_{55} pair, (3) intercalation of G_{57} between the 18-55 and 19-56
base pairs and stacking of m^1A_{58} on G_{18} to form the stack G_{19}-G_{57}-
G_{18}-m^1A_{58}, and (4) cross-linking of loop IV by a reverse Hoogsteen
base pair of m^1A_{58}-T_{54}. The remainder of the region joining the two
helices is stabilized by the following layer of tertiary interactions.
They are from top to bottom as follows.

1. G_{15}-C_{48} in a reverse, or trans, Watson-Crick base pair.

2. U_8-A_{14} by a reverse Hoogsteen base pair.

3. m^7G_{46}-G_{22}-C_{13} triple in which the m^7G_{46} pairs with the G_{22}
     of the G_{22}-C_{13} base pair.

4. A_9-A_{23}-U_{12} base triple in which A_9 pairs with A_{23} of
     A_{23}-U_{12} base pair.

5. $G_{45}-m^2G_{10}-C_{25}$ base triple in which G_{45} pairs with the m^2G_{10} of the $G_{10}-C_{25}$ base pair.

6. $G_{26}-A_{44}$ non-co-planar base pair.

The invariant base pair $Y_{11}-R_{24}$ is stacked between the $A_9-A_{23}-U_{12}$ and $G_{45}-m^2G_{10}-C_{25}$ base triples.

These interactions involve most but not all of the invariant and semi-invariant bases (see above) and thus explain in large measure their existence. For example, the $R_{15}-Y_{48}$ invariance mentioned above can now be understood in terms of a tertiary base pair. The invariants and semi-invariant bases *not* so occupied are A_{21}, Y_{32}, U_{33}, R_{37}, and Y_{60}. A_{21} forms a single H-bond with the O_2' of U_8 which *may* explain its invariance (77, 78), and R_{37} plays an important role in codon-anticodon recognition (see below). Y_{32} and U_{33} are involved in maintaining the conformation of the anticodon loop in which the three anticodon bases are stacked on the two 3' bases of the loop forming a stacked continuation of stem b. Thus only Y_{60} remains unexplained. It is also not clear why $Y_{11}-R_{24}$ is invariant since the reverse pair, $R_{11}-Y_{24}$, should also stack between the two triples, yet it is found only in prokaryote (and bean chloroplast) initiator tRNA.

Other tRNA sequences can be fitted to this structure with little difficulty. Obviously those tRNAs with only 3 base-pairs in stem a must sacrifice the 46-22-13 triple, but a 46-22 interaction may still be possible. A long arm III does not interfere as that loop extends out in a direction away from the rest of the molecule. In general, substitution of other bases in the tertiary interaction pairs and triples described above, either as exceptions to the invariant ones, or as one of the variable bases, does not disturb the H-bonding interactions described. This is because there are usually compensating changes in more than one residue which allows alternative base-pairs of equivalent stereochemistry to form (3, 36, 66, 68, 77, 78). This is not always true in detail, of course. Some tRNAs will be unable to form all of the tertiary interactions described here, and others will no doubt devise different ones to achieve the same end result.

This structure may also explain why all of the variations of tRNA size is confined to the two regions of loop I and arm III. Since the rest of the molecule is so involved in either H-bonding or stacking interaction, these are the only places where variation *can* occur.

The yeast $tRNA_i^{Met}$ structure at the current level of resolution, 4Å. differs from the $tRNA^{Phe}$ structure only in the conformation of the anticodon loop which is slightly more skewed (79). The tertiary structural interactions appear to be preserved in those cases where the bases are invariant or coordinately changed. In particular, the base pair, $A_{54}-m^1A_{58}$ is structurally equivalent to the $T_{54}-m^1A_{58}$ of $tRNA^{Phe}$ (84).

The *E. coli* tRNA$_f^{Met}$ structure is also similar to that of tRNA^{Phe}, but with three significant differences (80). First, since tRNA$_f^{Met}$ has an additional base at 17:1, the folding of the region 16-17:1 is different. Surprisingly, the extra residue is folded in toward the center of the molecule rather than outward toward the solvent. Perhaps this is why there are no examples of tRNAs with more than three residues in this region. There may be no additional space to accommodate them in the interior of the tRNA. Second, U$_{33}$ is turned outward to the solvent instead of in toward the anticodon loop. Third like yeast tRNA$_i^{Met}$, the anticodon loop is skewed to one side. Otherwise, the tertiary interactions appear similar to yeast tRNA^{Phe} although the analysis must proceed to higher resolution before these interactions can be clearly seen.

The yeast tRNA^{Asp} has been obtained in two crystal forms. In one form, loops I and IV interact, but in the other they do not. This may suggest how the two loops may open during ribosomal binding but nothing is known so far. U$_{33}$ is turned toward loop II as in yeast tRNA^{Phe} and the overall structure is very similar.

In summary, while there are small structural differences between these other tRNAs and tRNA^{Phe}, there is no explanation for why they occur, especially the anticodon loop changes, nor is there anything other than speculation (80) about possible functional roles that they may play.

Effect of modified bases on structure. It is noteworthy that none of the tertiary interactions described above require the participation of a modified base. In fact, just the opposite is true. Those bases which are modified at normal H-bonding positions, such as m^{1}A, ^{4}S, and m^{1}G form unusual H-bonds which do not use these atoms. Other base modifications such as in m^{1}G and m^{2}G do not interfere with normal base pairing. Even in the case of Ψ, where synthesis requires the breaking and remaking of a glycosidic bond at the polynucleotide level, the extra H-bond donor at N$_5$ is not used for structural interactions. One exception is the case of s^{2}T. This residue has been shown to stabilize loop I-IV interactions (85).

The main effect of base modification appear to be an increase in the solvent accessible surface of the molecule. As a result of the extensive system of base stacking and H-bonding, over 90% of the bases are inaccessible to solvent. The base modifications increase the overall exposure area of the bases by up to 20%, suggesting that the modifications exist for recognition by various proteins (36). Indeed, it was long ago noted by Crick that tRNA with all its modified bases appeared to be an RNA molecule trying to pass itself off as a protein.

Structure in solution. There is little doubt now that the tRNA tertiary structure found in crystals also exists in solution. Many

this topic. The reader is referred to the monograph by Soll et al.
(7) and especially the review chapters on tRNA biosynthesis (217) and
tRNA gene organization (218) for a detailed analysis. tRNA genes in
eukaryotes are amplified several hundred to several thousand times.
They are not associated with the nucleolar organizer as are the rRNA
genes, but rather are disperse among many chromosomes. In some
species, like *Xenopus* multiple gene copies appear to be in tandem
while in other cases, as in yeast, they are dispersed. Most
eukaryotic tRNA precursors which have been identified are monomeric,
although dimers have also been detected (219, 220). The precursors
contain a small number of extra nucleotides at each end, and lack the
common CCA end which is added later. Some, but not all, precursor
tRNAs of eukaryotes contain an intervening sequence of 14-34 bases
inserted at the 3' side of the anticodon loop junction with its stem.
As in eukaryotic mRNAs, this sequence is removed and the two newly-
made ends of the tRNA spliced to produce the mature molecule (221,
222). Although it had been thought that splicing, CCA addition, and
some removal of the termina extra nucleotides occurred in the cytoplasm, recent experiments indicate that all these events occur in
the nucleus at least in *Xenopus* oocytes (223).

The minor bases are all made by modification of their parent in
the precursor polynucleotide chain except for the base Q which is
inserted by base exchange with G (53). Earlier studies had not
indicated any specific minor base requirement in any of the processing
steps nor a restriction to any particular precursor stage, but the
same recent *Xenopus* experiments (223) again show that this is not the
case. At least four modifications occur in the nucleus at defined
stages of the processing sequence. The only modification that definitely occurs in the cytoplasm is Gm formation.

ACKNOWLEDGEMENT

This review could not have been written without the assistance
of the transcriber of these words, Barbara Kerr. She has my heartfelt thanks for her extraordinary skill, dedication, and endurance.

REFERENCES

1. Crick, F.H.C. (1957), Biochem. Soc. Symp., 14, 25-26.
2. Ofengand, J. (1977), In *Molecular Mechanisms of Protein Biosynthesis*, (Eds. H. Weissbach and S. Pestka), pp. 7-79, Academic
   Press, New York.
3. Goddard, J.P. (1977), Progr. Biophys. Molec. Biol., 32, 233-308.
4. Schimmel, P.R. and Soll, D. (1979), Ann. Rev. Biochem., 48, 601-
   648.
5. Altman, S. (Ed.) (1978), *Transfer RNA*, M.I.T. Press, Cambridge,
   Massachusetts.

edly do occur, see for example (98), but gross changes in tertiary
structure appear unlikely.

Two kinds of conformational changes in tRNA have been proposed
to take place when tRNA binds to the ribosome. The first is related
to the postulated interaction of loop IV with ribosomal RNA at the
A site (reviewed in ref. 82). Since loops I and IV are normally
strongly complexed, such an interaction presupposes a separation of
these two loops on binding to the A site. It has been suggested that
codon-anticodon recognition such as occurs at the A site might be the
trigger that causes loops I and IV to separate, and evidence for (98)
and against (90b, 93, 99, 100) this proposal has been obtained. At
the present time, it remains only an attractive possibility. The
evidence for interaction of loop IV, in particular the common GTΨCR
sequence, with ribosomal RNA at the A site seems somewhat stronger.
This question is discussed in some detail in (82). Suggestive evidence for opening of the contacts between loops I and IV at the P site
have been obtained recently by slow ^{3}H-exchange studies (100a). It
remains to be seen whether such conformational changes occur at the
A site.

The second conformational change frequently proposed is in the
anticodon loop. It has been proposed, for a variety of reasons by
different authors, that the anticodon stack on the 3' side of loop II
flips to an alternate state in which the anticodon stacks with the
5'-bases of the loop as the tRNA moves through its ribosome binding
cycle (78, 101, 102). Evidence for such a 5'-stack in a solution
complex of tRNA^{Phe} and the codon-containing pentamer UUCAG exists
(90b). However, recent studies by Kuechler et al. (103) tend to
discredit this notion as applied to ribosomal complexes. By cross-
linking the anticodon to the messenger RNA in both P and A sites,
these authors obtained evidence that the conformation of the anticodon
loop is the same as in the crystal, i.e., a 3'-stack, and that it does
not change when the tRNA moves from the A to the P site.

It would be surprising if small conformational changes in the
tRNA did not occur when it binds to an object as large as the ribosome, particularly since multiple contact surfaces must be involved.
In fact, a recent study shows that codon-dependent conformational
changes in the anticodon loop do occur (104), and tritium exchange
studies also suggest that small changes in conformation occur (90,
100a). Nevertheless, there is so far no evidence for a major structural re-organization of tRNA as it passes through the ribosomal
cycle of protein synthesis (105).

STRUCTURE OF AMINOACYL-tRNA SYNTHETASES

Multiplicity and Cellular Localization

There is only one synthetase in the prokaryotic cell for each
amino acid (2) and the same is probably also true in eukaryotic
cytoplasm although it is less certain because of the existence of
organelle-specific synthetases both in chloroplasts and in mitochondria. It is not clear whether organelles have their own complete set
of organelle-specific synthetases, or import some cytoplasmic enzymes
also. In all case, all indications are that all of the synthetases
are made in the cytoplasm using nuclear genes, both in the case of
yeast (73, 106, 107) and chloroplasts (108). The extent of cross-
reaction between cytoplasmic and organelle tRNAs and synthetases can
vary all the way from no to complete cross-reaction just as do inter-
species heterologous reactions, with as little discernable pattern
(59).

Protein Structure

Subunit Structure. Despite the fact that all synthetases catalyze the same reaction, they differ markedly in their physical properties. Eukaryotic synthetases range in size from 74 to 287 kilodaltons, and consist of from 1 to 4 subunits. The three subunit
structure classes are α_1, α_2, and $\alpha_2\beta_2$. However, as seen in Table 5,
the variation is largely related to amino acid specificity. Except
for E^{Arg}, E^{Leu}, E^{Phe}, and E^{Trp}, synthetases from different sources
for a given amino acid are constant in subunit structure and very
similar in size. (Yeast E^{Met} and E^{Tyr} are size exceptions.) Even
E^{Arg}, E^{Leu} (bovine mammary), and E^{Trp} only represent an apparent
multimerization of the basic subunit, and could be due to experimental artifacts. Likewise the $\alpha\beta$ structure of human placenta E^{Phe}
could be due to a failure to dimerize. Comparison of this Table with
the list of prokaryotic synthetases in (4, 109) confirms this correlation although there are apparent occasional exceptions. However,
as the size as well as the ability to self-associate is strongly
affected by mild proteolysis during isolation (2, 4), the occasional
deviation from the general structure for a given amino acid should be
viewed with caution. The α_2 structure of E^{Leu} (wheat germ) may be due
to such a proteolysis, as was recently found for the yeast E^{Leu} (110).

Although only 15 amino acids are represented in Table 5, examples are known for all amino acids, if prokaryotic synthetases are
included. In all cases, only the same three subunit structure classes
are used. Most synthetases fall into either the α_1 or α_2 class.
E^{Gly} is the only additional synthetase to be added to the $\alpha_2\beta_2$ class.

Closer inspection of the table reveals that the $_1$ class subunit
(73-130 kd) is approximately twice the size of the α_2 class subunit

Table 5. Subunit Structure of Eukaryotic AA-tRNA Synthetases

Amino Acid	(Source)	Type	MW kilo-daltons	Amino Acid	(Source)	Type	MW kilo-daltons
Ala	(yeast)	α_1	128	Phe	(wheat germ)	$\alpha_2\beta_2$	260
Arg	(N. crassa)	α_1	85			$\alpha=61;\beta=69$	
	(yeast)	α_1	73		(lupin seed)	$\alpha_2\beta_2$	260
	(wheat germ)	α_1	73			$\alpha=59;\beta=75$	
	(lupin seed)	α_2	140		(rat liver)	$\alpha_2\beta_2$	287
Asp	(yeast)	α_2	120			$\alpha=69;\beta=75$	
Cys	(rat liver)	α_2	240		(human placenta)	$\alpha\beta$	130
His	(rabbit retic.)	α_2	122			$\alpha=55;\beta=72$	
Ileu	(yeast)	α_1	124	Ser	(yeast)	α_2	120
Leu	(S. cerev.)	α_1	120		(lupin seed)	α_2	110
	(T. utilis)	α_1	128		(hen liver)	α_2	122
	(N. crassa)	α_1	110	Thr	(yeast)	α_2	160
	(N. crassa mito.)	α_1	90		(rat liver)	α_2	170
	(E. gracilis)	α_1	110	Trp	(yeast)	α_2	110
	(E. gracilis chloro.)	α_1	105		(lupin seed)	α_4	200
	(T. pyriformis)	α_1	120		(bovine pancreas)	α_2	114
	(T. pyrif. mito.)	α_1	100		(human placenta)	α_2	116
	(wheat germ)	α_2	110		(buffalo brain)	α_3	155
	(T. molitor)	α_1	99	Tyr	(yeast)	α_2	80
	(bovine mammary)	α_2	182		(soybean cyto.)	α_2	122
Lys	(yeast)	α_2	138		(soybean chloro.)	α_2	86
	(rabbit retic.)	α_2	122		(rat liver)	α_2	140
Met	(yeast)	α_2	110		(hog pancreas)	α_2	118
	(wheat germ)[a]	α_2	165	Val	(yeast)	α_1	130
	(lupin seed)	α_2	170		(E. gracilis)	α_1	125
Phe	(yeast cyto.)	$\alpha_2\beta_2$	276		(E. gracilis chloro.)	α_1	126
		$\alpha=63;\beta=75$			(lupin seed)	α_1	125
	(yeast mito.)	$\alpha_2\beta_2$	260				
		$\alpha=57;\beta=72$					

Data extracted from (4,109) except for yeast E^{Leu} (110); yeast E^{Met} (111); yeast mitochondria E^{Phe} (112); and rat liver E^{Thr} (113).

[a]Rosa and Sigler (116) find two α_1 forms of 105 and 70 kilodaltons, respectively.

(40-70 kd), except for E^{Met}_{Cys} and E^{Thr} which appear to be made up of 2 α_1 class subunits. (E^{Cys} is an exception.) This has raised the possibility that the α chain is actually made up of two primordial subunits of the α_2 size which have become covalently joined by gene duplication and fusion. Two lines of evidence support this concept. First, in a number of instances, very mild proteolysis cleaves α_1 synthetases into two fragments, which in combination, sometimes retain partial enzymatic activity (2, 4). Second, a number of α_1 class enzymes show internal sequence repeats with one set of sequences appearing in each of the two large fragments that are generated by mild proteolysis (2, 4). Similar studies on α_2 enzymes have failed to show sequence repeats, except for E^{Met}. However, as this enzyme seems to be made up of 2 α_1 type subunits, sequence repeats are to be expected. Yeast E^{Phe} also showed sequence repeats in both of its subunits, but here too each subunit approaches the size of an α_1 class enzyme (117). In summary, it appears as if all synthetases were made from a primordial subunit of 40-70 kilodaltons. In the α_1 class, the gene apparently duplicated, fused, and underwent partial mutation. In the α_2 class, association only occurs at the protein level. In the $\alpha_2\beta_2$ class, both processes have occurred.

The α_2 and $\alpha_2\beta_2$ enzymes have 2 substrate binding sites, and the α_1 class is generally thought to have but one site (2, 4). However, at least one α_1 enzyme, *E. coli* E^{Val}, has been shown to have two binding sites (118) and a mechanism has been proposed by which 2 binding sites could enhance catalysis through site interaction (119). This proposal might explain why so many synthetases are functional dimers, although it cannot be true in all cases. E^{Cys} from *B. stearothermophilus*, a 54 kd monomer, has only one substrate binding site and does not dimerize even under aminoacylation conditions (120).

Tertiary structure. The ultimate understanding of the structure of synthetases will come from X-ray crystallographic analysis. Three enzymes are currently under study, two from prokaryotes (121, 122) and one from yeast (123). Study of the prokaryotic enzymes has already shown that the predictions about structure from the presence or absence of internal repeated sequences is correct. The E^{Met} monomer consists of two parts separated by a large cleft (121), while the E^{Tyr} dimer shows no such domain structure (122). X-ray analysis of the yeast E^{Asp} only shows so far that it is elongated (123). Of much more interest is the recent announcement that co-crystals of tRNA^{Asp} and E^{Asp} have been obtained that are suitable for X-ray diffraction work (Giege, R., personal communication). Since the crystal structure of tRNA^{Asp} is already known (see above), the availability of diffracting co-crystals should make it possible for the first time to see exactly how a tRNA and its *cognate* synthetase fit together.

Multi-Enzyme Complexes

While in prokaryotes, synthetases exist free or in complex with tRNA (2), in eukaryotes, many synthetases are found as components of high molecular weight complexes, 1-5 x 10^6 daltons. These complexes have been reported now for a wide variety of tissues including rabbit reticulocytes, mouse liver, mouse embryo, chicken embryo, Friend leukemia cells, HeLa cells, CHO cells, sheep liver and bovine brain, to cite some recent examples (124-127). The size of the complex has been variously reported as 1 x 10^6 (126), 5 x 10^6 (127), "22-35S" (125) or "18-25S" (124) and there appears to be neither lipid nor carbohydrate present (128). The synthetase composition has also varied, although the enzymes for Arg, Gln, Glu, Leu, and Lys are always there. Thus, activity for Asp, Ile, Met, Pro, and Val was found in addition in six different cell types (124). Asp, Ile, Met, Pro, and Thr activity was additionally found in CHO cells (125). The additional activities were Ile and Met in sheep liver (126). Asp and Val were the additions in bovine brain (127).

A brief inspection of the above shows that the same cast of characters keeps appearing, although with minor variations and additions. E^{Phe}, which is *not* found in these macromolecular complexes, tends to be associated with the ribosomes (124). In sheep liver, all of the cellular activity for the listed amino acids is found in the complex (126), as is the Glu activity in bovine brain (127), but in other cases, this is true to a lesser extent, each tissue and amino acid activity giving somewhat different results. Some synthetases, such as those for Ala, Ser, and Trp are almost never found in high molecular weight complexes, although in the early reports (cited in ref. 2), it was claimed that all synthetases were present. It may be that the current methods used for isolation are too harsh, resulting in the loss of some synthetases. Thus, the relatively constant content of amino acid specific activities mentioned above may merely be a result of the intrinsically greater association ability of these particular enzymes. On the other hand, there is no evidence that a given complex is made up of all the activities detected, or that any particular stoichiometry is involved. The complexes could be very heterogeneous in composition, each containing different groups of enzymes. What is badly needed here are studies to show how many other activities are present, and in what stoichiometry, in a complex selected out by some affinity tehnique for having a particular synthetase.

The physiological reason for the existence of synthetase in these complexes is obscure. It has been noted that when CHO cells are arrested in G_1 phase, a decrease occurs in four of the 10 synthetases present in the 30S complex. The decrease was not too large, but it was highly specific, leading the authors (125) to speculate on a role for these complexes in regulation of the cell cycle. In a more recent study, the complexes from bovine brain which contain the

synthetases for glutamine and glutamic acid were shown to maintain
an internal pool of these two amino acids, but not of any of the
other amino acids represented by the synthetases in the complex (129).
These authors speculate that such complexes might be a way of insu-
lating the protein synthetic apparatus from large fluctuations in
amino acid supply, in cells where amino acids play particular regulatory roles. However, the wide distribution of these complexes in
many kinds of tissue makes it unlikely that this explanation will be
generally true.

SPECIFICITY OF AMINOACYLATION

Importance of Specificity

Proof of the Adaptor Hypothesis, namely that when attached to
tRNA, the amino acid no longer has any effect on codon selection, has
long been available (cited in ref. 2). The consequence of this fact
is that AA-tRNA synthetases must be highly accurate both in their
selection of an amino acid and of a tRNA, since any error in either
recognition process will inevitably result in the incorporation of
wrong amino acids into proteins. An *upper* limit to the degree of
error tolerated in *finished* proteins was obtained for the misincor-
poration of valine in place of isoleucine in ovalbumin and globin.
This value, which includes transcriptional and translational errors
is $< 3 \times 10^{-4}$ (130). The demands on synthetases are therefore much
greater than for many other enzymes. Ordinarily, an error in the
recognition of a substrate will most likely be corrected by the enzyme
catalyzing the next step of the reaction which will reject the
incorrect substrate, but in this case such a separate error-correction
mechanism does not exist. Error-correction, editing, or proofreading, as it is variously called, with respect to amino acid specificity appears to occur before the AA-tRNA and synthetase separate
from each other (see below).

Synthetase specificity can be conveniently divided into two
parts, namely specificity for the small ligand, the amino acid, and
for the large ligand, the tRNA.

Amino Acid Recognition

General specificity. The general recognition feature of
synthetases for amino acids can be summed up as follows (2). A free
α-amino group is essential both for substrate or inhibitor activity.
Also, N-substituted aminoacyl-tRNAs do not undergo deacylation in the
presence of AMP and PP (the reverse of equations 1 and 2). The
carbonyl group, on the other hand, is unnecessary, and amino acid
amides or alcohols are excellent competitive inhibitors. Synthetases

are L-isomer specific, except for E^{Tyr} from *E. coli* and *B. Subtilis*.
Amino acid analogs may or may not be active, or may be active in
activation (equation 1) but not in transfer to tRNA (equation 2). In
general, naturally occurring analogs likely to be in the environment
of a given synthetase are inactive. However, a synthetase from a
different species which does not normally encounter such an analog
may react with it. Synthetic analogs may be unreactive, reactive in
equation 1 but not in 2, or completely active. This depends on the
editing properties of the synthetase in question and on the structure
of the analog.

    <u>Editing Mechanisms</u>. The selection of a given amino acid from
among the other naturally occurring ones of similar structure is a
formidable task, and the way synthetases do this has been the subject
of much study and speculation. It now appears that the way this high
selectivity is achieved is by a "double-sieve" editing mechanism (131,
132). First, larger amino acids are excluded from the aminoacylation
site by the usual steric mechanisms. Measurements of the kinetic
constants for ATP-PP exchange (equation 1, the activation reaction)
for isoleucine and valine with E^{Val} from *E. coli*, *B. stearothermophilus*, yeast, and yellow lupin seeds have shown that the synthetases
can discriminate adequately ($<10^{-4}$) against an amino acid just one
methylene group larger (132, 132a). When challenged with *unnatural*
amino acids, however, as in the case of ethionine and E^{Met} (133), the
discrimination ratio, 3×10^{-2}, is much less. O-methyl threonine is
also not very well discriminated (ratio of 3×10^{-3}) against by E^{Val}
of yeast (134) and in both cases, the unnatural amino acid remains
stably attached to the tRNA.

    Amino acids smaller than the specific substrates are only par-
tially excluded, the rates of activation (AA-AMP formation) decreasing
with decreasing size because of the poorer fit (132, 133). For amino
acids much smaller, the lack of specific contact areas suffices to
prevent misactivation. However, for amino acids only slightly
smaller, such as the pair isoleucine-valine, valine-α aminobutyrate,
or methionine-homocysteine, the discrimination ratio is not sufficiently low to account for the accuracy of finished proteins. In
particular, it has been estimated that 1 isoleucine in 16 would be
replaced by valine in the absence of an editing mechanism (135). In
these cases, a special hydrolytic mechanism is used. The incorrect
amino acid is transferred to the tRNA (131, 136, 137), and then
rapidly hydrolyzed while the AA-tRNA is still enzyme-bound (137).
This appears to occur at an amino acid binding site distinct from the
original amino acid recognition site, since enzyme-bound AA-AMP does
not interfere with this process (137). This second site is presumed
to be slightly smaller than the first, so that the correct amino acid
does not enter and is not hydrolyzed. Consistent with this hypothesis
is the experimental fact that the well-known AMP and PP-independent
hydrolytic activity of synthetases (reviewed in ref. 2) is much
faster for those cognate tRNAs misacylated with smaller or isosteric

amino acids than for either correctly aminoacylated tRNA or for
cognate tRNA misacylated with larger amino acids (132, 134, 136).

Naturally occurring amino acids isosteric with the correct one
require some special property either of the aminoacylation site or of
the hydrolytic site for exclusion. Thus, E^{Cys} is highly specific for
the activation of cysteine, and does not recognize similarly sized
amino acids like serine or alanine, presumably because it is not
difficult to build a binding site uniquely for cysteine (138).
Consequently, it has no need for a hydrolytic capacity for
misacylated-tRNA^{Cys} and experimentally has none. On the other hand,
threonine which is isosteric with valine, is mis-activated by E^{Val}
which apparently cannot so easily distinguish the extra OH of
threonine from valine. In this case the Thr-tRNA is edited by hydrolysis of the Thr-tRNA^{Val} (137). In some instances, editing can also
occur at the level of the enzyme-bound aminoacyl adenylate (132a).

Use of tRNAs with deoxy-A at the 3'-end (see Aminoacylation
above) has revealed another interesting feature of this editing
mechanism. In a number of cases, the absence of one of the 3'-
terminal hydroxyls blocks the hydrolytic editing function and allows
the incorrect AA-tRNAs to accumulate (134, 139). This has prompted
Cramer and his colleagues to propose that the OH group is a direct
participant in the hydrolytic reaction (134). However, this cannot
always be the case since at least one example exists of a deoxy
tRNA^{Met} which does hydrolytically edit (133) and a slow but significant hydrolysis of Thr-deoxy tRNA^{Val} by E^{Val} of yeast also occurs
(134).

It has also been noted that the two synthetases, E^{Cys} and E^{Tyr},
which most clearly do not have any specificity for the position of
aminoacylation (Table 1) also do not have any editing function,
apparently because they are sufficiently specific without one (138,
139). This is in keeping with the notion that there are two amino
acid binding sites on synthetases with editing capacity, one for
aminoacylation which requires one hydroxyl, and one for hydrolytic
editing, which in some way requires the other hydroxyl. Assuming
that by chance, a given synthetase evolved to a specificity for one
hydroxyl, the requirement of editing for the other one would preclude
the possibility of reversing specificity. This could explain why
only those amino acids which do not require editing are non-specific
for their position of aminoacylation and why the selection of 2' or
3' for the other amino acids is about equally distributed. The
examples of E^{Met} and E^{Val} mentioned above which are hydroxyl-specific
in aminoacylation and which can edit without the "other" hydroxyl do
not disprove this notion, but serve to show that the second hydroxyl
is not an absolute requirement in all cases.

tRNA Recognition

Overview. Probably no area of tRNA and synthetase research has received as much attention as the mechanism of recognition of a tRNA by its cognate synthetase. This is due not only to the intrinsic fascination of understanding how these two macromolecules recognize each other, and select against non-cognate but very similar molecules, but also to the fact that this easily-studied system serves as the paradigm for all nucleic acid-protein recognition systems. Unfortunately, despite the many man-years of effort devoted to this problem, there is still no exact understanding of how this mutual recognition occurs.

In principle, recognition could occur in one of several ways. (1) A specific sequence of bases in the tertiary structure of tRNA is the recognition site. All tRNAs use the same positions, specificity being achieved by sequence variation. (2) The recognition site consists of multiple sets of sequences as in (1) which need not be adjacent in either the primary or tertiary structure, but all tRNAs use the same regions; (3) and (4), as in (1) and (2) but with the proviso that each tRNA-synthetase pair may use a different set of sites; (5) the detailed tertiary structure of the tRNA or part of it determines the specificity, and each synthetase may sense different parts of the structure. In this connection, it should be noted that tRNAs can be denatured (56), and thus that some elements of tertiary structure are clearly necessary.

Since this subject has been reviewed in detail many times (2-4, 66, 140-142), the following sections will only summarize the approaches used and the conclusions reached without repeating the citations of the original papers. Reference to specific work will only be made where appropriate.

tRNA-synthetase interaction. Physical complex formation measures the specificity of interaction independent of reaction catalysis and thus allows this property to be studied in the absence of other ligands. Many methods, both non-equilibrium such as gel chromatography and centrifugation to mention but two, and equilibrium methods like fluorescence quenching have been used. While in almost all cases, the homologous cognate interaction was strongest, considering the high degree of specificity for acylation, some non-cognate interactions were found to be surprisingly strong, from 10 to 20% of the cognate ones, and sometimes even stronger.

Similar results were obtained by kinetic methods. Comparison of K_m and V_{max} for a number of cognate and non-cognate acylation reactions has shown that the K_m for non-cognate tRNAs is in many cases as low or lower than for the cognate tRNA. However, the non-cognate V_{max} is vastly reduced (ca 10^{-4}) compared to the cognate

reaction. Thus both physical and kinetic methods agree that complex
formation between tRNA and a synthetase is not highly selective.
Rather, kinetic control over the catalytic event of aminoacylation
appears to be the selective step.

This is supposed to occur by a series of conformational changes
in the $E^1 \cdot tRNA^1$ complex which properly positions the 3'-end of the
tRNA at the catalytic center only when the correct tRNA is bound.
Evidence has been obtained by fast kinetic techniques for a rapid
non-specific binding step occurring with both cognate and non-cognate
tRNAs which is followed by a slower conformational change that is only
detected with the cognate tRNA (summarized in ref. 4). This second
step requires a correct 3'-end since it is not detected when the 3'-A
is removed (143, 144) and its characteristics are altered when the
3'-A is modified (145). Similar conclusions about multiple conformational changes and the "triggering" role of the 3'-A have also been
reached from steady-state measurements (146, 147).

Thus the high degree of specificity in homologous aminoacylation
is now understood not simply in terms of specificity of complex
formation, but as an interplay of (a) specificity or "triggering" in
the catalytic step or step(s) and (b) competition at each of these
steps between $tRNA^A$ and $tRNA^A$ for synthetase^A and also between
synthetase^A and synthetase^B for $tRNA^B$ (147).

Molecular basis for specific recognition. A rich variety of
methods have been used to probe this question. The sequence homology approach was one of the first used. In this approach, sequences
of tRNAs capable of being acylated by a given synthetase are compared,
and common regions located. The difficulties with this approach which
ignores tRNA tertiary structure is well illustrated in ref (68)
pp. 29-31. Nevertheless, a number of useful correlations have been
shown to exist between this method and other techniques (66).

The remainder of the methods used fall into two classes. In
one, the tRNA is modified, either by chemical treatment, nuclease
action including the removal of selected parts of the molecule,
mutation, or base analog substitution, and the aminoacylation
activity then tested. There are two main pitfalls with this type of
approach. First, *any* perturbation, even a base change, in the
natural structure may have an effect distant from the site of action,
i.e. it may break a tertiary interaction, making it difficult to
know the actual cause of the inactivation event. Thus, the most
meaningful results probably come from *retention* of activity since the
modified areas are clearly unnecessary. Second, the frequent failure
to distinguish between K_m and V_{max} effects makes it difficult to know
whether the general binding site is affected or the specific determinants. Only rarely has the *specificity* of aminoacylation been tested.

The results of a large number of chemical modification studies have been recently summarized (3, 66). A useful compilation of a series of carefully controlled modification studies on *E. coli* tRNA$_f^{Met}$ which illustrates the possibilities and limitations of the method is given by Schulman (148). The conclusion from these studies is that a single C to U change in as diverse locations are stem e or loop II can block aminoacylation, although other C to U changes, even those nearby, do not.

Nuclease treatment and the use of tRNA fragments to determine essential or non-essential areas has been another approach (2, 4). In some cases, large fractions of the tRNA could be eliminated with retention of aminoacylation activity. However, the results varied widely from one synthetase–tRNA pair to another. The main conclusion to be drawn from these studies is that different synthetases recognize different parts of the tRNA. For a recent example of the complexities involved, see (149). The logical extension of such "dissected molecule" studies has been to replace excised sections of tRNA with chemically synthesized oligonucleotides which can then be ligated to the rest of the molecule (150). In this way it should be possible to examine the effects of specific base changes at will.

Another way to look at base changes has been to select mutant tRNAs (2, 4). This has only been possible in *E. coli* where mutants of the tyrosine suppressor tRNA which insert a different amino acid could be selected. All were due to a single base change in stem e, and all could then be acylated with glutamine in addition to tyrosine. These studies are fundamentally different from all the others cited above since here a single base change resulted in an *altered* specificity. Moreover, there is no simple structural relationship between these two tRNAs. tRNA^{Tyr} has a large arm III while tRNA^{Gln} has the more common short one. Other similar experiments showed that a single C to U change in the anticodon of tRNA^{Trp} could also allow E^{Gln} recognition. Recently, these results were confirmed by *in vitro* bisulfite-catalyzed mutation, and it was directly shown that the altered tRNA could now be recognized by both E^{Gln} and E^{Trp} (151). Although there is no simple rationale that can account for these results (however, see ref. 151a for a recent speculation), it does seem clear that both the anticodon region and the acceptor end *can* be important recognition areas.

Base analog substitutions have not been much used because of the difficulty of replacing normal bases with synthetic analogs. In one approach, all of the uridine and uridine-derived minor bases of tRNA were replaced by 5-fluorouridine, yet all but a few species had normal acylation kinetics (2, 151b). The strongest effect was on tRNA^{Lys} which should have three fU residues in its anticodon. Since most tRNAs were unaffected, none of the replaced residues in those species were essential, or if they were, the 5-fluoro analog was just as good (2).

The second major approach has been to look at the physical contact areas between tRNA and synthetase, in the belief that the contact sites must include the recognition regions. Contact sites have been studied by protection experiments, direct crosslinking, and by a unique chemical reaction. In protection experiments, complex formation is used to protect the tRNA from a nuclease, from complexation with a specific small oligonucleotide probe for one or another single-stranded region, or from ^{3}H-exchange of the purine C-8 with water. Most of these studies have been carried out by Schimmel and his colleagues and have been recently summarized (4,140). The earlier work is reviewed in (2). More recent nuclease protection experiments are described by Ebel (142). Direct crosslinking studies have been carried out by UV irradiation of synthetase-tRNA complexes (2, 4, 140, 152-154) to obtain general crosslinking sites, or by chemical crosslinking of a lysine residue of the synthetase to the periodate-oxidized 3'-end of the tRNA (153, 155).

From the results of all of these direct contact studies, as well as from the tRNA modification experiments, it seems clear that no unitary recognition mechanism exists which is applicable in its entirety to all tRNA-synthetase pairs. Nevertheless, some general conclusions can be drawn (see also ref. 142). The weight of the evidence indicates that the amino acid acceptor stem (stem e), stem a and the residues joining stems a and e, and the anticodon loop region (loop II), all of which are located approximately on the right-hand side of the tRNA "L" in Fig. 5, are involved in both contact with, and recognition by, the synthetase. All three regions may not be required simultaneously, however, nor in all tRNAs. In addition, other elements may be important in some cases. For example, in the methionine system, the stem a interaction was not detected. Instead, parts of stem d and loop IV were found to be contact regions for E^{Met} of E. coli using both prokaryotic and eukaryotic tRNA$_i^{Met}$ as well as E. coli tRNA$_m^{Met}$ (152).

Since the distance from the anticodon to the amino acid is about 80Å, some synthetases may not be large enough or elongated enough to span that distance. Thus, the anticodon area may not always be a recognition region, and indeed some of the experiments cited above show that the anticodon can be modified or deleted without effect. Other experiments also cited above show the opposite effect. These conflicting results might be explained in terms of either the size of the synthetase involved or the location of the tRNA binding site.

Additional evidence for involvement of the regions between stems e and a comes from recent work by Schimmel (140, 156). These workers observed several peculiar reactions between U$_8$ of tRNA^{Ileu} and its cognate synthetase, which do not occur with non-cognate enzyme. These reactions include (a) an abnormally high ^{3}H exchange into U$_8$, (b) specific inhibition of aminoacylation by the trinucleotide, UAG, which is the sequence around U$_8$, (c) an abnormally high ^{3}H exchange into the U of this trinucleotide, (d) direct binding of UAG to E^{Ileu}, and (e) glycosidic bond cleavage of 5-bromouridine, other 5-halouridines, and even of uridine itself. The U_8 tritium exchange was also seen with $tRNA^{Tyr}-E^{Tyr}$ and with unfractionated tRNA and synthetases. The glycosidic bond cleavage reaction was found with four other synthetases (156). Moreover, treatment of E^{Ileu} with 5-bromouridine at 0° inhibited the enzyme for subsequent amino-acylation (156), but this inhibition could be blocked by isoleucine, ATB, and $tRNA^{Ileu}$, or reversed by thiol treatment (157). Consistent with these results, a covalent adduct of 5-bromouridine and E^{Ileu} has been found (157). Of even more interest, however, is the further observation that treatment of E^{Ileu} with 6-hydroxyuridine leads to formation of a 1:1 covalent complex between enzyme and uridine which is catalytically inactive, and that $tRNA^{Ileu}$ protects the synthetase (Koontz, S.W. and Schimmel, P.R., personal communication).

This interaction cannot be a unique specificity site although it requires a cognate tRNA-E pair, since several synthetases show this property. Moreover, U (or ^4S) is present at position 8 in all known tRNAs (one exception is Ψ_8, see Table 2). Schimmel (140) has suggested that U_8 might be a kind of benchmark for the synthetase once recognition has occurred. That is, by anchoring to the tRNA at this point the proper placement of the 3'-end of the tRNA at the catalytic center would be assured.

So far this discussion has centered around the tRNA. The synthetase is also a partner in recognition but so far little is known. Crosslinking studies with yeast E^{Phe} (153) have revealed that the tRNA, as well as the ATP and amino acid, all have contact sites on the α (smaller) subunit in this $\alpha_2\beta_2$ enzyme, but not more is known. Neutron scattering studies indicate that in yeast E^{Val}, the tRNA is located nearer to the center of mass of the complex than is the protein, and that the elongated enzyme contracts when it is complexed with tRNA (158). These features appear to be characteristic of both strong cognate and of strong non-cognate complexes. Finally, preliminary studies indicate that mutant synthetases may have altered specificity for tRNA (141).

tRNA Proofreading. When the AMP and PP independent hydrolysis of AA-tRNA by synthetase was first discovered, it was suggested that this might be an editing mechanism at the tRNA level. That is, if $tRNA^B$ were acylated with amino acid^A by E^A, it could be removed by E^B. This is not likely to occur in vivo for two reasons. First, the deacylation rates are only rapid for incorrect amino acids smaller or isosteric with the correct amino acid (132, 134, 136), and second the in vivo concentration of EFTu and probably also of EF-1, is high enough to sequester all AA-tRNA released by E^A before it can be complexed with E^B (131).

Role of minor bases. An obvious candidate for a major role in

determining the specificity of tRNA-synthetase interactions are the
many unusual nucleotides found in tRNA (Table 4). This hypothesis
stimulated the development of a numebr of chemical reactions to
modify specific minor bases in order to probe this question. The
end result of these studies as well as those in which the post-
transcriptional modification reactions leading to the minor bases
were blocked is that none of these bases influence aminoacylation to
any great extent. Much of the earlier work has been reviewed (2).
In summary, Ψ, rT, ^4S, hU, I, cmo^5U, mnm^5s^2U, i^6A, ms^2i^6A, yW, and
m^1G are not required. More recent studies confirm that yW (159),
mnm^5s^2U (160), i^6A (161), and ms^2i^6A (162) are not required. In
addition, Gm (50, 162), m^2C (163), m^1G (160), t^6A (164), mcm^5s^2U
(165), s^2C_{32} (166), and Q (cited in ref. 167) are not needed. So far
only two examples are known in which a modification has had an effect.
Conversion of G_{10} to m^2G_{10} in E. coli $tRNA^{Phe}$ markedly increased the
V_{max} for aminoacylation (168) and a mutant B. subtilis $tRNA_f^{Met}$ with
G in place of m^1G_{46} had only one-half of the wild-type V_{max} (162).
On the other hand, reduction of m^1G in $tRNA^{Phe}$ and $tRNA^{Val}_{max}$
to the uncharged 8-hydro analog did not inhibit aminoacylation (169).

CODON-ANTICODON RECOGNITION

Codon Translation in the Cytoplasm

      Role of N_{34}. One of the early observations during the decipher-
ing of the genetic code was that a single tRNA species could recognize more than one codon. In order to account for this, Crick (65)
proposed the Wobble Hypothesis, since amply verified experimentally,
which stated that at the 5'-end of the anticodon, G·U, I·U, and I·A
base pairs are found in addition to the standard G·C, U·A, and U·I
pairs. This allows the 64 assigned codons to be read by a minimum
of 32 tRNAs. This follows directly from the fact that 2 RNAs are
needed for each of the 16 sets of codons of the form XYN, plus 1
initiator tRNA and minus 1 tRNA for the terminator codon pairs UA^A_G.
However, more than 32 tRNAs are used. This is because the tRNA for
the five amino acids (Arg, Gln, Glu, Leu, Lys) whose codons are XY^A_G,
do not use U_{34} corresponding to the Wobble rules. Instead, in all
cases, where information is available (except $tRNA^{Leu}_3$), a modified
U such as mcm^5s^2U, mcm^5U, or mnm^5s^2U is used. This residue does not
pair with G, so that a second C_{34} containing tRNA is needed.
Nishimura (53) has suggested that the reason an unmodified U_{34} is not
allowed in these cases is that it might too readily mispair with
XYU, which would be lethal since those codons stand for a different
amino acid. Presumably the cell has not been able to design a base
capable of pairing with both G and A but not with U, and therefore
has had to resort to the expedient of using two separate tRNAs.

      E. coli has tried to do this and failed. The $tRNA^{Leu}$ recog-
nizing the UUR codon series has instead of a modified Y_{34}, a

modified A_{34} (172). How this recognizes A and G is mysterious in its
own right, but the point is that it also can weakly misread UUU.
In fact, the well-known miscoding of poly(U) for leucine incorporation
into protein turns out to be entirely due to this particular species
of $tRNA^{Leu}$ (173).

Sundaralingam (174) has suggested a mechanism for the mispairing
of unmodified U_{34} with pyrimidines. He found that the G·U pair stacks
far better on a standard Watson-Crick pair when it is at the 5'-side
of the U than when it is at the 3' side. Therefore, a G·U_{34} pair
would not be able to stack well since the adjacent base pair is on
the 3'-side. The U_{34} may then have sufficient freedom to engage in
U_{34}·Y pairing.

The complementary codons, XY^U_C, for Asp, Asn, Cys, His, Ileu, Phe,
Ser, and Tyr are recognized by a single anticodon using G_{34}, as
expected. However, in both prokaryotes and higher eukaryotes, G_{34} is
modified to Q_{34}, but only when the adjacent anticodon base is U_{35}.
Since this does not occur in yeast, the Q modification cannot be
essential for codon recognition. In hemoglobin synthesis, $tRNA^{His}$
with G_{34} reads both His codons as well as the tRNA containing Q_{34}
(175). Other roles for the Q modification such as in differentiation
are discussed in (53).

The remaining codons are all sets of four of the form XYN,
specifying a single amino acid. There are eight such sets, for Ala,
Arg, Gly, Leu, Pro, Ser, Thr, Val. Now if the same device were to
be used to decode the XY^A_G half of the set as before, three tRNAs would
be needed per set. Instead, two alternative methods are used to
decode the four codons with only two tRNAs. Valine illustrates the
situation. In prokaryotes, one tRNA with a modified U_{34} can read the
set GU^A_G, while a second tRNA with G_{34} reads GU^C_U. In eukaryotes, one
tRNA with I_{34} reads GU^C_U, while the other tRNA with C_{34} reads GUG.
These modified U residues, mo^5U and cmo^5U, are altered at the C-5
position, like the ones discussed above, but they allow pairing with
A, G, and U while the others are strictly for A. A possible explanation for this, based on a detailed structural analysis of the anticodon loop, is that the 5-substituents of the restrictive U residues
interfere with the movement required for U·G pairing, while the cmo-
and mo- groups do not (L76). Why prokaryotes use one method, and
eukaryotes another, is unknown. The capacity to form I_{34} exists in
E. coli since it is used for the Arg codons CG^U_A.

It is not known why a single tRNA with an unmodified U_{34} is not
used in these cases, where mispairing with U and C would not matter
since all four codons specify the same amino acid. It would be more
economical, but perhaps it has not yet been possible for most cells
to devise an enzyme capable of modifying the U_{34} of the tRNAs for
Arg^1, Gln, Glu, Leu^1, and Lys, but not those for Ala, Arg^2, Leu, Pro,

Ser, Thr, and Val.

The Gly anticodons are an exception. There is both a G_{34} and C_{34} containing mammalian $tRNA^{Gly}$ (Table 2), implying that a third tRNA exists to decode GGA, and three such tRNAs are known in *E. coli* (57). It is not clear why a two tRNA system is not used. In *Mycoplasma*, there is a single $tRNA^{Gly}$ with U_{34} for translation of all 4 codons (ref. 40 of Table 2). Another exception is yeast $tRNA^{Leu}$ with anti-codon UAG (Table 2). This tRNA and the above-mentioned $tRNA^{Gly}$ are unusual in being the only cytoplasmic tRNAs functional in protein synthesis with an unmodified U_{34}. The $tRNA^{Leu}$ also has the unusual ability to translate all six Leu codons (177).

The Ile codons, AUN, are a special case since the AUG has been taken over by Met. In eukaryotes, AUC_A is read by an I_{34}-containing tRNA, but in *E. coli* for some reason, a G_{34}-containing tRNA is used for AU^U_C, and a specific C^*_{34}-containing tRNA is used to read AUA (Table 2). Presumably, the C^* modification is better able to pair with A while avoiding a $C^* \cdot G$ pair, than a modified U. Pairing with G must be avoided or isoleucine will go where methionine should be. Alternatively, since the $tRNA^{Ile}$ reading AUA is present in only very small amounts relative to $tRNA_m^{Met}$, it is possible that the C^* modi- fication allows more effective competition with $tRNA_m^{Met}$ for AUA than does the U^* modification. Acetylation of C_{34} in $tRNA_m^{Met}$ also modifies this competition. Removal of the acetyl group by bisulfite treatment allows misreading of AUA (178). Acetylation is not needed in eukaryotic $tRNA_m^{Met}$ because the Ile codons are read by a single I_{34}-containing tRNA (53), which is a better competitor both because of I-A pairing and because there is more of this tRNA.

The last point to note is that A_{34} is never found in a mature tRNA. It is found in several DNA sequences and *in vitro* transcrip- tion products of tRNAs expected to contain I_{34}. Presumably, deami- nation of A occurs as a late maturation event. The most reasonable explanation for the lack of A_{34} is that a second tRNA would be required for the XYC codon, so that it is more economical to use G or a derivative.

Role of N_{37}. Three **classes** of modified bases are found next to the 3'-side of the anticodon. (1) In most cases, but not all, when the adjacent base is A_{36}, N_{37} is i^6A, ms^2i^6A, or yW. These are all bulky hydrophobic bases. (2) In most cases where U_{36} is present, N_{37} is t^6A, mt^6A, on ms^2t^6A. These are also bulky but hydrophilic. The only known exception is prokaryote $tRNA_f^{Met}$ with A_{37} even though the organism possesses the modifying enzyme (cited in ref. 2). (3) When C_{36} ir G_{36} are present, methylated G_{37}, I_{37} or A_{37}, or unmodi- fied A_{36} are found. Evidence exists that the 1st and 2nd class of modifications are necessary for optimal ribosomal decoding to occur (161, 164, 179, 180), but why they should be required is unknown. On the other hand, other experiments show that tRNAs lacking i^6A can make proteins normally (53).

It has been suggested that N_{37} modifications are designed to modulate the strength of the codon-anticodon interaction so that all tRNAs, irrespective of the number of G-C base-pairs will form a codon-anticodon complex of approximately equal strength. While there is evidence for such a modulation effect (181), it is not clear why the absence of the modified bases in mutants should have such a deleterious effect on protein synthesis, if their only function is to fine-tune codon recognition.

Another role for t^6A is suggested by the G·U pair stacking effect mentioned above. It is known that prokaryote $tRNA_f^{Met}$ can read GUG as well as AUG, indicating a G·U wobble in the N_{36} position. This is understandable from Sundaralingam's results (174) since in this case, the G·U pair can stack on the middle anticodon base pair. Other tRNAs with U_{36}, in particular eukaryotic $tRNA_i^{Met}$, do not wobble (181a). The phenomenological reason for this appears to be that they are all followed by t^6A_{37} or mt^6A_{37} (53), which has been shown to have the possibility of contacting and modifying the base-pairing potential of U_{36} (182).

Finally, it should be noted that all modifications of N_{37} block standard base-pairing, while all modifications to N_{34} do not. This suggests that an additional function of N_{37} may be to define the reading frame on the 3' side of the anticodon, which is stacked on N_{37} and N_{38}. The 5'-side, being defined by the U-turn between U_{33} and N_{34}, may not need any special mechanism.

Effect of other sites. There is one example of the long-range effect of base modifications elsewhere, albeit in a prokaryote. This is the well-known change of $G_{24}·U_{11}$ in stem a of E. coli $tRNA^{Trp}$ to $A_{24}·U_{11}$. Changing this G U pair to an A·U pair allows the $tRNA^{Trp}$ anticodon, CCA, to translate the termination codon, UGA, in addition to UGG, and also to translate the cysteine codons, UG_C^U, in the absence of the normal $tRNA^{Cys}$ (183). The anticodon loop appears normal since both UGA and UGG are bound with the same affinity, but other evidence indicates that the tertiary structure in the neighborhood of stem a is perturbed. That this can effect codon recognition is shown by the fact that UV-crosslinking of $^4S_8$ to C_{13} (184) in the $A_{24}·U_{11}$ suppressor tRNA causes the loss of in vivo translation of UGA, i.e., the tRNA is no longer a suppressor. These results show rather clearly that modification of the tRNA elsewhere than in the anticodon can affect codon recognition. The most likely explanation is that the first modification increased the overall affinity of the tRNA for the ribosome so that two base-pairs suffice in the codon-anticodon complex, while the second modification reversed the increased affinity. In this connection, note that the crosslinked C_{13} is only two bases away from the U_{11} involved in the modified base-pairs. A similar situation may exist in rabbit reticulocytes where a minor tRNA^{Trp} species has been shown to read UGA (185), although the nature of the modification is not known in this case.

The second example is of a tRNA^{Arg} modified at s^2C_{32}, two residues away from the anticodon base I_{34} (166). Whereas the tRNA containing either s^2C_{32} or C_{32} can recognize the codons CGA, as well as CGG, alkylation of the s^2C residue which protonates N_1, also restricts codon recognition to mainly only CGU. Presumably, the positive charge on the s^2C_{32} ring in some way perturbs all of the I wobble pairs except the most stable one, I·U, but how this occurs is unknown.

Codon Translation in Mitochondria

Role of N_{34}. As pointed out earlier, mitochondrial tRNAs of yeast and mammals have only 23-24 distinct tRNAs. Since 32 tRNAs are the minimum required according to the Wobble Hypotehsis (see above), and cytoplasmic tRNAs are not used, either codon usage in mitochondria is restricted, degenerate codons are recognized by a mechanism other than "wobbling", or a modified genetic code is used. In fact, all three methods are used. Recent work has shown that in yeast (60), N. crassa (ref. 1 of Table 2), and mamalian (63) mitochondria, a single tRNA with U_{34} is used in all of the cases where a set of four codons XYN codes for the same amino acid. This corresponds to U_{34} pairing with U and C in addition to A and G. Such pairing, while unusual, has been described previously (ref. 40 of Table 2, see also 181). Thus mitochondria use one tRNA for this series of codons instead of the two or more used in the cytoplasm (see above). For those codons XY_G^A where XY_C^U stand for a different amino acid, mito-chondria use a single tRNA with a modified U_{34}. This has been explicitly shown only for N. crassa (ref. 1 of Table 2), but is probably also true for the other species where the DNA sequence can only show the presence of a U (60, 63). It appears that this unknown modification of U (ref. 1 of Table 2) can allow, at least on the mitochondrial ribosome, both A and G pairing but with exclusion of U and C interaction. This is something that neither prokaryotic nor eukaryotic cytoplasm has been able to do (see above). As a result of these two simple modifications of the genetic code, only 24 tRNAs are now required.

There are two further points to note. tRNA^{Trp} also has a modified U_{34} in its anticodon. This has been directly shown for N. crassa (ref. 30 of Table 2) and is inferred from the DNA sequence of the yeast tRNA (ref. 31 of Table 2). Consequently, UGA codes for Trp instead of for termination, as had been inferred previously both for yeast (186, 187) and for human (71) mitochondria, and directly demonstrated for a yeast mRNA (187a). Sharing of the AUN codon set by Ile and Met is handled differently in yeast and in mammals. In yeast tRNA^{Ile} has G_{34} or more probably G_{34}^*, but is known to read AUA in addition to AU_C^U (60). Recall that in E. coli, AUA is read by UAC* (ref. 7 of Table 2), and that an R.A* pair occurs in a tRNA^{Leu} (172). tRNA^{Met} is normal with C$_{34}$. In mammals, tRNE^{Ile} has G$_{34}$, which might be modified, since only the sequence of the gene is known, but tRNA$^{Met}_{Met}$ is thought to have U$_{34}$, unlike all other tRNA^{Met} species. This would allow it to read $^{G}_{34}$AU$^{G}_{A}$, and there is some evidence for this (71).

Role of N$_{37}$. There is no information about any special features of N$_{37}$ modifications, except for the case of *N. crassa* tRNA^{Ala}. This tRNA has U$_{37}$ in the position which up to now has always been a purine.

Changes in the code. In yeast mitochondria, the codon CUA is read as Thr, not Leu, by an unusual tRNA with an extra base in its anticodon loop (ref. 29 of Table 2). CUU is also read as Thr, and it is likely that the entire codon set CUN has been changed from Leu to Thr, as only one tRNA^{Leu} (for UU$^{A}_{G}$) has been found in yeast (60). This is not the case in *N. crassa* or mammals where CUN stands for Leu (63, ref. 1 in Table 2).

There is a tRNA^{Arg} in yeast with A$_{34}$-C$_{35}$-G$_{36}$ which could read the codon CGU, but so far the CGN series has not been used (60). In mammals, there is a U$_{34}$C$_{35}$G$_{36}$-tRNA^{Arg} for the set CGN, but no tRNA^{Arg} for the codons AG$^{A}_{G}$. On the basis of analysis of the genes for cytochrome oxidase subunits II and IV, the AG$^{A}_{G}$ codons were suggested to be terminators (63), whereas in yeast, the AG$^{A}_{G}$ codons are used for Arg.

Codon Translation *in vivo*

Suppression. In yeast, UAA, UAG, and UGA codons are used for termination, as in prokaryotes. This has been shown by the isolation of appropriate suppressor tRNAs and analysis of the anticodon changes (188, 189). In higher eukaryotes, suppression can also be shown to occur both by the supplementation of *in vitro* systems with yeast suppressor tRNAs (190-192) and by polyamine additions (192) or an increase in the Mg concentration (190). In the latter cases, suppression occurs by translation of the terminator codons by a normal tRNA (190, 192), as was also mentioned above in the case of a minor tRNA^{Trp} species in reticulocyte lysates (185). This may be an important control mechanism in higher eukaryotes.

Mistranslation and "two out of three". Mistranslation, that is failure to follow the coding rules outlined above, has been shown to occur when CHO cells are deprived of certain AA-tRNAs either by use of synthetase ts mutants, or by amino acid analogs (193). In these cases, the data suggest that misrecognition is occurring at the 3'-codon base (N$_{34}$ in the anticodon). The same phenomenon has also been described in bacterial cells (194 and references therein).

Another kind of miscoding at the N$_{34}$ anticodon position that has so far been confined to bacterial cells has been termed "two out of three" reading by Lagerkvist (195 and references therein). In his *in vitro E. coli* system, $tRNA^{Val}$ and $tRNA^{Ala}$ isoacceptors respond to codons they should not recognize, although $tRNA^{Lys}$ does not. Thus, this type of misreading does not result in amino acid substitutions, in contrast to the effect of specific AA-tRNA deprivation described above. A similar but more restrictive result has been obtained in a study of $tRNA^{Leu}$ (196). In contrast, an *in vivo* study, again in *E. coli*, failed to find evidence for "two out of three" reading in the case of $tRNA^{Gly}$ (197).

Other aspects. Additional situations known to affect codon recognition are ribosomal frameshifting (198) and the effect of codon context on codon recognition (199). Both phenomena have so far only been shown to occur in prokaryotes, although the latter can be expected to be of more general occurrence. Context effects have been ascribed to secondary structural effects of the adjacent codon in the mRNA, but the more likely explanation is that they are related to interaction of the two tRNAs specified by the two adjacent codons when they are simultaneously bound at the A and P sites of the ribosome. It is not difficult on structural grounds to imagine that certain modified bases at N_{37} in the tRNA at the A site might interact with the modified bases at N_{34} in the P site tRNA and thus certain tRNA pairs might be incompatible, or possibly even particularly compatible.

The role of the ribosome in maintaining the fidelity of codon translation has been summarized recently by Yarus (200). He has emphasized that the low error rate of codon recognition cannot be explained on physical-chemical grounds, and that the ribosomal structure itself must play an important, although as yet undefined, role.

tRNA RECOGNITION BY THE EUKARYOTIC PROTEIN SYNTHESIS SYSTEM

Initiation and Elongation Factors

Very little is known about specific tRNA recognition features in eukaryotic initiation and elongation factors. As mentioned earlier, initiation factor eIF-2 can distinguish between the unformylated forms of prokaryotic $tRNA_f^{Met}$ and eukaryotic $tRNA_i^{Met}$, and it can also distinguish eukaryotic $tRNA_m^{Met}$ from $tRNA_i^{Met}$ (75). The latter recognition process is of course essential for proper initiation. The assumption that it is the GAUCG sequence in loop IV which is recognized has been clouded by the finding (mentioned earlier) that silkworm $tRNA^{Ala}$ which has no known initiation function has almost the identical sequence in loop IV.

Elongation factor 1 forms a ternary complex with AA-tRNA and GTP just as in prokaryotes, which is used to transfer AA-tRNA to the ribosomal A site. As in prokaryotes (2), neither deacylated nor

N-acylated AA-tRNAs can substitute for AA-tRNA (201). Weak complex formation occurs with Met-tRNA$_f^{Met}$, but it is clearly distinguishable from Met-tRNA$_m^{Met}$ or other elongator AA-tRNAs (2).

Ribosomes

The main recognition feature in tRNA for ribosomal binding in both prokaryotes and eukaryotes is considered to be the GTΨCR sequence in loop IV. It is usually thought to interact with 5S RNA by forming complementary base-pairs with the sequence CGAAC found in nearly the same position, residues 42-49, in all prokaryotic 5S RNA sequences (201a), although there is some evidence which contradicts this hypothesis (reviewed in 82). Eukaryotic 5S RNA does not have this sequence, except in plants. However, 5.8S RNA, a characteristic component of the large subunit of eukaryotic ribosomes, has two complementary sequences at residues 42-45 and 101-107 (201a), and is thought to be responsible for binding tRNA to eukaryotic ribosomes. Other homologies between 5S RNA of prokaryotes and 5.8S RNA of eukaryotes have been pointed out elsewhere (202). The 5S RNA of eukaryotes, on the other hand, has been postulated to specifically recognize eukaryotic initiator tRNA because, except for plants, it has the sequence YGAUC complementary to loop IV of eukaryotic initiator tRNA in the same region where the GAAC sequence is found in prokaryotic 5S RNA. The argument that 5S RNA in eukaryotes interacts with initiator tRNA, and that 5.8S RNA binds elongator tRNA is put forth in some detail in (2).

Plant 5S RNA is peculiar because it not only has a GAAC sequence located where it is found in prokaryotes, but it also has a GAUC sequence about 10 residues to the 5' side of the GAAC. This second GAUC sequence, between residues 30-34, is also found in all other eukaryotic 5S RNAs, except for yeasts and *Drosophila*. In these species, there is only the single GAUC at residues 41-44. Unfortunately for the hypothesis that 5S RNA recognizes initiator tRNA, the tRNA$_i^{Met}$ of both wheat germ and bean do not have the prokaryotic GTΨCR sequence in loop IV, but rather a normal eukaryotic GAΨC, despite the presence of GAAC in their respective 5S RNAs. Thus, the true role of 5S and 5.8S RNAs in ribosomal recognition of tRNA is still unclear.

A related issue is the effect of the presence or absence of rT_{54} on eukaryotic protein synthesis. Dudock showed that the rate of wheat germ protein synthesis using a natural mRNA and wheat germ tRNA^{Gly} which normally has U_{54} was reduced in half when the U_{54} was converted to rT_{54} by an *E. coli* methylating enzyme (203). He also showed that the effect could be abolished by addition of spermine. Subsequently, Roe did the reciprocal experiment. He compared submethylated tRNA^{Phe} with rT-containing tRNA^{Phe} in an *in vitro* rat liver system with polyU as mRNA. When the normal tRNA^{Phe} was fully rT_{54}, it was *more* active than the partially U_{54}-containing tRNA^{Phe}

(204). The effect of spermine was not tested. Similar results have been described *in vivo*. Kersten has found a preference for rT_{54}-containing tRNAs vs. U_{54}-containing species on polysomes in the slime mold by analyzing the content of polysome-bound tRNAs as compared to total tRNA (205). These reports suggest that *in vivo* as well as *in vitro* , methylation of U_{54} can serve to modulate the activity of a tRNA in ribosomal protein synthesis. This may be related to the loop IV hypothesis, but especially in view of the spermine effect and the reverse effect of methylation in wheat germ ^{Gly}, it seems more likely that methylation or the lack of it is used to "fine-tune" tRNA to its optimal structure for proper functioning. Some tRNAs may need methylation and some may be better off without it.

Recognition between tRNA and ribosomes from the viewpoint of the ribosome is dealt with in the chapter by I. Wool (this volume). In the context of tRNA discrimination, it is worth noting here that the rat liver ribosomal proteins L6, L35a, and S15 were able to form specific complexes with yeast elongator tRNAs, but only two of them, L6 and L35a, could also complex with yeast $tRNA_i^{Met}$ (205a). Moreover, L6 was also able to form complexes with both $5S$ and $5.8S$ RNA (205b).

OTHER FUNCTIONS OF tRNA

tRNA-Like Structures in Viral RNA

Ten years ago, it was reported that the RNA of TYMV could be aminoacylated at its 3'-end with valine. Subsequently, this phenomenon has been extended to a number of plant viruses as well as to animal viruses of the picornavirus group. The subject has been reviewed in detail recently (206) so that only a brief summary will be given here. All of the following statements have not been examined with each of the known viral RNAs, but no contradictions are known.

Both the viral RNA as well as tRNA-sized fragments from the 3'-end can be aminoacylated. Synthetases from *E. coli*, yeast, plants, and mammals can catalyze the incorporation as can *Xenopus* oocytes. In all cases tested, the host enzyme can also catalyze the reaction. Esterification is via a normal AA-tRNA bond (Fig. 1) to the 3'-terminal A of a CCA-terminus on the RNA, either naturally present or added by CCA-pyrophosphorylase. With TYMV RNA and yeast E^{Val}, the K_m and V_{max} were very similar to that using yeast $tRNA^{Val}$. There is specificity for the amino acid attached, which depends on the particular viral RNA. So far, His, Tyr, and Val have been found to acylate plant viral RNA, and His and Ser the picornavirus RNAs. Additional enzymes that can recognize the 3'-end of viral RNA as a tRNA-like structure are peptidyl-tRNA hydrolase, wheat germ EF-1, and *E. coli* U_{54} methylase. Aminoacylated viral RNA cannot function in any *in vitro* ribosomal system yet tried.

The complete primary structure of the 3'-regions of TYMV and BMV

RNAs have been determined as well as 48 residues of TMV RNA. While
they can be folded up into a structure with stems and loops, they
only crudely resemble a tRNA cloverleaf. Moreover, they have no
modified nucleotides at all. These facts reinforce two points made
earlier in this review. First, aspects of *tertiary* structure determine synthetase recognition, and second, minor bases have essentially
no role to play in either the specific or general aspects of aminoacylation. It would seem that a study of the interaction of the
3'-fragments of TYMV RNA with EV^{al} of yeast by the various methods
cited earlier in this review should be very helpful for understanding
the problem of tRNA-synthetase recognition, particularly since the
tRNA-like structure is so aberrant. However, no such studies have
yet been carried out.

Why these structures are present in viral RNA is not known.
Haenni and Chapeville (206) speculate that their role may be to
inhibit host protein synthesis possibly by competing for one of the
host isoacceptor tRNAs not needed for virus synthesis. The virus
could afford to sacrifice the 3'-portion of some of its RNA by an
RNA processing activity in the host in order to generate a specific
host inhibitor.

<u>Primer for Reverse Transcriptase</u>

Replication of the retroviruses requires a tRNA primer for
initiation of synthesis of the DNA strand complementary to the virion
RNA (207). For the avian sarcoma/leukosis viruses, (ASV/ALV), the
primer is the host cell tRNA^{Trp} and for murine leukemia virus (MuLV),
it is tRNA^{Pro}. Recently, tRNA$_3^{Lys}$ has been shown to be the primer for
mouse mammary tumor virus. Priming occurs by reverse transcriptase-
induced hybridization of the 3'-region of the tRNA, residues 60-75
for tRNA^{Trp} and residues 58-75 for tRNA^{Pro}, to a region near to the
5'-end of the viral RNA. Priming occurs by addition to the 3'-end
of the tRNA.

In these viruses, the primer tRNA as well as a number of other
tRNAs can be found within the virion. These are a specific subset
of the host cell tRNA population which is relatively strongly bound
to the viral reverse transcriptase. The evidence for this is (a)
there are about the same number of tRNAs as enzyme molecules in the
virion, (b) mutant viruses lacking reverse transcriptase also lack
the tRNAs, and (c) binding of the same subset of tRNAs can be duplicated using pure enzyme *in vitro* (207). Only one of these specific
tRNAs is used as the primer, even though the others are also in the
virion.

Three distinct events must occur for tRNA to function as a
primer. First, the enzyme must recognize and bind strongly to a
specific tRNA species. Second, the enzyme must unwind and hybridize
stems <u>e</u> and <u>d</u> to the viral RNA. Third, the enzyme must initiate DNA

synthesis by addition of deoxynucleotides to the 3'-end of the tRNA. The structural features of tRNA required for each of these steps are examined here.

Strong binding to the reverse transcriptase requires all of the tRNA sequence (208) except for residues 67-76 (209), the 3'-terminal nonamer. Aminoacylation, opening of the 3'-terminal ribose ring by periodate oxidation, or even removal of the 3'-terminal A, does not affect binding (207), nor does cleavage at the anticodon so long as the two halves are not separated (207, 208). On the other hand, there is a strong tRNA specificity, and a requirement for a particular conformation. Only $tRNA^{Trp}$ and $tRNA_4^{Met}$ of chicken cells can bind with high affinity to the ASV/ALV enzyme, the complex of $tRNA^{Trp}$ with its hybridized fragment of viral RNA does not bind, and only a fraction of the $tRNA^{Trp}$ of the cell can bind to the enzyme with high affinity (208, 209a). In this latter case, the sequence of the $tRNA^{Trp}$ which binds to the enzyme and that which does not bind were carefully examined and shown to be identical (209a), demonstrating that the binding ability is conformation-dependent in some as yet unexplained manner. Interestingly, the enzyme-bound conformer was also the one which could be misaminoacylated with methionine (209a).

Only the 3'-terminal region, 60-75 in $tRNA^{Trp}$ and 58-75 in $tRNA^{Pro}$ are involved in base-pairing with the viral RNA (207). By contrast, initiation of DNA synthesis requires residues 50-76 (208), that is almost all of arm IV plus the 3'-side of stem e. Note that initiation of DNA synthesis must include binding to the reverse transcriptase. Clearly stable binding, as described above, is not a prerequisite for efficient initiation of DNA synthesis.

It is interesting to note that although only part of arm IV is actually hybridized to the viral RNA, the entire loop and double-stranded stem is needed for initiation. This may reflect a requirement for the loop-stem structure, for the specific sequence $\Psi_{54}\Psi_{55}$ which is present in both $tRNA^{Pro}$ and $tRNA^{Trp}$ primers, or for both elements. In this connection, it would be interesting to know the loop IV sequence of the $tRNA_3^{Leu}$, $tRNA^{Asp}$, and $tRNA^{Gly}$ species implicated as being retrovirus primers (210). Presumably, the tRNA sequence and the viral RNA initiation site co-evolved in order to maximize base-pairing, but why these particular tRNAs were selected as primers, why tRNAs are used as primers at all, and how priming of the second or (+) DNA strand is accomplished, are all unknown areas.

Aminoacyl-tRNA Protein Transferase

An enzyme exists in all nucleated eukaryotic cells which cata-lyze the transfer of Arg from Arg-tRNA to the NH_2-terminus of proteins or peptides (211). The enzyme can use Arg-tRNA from E. coli or rat liver and probably from any source. As all isoacceptors are utiliz-able, it is likely that any tRNA or 3'-fragment of tRNA would be

acceptable, but such donor specificity studies have not been carried
out. The α-NH$_2$ group of the Arg must not be blocked. The acceptor
amino acid can only be Glu, Asp, or Cys, and only when they are
exposed residues in the proteins or peptide. In some cases, stoichiometric addition of Arg is found. The acceptor can also be as small as
a dipeptide, if it has the right N-terminus and is made from L-amino
acids, but larger peptides are better. The enzyme is *not* required
for cell viability as yeast mutants totally lacking the activity will
still grow. Its function is totally obscure.

A similar enzyme with different donor and acceptor amino acid
specificity has also been found in gram negative bacteria. It is
also not necessary for viability, and has no known function.

Regulatory Functions

Unlike the situation in bacteria, it has been difficult to
clearly establish regulatory functions for AA-tRNA in the metabolism
of eukaryotic cells. Most such studies have used either temperature-
sensitive mutants in an AA-tRNA synthetase or an amino acid analog
to block AA-tRNA synthesis, followed by an analysis of the result
according to the bias of the investigator. In one such series of
studies in CHO cells, earlier indications by the same group that Asp-
tRNA was involved in the derepression of asparagine synthetase was
shown to be incorrect. In this case, although asparagine synthetase
activity did vary according to the *in vivo* level of Asp-tRNA induced
either by removal of Asp or use of a ts mutant, the same effect was
found when ts mutants of E^{Leu}, E^{Met}, or E^{Lys} were used (212). Clearly
there *is* regulation of enzyme activity levels by the concentration of
AA-tRNA or some consequence of it, but it is not specific to Asp-tRNA.
It was also not related to a general decreased rate of protein synthesis, as shown by the use of antibiotic inhibitors. In another
study, decreasing the rate of protein synthesis in CHO cells by either
the use of histidinol to block E^{His} activity, or of E^{His} ts mutants
in combination with histidinol, led to an increased rate of protein
degradation (213). Inhibition of protein synthesis by cycloheximide
did not stimulate degradation, leading the authors to suggest a
functional connection between general AA-tRNA levels and the regulation of protein degradation.

No mechanism or direct connection between AA-tRNA and the effects
observed were shown in either of these studies, and it must be concluded that while regulation in this way is an attractive hypothesis
in view of the prokaryotic examples (214-216), there is so far no
evidence to support it.

tRNA BIOSYNTHESIS

Since the subject of tRNA biosynthesis is not properly part of
this review, the following paragraphs will only briefly summarize

this topic. The reader is referred to the monograph by Soll et al. (7) and especially the review chapters on tRNA biosynthesis (217) and tRNA gene organization (218) for a detailed analysis. tRNA genes in eukaryotes are amplified several hundred to several thousand times. They are not associated with the nucleolar organizer as are the rRNA genes, but rather are disperse among many chromosomes. In some species, like *Xenopus* multiple gene copies appear to be in tandem while in other cases, as in yeast, they are dispersed. Most eukaryotic tRNA precursors which have been identified are monomeric, although dimers have also been detected (219, 220). The precursors contain a small number of extra nucleotides at each end, and lack the common CCA end which is added later. Some, but not all, precursor tRNAs of eukaryotes contain an intervening sequence of 14-34 bases inserted at the 3' side of the anticodon loop junction with its stem. As in eukaryotic mRNAs, this sequence is removed and the two newly-made ends of the tRNA spliced to produce the mature molecule (221, 222). Although it had been thought that splicing, CCA addition, and some removal of the termina extra nucleotides occurred in the cytoplasm, recent experiments indicate that all these events occur in the nucleus at least in *Xenopus* oocytes (223).

The minor bases are all made by modification of their parent in the precursor polynucleotide chain except for the base Q which is inserted by base exchange with G (53). Earlier studies had not indicated any specific minor base requirement in any of the processing steps nor a restriction to any particular precursor stage, but the same recent *Xenopus* experiments (223) again show that this is not the case. At least four modifications occur in the nucleus at defined stages of the processing sequence. The only modification that definitely occurs in the cytoplasm is Gm formation.

ACKNOWLEDGEMENT

This review could not have been written without the assistance of the transcriber of these words, Barbara Kerr. She has my heartfelt thanks for her extraordinary skill, dedication, and endurance.

REFERENCES

1. Crick, F.H.C. (1957), Biochem. Soc. Symp., 14, 25-26.
2. Ofengand, J. (1977), In *Molecular Mechanisms of Protein Biosynthesis*, (Eds. H. Weissbach and S. Pestka), pp. 7-79, Academic Press, New York.
3. Goddard, J.P. (1977), Progr. Biophys. Molec. Biol., 32, 233-308.
4. Schimmel, P.R. and Soll, D. (1979), Ann. Rev. Biochem., 48, 601-648.
5. Altman, S. (Ed.) (1978), *Transfer RNA*, M.I.T. Press, Cambridge, Massachusetts.

6.   Schimmel, P.R., Soll, D. and Abelson, J.N. (Eds.) (1979),
     Transfer RNA: Structure, Properties, and Recognition, Cold
     Spring Harbor Laboratory, Cold Spring Harbor, New York.

7.   Soll, D., Abelson, J.N. and Schimmel, P.R. (Eds.) (1980),
     Transfer RNA: Biological Aspects, Cold Spring Harbor Laboratory,
     Cold Spring Harbor, New York.

8.   Chousterman, S., Herve, G. and Chapeville, F. (1966), Bull. Soc.
     Chim. Biol., 48, 1295-1303.

9.   Schofield, P. and Zamecnik, P.C. (1968), Biochem. Biophys. Acta,
     155, 410-416.

10. Ofengand, J., Chladek, S., Robilard, G. and Bierbaum, J. (1974),
     Biochemistry, 13, 5425-5432.

11. Sprinzl, M. and Cramer, F. (1979), Progr. Nucleic Acid Res. Mol.
     Biol., 22, 1-69.

12. Julius, D.J., Fraser, T.H. and Rich, A. (1979), Biochemistry, 18
     604-609.

13. Hecht, S.M. (1979), In *Transfer RNA: Structure, Properties, and
     Recognition,* (Eds. P.R. Schimmel, D. Söll and J.N. Abelson), pp.
     345-360, Cold Spring Harbor Laboratory, Cold Spring Harbor, New
     York.

14. Sprinzl, M. and Cramer, F. (1975), Proc. Natl. Acad. Sci. USA,
     72, 3049-3053.

15. Chinault, A.C., Tan, K.H., Hassur, S.M. and Hecht, S.M. (1977),
     Biochemistry, 16, 766-776.

16. Fraser, T.H. and Rich, A. (1975), Proc. Natl. Acad. Sci. USA, 72
     3044-3048.

17. Zachau, H.G. and Feldman, H. (1965), Prog. Nucleic Acid Res.,
     4, 217-230.

18. Alford, B. and Hecht, S.M. (1978), J. Biol. Chem., 253, 4844-
     4850.

19. Igloi, G.L. and Cramer, F. (1978), FEBS Lett., 90, 97-102.

20. Alford, B.L. and Hecht, S.M. (1979), J. Biol. Chem., 254, 6873-
     6875.

21. Loftfield, R.B. (1972), Prog. Nucleic Acid Res. Mol. Biol., 12,
     87-128.

22. Lovgren, T.N.E., Heinonen, J. and Loftfield, R.B. (1975), J.
     Biol. Chem., 250, 3854-3860.

23. Fersht, A.R., Gangloff, J. and Dirheimer, G. (1978), Bio-
     chemistry, 17, 3740-3746.

24. Fasiolo, F. and Fersht, A.R. (1978), Eur. J. Biochem., 85, 85-88.

25. Mulvey, R.S. and Fersht, A.R. (1978), Biochemistry, 17, 5591-
     5597.

26. Fersht, A.R. and Kaethner, M.M. (1976), Biochemistry, 15, 818-
     823.

27. Kern, D. and Lapointe, J. (1980), J. Biol. Chem., 255, 1956-1961.

28. Thiebe, R. (1978), Nucleic Acids Res., 5, 2055-2071.

29. Lui, M., Chakraburtty, K. and Mehler, A.H. (1978), J. Biol.
     Chem., 253, 8061-8064.

30. Kim, J.J.P., Chakraburtty, K. and Mehler, A.H. (1977), J. Biol.
     Chem., 252, 2698-2701.

31. Lagerkvist, U., Akesson, B. and Branden, R. (1977), J. Biol.
     Chem., 252, 1002-1006.
32. Thiebe, R. (1977), FEBS Lett., 79, 212-214.
33. Lovgren, T.N.E., Petersson, A. and Loftfield, R.B. (1978), J.
     Biol. Chem., 253, 6702-6710.
34. Hyafil, F. and Blanquet, S. (1977), Eur. J. Biochem., 74, 481-
     493.
35. Igarashi, K., Eguchi, K., Tanaka, M. and Hirose, S. (1978), Eur.
     J. Biochem., 82, 301-307.
36. Kim, S.-H. (1979), In *Transfer RNA: Structure, Properties and
     Recognition*, (Eds. P.R. Schimmel, D. Söll and J.N. Abelson),
     pp.83-110, Cold Spring Harbor Laboratory, Cold Spring Harbor,
     New York.
37. Rich, A., Quigley, G.J., Teeter, M.M., Decruix, A. and Woo, N.
     (1979) In *Transfer RNA: Structure, Properties, and Recognition*,
     (Eds, P.R. Schimmel, D. Söll and J.N. Abelson), pp. 101-113,
     Cold Spring Harbor Laboratory, Cold Spring Harbor. New York.
38. Jabubowski. H. (1980). FEBS Lett.. 109. 63-66.
39. Igloi. C.L. and Cramer. F. (1978). In *Transfer RNA: Structure
     Properties and Recognition*. (Ed. S. Altman) pp. 294-349. M.I.T.
     Press. Cambridge. Massachusetts.
40. Godeau. J.-M. (1980), Eur. J. Biochem., 103, 169-177.
41. Merault, G., Graves, P.V., Labouesse, B. and Labouesse, J. (1978)
     Eur. J. Biochem., 87, 541-550.
42. Yarus, M. and Berg, P. (1969), J. Mol. Biol., 42, 171-189.
43. Delorenzo, F. and Ames, B.N. (1970), J. Biol. Chem., 245 1710-
     1716.
43a. Blanquet. S.. Iwatsubo. M. and Waller. J.P. (1973). Eur. J.
     Biochem.. 36. 213-226.
44. Bonnet. J. and Ebel. J.P. (1974), FEBS Lett., 39, 259-262.
45. Furano, A.V. (1976), Eur. J. Biochem., 64, 597-606.
46. Jacobson, G.R., Takacs, B.J. and Rosenbusch, J.P. (1976),
     Biochemistry, 15, 2297-2303.
47. Pingoud, A. and Urbanke, C., (1980), Biochemistry, 19, 2108-2112.
48. Knowlton, R.G. and Yarus, M. (1980), J. Mol. Biol., 139, 721-
     732.
48a. Slobin, L.I. (1980), Eur. J. Biochem., in press.
49. Weber, L. and Berger, E. (1976), Biochemistry, 15, 5511-5519.
50. Rogg, H., Muller, P., Keith, G. and Staehelin, M. (1977), Proc.
     Natl. Acad. Sci. USA, 74, 4243-4247.
51. Kitchingham, G.R. and Fournier, M.J. (1977), Biochemistry, 16,
     2213-2220.
52. Vold, B.S. (1978), J. Bact., 135, 124-132.
53. Nishimura, S. (1979), In *Transfer RNA: Structure, Properties,
     and Recognition*, (Eds. P.R. Schimmel, D. Söll and J.N. Abelson),
     pp. 59-79, Cold Spring Harbor Laboratory, Cold Spring Harbor,
     New York.
54   Nishimura, S. (1978), In *Transfer RNA*, (Ed. S. Altman), pp. 168-
     195, M.I.T. Press, Cambridge, Massachusetts.
54a Nishimura, S. (1979), Gann Monograph on Cancer Research, 24,
     245-262.

55. Petrissant, G. and Favre, A. (1972), FEBS Lett., 23, 191-194.
56. Kowalski, S. and Fresco, J.R. (1971), Science, 172, 384-385.
57. Sprinzl, M., Grueter, F., Spelzhaus, A. and Gauss, D.H. (1980), Nucleic Acids Res., 8, r1-r22.
58. Dirheimer, G., Keith, G., Sibler, A.-P. and Martin, R.P. (1979), In *Transfer RNA: Structure, Properties, and Recognition*, (Eds. P.R. Schimmel, D. Söll and J.N. Abelson), pp. 19-41, Cold Spring Harbor Laboratory, Cold Spring Harbor, New York.
59. Barnett, W.E., Schwartzbach, S.D. and Hecker, L.I. (1978), Progr. Nucleic Acid Res. Mol. Biol., 21, 143-179.
60. Bonitz, S.G., Berlani, R., Coruzzi, G., Li, M., Macino, G., Nobrega, F.G., Nobrega, M.P., Thalenfeld, B.E. and Tzagoloff, A. (1980), Proc. Natl. Acad. Sci. USA, 77, 3167-3170.
61. Martin, N.C. and Rabinowitz, M. (1978), Biochemistry, 17, 1628-1634.
62. Wesolowski, M. and Fukuhara, H. (1979), Mol. gen. Genet., 170, 261-275.
63. Barrell, B.G., Anderson, S., Bankier, A.T., Debruijn, M.H.L., Chen, E., Coulson, A.R., Drouin, J., Eperon, I.C., Nierlich, D.P., Roe, B.A., Sanger, F., Schreier, P.H., Smith, A.J.H., Staden, R. and Young, I.G. (1980), Proc. Natl. Acad. Sci. USA, 77, 3164-3166.
64. Martin, R.O., Schneller, J.-M., Stahl, A.J.C and Dirheimer, G. (1979), Biochemistry, 18, 4600-4605.
65. Crick, F.H.C. (1966), J. Mol. Biol., 19, 548-555.
66. Singhal, R.P. and Fallis, P.A.M. (1979), Progr. Nucleic Acid Res. Mol. Biol., 23, 228-290.
67. Rich, A. and Rajbhandary, U.L. (1976), Ann. Rev; Biochem., 45, 805-860.
68. Clark, B.F.C. (1978), In *Transfer RNA*, (Ed. S. Altman), pp. 14-47, M.I.T. Press, Cambridge, Massachusetts.
69. Clark, B.F.C. (1980), In *Ribosomes: Structure, Function and Genetics*, (Ed. G. Chambliss, G. Craven, J. Davies, L. Kahan and M. Nomura), pp. 413-444, University Park Press, Baltimore, Maryland.
70. Eperon, I.C., Anderson, S., and Nierlich, D.P. (1980), Nature, 286, 460-467.
71. Barrell, B.G., Bankier, A.T. and Drouin, J. (1979), Nature, 282, 189-194.
72. Debruijn, M.H.L., Schreier, P.H., Eperon, I.C., Barrell, B.G., Chen, E.Y., Armstrong, P.W., Wong, J.F.H., and Roe, B.A. (1980), Nucleic Acids Res., 8, in press.
72a. Taylor, R.H., Varricchio, F. and Dubin, D.T. (1980), Biochim. Biophys. Acta, 607, 521-526.
73. Borst, P. and Grivell, L.A. (1978), Cell, 15, 705-723.
74. Friedman, S. (1972), Biochemistry, 11, 3435-3443.
75. Filipowicz, W., Sierra, J.M. and Ochoa, S. (1975), Proc. Natl. Acad. Sci. USA, 72, 3947-3951.
76. Deutscher, M. (1973), Progr. Nucleic Acid Res. and Mol. Biol., 13, 51-92.

77. Kim, S.-H. (1978), Adv. Enzymol., 46, 279-315.
78. Kim, S.-H. (1978), In *Transfer RNA*, (Ed. S. Altman), pp. 248-293, M.I.T. Press, Cambridge, Massachusetts.
79. Schevitz, R.W., Podjarny, A.D., Krishnamachari, N., Hughes, J.J. and Sigler, P.B. (1979), Nature, 278, 188-190.
80. Woo, N.H., Roe, B.A. and Rich, A. (1980), Nature 286, 346-351.
81. Moras, D., Comarmond, M.B., Fischer, J., Weiss, R. Thierry. J.C., Giege, R., Dietrich, A. and Ebel, J.P. (1980), Abstracts EMBO-FEBS tRNA Workshop, Strasbourg, France.
82. Ofengand, J. (1980), In *Ribosomes: Structure, Function, and Genetics*, (Eds. G. Chambliss, G. Graven, J. Davies, L. Kahan and M. Nomura), pp. 497-529, University Park Press, Baltimore, Maryland.
83. Holbrook, S.R., Sussman, J.L., Warrant, R.W. and Kim, S.-H. (1978), J. Mol. Biol., 123, 631-660.
84. Schevitz, R.W., Podjarny, A.D., Gross, M., Sussman, J.L. and Sigler, P.B. (1980), Abstracts EMBO-FEBS tRNA Workshop, Strasbourg, France.
85. Davanloo, P., Sprinzl, M., Watanabe, K., Albanai, M. and Kersten, H. (1979), Nucleic Acids Res., 6, 1571-1581.
86. Reid, B.R., McCollum, L., Ribeiro, N.S., Abbate, J. and Hurd, R.E. (1979), Biochemistry, 18, 3996-4005.
87. Hurd, R.E. and Reid, B.R. (1979), Biochemistry, 18, 4005-4011.
88. Hurd, R.E., Azhderian, E. and Reid, B.R. (1979), Biochemistry, 18, 4012-4017.
89. Hurd, R.E. and Reid, B.R. (1979), Biochemistry, 18, 4017-4024.
90. Schimmel, P.R. and Redfield, A,G. (1980), Ann. Rev. Biophys. Bioeng., 9, 181-221.
90a. Wrede, P., Woo, N.H. and Rich, A. (1979), Proc. Natl. Acad. Sci. USA, 76, 3289-3293.
90b. Geerdes, H.A.M., Van Boom, J.H. and Hilbers, C.W. (1980), J. Mol. Biol., 142, 195-217.
91. Negishi, K., Nishimura, S., Harada, F. and Hayatsu, H. (1979), Nucleic Acids Res., 6, 899-914.
92. Potts, R.O., Wang, C.-C., Fritzinger, D.C., Ford, JR., N.C. and Fournier, M.J. (1979), In *Transfer RNA: Structure, Properties, and Recognition*, (Eds. P.R. Schimmel, D. Soll and J.N. Abelson), pp. 207-220, Cold Spring Harbor Laboratory, Cold Spring Harbor, New York.
93. Davanloo, P., Sprinzl, M. and Cramer, F. (1979), Biochemistry, 18, 3189-3199.
94. Shulman, R.G., Hilbers, C.W., Soll, D.G. and Yang, S.K. (1974), J. Mol. Biol., 90, 609-611.
95. Shulman, R.G., Hilbers, C.W. and Miller, D.L. (1974), J. Mol. Biol., 90, 601-607.
96. Schoemaker, H.P. and Schimmel, P.R. (1976), J. Biol. Chem., 251, 6823-6830.
97. Jekowsky, E., Miller, D.L. and Schimmel, P.R. (1977), J. Mol. Biol., 114, 451-458.

98. Moller, A., Wild, U., Riesner, D. and Gassen, H.G. (1979), Proc. Natl. Acad. Sci. USA, 76, 3266-3270.

99. Wagner, R. and Garrett, R.A. (1979), Eur. J. Biochem., 97, 615-621.

100. Vlassov, V.V. and Mamayev, S.V. (1980), FEBS Lett., 113, 65-67.

100a. Farber, N. and Cantor, C. (1980), Proc. Natl. Acad. Sci. USA, 77, 5135-5139.

101. Woese, C. (1970), Nature, 226, 817-820.

102. Lake, J.A. (1977), Proc. Natl. Acad. Sci. USA, 74, 1903-1907.

103. Matzke, A.J.M., Barta, A. and Kuechler, E. (1980), Proc. Natl. Acad. Sci. USA, 77, 5110-5114.

104. Ofengand, J. and Liou, R. (1981), Biochemistry, 20, in press.

105. Cantor, C.R. (1979), In *Transfer RNA: Structure, Properties, and Recognition*, (Eds. P.R. Schimmel, D. Soll and J.N. Abelson), pp. 363-392, Cold Spring Harbor Laboratory, Cold Spring Harbor, New York.

106. Schneller, J.M., Schneller, C., Martin, R. and Stahl, A.J.C. (1976), Nucleic Acids Res., 3, 1151-1165.

107. Beauchamp, P., Horn, E.W. and Gross, S.R. (1977), Proc. Natl. Acad. Sci. USA, 74, 1172-1176.

108. Hecker, L.I., Egan, J., Reynolds, R.J., Nix, C.E., Schiff, J.A. and Barnett, W.E. (1974), Proc. Natl. Acad. Sci. USA, 71, 1910-1914.

109. Joachimiak, A. and Barciszewski, J. (1980), FEBS Lett., 119, 201-211.

110. Lin, C.-S., Irwin, R. and Chirikjian, J.G. (1979), Nucleic Acids Res., 6, 3651-3660.

111. Sternbach, H., Hellmann, K.P. and Cramer, F. (1980), Abstracts EMBO-FEBS tRNA Workshop, Strasbourg, France.

112. Diatewa, M. and Stahl, A.J.C. (1980), Biochem. Biophys. Res. Commun., 94, 189-198.

113. Dignam, J.D., Rhodes, D.G. and Deutscher, M.P. (1980), Biochem., 19, 4978-4984.

116. Rosa, M.D. and Sigler, P.B. (1977), Eur. J. Biochem., 78, 141-151.

117. Robbe-Saul, S., Fasiolo, F. and Boulanger, Y. (1977), FEBS Lett., 84, 57-62.

118. Mulvey, R.S. and Fersht, A. (1977), Biochemistry, 16, 4005-4013.

119. Fersht, A. (1975), Biochemistry, 14, 5-13.

120. Bruton, C.J. and Cox, L.A.-M. (1979), Eur. J. Biochem., 100, 301-308.

121. Zelwer, C., Risler, R. and Monteilheit, C. (1976), J. Mol. Biol. 102, 93-101.

122. Irwin, M.J., Nyborg, J., Reid, B.R. and Blow, D.M. (1976), J. Mol. Biol., 105, 577-586.

123. Dietrich, A., Giege, R., Comarmond, M.B., Thierry, J.C. and Moras, D. (1980), J. Mol. Biol., 138, 129-135.

124. Ussery, M.A., Tanaka, W.K. and Hardesty, B. (1977), Eur. J. Biochem., 72, 491-500.

125. Enger, M.D., Ritter, P.O. and Hampel, A.E. (1978), Bio-
     chemistry, 17, 2435-2438.
126. Kellermann, O., Brevet, A., Tonetti, H. and Waller, J.-P. (1979),
     Eur. J. Biochem., 99, 541-550.
127. Vadeboncoeur, C. and Lapointe, J. (1980), Brain Res., 188, 129-
     138.
128. Agris, P.F., Setzer, D. and Gehrke, C.W. (1977), Nucleic Acids
     Res., 4, 3803-3819.
129. Vadeboncoeur, C. and Lapointe, J. (1980), Eur. J. Biochem., in
     press.
130. Loftfield, R.B. and Vanderjagt, D. (1972), Biochemical Journal,
     128, 1353-1356.
131. Fersht, A.R. (1979), In *Transfer RNA: Structure, Properties,
     and Recognition*, (Eds. P.R. Schimmel, D. Soll and J.N. Abelson),
     pp. 247-254, Cold Spring Harbor Laboratory, Cold Spring Harbor,
     New York.
132. Fersht, A.R. and Dingwall, C. (1979), Biochemistry, 18, 2627-
     2631.
132a. Jakubowski, H. (1980), Biochemistry, 19, 5071-5078.
133. Fersht, A.R. and Dingwall, C. (1979), Biochemistry, 18, 1250-
     1256.
134. Igloi, G.L., Van der Haar, F. and Cramer, F. (1977), Bio-
     chemistry, 16, 1696-1702.
135. Mulvey, R.S. and Fersht, A.R. (1977), Biochemistry, 16, 4731-
     4737.
136. Fersht, A.R. and Dingwall, C. (1979), Biochemistry, 18, 1238-
     1244.
137. Fersht, A.R. and Kaethner, M.M. (1976), Biochemistry, 15, 3342-
     3346.
138. Fersht, A.R. and Dingwall, C. (1979), Biochemistry, 18, 1245-
     1249.
139. Igloi, G.L. Von der Haar, F. and Cramer, F. (1978), Biochemistry,
     17, 3459-3468.
140. Schimmel, P.R. (1979), Adv. Enzymol., 49, 187-222.
141. Schimmel, P.R. (1979), In *Transfer RNA: Structure, Properties,
     and Recognition*, (Eds. P.R. Schimmel, D. Soll and J.N. Abelson),
     pp. 297-310, Cold Spring Harbor Laboratory, Cold Spring Harbor,
     New York.
142. Ebel, J.-P., Renaud, M., Dietrich, A., Fasiolo, F., Keith, G.,
     Favorova, O.O., Vassilenko, S., Baltzinger, M., Ehrlich, R.,
     Remy, P., Bonnet, L., and Giege, R. (1979), In *Transfer RNA:
     Structure, Properties, and Recognition*, (P.R. Schimmel, D. Soll,
     and J.N. Abelson), pp. 325-343, Cold Spring Harbor Laboratory,
     Cold Spring Harbor, New York.
143. Krauss, G., Riesner, D. and Maass, G. (1977), Nucleic Acids
     Res., 4, 2253-2262.
144. Krauss, G., Coutts, S.M., Riesner, D. and Maass, G. (1978),
     Biochemistry, 17, 2443-2449.
145. Krauss, G., Von der Haar, F., and Maass, G. (1979), Biochemistry,
     18, 4755-4761.

146. Von der Haar, F. and Cramer, F. (1978), Biochemistry, 17, 3139-3145.
147. Von der Haar, F. and Cramer, (1978), Biochemistry, 17, 4059-4514.
148. Schulman, L.H. (1979), In *Transfer RNA: Structure, Properties and Recognition*, (Eds. P.R. Schimmel, D. Soll and J.N. Abelson), pp. 311-324, Cold Spring Harbor Laboratory, Cold Spring Harbor, New York.
149. Renaud, M., Ehrlich, R., Bonnett, J.H. and Remy, P. (1979), Eur. J. Biochem., 100, 157-164.
150. Ohtsuka, E., Nishikawa, S., Fukumoto, R., Uemura, H., Tanaka, T., Nakagawa, E., Miyake, T. and Ikehara, M. (1980), Eur. J. Biochem., 105, 481-487;
151. Iwata, K., Yagura, T., Takeishi, K. and Seno, T. (1980), Biochim. Biophys. Acta., 606, 262-273.
151a. Wright, H.T. (1980), FEBS Lett., 118, 165-171.
151b. Ramberg, E.S., Ishaq, M., Rulf, S., Moeller, B. and Horowitz, J. (1978), Biochemistry, 17, 3978-3985.
152. Rosa, J.J., Rosa, M.D. and Sigler, P.B. (1979), Biochemistry, 18, 637-647.
153. Baltzinger, M., Fasiolo, F. and Remy, P. (1979), Eur. J. Biochem. 97, 481-494.
154. Renaud, M., Dietrich, A., Giege, R. Remy, P. and Ebel, J.-P. (1979), Eur. J. Biochem., 101, 475-483.
155. Hountondji, C., Fayat, G. and Blanquet, S. (1979), Eur. J. Biochem., 102, 247-250.
156. Koontz, S.W. and Schimmel, P.R. (1979), J. Biol. Chem., 254, 12277-12280.
157. Koontz, S.W. and Schimmel, P.R. (1980), Abstracts EMBO-FEBS tRNA Workshop, Strasbourg, France.
158. Zaccai, G., Morin, P., Jacrot, B., Moras, D., Thierry, J.-C. and Giege, R. (1979), J. Mol. Biol., 129, 483-500.
159. Krauss, G., Peters, F. and Maass, G. (1976), Nucleic Acids Res. 3, 631-639.
160. Bjork, G.R. and Kjellin-Straby, K. (1978), J. Bacteriol., 133, 508-517.
161. Laten, H., Gorman, J. and Bock, R.M. (1980), in *Transfer RNA: Biological Aspects*, (Eds. D. Soll, J.N. Abelson and P.R. Schimmel), pp. 395-406, Cold Spring Harbor Laboratory, Cold Spring Harbor, New York.
162. Hoburg, A., Aschhoff, H.J., Kersten, H., Manderschied, U. and Gassen, H.G. (1979), J. Bacteriol., 140, 408-414.
163. Harris, J.S. and Randerath, K. (1978), Biochim. Biophys. Acta, 521, 566-575.
164. Miller, J.P., Hussain, Z. and Schweizer, M.P. (1976), Nucleic Acids Res., 3, 1185-1201.
165. Sen, G.C. and Ghosh, H.P. (1976), Nucleic Acids Res., 3, 523-535.
166. Kruse, T.A., Clark, B.F.C. and Sprinzl, M. (1978), Nucleic Acids Res., 5, 879-892.

167. Yokoyama, S., Miyazawa, T., Iitaka, Y., Yamaizumi, Z., Kasai, H. and Nishimura, S. (1979), Nature, 282, 107-109.
168. Roe, B., Michael, M. and Dudock, B. (1973), Nat. New Biol., 246, 135-138.
169. Arcari, P. and Hecht, S.M. (1978), J. Biol. Chem., 253, 8278-8284.
170. Langdon, S.P. and Lowe, G. (1979), Nature, 281, 320-321.
171. Kovaleva, G.K., Moroz, S.G., Favorova, O.O. and Kisselev, L.L. (1978), FEBS Lett., 95, 81-84.
172. Yamaizumi, Z., Kuchino, Y., Harada, F., Nishimura, S. and McCloskey, J.A. (1980), J. Biol. Chem., 255, 2220-2225.
173. Grosjean, H., Ballivian, L., DeHenau, S. and Bollen, A. (1979), Arch. Int. Physiol. Biochim., 87, 415-417.
174. Mizuno, H. and Sundaralingam, M. (1978), Nucleic Acids Res., 5, 4451-4461.
175. McNamara, A.L. and Smith, D.W.E. (1978), J. Biol. Chem., 253, 5964-5970.
176. Hillen, W., Egert, E., Lindner, H.J. and Gassen, H.G. (1978), FEBS Lett., 94, 361-364.
177. Weissenbach, J., Dirheimer, G., Falcoff, R., Sanceau, J. and Falcoff, E. (1977), FEBS Lett., 82, 71-76.
178. Stern, L. and Schulman, L.H. (1978), J. Biol. Chem., 253, 6132-6139.
179. Eisenberg, S.P., Yarus, M. and Soll, L. (1979), J. Mol. Biol., 135, 111-126.
180. Janner, F., Vogeli, G. and Fluri, R. (1980), J. Mol. Biol., 139, 207-219.
181. Grosjean, H.J., DeHenau, S. and Crothers, D.M. (1978), Proc. Natl. Acad. Sci. USA, 75, 610-614.
181a. Sherman, F., McKnight, G. and Stewart, J.W. (1980), Biochim. Biophys. Acta, 609, 343-346.
182. Parthasarathy, R., Ohrt, J.M. and Chheda, G.B. (1977), Biochemistry, 16, 4999-5008.
183. Buckingham, R.H. and Kurland, C.G. (1980), in *Transfer RNA: Biological Aspects*, (Eds. D. Soll, J.N. Abelson and P.R. Schimmel), pp. 421-426, Cold Spring Harbor Laboratory, Cold Spring Harbor, New York.
184. Favre, A., Yaniv, M. and Michelson, A.M. (1969), Biochem. Biophys. Res. Commun., 37, 266-271.
185. Geller, A.I. and Rich, A. (1980), Nature, 283, 41-46.
186. Fox, T.D. (1979), Proc. Natl. Acad. Sci. USA, 76, 6534-6538.
187. Macino, G., Coruzzi, G., Nobrega, F.G., Li, M. and Tzagoloff, A. (1979), Proc. Natl. Acad. Sci. USA, 76, 3784-3785.
187a. Deronde, A., Van Loon, A.P.G.M. and Grivell, L.A. (1980), Nature, 287, 361-363.
188. Piper P.W. (1980), In *Transfer RNA: Biological Aspects*, (Eds. D. Soll, J.N. Abelson and P.R. Schimmel), pp. 379-394, Cold Spring Harbor Laboratory, Cold Spring Harbor, New York.

189 . Kohli, J., Altruda, F., Kwong, T., Rafalski, A., Wetzel, R. and
      Soll, D. (1980), In *Transfer RNA: Biological Aspects*, (Eds.
      D. Soll, J.N. Abelson and P.R. Schimmel), pp. 407-419, Cold
      Spring Harbor Laboratory, Cold Spring Harbor, New York.

190 . Pelham, H.R.B. (1978), Nature, 272, 469-471.

191 . Philipson, L. (1979), In *Nonsense Mutations and tRNA Suppressors*,
      (Eds. J.E. Celis and J.D. Smith), pp. 313-319, Academic Press,
      London.

192 . Morch, M. and Benicourt, C. (1980), Eur. J. Biochem., 105, 445-
      451.

193 . Parker, J., Pollard, J.W., Friesen, J.D. and Stanners, C.P.
      (1978), Proc. Natl. Acad. Sci. USA, 75, 1091-1095.

194 . Parker, J. and Friesen, J.D. (1980), Molec. gen. Genet., 177,
      439-445.

195 . Samuelsson, T., Elias, P., Lustig, F., Axberg, T., Folsch, G.,
      Akesson, B. and Lagerkvist, U. (1980), J. Biol. Chem., 255,
      4583-4588.

196 . Goldman, E., Holmes, W.M. and Hatfield, G.W. (1979), J. Mol.
      Biol., 129, 567-585.

197 . Murgola, E.J. and Pagel, F.T. (1980), J. Mol. Biol., 138, 833-
      844.

198 . Atkins, J.F., Gexteland, R.F., Reid, B.R. and Anderson, C.W.
      (1979), Cell, 18, 1119-1131.

199 . Bossi, L. and Roth, J.R. (1980), Nature, 286, 123-127.

200 . Yarus, M. (1979), Progr. Nucleic Acid Res. and Mol. biol., 23,
      195-225.

201 . Weissbach, H. and Ochoa, S. (1976), Ann. Rev. Biochem., 45, 191-
      216.

201a . Erdmann, V.A. (1980), Nucleic Acids Res., 8, r31-r47.

202 . Wrede, P. and Erdmann, V.A. (1977), Proc. Natl. Acad. Sci. USA,
      74, 2706-2709.

203 . Marcu, K. and Dudock, B.S. (1976), Nature, 261, 159-162.

204 . Roe, B.A. and Tsen, H.-Y. (1977), Proc. Natl. Acad. Sci. USA,
      74, 3696-3700.

205 . Dingermann, T., Pistel, F., and Kersten, H. (1980), Eur. J.
      Biochem., 104, 33-40.

205a . Ulbrich, N., Wool, I.G., Ackerman, E. and Sigler, P.B. (1980),
      J. Biol. Chem., 255, 7010-7016.

205b . Ulbrich, N., Lin, A., Todokoro, K. and Wool, I.G. (1980), J.
      Biol. Chem., 255, 797-801.

206 . Haenni, A.-L. and Chapeville, F. (1980), In *Transfer RNA: Bio-
      logical Aspects*, (Eds. D. Soll, J.N. Abelson and P.R. Schimmel),
      pp. 539-556, Cold Spring Harbor Laboratory, Cold Spring Harbor,
      New York.

207 . Dahlberg, J.E. (1980), In *Transfer RNA: Biological Aspects*,
      (Eds. D. Soll, J.N. Abelson and P.R. Schimmel), pp. 507-516,
      Cold Spring Harbor Laboratory, Cold Spring Harbor, New York.

208 . Cordell, B., Swanstrom, R., Goodman, H.M. and Bishop, J.M.
      (1979), J. Biol. Chem., 254, 1866-1874.

209. Baroudy, B.M. and Chirikjian, J.G. (1980), Nucleic Acids Res.,
     $\underline{8}$, 57-66.
209a. Cordell, B., Denoto, F.M., Atkins, J.F., Gesteland, R.F.,
     Bishop, J.M. and Goodman, H.M. (1980), J. Biol. Chem., $\underline{255}$,
     9358-9368.
210. Yang, W.K. and Hwang, D.L.R. (1980), In *Transfer RNA: Bio-
     logical Aspects*, (Eds. D. Soll, J.N. Abelson and P.R. Schimmel),
     pp. 517-537, Cold Spring Harbor Laboratory, Cold Spring Harbor,
     New York.
211. Soffer, R.L. (1980), In *Transfer RNA: Biological Aspects*, (Eds.
     D. Soll, J.N. Abelson and P.R. Schimmel), pp. 517-537, Cold
     Spring Harbor Laboratory, Cold Spring Harbor, New York.
212. Andrulis, I.L., Hatfield, G.W. and Arfin, S.M. (1979), J. Biol.
     Chem., $\underline{254}$, 10629-10633.
213. Scornik, O.A., Ledbetter, M.L.S. and Malter, J.S. (1980), J.
     Biol. Chem., $\underline{255}$, 6322-6329.
214. Oxender, D.L., Zurawski, G. and Yanofsky, C. (1979), Proc. Natl.
     Acad. Sci. USA, $\underline{76}$, 5524-5528.
215. Keller, E.B. and Calvo, J.M. (1979), Proc. Natl. Acad. Sci. USA,
     $\underline{76}$, 6186-6190.
216. Johnston, H.M., Barnes, W.M., Chumley, F.G., Bossil, L. and
     Roth, J.R. (1980), Proc. Natl. Acad. Sci. USA, $\underline{77}$, 508-512.
217. Mazzara, G.P. and McClain, W.H. (1980), In *Transfer RNA: Bio-
     logical Aspects*, (Eds. D. Soll, J.N. Abelson and P.R. Schimmel),
     pp. 3-27, Cold Spring Harbor Laboratory, Cold Spring Harbor,
     New York.
218. Abelson, J.N. (1980), In *Transfer RNA: Biological Aspects*,
     (Eds. D. Soll, J.N. Abelson and P.R. Schimmel), pp. 211-219, Cold
     Spring Harbor Laboratory, Cold Spring Harbor, New York.
219. Schmidt, O., Mao, J., Ogden, R., Beckmann, J., Sakano, H.,
     Abelson, J., and Soll, D. (1980), Nature $\underline{287}$, 750-752.
220. Mao, J., Schmidt, O., and Soll, D. (1980), Cell, $\underline{21}$, 509-516.
221. Abelson, J.N. (1979), Ann. Rev. Biochem., $\underline{48}$, 1035-1069.
222. Johnson, J.D., Ogden, R., Johnson, P., Abelson, J., Dembeck, P.
     and Itakura, K. (1980), Proc. Natl. Acad. Sci. USA, $\underline{77}$, 2564-
     2568.
223. Melton, D.A., DeRobertis, E.M. and Cortese, R. (1980), Nature,
     $\underline{284}$, 143-148.

APPENDIX

Table 2.   tRNA Sequences

	Aminoacyl Stem									D Stem				D Loop							20₁	20₂		D Stem					Anticodon Stem				
	1	2	3	4	5	6	7	8	9	10	11	12	13	14	15	16	17	18	19	20	20₁	20₂	21	22	23	24	25	26	27	28	29	30	31
ALANINE (1) N. crassa mito.	G	G	G	G	G	G	U	A	U	G	U	A	U	A	A	D	U	G	G	D			A	A	G	U	A	C	A	G	C	A	A	U
ARGININE (2) Drosophila*	G	G	U	C	C	U		G	U	G	C	G	C	A	A	U		G	G	A	U		A	A	C	G	G	C	G	▼	C	U	G	A
CYSTEINE (3) S. cerev. mito.*	G	G	A	G	A	U		G	U	U	U	U	U	U	A	A		G	G	U	U		A	A	A	C	A	C	C		U	U	A	G	A
GLUTAMIC ACID (4) Drosophila 4	U	C	C	C	A	U		A	U	G	U	C	U	A	G	D		G	G	U	D		A	A	G	G	A	U	U	C	A	U	U	G	G
GLYCINE (5) Human Placenta (GCC)	G	C	A	Um	U	G		G	U	G	U	U	U	A	G	U		G	G	D			A	A	G	A	A	U	U	m²₂G	C	U	C	G	C
(6) Human Placenta (CCC)	G	C	G	C	C	G		C	U	G	U	G	G	A	G	U		G	G	D			A	A	U	C	A	U	G	m²₂G	C	A	A	G	A
ISOLEUCINE (7) E. coli 2	G	G	C	C	C	C	U⁴S	A	G	C	U	ac⁴C	U	A	G	U		Gm	G	D			A	A	G	A	G	C	A	A	G	C	G	G	A
LEUCINE (8) Bovine Liver	G	G	U	A	G	C		G	U	m²₂G	C	C	G	A	G	C		G	G	D	C	Y	A	A	A	G	G	C	m²₂G	C	U	U	G	A
(9) Rat Hepatoma	G	U	C	A	G	G		A	U	m²₂G	C	C	G	A	A	U		G	G	D	C	Y	A	A	G	A	C	m²₂G	C	C	A	G	G	A
(10) Bean Chloroplast 1	G	G	G	G	A	U		A	U	G	G	C	G	A	A	U	U	Gm	G	D	A		A	G	A	C	U	A	C	C	G	G	A	A
(11) Bean Chloroplast 2	G	G	C	U	U	G		A	U	G	G	U	G	A	A	U	U	Gm	G	D	A		A	G	A	C	A	U	C	C	G	A	G	R
(12) Bean Chloroplast 3	G	C	C	C	G	U		A	U	A	U	G	G	A	A	U	U	Gm	G	D	A		A	G	A	C	A	U	C	C	U	G	C	U
(13) N. crassa mito. 1	A	U	C	C	G	A		G	U	A	U	G	G	A	A	D		G	G	D	A		A	G	A	C	A	U	A	A	C	A	U	G	G
(14) N. crassa mito. 2	A	U	A	G	G	U		G	U	C	U	U	G	A	A	D		G	G	D	A		A	G	A	C	A	G	U	U	C	A	U	C	C
METHIONINE-Initiator (15) B. subtilis	C	G	C	G	G	G		U	A	G	A	G	C	A	G	U	U	C	G	G	D		A	A	G	C	U	C	G	m²₂G	C	A	G	G	G
(16) T. utilis	A	G	C	G	U	C		U	G	G	C	G	C	A	G	D		G	G	A			A	A	G	C	U	U	C	G	C	A	G	G	G
(17) Yeast mito.	U	G	C	A	A	U	m¹G	A	U	A	U	G	U	A	A	D		G	G	U	D		A	A	C	A	U	A	m²₂G	U	U	U	A	G	G
(18) Bean Cytoplasm	A	U	C	A	G	A		G	U	G	C	G	C	A	G	C		G	G	A			A	G	C	G	U	G	m²₂G	G	U	G	G	G	G

Table 2. tRNA Sequences (cont.)

	Aminoacyl Stem									D Stem								D Loop							D Stem				Anticodon Stem				
	1	2	3	4	5	6	7	8	9	10	11	12	13	14	15	16	17	18	19	20	20₁	20₂	21	22	23	24	25	26	27	28	29	30	31
METHIONINE-Initiator																																	
(19) Bean Chloroplast	C	G	C	G	G	A	G	U	A	G	A	G	C	A	A	C	U	Gm	G	D			A	G	C	U	U	C	C	A	A	G	G
(20) S. obliquus	A	G	C	U	G	A	G	U	m¹G	m²G	C	G	C	A	G	D		G	G	A			A	G	C	U	G	m²G	A	ψ	G	G	G
(21) Starfish oocyte	A	G	C	A	G	A	G	U	G	m²G	C	G	G	A	G	U		G	G	A			A	G	C	U	G	G	C	U	U	G	G	G
METHIONINE-Elongator																																	
(22) S. obliquus	G	C	C	U	G	C	U	U	A	G	C	U	C	A	G	U	D	Gm	G	C		C	A	G	C	U	C	A	ψ	C	C	C	ψ
PHENYLALANINE																																	
(23) Bombyx mori	G	C	C	C	A	A	A	U	A	m²G	C	U	C	A	G	D	D	G	G	G			A	G	A	G	C	m²G	ψ	ψ	A	G	A
(24) Drosophila	G	C	C	G	A	A	A	U	A	m²G	C	U	C	A	G	D	D	G	G	G			A	G	A	G	C	m²G	ψ	ψ	A	G	A
(25) N. crassa cyto.	G	C	G	G	G	U	U	U	A	m²G	C	U	C	*A	G	D	D	G	G	G			A	G	A	G	C	m²G	ψ	C	A	G	A
PROLINE																																	
(26) Chick 1(2)	G	G	C	Um	C	G	U	U	G	G	C	U	ψ	A	G	G		G	G	D			A	U	C	A	C	U	C	A	G	C	C
SERINE																																	
(27) Yeast mito.*	G	G	A	U	G	G	U	U	G	A	C	U	G	A	G	U		G	G	U	U	U	A	A	G	U	U	U	U	U	A	U	U	G
THREONINE																																	
(28) N. crassa mito.	G	C	C	U	G	G	U	A	G	G	C	A	U	A	A	A		A	A	D			A	A	A	C	A	A	A	U	U	U	G	ψ
(29) Yeast mito. 1*	G	U	A	A	A	U	A	U	A	A	U	U	U	A	A	U		G	G	U			A	A	A	C	U	G	A	U	U	C	G	U
(29a) Yeast mito. 2*	G	U	U	A	U	A	U	U	A	G	C	U	U	A	U	U	U	G	G	U			A	G	A	C	U	G	A	U	U	C	G	U
TRYPTOPHAN																																	
(30) N. crassa mito.	A	A	G	A	G	U	A	U	A	G	U	U	U	A	G	D	D	G	G	D			A	A	A	C	A	A	A	A	A	U	G	A
(31) Yeast mito.*	A	A	G	G	A	U	A	U	A	G	U	U	U	A	U	U		G	G	U			A	A	A	C	A	A	G	U	U	G	A
TYROSINE																																	
(32) S. pombe	C	U	C	C	U	G	A	U	G	G	U	G	ψ	A	G	D		G	G	D			A	U	C	A	C	A	(ψ)	C	C	C	G	G
VALINE																																	
(33) N. crassa mito.	G	A	G	A	G	A	U	U	A	G	C	U	C	A	G	U		G	G	D			A	G	A	G	C	A	C	C	U	C	G	U
(34) Yeast mito.*	A	G	G	A	G	A	U	U	A	G	C	C	U	A	U	A		G	G	U			A	U	A	G	C	A	A	U	U	C	G	U

Table 2. tRNA Sequences (cont.)

	Anticodon Loop							Anticodon Stem									Extra Arm											
	32	33	34	35	36	37	38	39	40	41	42	43	44	45	46	47	47₁	47₂	47₃	47₄	47₅	47₆	47₇	47₈	47₉	47₁₀	47₁₁	48
ALANINE (1) N. crassa mito.	C	U	U	G	C	U	C	A	Ũ	U	G	C	U	U	G													U
ARGININE (2) Drosophila*	C	U	A	C	G	G	A	A	U	C	A	G	A	A	G	A												U
CYSTEINE (3) S. cerev. mito.*	U	U	G	C	A	A	A	A	U	C	U	A	C	U	U	A												U
GLUTAMIC ACID (4) Drosophila 4	Om	U	U	C	A	C	m⁵C	C	C	A	G	A	G	G														C
GLYCINE (5) Human Placenta (GCC)	C	U	G	C	C	A	C	G	C	G	G	G	A	G	G													m⁵C
(6) Human Placenta (CCC)	Um	U	C	C	C	A	U	U	C	U	U	G	C	G	A(acp³)													C
ISOLEUCINE (7) E. coli 2	C	U	C	A	U	A	A	Ψ	C	C	A	G	U	U	G	G	U											C
LEUCINE (8) Bovine Liver	Ψm	U	I	A	G	G	C	C	C	A	G	U	U	C	Y	C	Y	U	C	G	G	G	G	G				m⁵C
(9) Rat Hepatoma	C	U	Om	A	A	A	G	Ψ	C	U	G	G	U	Y	C	C	Y	U	C	G	G	A	G					m⁵C
(10) Bean Chloroplast 1	C	U	U	A	G	G	A	Ψ	C	C	G	U	C	G	A	C	U	U	A	A	U	A	G	C	A	U		U
(11) Bean Chloroplast 2	C	U	Om	A	A	U	A	U	C	U	C	U	A	U	C		A	A	A	G	A	G	C	G				U
(12) Bean Chloroplast 3	C	U	U	A	G	G	A	A	G	C	A	G	U	G	G	C	A	G	A	G	C	A						U
(13) N. crassa mito. 1	C	U	Ũ	A	A	A	A	U	C	A	U	G	U	G	G	C	U	U	C	A	A	G	C	U	G			U
(14) N. crassa mito. 2	Ψ	U	U	A	G	G	C	C	G	G	A	A	U	G	G	U	U	A	A	A	A	C	U	G				U
METHIONINE-Initiator (15) B. subtilis	C	U	C	A	U	A	A(t⁶)	C	C	G	A	A	A	A	G	G(m⁷)												U
(16) T. utilis	C	U	C	A	U	A	A	C	C	C	U	G	A	U	G	D												C
(17) Yeast mito.	G	U	C	A	U	G	A(t⁶)	C	C	U	A	A	U	U	A(m⁷,A)													m⁵C
(18) Bean Cytoplasm	C	C	C	A	U	A	A	C	C	A	C	A	G	G	D													C

Table 2. tRNA Sequences (cont.)

	Anticodon Loop							Anticodon Stem					44	45	46	47	Extra Arm (47:1–47:13)	48
	32	33	34	35	36	37	38	39	40	41	42	43					47:1 … 47:13	
METHIONINE-Initiator																		
(19) Bean Chloroplast	C	U	C	A	U	A	A	C	C	U	U	G	A	A	m7G	acp3		m5C
(20) S. obliquus	C	U	C	A	U	t6A	A	C	C	C	A	U	G	G	m7G	U		C
(21) Starfish oocyte	C	C	C	A	U	t6A	A	C	C	C	A	G	G	A	G	D		C
METHIONINE-Elongator																		
(22) S. obliquus	ψ	U	C	A	U	m2A	C	G	C	G	G	A	A	A	m7G	D		C
PHENYLALANINE																		
(23) Bombyx mori	Cm	U	Gm	A	A	yW	A	Y	C	U	A	A	A	G	m7G	D		C
(24) Drosophila	Cm	U	Gm	A	A	yW	A	Y	C	m5U	A	A	A	G	m7G	D		m5C
(25) N. crassa cyto.	Cm	U	Gm	A	A	yW	A	Y	C	U	G	A	A	G	m7G	D		m5C
PROLINE																		
(26) Chick 1(2)	Um(ψ)	U	I	G	G	G	Y(N)	G	C	G	G	A	A	G	G	D		C
SERINE																		
(27) Yeast mito.*	U	U	U	G	A	G	C	U	A	U	C	A	A	U	U	A	U C U U U A U G G C U A C	C
THREONINE																		
(28) N. crassa mito.	U	U	U	G	U	A	A	Y	C	A	A	U	A	A	G	A		A
(29) Yeast mito. 1*	UU	U	U	A	G	G	t6U	G	C	A	U	A	A	U	U	A		U
(29a) Yeast mito. 2*	U	U	U	G	U	A	A	U	C	G	A	A	G	G	U			U
TRYPTOPHAN																		
(30) N. crassa mito.	C	U	U	ŭ	C	A	*A	C	U	U	U	A	A	A	U			U
(31) Yeast mito.*	U	U	U	C	A	A	A	U	C	A	A	U	C	A	U			m5C
TYROSINE																		
(32) S. pombe	C	U	G	Y	A	A	i6A	C	C	G	G	U	m7G	G	U			m5C
VALINE																		
(33) N. crassa mito.	U	U	U	A	C	A	C	A	C	G	G	A	A	G	G	U		U
(34) Yeast mito.*	U	U	U	A	C	A	C	A	C	G	A	A	A	G	A	U		U

Table 2.   tRNA Sequences (cont.)

	TΨ Stem					TΨ Loop							TΨ Stem					Aminoacyl Stem										
	49	50	51	52	53	54	55	56	57	58	59	60	61	62	63	64	65	66	67	68	69	70	71	72	73	74	75	76
ALANINE																												
(1) N. crassa mito.	m5C	A	A	G	G	T	Ψ	C	A	A	A	U	C	C	U	U	G	U	U	A	U	C	U	C	A	C	C	A
ARGININE																												
(2) Drosophila*	C	C	A	G	G	U	U	U	C	G	A	C	C	C	U	G	G	G	C	A	G	G	A	U	C	G		
CYSTEINE																												
(3) S. cerev. mito.*	A	A	G	A	G	U	U	U	C	G	A	U	U	C	U	C	U	C	A	U	C	U	C	U	U			
GLUTAMIC ACID																												
(4) Drosophila 4	C	C	C	G	G	T	Ψ	C	G	A	U	U	C	C	C	G	G	U	A	U	G	G	G	A	A	C	C	A
GLYCINE																												
(5) Human Placenta (GCC)	m5C	m5C	C	G	G	T	Ψ	C	G	mA	U	U	C	C	C	G	G	C	C	A	A	U	G	C	A	C	C	A
(6) Human Placenta (CCC)	m5C	m5C	C	G	G	T	Ψ	C	G	mA	U	U	C	C	C	G	G	G	C	C	A	A	U	C	A	C	C	A
ISOLEUCINE																												
(7) E. coli 2	G	C	U	G	G	T	Ψ	C	A	A	G	U	C	C	A	G	C	A	G	G	G	G	C	C	A	C	C	A
LEUCINE																												
(8) Bovine Liver	G	U	G	G	G	T	Ψ	C	G	mA	A	U	C	C	C	A	C	C	U	U	G	U	G	C	A	C	C	A
(9) Rat Hepatoma	G	U	G	G	G	T	Ψ	C	A	A	U	C	C	C	A	C	U	U	G	A	C	C	C	A	A	C	C	A
(10) Bean Chloroplast 1	G	A	A	G	G	T	T	C	A	A	G	U	C	C	U	C	U	U	C	A	A	G	C	C	A	C	C	A
(11) Bean Chloroplast 2	G	G	A	G	G	T	T	C	A	A	G	U	C	C	U	G	A	U	C	A	A	G	G	C	A	C	C	A
(12) Bean Chloroplast 3	C	U	C	G	G	T	T	C	A	A	G	U	C	C	U	G	U	U	C	A	U	C	A	U	A	C	C	A
(13) N. crassa mito. 1	G	A	A	G	G	T	Ψ	C	A	A	A	U	C	C	U	U	U	G	C	A	U	C	A	U	A	C	C	A
(14) N. crassa mito. 2	A	C	A	A	G	T	T	C	A	A	A	U	C	U	U	G	U	U	A	U	C	A	U	U	A	C	C	A
METHIONINE-Initiator																												
(15) B. subtilis	G	C	A	G	G	T	Ψ	C	A	A	A	U	C	C	U	G	C	C	C	C	C	C	A	C	A	C	C	A
(16) T. utilis	m5C	C	U	U	G	G	A	U	C	G	A	A	U	C	C	G	G	C	G	A	U	U	U	U	A	C	C	A
(17) Yeast mito.	A	U	A	C	G	T	Ψ	C	A	A	U	U	C	C	G	U	U	A	U	U	U	G	A	Ψ	A	C	C	A
(18) Bean Cytoplasm	C	C	A	G	G	A	Y	C	G	A	A	A	C	C	U	GmG	C	C	U	C	U	G	A	U	A	C	C	A

Table 2. tRNA Sequences (concluded)

	Tψ Stem					Tψ Loop							Tψ Stem					Aminoacyl Stem										
	49	50	51	52	53	54	55	56	57	58	59	60	61	62	63	64	65	66	67	68	69	70	71	72	73	74	75	76
METHIONINE-Initiator (19) Bean Chloroplast	A	C	G	G	G	T	ψ	C	A	A	A	U	C	C	C	G	U	C	U	C	C	G	C	A	A	C	C	A
(20) S. obliquus	A	C	A	G	G	A	U	C	G	A	A	A	C	C	C	U	G	C	U	C	A	G	C	U	A	C	C	A
(21) Starfish oocyte	C	G	A	G	G	A	ψ	C	G	A	A	A	C	C	U	C	G	C	U	C	U	G	C	U	A	C	C	A
METHIONINE-Elongator (22) S. obliquus	A	C	U	A	G	T	ψ	C	G	A	A	U	C	U	A	G	U	U	A	G	C	A	G	C	N	C	C	A
PHENYLALANINE (23) Bombyx mori	m5C	C	U	G	G	T	ψ	C	G(A)	A	U	C	C	C	G	G	G	U	U	U	C	G	G	C	C	C	C	A
(24) Drosophila	C	C	C	G	G	T	ψ	C	A	A	U	C	C	C	A	C	C	U	U	U	C	C	G	C	C	C	C	A
(25) N. crassa cyto.	G	m5U	G	U	G	T	ψ	C	G	A	A	U	C	A	C	A	G	A	A	A	C	C	G	C	C	C	C	A
PROLINE (26) Chick 1(2)	C	m5C	G	G	G	ψ	ψ	C	A	A	U	U	C	C	C	C	G	A	C	G	A	G	C	C	C	C	C	A
SERINE (27) Yeast mito.*	G	U	A	G	G	U	U	C	A	A	A	U	C	C	U	A	C	A	U	C	A	U	C	C	G			
THREONINE (28) N. crassa mito.	G	C	A	A	G	T	G	C	G	A	U	A	C	U	U	G	C	C	A	C	U	G	G	ψ	U	C	C	A
(29) Yeast mito. 1*	C	U	A	A	G	U	U	C	A	A	A	U	C	U	U	A	G	U	A	U	U	A	A	C	A	C	C	A
(29a) Yeast mito. 2*	U	G	G	G	G	U	U	C	A	A	A	U	C	C	C	U	A	A	U	A	U	A	C	C	A	C	C	A
TRYPTOPHAN (30) N. crassa mito.	C	U	U	A	G	T	ψ	C	G	A	G	U	C	U	A	A	G	U	A	C	U	C	U	U	G	G	C	A
(31) Yeast mito.*	A	G	G	A	G	U	U	C	G	A	A	U	C	U	C	U	U	U	U	A	C	U	U	U	G	G		
TYROSINE (32) S. pombe	G	C	U	A	G	T	ψ	C	G	A	U	U	C	U	G	G	C	U	C	A	G	G	A	G	A	C	C	A
VALINE (33) N. crassa mito.	G	G	G	U	G	T	ψ	C	G	A	A	U	C	A	C	C	C	A	U	U	U	U	U	C	C	C	C	A
(34) Yeast mito.*	A	U	A	G	G	U	U	C	G	A	A	C	C	C	U	A	U	U	U	C	U	C	U	ψU	A			

Table 2. Legend (continued)

Published tRNA sequences as of August 1, 1980 that are not listed in (57). Mammalian mitochondrial tRNAs are not included. See text for further discussion. Minor nucleotide abbreviations are as given in Table 4. *DNA sequence of the tRNA gene. Not listed in the table are the tRNA gene sequences for Drosophila tRNA^{Asn} and tRNA^{Ile} (35), and for *Xenopus* tRNA^{Tyr} (36). More recently, the DNA sequence for yeast mitochondria tRNA^{Ala}, tRNA^{Arg}, tRNA^{Asp}, tRNA^{Gly}, tRNA^{Lys}, and tRNA^{Ser}. (37), and the tRNA sequences of *B. subtilis* tRNA$_m^{Met}$ (38), spinach chloroplast tRNA^{Thr} (39) and tRNA$_f^{Met}$ (40), Drosophila tRNA^{His} (41), and *Mycoplasma* tRNA^{Gly} (42) have appeared. References are as follows:

1, 13, 14, 28, 30, 33) Heckman, J.E., Sarnoff, J., Alzner-DeWeerd, B. Yin, S., and RahBhandary, U.L. (1980), Proc. Natl. Acad. Sci. USA, 77, 3159-3163.

2) Silverman, S., Schmidt, O., Soll, D. and Hovemann, B. (1979), J. Biol. Chem., 254, 10290-10294.

3) Bos, J.L., Osinga, K.A., Van der Horst, G. and Borst, D. (1979), Nucleic Acids Res., 6, 3255-3266.

4) Altwegg, M. and Kubli, E. (1980), Nucleic Acids Res., 8, 215-223.

5) Gupta, R.C., Roe, B.A. and Randerath, K. (1979), Nucleic Acids Res., 7, 959-970.

6) Gupta, R.C., Roe, B.A. and Randerath, K. (1980), Biochemistry, 19, 1699-1705.

7) Kuchino, Y., Watanabe, S., Harada, F. and Nishimura, S. (1980), Biochemistry, 19, 2085-2089.

8) Pirtle, R., Kashdan, M., Pirtle, I. and Dudock, B. (1980), Nucleic Acids Res., 8, 805-815.

9) Randerath, E., Gupta, R.C., Morris, H.P. and Randerath, K. (1980) Biochemistry, 19, 3476-3483.

10, 11, 12) Osorio-Almeida, M.L., Guillemaut, P., Keith, G., Canady, J. and Weil, J.H, (1980), Biochem. Biophys. Res. Commun., 92, 102-108.

15) Yamada, Y., Kuchino, Y. and Ishikura, H. (1980), J. Biochem., 87, 1261-1269.

16) Yamashiro-Matsumara, S. and Takemura, S. (1979), J. Biochem., 86, 335-346.

17) Canady, J., Dirheimer, G. and Martin, R.P. (1980), Nucleic Acids Res., 8, 1445-1457.

18, 19) Canady, J., Guillemaut, P. and Weil, J.-H. (1980), Nucleic Acids Res., 8, 999-1008.

20) Olins, P.O. and Jones, D.S. (1980), Nucleic Acids Res., 8, 715-729.

21) Kuchino, Y., Kato, M., Sugisaki, H. and Nishimura, S. (1979), Nucleic Acids Res., 6, 3459-3469.

22) Jones, D.S. (1980), Abstracts EMBO-FEBS tRNA Workshop, Strasbourg, France.

23) Keith, G. and Dirheimer, G. (1980), Biochem. Biophys. Res.
     Commun., 92 , 109-115.
24) Altwegg, M. and Kubli, E. (1979), Nucleic Acids Res., 7, 93-105.
25) Alzner-DeWeerd, B., Hecker, L.I., Barnett, W.E. and RajBhandary,
     U.L. (1980), Nucleic Acids Res., 8, 1023-1032.
26) Harada, F., Peters, G.C. and Dahlberg, J.E. (1979), J. Biol.
     Chem., 254, 10979-10985.
27) Miller, D.L., Martin, N.C., Pham, H.D. and Donelson, J.E. (1979)
     J. Biol. Chem., 254, 11735-11740.
29, 34) Li, May and Tzagoloff, A. (1979), Cell, 18, 47-53.
29a) Berlani, R.E., Pentella, C., Macino, G. and Tzagoloff, A. (1980),
     J. Bact., 141, 1086-1097.
31) Martin, N.C., Pham, H.D., Underbrink-Lyon, K., Miller, D.L. and
     Donelson, J.E. (1980), Nature, 285, 579-581.
32) Vogeli, G. (1979), Nucleic Acids Res., 7, 1059-1065.
35) Hovemann, B., Schmidt, O., Yamada, H., Silverman, S., Mao, J.,
     De Franco, D. and Soll, D. (1980), In Transfer RNA: Biological
     Aspects, (D. Soll, J.N. Abelson and P.R. Schimmel), pp. 325-338,
     Cold Spring Harbor Laboratory, Cold Spring Harbor, New York.
36) Muller, F. and Clarkson, S.G. (1980), Cell, 19, 345-353.
37) Bonitz, S.G. and Tzagoloff, A. (1980), J. Biol. Chem., 255, 9075-
     9081.
38) Yamada, Y. and Ishikura, H. (1980), Nucleic Acids Res., 8, 4517-
     4520.
39) Kashadan, M.A., Pirtle, R.M., Pirtle, I.L., Calagan, J.L.,
     Vreman, H.J. and Dudock, B.S. (1980), J. Biol. Chem., 255, 8831-
     8835.
40) Calagan, J.L., Pirtle, R.M., Pirtle, I.L., Kashdan, M.A., Vreman
     H.J. and Dudock, B.S. (1980), J. Biol. Chem., 255, 9981-9984.
41) Altwegg, M. and Kubli, E. (1980), Nucleic Acids Res., 3259-3262.
42) Kilpatrick, M.W. and Walker, R.T. (1980), Nucleic Acids Res., 8,
     2783-2786.

THE STRUCTURE OF EUKARYOTIC RIBOSOMES

Ira G. Wool

The Department of Biochemistry
University of Chicago
Chicago, Illinois 60637 (U.S.A.)

INTRODUCTION

I begin with a litany of the gospel. The grand design in research on eukaryotic ribosomes is to know the structure and the function of the organelle: to be able to specify the coordinates, and to define the activity, of each of the molecules in the particle. A subsidiary aim is to compare the structure of eukaryotic and prokaryotic ribosomes. There are significant differences between the two classes of particles. Those differences are the basis of what has been called the "central dilemma". Eukaryotic ribosomes are appreciably larger: they contain a greater number of proteins, seventy-eighty rather than fiftytwo, and they have an extra molecule of RNA as well. Moreover, the proteins and nucleic acids are, on the average, larger. The difference in size is a paradox since eukaryotic ribosomes perform the same general function, namely to catalyze the synthesis of protein; more to the point, they appear to employ appreciably the same partial reactions, although there may be real differences in the means by which the initiation of peptide synthesis is accomplished. One would like to know the nature of the evolutionary pressure for the accretion of the extra proteins in eukaryotic ribosomes- whether there are functions of the particle still be to uncovered. It has not escaped attention that knowledge of the conserved features may be of help in understanding the function of both eukaryotic and prokaryotic ribosomes.

This review supplements two on the same subject that have appeared recently (1, 2). Several of the topics dealt with in greater detail before are epitomized here; on the other hand there is more information now on the identity of the proteins that bind to ribosomal RNA and that topic is given more attention. It should be noted that the emphasis is on mammalian (specifically rat liver) ribosomes. One assumes that any one of the set of eukaryotic ribosomes will be prototypical of the group. By a concatenation of events that no longer has rhyme or reason we chose rat liver, and now, for good and for bad, in sickness and in health, we are stuck with them.

GENERAL CHARACTERISITCS OF EUKARYOTIC RIBOSOMES

Eukaryotic ribosomes are composed of two subunits, which together contain four molecules of RNA and perhaps as many as eighty proteins. The sedimentation coefficient of the monomer and subparticles varies between species (as do other of the characteristics) but are close to 80 for the ribosome and 60 and 40 for the subunits (3); those numbers are used to designate the particles.

The small subunit of rat liver ribosomes has a sedimentation co-efficient of 36.9 (4). The particle contains one molecule of 18S RNA of molecular weight 0.7×10^6 (5,6) and about thirty proteins whose combined mass is about 0.74×10^6 daltons which is almost exactly what has been measures (4). The particle might be expected then to have nearly equal amounts of protein and RNA, whereas the chemical determination (7) reveals it to contain only 45% RNA. The buoyant density of the small subparticle is 1.515, from which the percent RNA can also be calculated to be 45 (7). The values obtained by others (8, 9) in a different way are similar. The extinction coefficient for the 40S subparticle is 94.5, and the diffusion coefficient 2.0×10^{-7} (4).

The large subunit of rat liver ribosomes has a sedimentation co-efficient of 56.3 and is composed of one molecule each of 5S, 5.8S,and 28S RNA, and 45 to 50 proteins. The molecular weight of 28S RNA is given as 1.7×10^6 (5, 6) and the molecular weights of 5S and 5.8S RNA, calculated from the sequence (10, 11) is approximately 39,000, and 51,000, respectively. The total mass of RNA in the particle is about 1.79×10^6 daltons. The mass of the protein is approximately 1.05×10^6 The expected molecular weight of the particle then is 2.84×10^6, and the actual measuremenet is 2.9×10^6 (4) which, given the uncertainties, is in reasonable agreement; a similar value had been obtained before (8). The 60S subunit has a buoyant density of 1,600 (7); the RNA content from the buoyant density is 59.4%,and the chemical determination gave the same value (7). Thus the large sub-article has an appreciably greater proportion of RNA than the 40S sub-unit.

Eukaryotic ribosomes are by no means a uniform groum; "80S ribo-somes" actually range in aggregate mass from about 3.9 for plants to 4.55×10^6 daltons for mammals (12-14). The change in ribosome mass is the result of an increase in the size of the large subunit from 2.4 to 3.05×10^6 daltons; the increase is in the mass of both RNA and protein. The 40S particle, on the other hand, has not changed appreciably in size during eukaryotic evolution.

There are detailed electron microscopic studies of eukaryotic ribosomes (15-18). Observers (17, 18) have been struck by the general similarity in the morphology of eukaryotic and prokaryotic ribosomal subunits, despite the obvious larger size of the former. The most prominent features of rat liver ribosomes are the division of the small subunit into two unequal parts; the asymmetric apposition of the small subunit to the large in ribosome couples which confers a handedness to the structure; and a notch in the profile of the large subparticle (16). The 40S subunit approximates a curved and flattened prolate elipsoid with dimensions of 230 x 140 x 115 Å (15); it is divided into "head" (one-third) and "body" (two-thirds) by a transverse partition seen in electron micrographs as a dense line 80 Å from one end of the subunit. The 60S subunit is seen either as a rounded object, or as asymmetric and skiff-shaped with a pointed end opposite to a blunt end (16).

A particularly promising approach to the solution of the structture of eukaryotic ribosomes is through three-dimensional reconstruction from electron micrographs of naturally occurring crystalline arrays which develop in the oocytes of lizards during hibernation (19, 20), or which can be induced in chicken embryos by cooling (16, 21-23).

ISOLATION AND CHARACTERIZATION OF EUKARYOTIC RIBOSOMAL PROTEINS

There is no way to arrive at a solution of the structure of an organelle without knowing the chemistry of the constituents. So several years ago we undertook to isolate and characterize the proteins in rat liver ribosomes. The enterprise has been successful and we feel some pride of accomplishment. Still, we do not reckon those years to have been chock-a-block with excitement; we do not hold annual reunions to celebrate and trade fond memories. Sad to say, the purification of proteins is not the stuff of melodrama, although it can be at times tragicomic. What is required for the isolation of rat liver ribosomal proteins is a strong conviction of the importance of the exercise, good management, hard work, a small amount of ingenuity, and a stupendous number of rat livers. One needs a large number of livers since the only efficient step in the purification is the separation of the livers from the rat. While the methods employed were in general similar to those that had served for the purification of prokaryotic *E. coli* ribosomal proteins (24-27), the effort was facilitated by the development of two ancillary procedures: a micro-method for the rapid analysis of small samples of ribosomal proteins by two-dimensional polyacrylamide gel electrophoresis (28); and an efficient means of fractionating ribosomal subunit proteins into smaller subsets (29).

Rat liver ribosomal subunit proteins were separated into groups

by stepwise elution from carboxymethylcellulose with LiCl at pH 6.5
(29). The proteins in the fractions were resolved by ion-exchange
chromatography on carboxymethylcellulose, or on phosphocellulose,
or for the small subset of acidic proteins on DEAE-cellulose (30-34).
Some of the proteins were purified entirely by ion-exchange chromatography, others required gel filtration as well. In that manner
thirtyfour proteins were puried from extracts of 40S subunits (30, 33,
34), fifty from the 60S subparticle (31, 32, 33a, 34), a total of
eightyfour (Table 1). Earlier Terao and Ogata (35) had isolated
twelve proteins from the 40S subparticle of rat liver ribosomes, and
Westermann and Bielka (36) the preparation and characterization of
twentyfour of the proteins. Some rat liver ribosomal proteins have
been prepared by elution from plugs cut out of two-dimensional gels
(37). A start has been made on the isolation of the proteins from
the ribosomes of other eukaryotic species: from chicken liver (38);
from yeast (39, 40); from Drosophila (41); and from wheat germ (42).

     The molecular weight of the purified proteins was determined by
electrophoresis in gels containing sodium dodecyl sulfate (Table 1).
The number average molecular weight for the thirtyfour 40S ribosomal
subunit proteins is about 22,000 (the range is 11,200 to 39,000).
The number average molecular weight for the fifty proteins isolated
from 60S subparticles is 21,000 (the range is 11,500 to 41,800).
The average molecular weight of an eukaryotic (rat liver) ribosomal
protein is about 21,300; the average for $E.$ $coli$ ribosomal proteins
is approximately 18,000. Although there are some discepancies, the
results, in general, do not differ greatly from values reported
earlier (35, 36, 43-48), especially when one gives consideration to
the variety of methods that have been used.

     The amino acid composition of the isolated rat liver ribosomal
proteins tends to be similar, but in most cases they are definitely
unique (30-34).

     There are two useful procedures- genetic analysis and recon-
stitution of subunits- for testing whether a protein is a constituent
of a prokaryotic ribosome. Neither can be used in a determination of
the status of putative eukaryotic ribosomal proteins. It is vexing
that one still can not say how many proteins there are in eukaryotic
(mammalian) ribosomes. The original estimate was from inspection of
two-dimensional polyacrylamide gel plates (49-54). It was calculated
that there were about seventytwo; thirtyone in the 40S, and fortyone
in the 60S subunit (49). Eightyfour proteins have been isolated from
rat liver ribosomes (30-34). That is twelve more than was estimated
at first. Several proteins designated initially have not been isolated, indeed were not even encountered during the purification;
whereas other proteins which were not resolved by electrophoresis
were isolated. The proteins in the first subset were, in general,
only rarely seen on two-dimensional gels or stained lightly; the
proteins in the second group were, as a rule, difficult to resolve by

Table 1. Rat Liver Ribosomal Proteins

40S Subunit		60S Subunit	
Proteins	M_r (x 10^{-3})	Proteins	M_r (x 10^{-3})
(Sa)[a]	41.5	(La)	37.9
(Sc)	33.0	(Lb)	29.8
S1	39.0	(Lf)	14.6
S2	33.1	P1	16.1
S3	30.4	P2	15.2
S3a	32.0	P3	13.0
(S3b)	30.4	L3	37.8
S4	29.5	L4	41.8
S5	22.8	L5	32.5
(S5a)	21.5	L6	33.0
S6	31.0	L7	29.2
S7	22.2	L7a	28.7
S8	26.8	L8	28.4
S9	24.3	L9	24.7
S10	20.1	L10	24.2
S11	20.7	L11	21.3
S12	14.9	L12	18.7
S13	18.6	L13	26.3
S14	17.3	L13a	24.6
S15	19.6	L14	25.8
S15a	15.7	L15	24.5
(S16)	17.1	L16	18.7
S17	18.0	L17	22.1
S18	18.5	L18	24.5
S19	17.1	L18a	21.3
S20	16.5	L19	25.3
S21	12.3	L20	16.2
S23/24	18.8	L21	20.3
S25	17.0	L22	16.1
S26	16.5	L23	15.6
S27	14.5	(L23a)	18.0
S27a	12.8	(L25)	17.5
S28	11.3	L26	18.6
S29	11.2	L27	17.8
		L27a	18.0
		L28	17.8
		L29	20.5
		L30	14.5
		L31	15.6
		L32	17.2
		L33	15.6
		L34	15.8
		L35	17.5
		L35a	13.7
		L36	14.3
		L36a	16.2
		L37	15.4
		L37a	12.6
		L38	11.5
		L39	11.6

[a]The proteins in parentheses may not be components of the ribosome or may not be unique peptides.

In the 40S subunit the $\Sigma\ M_r$ = 746,000; number of proteins = 34; average M_r = 22,000. In the 60S subunit the $\Sigma\ M_r$ = 1,049,000; number of proteins = 50; average M_r = 21,000.

electrophoresis. Proteins enumerated but not isolated include: S22, S23 or S24, S30, and S31 from the small subunit; and L1, L2, L24, L40, and L41 from the large subparticle. The proteins which were isolated include a number from the 40S subunit (Sa, Sb, S3a, S3b, S5a, S15a, and S27a), and from the 60S subunit (La, Lb, Lf, P1, P2, P3, L7a, L13a, L18a, L23a, L27a, L35a, L36a, and L37a), that were not in the original classification and whose exact status is still uncertain.

We were from the first sensitive to the possibility that some of the isolated proteins might be chemically modified forms of another polypeptide. The chemical modification could either be physiological, as, for example, phosphorylation, acetylation, or methylation; or it could be advantitious, having occurred during purification. Since the proteins are exposed to urea during isolation, carbamylation of amino groups can occur even though the purification is done at acidic pH: carbamylation would change the net charge and later the migration of the protein during electrophoresis. A second change we were concerned about was proteolytic cleavage at a sensitive peptide bond. Of course, the enzyme would have to be specific since we assume most of the proteins are not affected.

The decision whether an isolated eukaryotic ribosomal protein is unique (which must be tentative until the sequence is available) has rested on the evaluation of the molecular weight, the amino acid composition, and the behavior on electrophoresis. Most of the proteins are distinct by all three criteria, however, there are exceptions. The exceptions are discussed in another place (1, 2).

While it is still not possible to give the exact number of proteins in eukaryotic ribosomes, there seems little doubt that it is considerably greater than the number in prokaryotic ribosomes. What purpose do the supernumerary proteins serve ? Some of the extra proteins may be for interaction with receptors in the endoplasmic reticulum (55-59), others may be involved in regulation, especially the regulation of translation of messenger RNA. In that regard it appears that the ribosome-mRNA recognition process, and the partial reactions in the initiation of protein synthesis, may be fundamentally different in eukaryotes (60-63, see Section II of this volume) and could require additional ribosomal proteins. The additional proteins may participate as well in the extraordinarily complicated process of ribosome biogenesis with its need for two-way traffic between the cytoplasm and nucleus: they could take part in assembly or in the transport of assembled subunits. Finally, it is conceivable that ribosomes have luxury proteins that serve no function. There are thiostrepton resistant mutants of *Bacillus megatariom* (64) and *Bacillus subtilis* (65) that lack ribosomal protein L11. While the former is sick the latter is not (66). Thus, if follows that *B. subtilis* L11 has no indispensable function in protein synthesis or at least it serves a role capable of being assumed by one or more of the remaining proteins.

One can look at the problem the other way around and ask why are prokaryotic ribosomes so small ? Perhaps, because they had to be streamlined. During rapid growth in rich medium as much as 30% of all cellular protein is ribosomal (67, 68) and the bacterium might have difficulty supporting a particle with as much protein as eukaryotic ribosomes have. The assumptions implicit in the suggestion are that ribosomes evolved only once- which seems reasonable: that prokaryotic ribosomes initially had a greater number of proteins, a number comparable to that in eukaryotic particles; and that prokaryotic ribosomes having had more opportunity to evolve responded to selective pressure by discarding proteins and reducing the size of those that were retained- the advantage being a greater number of smaller ribosomes performing the same functions as fewer larger particles. All that would have happended after divergence of primitive eukaryotes. The argument, however, does not change the fundamental nature of the problem, it merely alters the way it is put: what prokaryotic ribosomal proteins could be dispensed with without apparent loss of cunction; if the extra proteins could be discarded with impunity why have they been retained by eukaryotes ? It is possible that in the putative process, the function of several prokaryotic ribosomal proteins were combined, as seems to have happened with the intiation factors. Eukaryotes use a greater number of initiation factors than bacteria to catalyze a similar process (63, and chapters III and IV of this book).

PRIMARY STRUCTURE OF EUKARYOTIC RIBOSOMAL PROTEINS

Eukaryotic ribosomes have something of the order of 16, 000 amino acids. Determining the sequence of all would be a monumental task, at least by conventional methods, and its achievement clearly is not imminent. It is possible that more rapid automated methods for sequencing small amounts of protein will be developed and make the task feasible. It is also conceivable that it will become a routine matter to determine the structure of proteins by sequencing the genes that code for them. One would still need the means to prepare large amounts of eukaryotic ribosomal protein DNA; it would be necessary to know at least the amino-terminal and carboxy-terminal sequences of the protein; and it would be necessary to be able to recognize potential intervening DNA sequences. Nonetheless, one has the impression that if the sequence of all the proteins in eukaryotic ribosomes is ever to be had it will come from a combination of the analysis of DNA and of protein. What is being done for the time is to determine the amino-terminal sequence of proteins (69) since that gives the maximum information for the expenditure of practical amounts of time and effort, and the complete sequence of a selected few proteins that have been established to be important for function or might shed light on ribosome evolution.

One purpose in collecting the data is to determine the relatedness of individual eukaryotic ribosomal proteins since it is possible

that the number was increased from the fiftytwo in prokaryotes to the
approximately eighty in eukaryotes by gene duplication. A computer
comparison of the amino acid compotision of rat liver ribosomal
proteins suggests a number are significantly similar statistically
and hence supports the possibility (70). A second reason for the
analyses is to search for homologous ribosomal proteins by correlating
amino acid sequences from rat liver protesin with those from other
species. The assumtion is that an organelle like the ribosome that
is at once universal, essential, and complicated arose on a single
occasion, further that knowledge of the conserved features can not but
help in understanding how ribosomes function.

What is the evidence that there is in fact conservation of ribosomal proteins ? The large subunit of rat liver, and of other eukaryotic, ribosomes have been demonstrated by immunological techniques
to have acidic proteins that cross-react with *E. coli* L7/L12 (71,
72). At the time the rat liver proteins that cross-reacted were
designated L40/L41. The identity is no longer certain since those
proteins have not been isolated and characterized; it is possible they
are P1, or P2, or both (see later). At any rate antisera raised
against *E. coli* L7/L12 reacted with rat liver L40/L41, and anti-L40/
L41 cross-reacted with *E. coli* L7/L12. Moreover, antisera prepared
against the rat liver proteins inhibited the function of *E. coli*
ribosomes (72). Howard, Smith, and Gordon (73) obtained similar
results: antiserum raised against *E. coli* L7/L12 cross-reacted with
chicken liver ribosomal proteins and inhibited a number of EF-1
catalyzed reaction. However, the experiments are not without their
critics (74). What remains to be settled is the identity of the
eukaryotic proteins related to *E. coli* L7/L12 and the extent of the
sequence similarity. The issue is whether there is both structural
and functional homology, or only functional homology.

It is possible that other eukaryotic and prokaryotic ribosomal
proteins are related. Monovalent Fab fragments prepared from immuno-
globulins directed against *E. coli* small ribosomal subunit proteins
have been tested for their effect on poly(U)-directed synthesis of
polyphenylalanine by rat liver ribosomes (75). Of the eighteen Fabs
tested three, anti-S10, anti-S12, and anti-S14, inhibited polyphenylalanine synthesis. The presumption is that the prokaryotic *E. coli*
ribosomal proteins, S10, S12 and S14, possess antigenic determinants
also present in eukaryotic (rat liver) ribosomal proteins; the rat
liver proteins have not yet been identified. It is not known if the
rat liver and *E. coli* proteins are homologous. In the strictest
sense two proteins are homologous only if they are derived from a
common ancestral gene. We do not know that to be the case for the
prokaryotic and eukaryotic ribosomal proteins. Thus it is better to
say that they share common determinants. Since ribosomal proteins
are such an unusual set it is hard to know if shared determinants is
tantamount to homology.

The sequence of the amino-terminal region of eleven rat liver ribosomal proteins- S4, S6, S8, L6, L7a, L18, 127, L30, L37, L37a, and L39 -has been determined (69). The sequence studies confirm the homogeneity of the proteins, and a comparison of the sequences establishes that the individual proteins are unique- they contain no significant homologous sequences. The N-terminal regions of the rat liver proteins was compared with amino acid sequences in *S. cervisiae* and in *E. coli* ribosomal proteins. The protein L37 from rat liver (69) and Y55 from yeast ribosomes (76) are probably homologous; it is possible that rat liver L7a or L37a or both are related to *S. cervisiae* Y44 (77), although the similar sequences are at the amino-terminus of the rat liver proteins and in an internal region of Y44. A number of similarities in the sequences of rat liver and *E. coli* ribosomal proteins have been found. However, it is not yet possible to say whether they connote a common ancestry. The most interesting possibility is the relation of rat liver L37 and *E. coli* L16. There are similarities in the N-terminal region of rat liver L37 and two separate sequences (position 21-34 and 77-91) of *E. coli* L16 (69). The latter is likely to be involved in ribosomal peptidyl transferase activity. Thus it is possible that rat liver L37 serves a similar function.

It is difficult to judge the significance of the similarities in sequence between rat liver and *E. coli* ribosomal proteins. There are insufficient data for the eukaryotic ribosomal proteins to establish homologies. Moreover, there is a serious lack of criteria to assess the importance of limited similarities in sequence. Thus the analyses to date serve mainly as a guide to further studies of the sequence of rat liver proteins and in one instance (the similarity of *E. coli*-L16 and rat liver-L37) of the function of a protein. It seems certain that ascertaining homologies between eukaryotic and prokaryotic ribosomal proteins by comparison of primary sequences will be difficult since it is likely that during evolution conservative changes in amino acids, deletions, and recombinations will have occurred. Some of the similarities that have been noted, if they are the vestiges of real homologies rather than merely chance phenomena, would indicate those processes have happened. Immunological studies (1, 71, 72, 75) have shown that while prokaryotic and eukaryotic ribosomal proteins share determinants, cross-reaction is not common. Indeed, it has been demonstrated that antisera specific for chicken or rat liver ribosomal proteins recognize only about 20% of common determinants (78). It is by no means surprising then that sequence homologies between rat liver and *E. coli* ribosomal proteins are not prominent. Homologies might be more apparent if one knew and could compare the tertiary structure of these proteins. Nonetheless, intuition leads one to surmise that the ribosomes are likely to be related to their prokaryotic ancestors. There is also circumstantial evidence that the fundamental structure and organization of ribosomes has been conserved (1), even though the primary sequence of the components varies. To what extent those surmises on the evolution of ribosomes

are correct is most likely to come from further studies of the
sequence of the molecular constitutents.

The sequence of the rat liver large ribosomal subunit protein P3
has been determined (Fig. 1). P3 differs from P2 only in that it
lacks the C-terminal fragment of 10 amino acids; thus P3 may be a
proteolytic fragment of P2. P1/P2/P3 are a family of related acidic
phosphoproteins which may have the same function as *E. coli* L7/L12.
the primary structure of P3 (79) has been compared to that of yeast
(*Saccharomyces cerevisiae*) Y-L12 (80), to the brine shrimp (*Artemia
salina*) eL12 (81), to *E. coli* L7/L12.(82), and to *Halobacterium cuti-
rubrum* L20 (83) using a sophisticated computer program (84). There
is no doubt that rat liver P3, yeast Y-L12, and *Artemia* eL12 are
homologous. Thus this one protein has been conserved during evolution
from a fungus (yeast), to an invertebrate (the brine shrip), to mammals (rat). The comparison of the eukaryotic proteins with *E. coli*
L7/L12 gives results that are difficult to rationalize: rat liver P3
seems definitely homologous with *E. coli* L7/L12 (the number of
standard deviations of the real score above the random score is 3.4
where 3 standard deviations corresponds to a P value of < 0.001).
However, the standard deviation for the comparison of yeast Y-L12 with
E. coli L7/L12 is 2.8, and for *Artemia salina* and *E. coli* L7/L12 only
1.2. It is not clear why only rat liver is definitely homologous with
E. coli L7/L12 when the three eukaryotic proteins (P3, Y-L12, and
eL12) are themselves homologous.

RNA-PROTEIN INTERACTIONS IN EUKARYOTIC RIBOSOMES

The proteins and nucleic acids in ribosomes are arranged in a
specific three-dimensional configuration to form structural and
functional domains. Solution of the structure of those particles,
and an understanding of the relation of structure to function,
requires information on the interaction of the ribosomal proteins with
ribosomal RNA and with nucleic acid ligands like tRNA and mRNA that
associate with ribosomes. An attempt is being made to identify, by
affinity chromatography, the proteins that interact with the several
species of rRNA. The strategy is quite simple. One constructs an
affinity column by activating Sepharose with cyanogen bromide and
attaches an adipic acid dihydrazide spacer; the nucleic acid of
concern is oxidized with periodate and coupled to the dihydrazide
conjugate. A mixture of proteins is passed through the column and
the proteins that bind are eluted and identified by polyacrylamide
gel electrophoresis.

5S rRNA

In the initial experiments the proteins that bind to rat liver 5S
rRNA were characterized (85). The nucleic acid was oxidized with
periodate and coupled by its 3' terminus to Sepharose through an
adipic acid dihydrazide spacer (86, 87). Ribosomal subunit proteins

```
                                                    10                         20                          30
RL-P3     MET ARG TYR VAL ALA ser TYR LEU LEU ALA  ALA LEU GLY GLY ASN ser asn PRO SER ALA   lys ASP ILE LYS LYS ILE LEU ASP SER VAL
As-eL12   MET ARG VAL ALA ALA TYR LEU LEU LEU ALA  ALA LEU thr GLY ASN asp PRO THR ser        ala lys ILE glu LYS ILE LEU ser SER VAL
Y-L12     MET LYS TYR LEU ALA ALA TYR LEU LEU leu  val gln GLY GLY ASN ala ala PRO SER ALA    ALA ASP ILE LYS ala VAL val GLU SER VAL

                          40                         50                          60
RL-P3     GLY ILE GLU ala ASP ASP glu ARG lys leu  asn lys val ile SER glu leu can GLY LYS    asn ile GLU asp VAL ILE ALA gln GLY val
As-eL12   GLY ILE GLU GLU ser asn pro ser gln leu  lys val met asn glu LEU lys GLY LYS asp    leu ala leu ILE ALA glu glu ALA gln thr
Y-L12     GLY ala GLU val ASP glu ile ARG ile asn  gln leu ser SER LEU glu LYS LYS gly        ser glu ILE ILE ALA glu GLY gln

                          70                         80                          90
RL-P3     gly LYS leu ALA SER VAL PRO ala GLY GLY  ALA val ALA ala ser ALA ala pro gly ALA    ALA ala ala pro ALA GLY ser ala pro ALA
As-eL12   LYS leu ala ser met pro thr gly GLY ala  pro ala ALA ALA gly GLY ALA ALA ALA        ALA PRO ALA ALA GLU ala lys ala lys lys
Y-L12     LYS LYS phe ALA THR VAL PRO thr GLY GLY  ALA ser ser ALA ALA ALA GLY ALA ALA gly    ser PRO ALA ALA GLU glu gly ala ASP ala gln

                          100                        105            110                   111
RL-P3     ala GLU GLU LYS GLU SER GLU GLU           lys lys ASP  GLU                          ASP
As-eL12   lys GLU GLU lys LYS GLU GLU SER GLU GLU   glu GLU ASP ASP MET GLY PHE GLY LYS PHE   ASP
Y-L12     glu GLU lys GLU glu GLU GLU               ser ASP ASP ASP MET GLY PHE GLY LYS PHE   ASP
```

Fig. 1   Comparison of the primary structure of ribosomal protein P3 from rat liver (79), eL12 from *Artemia salina* (81), and Y-L12 from *Saccaromyces cerevisiae* (80).

were passed through the affinity column and those that were bound
specifically were eluted and identified by polyacrylamide gel electrophoresis. The eukaryotic 5S rRNA binding proteins revealed by this
procedure were L6 and L19 (Table 2); small amounts of L7, L23a, L27 or
L27a, L35a, and L39 also were bound to the affinity column. Of the
several minor binding proteins the most conspicuous was L7 (see
later). It was not certain if the proteins recovered in small
amounts were bound directly but weakly to 5S rRNA, or whether they
interacted with the major binding proteins (i.e. result from protein-
protein interactions), or whether they were merely fortuitous contaminants. The matter is taken up again later. No proteins of the
40S ribosomal subunit associated with immobilized 5S rRNA; nor did
the proteins of *E. coli* ribosomes bind to rat liver 5S rRNA.

Metspalu et al. (80) identified L6 and L18 as the major proteins
that associate with rat liver 5S rRNA attached to Sepharose. There is
reason to believe that the protein identified as L18 was in fact L19
(see later).

Experiments of this kind must be closely controlled, if for no
other reason, because basic ribosomal proteins can associate nonspecifically with nucleic acids. No 60S ribosomal proteins (nor any
other ribosomal proteins for that matter) were bound to the activated
Sepharose 4B-adipic acid dihydrazide conjugate lacking a nucleic acid,
indicating the proteins did not merrely associate by ion exchange
chromatography on unsubstituted hydrazide groups or on reactive
groups on the CNBr-activated Sepharose (85). It is more important
that no 40S ribosomal protein associated with immobilized 5S rRNA,
nor did the proteins of either the small (30S) or large (50S) subparticle of *E. coli* ribosomes, although they contain many basic
proteins. The results are reassuring in that they indicate that the
binding of L6, and L19 to 5S rRNA is likely to be specific, not merely
the chromatography of a basic protein on a nucleic acid.

Table 2. Rat Liver Ribosomal Proteins That Bind to Nucleic Acids

5S rRNA	5.8S rRNA	Elongator-tRNA	Initiator-tRNA
L6	L6	L6	L6
L7	L8	L35a	L35a
L19	L19	S15	
	S9		
	S13		

The failure of any *E. coli* ribosomal proteins to bind to rat liver 5S rRNA suggests that rat liver L6 and L19 are not structurally homologous with *E. coli* L5, L18, and L25- the proteins that bind to *E. coli* 5S rRNA (87, 89-91). The results conform to the finding of Wrede and Erdmann (92) that *E. coli* L18 and L25 do not bind to yeast 5S rRNA.

A 5S rRNA-protein complex (5S rRNP) can be extracted from the large subunit of eukaryotic ribosomes with EDTA (93-96). The single protein in the complex was identified as L5 (96) and it was assumed that it was one of, or the only, protein associated with 5S rRNA in the ribosome. No L5 was bound to immobilized 5S rRNA (85). The results of the two sets of observations can be reconciled in several ways. One possibility is that EDTA unfolded the ribosomal subparticle and caused the proteins to lose their specific locations on ribosomal RNA. EDTA will randomize the distribution of proteins in *E. coli* ribosomal subparticles (97). A second possibility is that the 5S rRNP particle prepared with EDTA preserves the physiological binding where- as affinity chromatography on 5S rRNA-Sepharose gives spurious results. The circumstances in which the chromatography was done - the ionic conditions were selected to favor detection of even weak binding of protein to 5S rRNA - and the control experiments mitigate against the possibility. There is another explanation and that is that the binding site for L5 is at or near the 3' end of 5S rRNA and that site is perturbed by periodate oxidation or coupling to the adipic acid dihydrazide or both.

Ionic strength is critical in the association of proteins with nucleic acids (98). At relatively low ionic strength, nonspecific binding can occur and at relatively high ionic strength authentic binding can be inhibited. The ionic strength of the binding buffer used in the affinity chromatography experiments described here was 0.38, close to that used by Traub and Nomura (i.e. 0.37) in the reconstitution of the small subunit of *E. coli* ribosomes (98).

It was apparent that it would be valuable to have an independent means to verify the results obtained by affinity chromatography. We have succeeded in that purpose (99). A sensitive nitrocellulose membrane filter assay, employed before by Spicer et al. (100) and by Spierer et al. (101), was adopted and used to characterize the inter- action of individual pure rat liver ribosomal proteins with (^{32}P)- labeled 5S rRNA. For that purpose, the 5S rRNA was treated with alkaline phosphatase to remove the 5'-terminal phosphate groups and then made radioactive with (γ- ^{32}P)ATP using polynucleotide kinase. 5S rRNA is not retained on millipore filters unless it associates with protein. The membrane filter assay has a number of advantages: one can use individual pure rat liver ribosomal proteins to determine if binding to the nucleic acid saturates in conditions that approxi- mate equilibrium; the molar ratio of protein and nucleic acid in the complex at saturation can be obtained; and the data can be used to calculate the apparent association constant (Ka'). The assay can also be used to determine the optimum temperature, pH, and concentrations of ions for formation of the complex.

An important criterion of the specificity of the association of a ribosomal protein with a ribosomal nucleic acid is that it be stoichiometric (102, 103): the existence of a specific binding site in a molecule of RNA is verified when the complex contains no more than one mol of protein even when the protein is present in excess. The rat liver ribosomal proteins that interact with 5S rRNA immobilized on Sepharose are L6 and L19 (85). However, it was not possible to determine by affinity chromatography if the binding of L6 and L19 was stoichiometric since the amount of 5S rRNA was generally 5- to 10-times greater than that of the protein. It follows that it was not possible to calculate the association constants for the interaction. Finally, small amounts of ribosomal proteins L7, L23a, L27/L27a, L35a, and L39 also were bound to 5S rRNA affinity columns (see above). We undertook then to characterize the formation of complexes of 5S rRNA with individual pure rat liver ribosomal proteins in the hope that the results would help to resolve those outstanding problems.

Radioactive 5S rRNA was incubated with increasing amounts of a single ribosomal protein (99). We confirmed that L6 and L19 bind to 5S rRNA, and found that binding reached saturation (Table 3). It was now possible to test whether the proteins (L7, L23a, L27/L27a, L35a, and L39) that had been found in small amounts in the eluate from 5S rRNA affinity columns had associated with the nucleic acid in a site-specific way. Of those proteins L23a, L27/L27a, L35a, and L39 did not bind to 5S rRNA at all even when present in 10 molar excess (Table 3). Thus it is likely that in the affinity chromatography experiments (85) they were contaminants or were present because they had interacted with authentic 5S rRNA binding proteins. The results with L7, which was also a minor binding protein in the affinity chromatography experiment (albeit the most prominent of the set), were quite different. L7 binds to 5S rRNA in significant amounts and, what is more important, the binding is saturable.

It is noteworthy that only very small amounts of L5 were bound to (^{32}P)-labeled 5S rRNA even when a 10-fold molar excess of the protein was added (99). Thus the failure to detect binding of L5 to a 5S rRNA affinity column is not likely to have been due to perturbation of the 3'-end of the nucleic acid by periodate oxidation, or by the coupling of the oxidized nucleic acid to the adipic acid dihydrazide, during construction of the affinity column.

No small (40S) ribosomal subunit proteins were bound to Sepharose-5S rRNA columns (85), nor did any of several individual pure proteins (S2, S9, S13, S15 and S21) bind to the nucleic acid when assessed by the filter assay (99), even though, once again, the

Table 3. Characteristics of Complexes Formed Between Rat Liver
Ribosomal Proteins and 5S rRNA

Complex	Ka' ($\times 10^{-5}M^{-1}$)		Protein / RNA	
	$4°C$	$22°C$	$4°C$	$22°C$
L6-5S rRNA	4.4	2.0	0.77	0.89
L7-5S rRNA	1.6	1.3	0.89	0.93
L19-5S rRNA	6.8	2.3	0.87	1.05

protein was added in great excess. The proteins that did not bind to
5S rRNA include S9 and S13 which associate with immobilized 5.8S rRNA
(104), and S15 which interacts with elongator-tRNAs on affinity
columns (105).

Metspalu et al.. (88) reported that L18 rather than L19 (85) was
bound to 5S rRNA immobilized on Sepharose. We have argued that the
discrepancy was the result of a failure to employ both one- and two-
dimensional electrophoresis, since L18 and L19 are hard to distinguish
by the latter procedure alone. The argument is strengthened by the
finding that pure L18 does not bind to (^{32}P)-labeled 5S rRNA.

The apparent association constant (Ka') for the interaction of
L6, L7, and L19 was determined by constructing a double reciprocal
plot of the data contained in the saturation binding curves (Table 3).
The analysis permits the determination from the intercept and from
the slope of the number of binding sites per mole of rRNA for each
protein, and the apparent association constant. The value for n (the
ratio of the moles of protein to the moles of rRNA in the complex),
which in every case approximates 1, establishes that the interaction
is specific (Table 3).

From the slope (the lines were fitted according to a least
squares regression) of the plot the apparent association constants
were calculated (Table 3). The association constants for the interaction of rat liver L6, L7, and L19 with 5S rRNA are surprisingly
similar, differing at most only by four-fold (Table 3), whereas the
binding of E. coli ribosomal proteins L5, L18, and L25 to bacterial
5S rRNA differs by two orders of magnitude (101). Moreover, the
strength of the binding of rat liver ribosomal proteins to 5S rRNA
is not only far less than that generally observed for protein-nucleic
acid interactions, but it is also less than that for formation of

E. coli 5S rRNA-protein complexes (101). It is, of course, possible
that the strength of the protein-rRNA interaction is greater when the
same proteins interact in the ribosome than when a single protein
binds to the nucleic acid. Indeed, it is likely that the association
of 5S rRNA with ribosomal proteins is substantially strengthened by
cooperative interactions involving many of the molecular constituents
of the subunit.

The nitrocellulose filter assay provides a valuable supplement to
affinity chromatography for the analysis of protein-nucleic acid
interactions in the ribosome. The advantage of affinity chromatography is that it permits a relatively rapid survey of all the ribosomal proteins to determine which interact with a species of nucleic
acid; the disadvantage is that one can not determine the stoichiometry of the binding, nor the apparent association constants, since
the experiment is done in nonequilibrium conditions. The filter
assay is a convenient, sensitive means for characterizing and quan-
titating the interaction of small amounts of single pure proteins with
nucleic acids.

5.8S rRNA

The proteins that were bound to immobilized 5.8S rRNA (107), a
nucleic acid that is associated by hydrogen bonds with 28S rRNA in the
60S subunit of eukaryotic ribosomes (106), were L19, L8, and L6 from
the 60S subunit; and S13, and S9 from the small subparticle (Table 2).
Small amounts of L14, L19a, L18, L27 or L27a, and L35a, and of S11,
S15, S23/24, and S26 also were bound to the affinity column. *E. coli*
ribosomal proteins did not bind to rat liver 5.8S rRNA. The finding
was somewhat surprising since *E. coli* ribosomal proteins L18, and L25,
associate specifically with yeast 5.8S rRNA (92). There is no satisfactory explanation for the apparently discordant results, although
it should be noted that *E. coli* ribosomal proteins do not bind to
yeast 5.8S rRNA immobilized on affinity columns (108).

Metspalu *et al.* (88) used affinity chromatography to identify L5,
L6, L7, and L18 as the ribosomal proteins that bind to rat liver 5.8S
rRNA rather than L5, L7, or L18. The discrepancies may be in part the
result of the conditions employed in the binding reaction, and in part
a reflection of the difficulty in making a precise identification of
the ribosomal proteins (c.f. ref. 107 for a discussion).

The rat liver ribosomal proteins L6, L7, L19, S9, and S13 have
been shown (110) by the millipore filter assay to bind stoichiometrically to 5.8S rRNA (Table 4).

The binding of the 40S ribosomal proteins S13 and S9 to 5.8S
rRNA, which is a component of the large subparticle, suggests that the
nucleic acid may participate in the formation of subunit couples (i.e.
80S ribosomes). Actually the theory has been tested (111): it has

Table 4.   Characteristics of Complexes Formed Between Rat Liver
           Ribosomal Proteins and 5.8S rRNA

| Complex | Ka' $(\times 10^{-5}\text{M}^{-1})$ | | Protein RNA | |
	4°C	22°C	4°C	22°C
L6-5.8S rRNA	1.3	1.6	1.04	0.93
L8-5.8S rRNA	2.6	1.5	1.05	1.17
L19-5.8S rRNA	18.0	14.2	0.89	0.92
S9-5.8S rRNA	0.9	0.2	1.16	1.22
S13-5.8S rRNA	6.0	1.6	1.04	1.27

been demonstrated that 5.8S rRNA immobilized on affinity columns forms
a complex with 40S subunits. The results suggest that 5.8S rRNA and/
or the proteins that associate with the nucleic acid participate in
forming 80S ribosomal couples. In addition, it follows that 5.8S rRNA
and the proteins that bind to it are on the surface of ribosomal subunits near the subunit interface, the region where the subunits abut,
a region known to be involved in transacting most of the business of
the ribosome.

L6 and L19 bind to 5S and to 5.8S rRNA (Table 2). If the results
obtained from affinity chromatography reflect protein-nucleic acid
association in the ribosome, then L6 and L19 must have at least two
domains for binding rRNA, one for 5S and a second for 5.8S. Since L6
also associates with tRNA (see later) it may well have a binding site
for that ligand too. We have now shown that L6 does in fact have
separate binding sites for 5S and 5.8S rRNA (112). A 5S rRNA affinity
column was constructed and pure L6 was bound to the nucleic acid; the
5S rRNA-L6 complex retained equimolar amounts of (^{32}P)-labeled 5.8S
rRNA indicating separate binding sites on L6 for the two nucleic
acids. The results further suggest that the two nucleic acids, 5S and
5.8S rRNA, and their associated proteins L6, L8, and L19 are in close
proximity in the 60S subunit. The domain may comprise a portion of
the A and P sites (see later).

The large subunit of eukaryotic ribosomes contain two small ribosomal nucleic acids, 5S and 5.8S, whereas prokaryotic ribosomes have
only 5S rRNA. There are reasons to suspect that the functionally
homologous nucleic acids are prokaryotic 5S and eukaryotic 5.8S (92,
113). The latter is part of the transcript of rDNA with 18 and 28S
rRNA, just as prokaryotic 5S rRNA is part of a transcription unit with
16 and 23S rRNA; on the other hand the genes for *eukaryotic* 5S rRNA

are located at a separate site in the nucleus. Moreover, there are
differences in the primary structure of prokaryotic and eukaryotic
5S rRNA (113-115). What is most significant is that the latter does
not have the sequence C-G-A-A-C that is in all prokaryotic 5S rRNAs,
and that is complementary to T-Ψ-C-G in loop IV of all elongator-tRNAs
and in prokaryotic initiator-tRNA (see preceding chapter, Table 2);
eukaryotic 5S rRNA does have the conserved sequence C-G-A-U that is
complementary to A-U-C-G of loop IV found only in eukaryotic initia-
tor-tRNAs (see reference 113 for a summary of the evidence and earlier
references). It is assumed that the T-Ψ-C-G of tRNA associates with a
complementary sequence in prokaryotic 5S (116, 117) and perhaps with
eukaryotic 5.8S rRNA (92, 113),during binding of aminoacyl-tRNA to the
ribosomal A site. The fragment T-Ψ-C-G inhibits factor-dependent
binding of aminoacyl-tRNA to prokaryotic (118, 119) and eukaryotic
ribosomes (120); the P site binding of eukaryotic initiator-tRNA is
not affected. A consistent construct is that in eukaryotes 5.8S rRNA
forms part of the ribosomal A site and participates in the binding of
elongator-tRNAs; that 5S rRNA forms part of the P site and takes part
in the binding of initiator-tRNA.

Putative A site and P site complexes have been constructed on
affinity columns (121). The putative A site complex contains 5.8S
rRNA and its associated proteins (L6, L8, L19, S19, and S13); the P
site complex has 5S rRNA and L6, L7, and L19. Both RNP complexes
retain pure isoaccepting species of tRNA, however, they do not
distinguish between elongator tRNAs (tRNA^{Phe}) and initiator tRNA
(tRNA$_f^{Met}$) - the initiator and elongator tRNAs bind equally well to the
5S and 5.8S RNP complexes. Moreover, the ternary complex EF-1.GTP.
phenylalanyl-tRNA binds to both RNP complexes also. Since neither
EF-1.GTP, nor puromycin, bind to the 5S and 5.8S RNP complexes, we
assume it is the tRNA (rather than the factor or the aminoacyl group)
that interacts with the ribosomal RNP complex. The failure of the
two complexes (5S and 5.8S RNP) to distinguish between elongator and
initiator tRNAs is a disappointment. We assume discrimination requires a binding site for the factors (EF-1 and eIF-2) as well as the
tRNA.

E. coli 5S rRNA

The E. coli large ribosomal subunit proteins L5, L18, and L25
have been shown by a variety of procedures (89-91), including affinity
chromatography (87), to bind independently and specifically to homologous 5S rRNA. We have confirmed the finding: when E. coli ribosomal proteins were chromatographed on 5S rRNA affinity columns L5,
L18, and L25 were bound (122).

A number of rat liver ribosomal proteins (L6, L7, L19, L35a, and
S9) bind to E. coli 5S rRNA. Lind et al. (123) also found that rat
liver ribosomal proteins bind to E. coli 5S rRNA but the identity of
the proteins was not fully documented. The finding is surprising

since no *E. coli* ribosomal proteins bind to rat liver 5S rRNA (85, 88, 123) nor to rabbit reticulocyte, nor yeast 5S rRNA (87). There is no immediate obvious explanation of the paradox. Still the idea that the function of prokaryotic 5S rRNA, which may be to participate in the binding of aminoacyl-tRNA, is split amongst two eukaryotic ribosomal nucleic acids, 5S and 5.8S, is attractive. The thesis is that bacterial 5S rRNA binds elongator-tRNAs, whereas eukaryotic 5S rRNA binds the initiator-tRNA and 5.8S rRNA the elongator-tRNAs. If that is the case then *E. coli* 5S rRNA might bind proteins that associate with either 5S or 5.8S rRNA, or both. Of the rat liver proteins that bind to *E. coli* 5S rRNA: L6 and L19 bind to both rat liver 5S and 5.8S rRNA; L35a binds to both also but only in trace amounts; L7 binds to 5S; and S9 binds to significant amounts only to 5.8S rRNA. Thus the pattern of binding of rat liver proteins to *E. coli* 5S rRNA overlaps the binding to rat liver 5S and 5.8S rRNA.

The similarities in the primary sequence of prokaryotic 5S rRNA and eukaryotic 5.8S rRNA are no greater than to be expected by chance (109) and hence is unlikely they share a common ancestor. However, it may be secondary structure rather than sequence that is relevant in the binding of proteins to rRNA; moreover, it may be possible to derive structurally similar binding sites from different primary sequences. It is relevant that there are similarities in the models for the secondary structure of prokaryotic and eukaryotic 5S rRNA (124). If the models reflect the native structure of 5S rRNA, the similarity could account for the biding of rat liver L6, L7, and L19 to *E. coli* and rat liver 5S rRNAs. *In situ* 5.8S rRNA is hydrogen-bonded to 28S rRNA (106); in some models 5.8S rRNA is given a secondary structure not entirely unlike that of 5S rRNA (125); that could account for the association with *E. coli* 5S rRNA of rat liver proteins (L6, L19, and S9) that bind to 5.8S rRNA. Nothing of the structure accounts for the failure of *E. coli* ribosomal proteins to bind to rat liver 5S rRNA in the reverse experiment.

The possibility that the binding of rat liver ribosomal proteins to *E. coli* 5S rRNA is a non-specific artifact needs to be considered. The possibility can not be excluded with certainty but does not seem likely. First, in exactly the circumstances in which the rat liver proteins bind to *E. coli* 5S rRNA on affinity columns, only three of 52 *E. coli* proteins (L5, L18, and L25) associate with the nucleic acid and they are the very proteins demonstrated to form a complex with the nucleic acid by a variety of other procedures (89-91). If the binding of *E. coli* ribosomal proteins to *E. coli* 5S rRNA reflects an authentic interaction, then the heterologous association may also derive from a fit between binding sites on the rRNA and the protein. It is important that the rat liver ribosomal proteins that bind to *E. coli* 5S rRNA (L6, L7, L19, L35a, and S9) are not a random set but rather a small subgroup of the total that includes most of the proteins that bind to rat liver 5S or 5.8S rRNA or both. Of the proteins that bind to rat liver 5S or 5.8S rRNA only L8 and S13 (5.8S

rRNA binding proteins) do not bind to *E. coli* 5S rRNA. We have in-
spected the properties of the binding protein - pI, number of basic
and acidic residues, hydrophobicity, aromatic amino acids - and do not
find anything to distinguish them from many ribosomal proteins that
do not bind to 5S or 5.8S rRNA. Thus the association is likely to be
the result of an affinity of the proteins for the nucleic acid rather
than of ion-exchange.

It is obvious that what would be helpful is the primary sequence
of the rat liver ribosomal proteins that bind to *E. coli* 5S rRNA and
some knowledge of the actual secondary structure of the small nucleic
acids in the large subunits of ribosomes. A comparison of the binding
of homologous and heterologous proteins to 5S rRNA may provide
important information on the nature of the rules that govern the
interaction.

tRNA

The binding of aminoacyl-tRNAs to the ribosome during protein
synthesis is not merely the result of codon-anticodon interaction:
for an exhaustive discussion of the question from the stand-point of
tRNA, see the preceding chapter.

The aminoacyl-tRNAs associate with ribosomal proteins (126, 127),
and perhaps ribosomal ribonucleic acids (113), as well. Knowledge of
the identity of the ribosomal proteins that interact with tRNA is
important for an understanding of the function of ribosomes, that is,
of the mechanism of protein synthesis. Experiments have been carried
out to determine by affinity chromatography the rat liver ribosomal
proteins that bind to tRNA (105). For that purpose either mixed yeast
elongator tRNAs (cytoplasmic tRNAs lacking $tRNA^{Met}_{f,m}$) or pure isoaccepting species of elongator tRNAs ($tRNA^{Met}_{m}$; $tRNA^{Phe}$), or
initiator tRNA ($tRNA^{Met}_{f}$) were immobilized on Sepharose columns and
the ribosomal proteins that formed a stable association with the
nucleic acids were identified by polyacrylamide gel electrophoresis.

The rat liver ribosomal proteins that bind to mixed elongator-
tRNAs and to pure isoaccepting species, $tRNA^{Met}_{m}$ and $tRNA^{Phe}$, are L6,
L35a, and S15. A number of proteins, of which L8 is perhaps the most
prominent, bind to elongator-tRNAs in small amounts.

Although it is not proven, the proteins L6, L35a, and S15 may
form a part of the ribosome surface that participates in the binding
of aminoacyl-tRNA during the elongation reaction of protein synthesis.
That domain may also encompass 5.8S rRNA and the proteins associated
with it -L6, L8, L19, S9, and S13 (see above). That L6 associates
with both nucleic acids would indicate that elongator-tRNAs bind to
the ribosome in the neighbourhood of 5.8S rRNA.

The rat liver ribosomal proteins that associate with $tRNA^{Met}_{f}$

immobilized on affinity columns are L6 and L35a. Once again the
assumption is made that the initiator-tRNA binds to the same proteins
that it interacts with during initiation of protein synthesis
although that has not been substantiated experimentally. If the
assumption is correct then L6 and L35a should be part of the ribosomal
surface designated as the P site. There are indications that the P
site includes 5S rRNA and the proteins associated with it – L6, L7,
and L19 (see above). It is not entirely surprising that two proteins
L6 and L35a can interact with elongator-tRNAs and initiator-tRNA.
First, the two proteins could each contribute two separate binding
sites; one each to the A site to bind elongator-tRNAs, and one each
to the P site to bind initiator-tRNA. This is consistent with the
requirement that the A and P sites be close enough (within 20 $\overset{\circ}{A}$) to
allow for peptide bond formation. Perhaps a more plausible explanation derives·from the observation that aminoacyl-tRNA binds successively, in a codon specific reaction, to the A and P sites during each
of the reiterative cycles of peptide bond formation (129, 130), hence
elongator-tRNAs may not bind exclusively to A site proteins in affinity experiments of the type we have carried out.

It is perhaps especially important that the small ribosomal subunit protein S15 associated with elongator-tRNAs but not with
initiator-tRNA. The finding suggests that the binding of Met-tRNA$_f$
to the 40S ribosomal subunit during formation of an initiation complex is specified by the initiation factor eIF-2 (63) and that the
interaction with ribosomal proteins does not occur until the 60S subunit is added to form an 80S initiation complex. At that stage the
Met-tRNA$_f$ would bind to L6 and L35a in the P site. S15 may be
exclusively in the A site.

CODA

A great deal has been learned of eukaryotic ribosomes in the last
decade. The proteins have been isolated and characterized and limited
sequence information is now available, moreover, it is likely that the
entire structure of some of the more important proteins will soon be
available. The structure of 5S and 5.8S rRNA from a number of eukaryotic species has been determined and the prospects are good that the
sequence of 18 and 28S rRNA will be known in the not too distant
future. Work has begun on the interaction of the proteins and nucleic
acids in the particle; that information will surely be valuable in
arriving at a solution of the structure of eukaryotic ribosomes.
Information is being accumulated on ribosomes from a variety of
eukaryotic species, from plants to humans, and from mitochondria and
chloroplast as well. The charting of the evolution of the particle
has predictably benefited from the data. Not too long ago one would
have guessed that the structure and function of eukaryotic ribosomes
would have to be resolved without assistance from a genetic analysis,
but there are now increasing numbers of authentic mutants in ribosomal
proteins. They are surely going to be useful. Perhaps the most

exciting advance is in the identification of the genes for ribosomal constituents. The analysis of the transcription units for rRNA is well underway and one suspects that the experiments will lead soon to an understanding of the regulation of their transcription and that the DNA will be sequenced, leading in turn to the sequence of 18 and 28S rRNA. A number of eukaryotic ribosomal protein genes have been identified: the extraordinary importance of that achievement can hardly be exaggerated. Not only is it likely to provide information on the organization and the regulation of the genes for the molecular constituents of the ribosome, but it should provide material for determining the sequence of the ribosomal proteins (from the DNA sequence), and for studying the regulation of transcription (by using the DNA as a template for coupled transcription and translation *in vitro*. In the circumstances it is not surprising that we know much more of ribosome biogenesis. Advances proceed on the supramolecular structure: there are now more refined three-dimensional reconstructions of eukaryotic ribosomes from electron micrographs of crystalline arrays; and immune-electron microscopy is being used to map the location of proteins on the surface of the particle. All that is lacking to overflow one's cup is a method for the reconstitution of ribosomal subunits from the components. One is entitled to be sanguine about prospects: clarly for eukaryotic ribosomes the future lies ahead. Some might go so far as to say the end is here even if it is not yet in.sight.

REFERENCES

1.   Wool, I.G (1979), Ann. Rev. Biochem. <u>48</u>, 719-754.
2.   Wool, I.G. (1980), In *Ribosomes: Structure, Function and Genetics* (Chamblis, G., Craven, G.R., Davies, J., Davis, K., Kahan, L., and Nomura, M., eds.) pp. 797-824,Univ.Park Press,Baltimore.
3.   Petermann, M.L. (1964), In *The Physical and Chemical Properties of Ribosomes,* Elsevier, New York.
4.   Blair, D., Hill, W.E., and Wool, I.G., unpublished data.
5.   Loening, U.E. (1968), J. Mol. Biol. <u>38</u>, 355-365.
6.   Weinberg, R.A., and Penman, S. (1970), J. Mol. Biol. <u>47</u>, 169-178.
7.   Sherton, C.C. and Wool, I.G. (1974), Mol. Gen. Genet. <u>135</u>, 97-112.
8.   Hamilton, M.G., Pavlovec, A., and Petermann, M.L. (1971), Biochemistry <u>10</u>, 3424-3427.
9.   Bielka, H., Welfle, H., Böttger, M., and Föster, W. (1968), Eur. J. Biochem. <u>5</u>, 183-190.
10.   Brownlee, G.G., Sanger, F., and Barrell, B.G. (1867), Nature, London, <u>215</u>, 735-736.
11.   Nazar, R.N., Sitz, T.O., and Busch, H. (1975), J. Biol. Chem. <u>250</u>, 8591-8597.
12.   Cammarano, P., Pons, S., Romeo, A., Gauldieri, M., and Gualerzi, C. (1972), Biochim. Biophys. Acta <u>281</u>, 571-596.
13.   Cammarano, P., Romeo, A., Gentile, M., Felsani, A., and Gualerzi, C. (1972), Biochim. Biophys. Acta <u>281</u>, 597 - 624.

14. Cammarano, P., Felsani, A., Gentile, M., Gualerzi, C., Romeo, A., and Wolf, G. (1972), Biochim. Biophys. Acta 281, 625-642.
15. Nonomura, Y., Blobel, G., and Sabatini, D.D (1971), J. Mol. Biol. 60, 303-323.
16. Lake, J.A., Sabatini, D.D., and Nonomura, Y. (1974), In *Ribosomes* (Nomura, M., Tissières, A., and Lengyel, P., eds.) pp. 543-557, Cold Spring Harbor Lab., Cold Spring Harbor, New York.
17. Emanuilov, I., Sabatini, D.D., Lake, J.A. and Freienstein,C. (1978), Proc. Nat. Acad. Sci. USA 75, 1389-1393.
18. Boublik, M., and Hellman, W. (1978), Proc. Nat. Acad. Sci. USA 75, 2829-2833.
19. Unwin, P.N.T., and Taddei, C. (1977), J. Mol. Biol. 114, 491-506.
20. Unwin, P.N.T. (1977), Nature, London, 269, 118-122.
21. Morimoto, T., Blobel, G., and Sabatini, D.D. (1972), J. Cell. Biol. 52, 338-354.
22. Morimoto, T., Blobel, G., and Sabatini, D.D. (1972), J. Cell. Biol. 52, 355-366.
23. Byers, B. (1967), J. Mol. Biol. 26, 155-167.
24. Hardy, S.J.S., Kurland, C.G., Voynow, P., and Mora, G. (1969), Biochemistry 8, 2897-2905.
25. Mora, G., Donner, D., Thammana, P., Lutter, L., and Kurland, C.G. (1971), Mol. Gen. Genet. 112, 229-242.
26. Hindennach, I., Stöffler, G. and Wittmann, H.G. (1971), Eur. J. Biochem. 23, 7-11.
27. Hindennach, I., Kaltschmidt, E., and Wittmann H.G. (1971), Eur. J. Biochem. 23, 12-16.
28. Lin, A., Collatz, E., and Wool, I.G. (1976), Mol. Gen. Genet. 144, 1-9.
29. Collatz, E., Lin, A., Stöffler, G., Tsurugi, K., and Wool, I.G. (1976), J. Biol. Chem. 251, 1808-1816.
30. Collatz, E., Wool, I.G., Lin, A., and Stöffler, G. (1976), J. Biol. Chem. 251, 4666-4672.
31. Tsurugi, K., Collatz, E., Wool, I.G., and Lin, A. (1976), J. Biol. Chem. 251, 7940-7946.
32. Tsurugi, K., Collatz, E., Todokoro, K., and Wool, I.G. (1977), J. Biol. Chem. 252, 3961-3969.
33. Collatz, E., Ulbrich, N., Tsurugi, K., Lightfoot, H.N,, Mackinlay W., Lin, A., and Wool, I.G. (1977), J. Biol. Chem. 252, 9071-9080
33a. Tsurugi, K., Collatz, E., Todokoro, K., Ulbrich, N., Lightfoot, H.N., and Wool, I.G. (1978), J. Biol. Chem. 253, 946-955.
34. Lin, A., Tanaka, T., and Wool, I.G. (1979), Biochemistry 18, 1634-1637.
35. Terao, K., and Ogata, K. (1972), Biochim. Biophys. Acta 285, 473-482.
36. Westermann, P., and Bielka, H. (1973), Mol. Gen. Genet. 126, 349-356.
37. Goerl, M., Welfle, H., and Bielka, H. (1978), Biochim. Biophys. Acta 519, 418-427.
38. Howard, G.A., Ramjoué, H.P.R., and Gordon, J. (1977), In *Translation of Natural and Synthetic Polypeptides* (Legocki, A.B., ed.) pp. 305-311, Univ. Argic., Poznan, Poland.

39. Higo, K., and Otaka, E. (1979), Biochemistry <u>18</u>, 4191-4196.

40. Itoh, T., Higo, K., and Otaka, E. (1979), Biochemistry <u>18</u>, 5787-5793.

41. Chooi, Y.W. (1980), Biochemistry <u>19</u>, 3469-3476.

42. Ting Shih, C.Y., Toivonen, J.E. and Craven, G.R. (1979), Eur. J. Biochem. <u>97</u>, 189-196.

43. Lin, A., and Wool, I.G. (1974), Mol. Gen. Genet. <u>134</u>, 1-6.

44. Howard, G.A., Traugh, J.A., Croser, E.A., and Traut, R.R. (1975), J. Mol. Biol. <u>93</u>, 391-404.

45. Terao, K., and Ogata, K. (1975), Biochim. Biophys. Acta <u>402</u>, 214-229.

46. Martini, O.H.W., and Gould, H.J. (1975), Mol. Gen. Genet. <u>142</u>, 317-331.

47. Issinger, O.G., and Beier, H. (1978), Mol. Gen. Genet. <u>160</u>, 297-309.

48. Welfle, H., Goerl, M., and Bielka, H. (1978), Mol. Gen. Genet. <u>163</u>, 101-112.

49. Sherton, C.C., and Wool, I.G. (1972), J. Biol. Chem. <u>247</u>, 4460-4467.

50. Welfle, H., Stahl, J., and Bielka, H. (1971), Biochim. Biophys. Acta <u>243</u>, 416-419.

51. Welfle, H., Stahl, J., and Bielka, H. (1972), FEBS Lett. <u>26</u>, 228-232.

52. Delaunay, J., Creusot, F., and Shapira, G. (1973), Eur. J. Biochem. <u>39</u>, 305-312.

53. Chatterjee, S.K., Kazemie, M., and Matthaei, H. (1973), Hoppe-Seylers Z. Physiol. Chem. <u>354</u>, 481-486.

54. Lastick, S.M., and McConkey, E.H. (1976), J. Biol. Chem. <u>251</u>, 2867-2875.

55. Blobel, G., and Sabatini, D.D. (1971), In *Biomembranes* (Manson, L.A., ed.), vol. <u>2</u>, pp. 193-196, Plenum Press, New York.

56. Sabatini, D.D., and Kreibich, G. (1976), In *The Enzymes of Biological Membranes* (Martinosi, A., ed.), vol. <u>2</u>, pp. 531-579, Plenum Press, New York.

57. Kreibich, G., Ulrich, B.L., and Sabatini, D.D. (1978), J. Cell. Biol. <u>77</u>, 464-487.

58. Kreibich, G., Freienstein, C.M., Pereyra, B.N., Ulrich, B.L., and Sabatini, D.D. (1978), J. Cell. Biol. <u>77</u>, 488-506.

59. Kreibich, G., Czako-Graham, M., Grebenau, R., Mok, W., Rodriguez-Boulan, E., and Sabatini, D.D. (1978), J. Supramol. Struct. <u>8</u>, 279-302.

60. Kozak, M. (1978), Cell. <u>15</u>, 1109-1123.

61. Kozak, M. (1979), J. Biol. Chem. <u>254</u>, 4731-4738.

62. Kozak, M. (1979), Nature, London, <u>280</u>, 82-85.

63. Trachsel, H., Erni, B., Schreier, M.H., and Staehelin, T. (1977), J. Mol. Biol. <u>116</u>, 755-767.

64. Cundliffe, E., Dixon, P., Stark, M., Stöffler, G., Ehrlich, R., Stöffler-Meilicke, M., and Cannon, M., J. Mol. Biol., in press.

65. Wienen, B., Ehrlich, R., Stöffler-Meilicke, M., Smith, I., Weiss, D., Vince, R., and Pestka, S. (1979), J. Biol. Chem. <u>254</u>, 8031-8041.

66. Pestka, S., Weiss, D., Vince, R., Wienen, B., Stöffler, G., and Smith, I., (1976), Mol. Gen. Genet. 144, 235-241.
67. Maaløe, O. (1969), Dev. Biol. 3, 33-58 (Suppl.).
68. Kjeldgaard, N.O., and Gausing, K. (1974), In *Ribosomes* (Nomura, M., Tissières, A., and Lengyel, P., eds.) pp. 369-392, Cold Spring Harbor Lab., Cold Spring Harbor, New York.
69. Wittmann-Liebold, B., Geissler, A.W., Lin, A., and Wool, I.G. (1979), J. Supramol. Struct. 12, 425-433.
70. Reisner, A.H., personal communication
71. Stöffler, G., Wool, I.G., Lin, A., and Rak, K.H. (1974), Proc. Nat. Acad. Sci. USA 71, 4723-4726.
72. Wool, I.G. and Stöffler, G. (1974), In *Ribosomes* (Nomura, M., Tissières, A., and Lengyel, P., eds.) pp. 417-460, Cold Spring Harbor Lab., Cold Spring Harbor, New York.
73. Howard, G.A., Smith, R.L., and Gordon, J. (1976), J. Mol. Biol. 106, 623-637.
74. Van Agthoven, A.J., Maasen, J.A., and Möller, W. (1977), Biochem. Biophys. Res. Comm. 77, 989-998.
75. Tanaka, T., Wool, I.G., and Stöffler, G. (1980), J. Biol. Chem. 255, 3832-3834.
76. Itoh, T., and Wittmann-Liebold, B., in preparation.
77. Itoh, T., and Wittmann-Liebold, B. (1978), FEBS Lett. 96, 399-402
78. Fischer, N., Stöffler, G., and Wool, I.G. (1978), J. Biol. Chem. 253, 7355-7360.
79. Lin, A., Wittman- Liebold, B., and Wool, I.G., unpublished data.
80. Itoh, T., and Osawa, S., submitted.
81. Amons, R., Pluijms, W., and Möller, W. (1979), FEBS Lett. 104, 85-89.
82. Terhorst, C., Möller, W., Laursen, R., and Wittmann-Liebold, B. (1973), Eur. J. Biochem. 34, 138-152.
83. Oda, G., Strøm, A.R., Visentin, L.P., and Yaguchi, M. (1974), FEBS Lett. 43, 127-130.
84. Dayhoff, M.O. (1976), In *Atlas of Protein Sequence and Structure* (Dayhoff, M.O., ed.) vol. 5, (suppl. 2) pp. 6. National Research Foundation, Washington, D.C.
85. Ulbrich, N., and Wool, I.G. (1978), J. Biol. Chem. 253, 9049-9052
86. Burrell, H.R., and Horowitz, J. (1975), FEBS Lett. 49, 306-309.
87. Burrell, H.R., and Horowitz, J. (1977), Eur. J. Biochem. 75, 533-544.
88. Metspalu, A., Saarma, M., Villems, R., Ustav, M., and Lind, A. (1978), Eur. J. Biochem. 91, 73-81.
89. Horne, J.R., and Erdmann, V.A. (1972), Mol. Gen. Genet. 119, 337-344.
90. Gray, P.N., Bellemare, G., Monier, R., Garrett, R.A., and Stöffler, G. (1973), J. Mol. Biol. 77, 133-152.
91. Yu, R.S.T., and Wittmann, H.G. (1973), Biochim. Biophys. Acta 324, 375-385.
92. Wrede, P., and Erdmann, V.A. (1977), Proc. Nat. Acad. Sci. USA 74, 2706-2709.
93. Blobel, G. (1971), Proc. Nat.. Acad. Sci. USA 68, 1881-1885.

94. Lebleu, B., Marbaix, G., Huez, G., Temmerman, J., Burney, A., and Chantrenne, H. (1971), Eur. J. Biochem. 19, 264-269.
95. Petermann, M.L., and Pavlovec, A. (1971), Biochemistry 10, 2770-2775.
96. Terao, K., Takahashi, Y., and Ogata, K. (1975), Biochim. Biophys. Acta 402, 230-237.
97. Newton, I., Rinke, J., and Brimacombe, R. (1975), FEBS Lett. 51, 215-218.
98. Traub, P., and Nomura, M. (1969), J. Mol. Biol. 40, 391-413.
99. Ulbrich, N., Todokoro, K., Ackerman, E., and Wool, I.G. (1980), J. Biol. Chem., in press.
100. Spicer, E., Schwarzbauer, J., and Craven, G.R. (1977), Nuc. Acids Res. 4, 491-499.
101. Spierer, P., Bogdanov, A.A., and Zimmermann, R.A. (1978), Biochemistry 17, 5394-5398.
102. Schaup, H.W., Green, M., and Kurland, C.G. (1970), Mol. Gen. Genet. 109, 193-205.
103. Garrett, R.A., Rak, K.H., Daya, L., and Stöffler, G. (1971), Mol. Gen. Genet. 114, 112-124.
104. Ulbrich, N., Lin, A., and Wool, I.G. (1979), J. Biol. Chem. 254, 8641-8645.
105. Ulbrich, N., Wool, I.G., Ackerman, E.J., and Sigler, P.B. (1980), J. Biol. Chem. 255, 7010-7016.
106. Pace, N.R., Walker, T.A., and Schroeder, E. (1977), Biochemistry 16, 5321-5328.
107. Ulbrich, N., Lin, A., and Wool, I.G. (1979), J. Biol. Chem. 255, 797-801.
108. Erdmann, V.A., personal communication.
109. Cedergren, R.J., and Sankoff, D. (1976), Nature, London, 260, 74-75.
110. Todokoro, K., Ulbrich, N., and Wool, I.G., unpublished data.
111. Villems, R., Saarma, M., Metspalu, A., and Toots, I. (1979), FEBS Lett. 107, 66-68.
112. Ulbrich, N., and Wool, I.G., unpublished data.
113. Erdmann, V.A. (1976), In *Progress in Nucleic Acid Research and Molecular Biology* (Cohn, E.W., ed.) vol. 18, pp. 45-90, Academic Press, New York.
114. Erdmann, V.A. (1978), Nucleic Acids Res; 5, r1-r14.
115. Erdmann, V.A. (1979), Nucleic Acids Res. 6, r29-r44.
116. Ofengand, J., and Henes, C. (1969), J. Biol. Chem. 244, 6241-6253
117. Schwarz, U., Menzel, H.M., and Gassen, H.G. (1976), Biochemistry 15, 2484-2490.
118. Richter, D., Erdmann, V.A., and Sprinzl, M. (1973), Nature New Biol. 246, 132-135.
119. Erdmann, V.A., Sprinzl, M., Richter, D., and Lorenz, S. (1974), Acta Biol. Med. Germ. 33, 605-608.
120. Grummt, F., Grummt, I., Gross, H.J., Sprinzl, M., Richter, D., and Erdmann, V.A. (1974), FEBS Lett. 42, 15-17.
121. Chan, Y.L., Ulbrich, N., Wool, I.G., Ackerman, E.J., and Sigler, P.B., unpublished data.
122. Ulbrich, N., Lin, A., and Wool, I.G. (1980), J. Biol. Chem. 255, 797-801.

123. Lind, A., Metspalu, A., Saarma, M., Toots, I., Ustav, M., and
     Villems, R. (1977), Bioorg. Chem. (USSR) 3, 1138-1140.
124. Hori, H. (1976), Mol. Gen. Genet. 145, 119-123.
125. Luoma, G.A., and Marshall, A.G. (1978), Proc. Nat. Acad. Sci.
     USA 75, 4901-4905.
126. Cantor, C.R., Pellegrini, M., and Oen, H. (1974), In *Ribosomes*
     (Nomura, M., Tissières, A., and Lengyel, P., eds.) pp. 573-585,
     Cold Spring Harbor Lab., Cold Spring Harbor, New York.
127. Stöffler, G. (1974), In *Ribosomes* (Nomura, M., Tissières, A., and
     Lengyel, P., eds.) pp. 615-667, Cold Spring Harbor Lab., Cold
     Spring Harbor, New York.
129. Wurmbach, P., and Nierhaus, K.H. (1979), Proc. Nat . Acad. Sci.
     USA 76, 2143-2147.
130. Lührmann, R., Eckhardt, H., and Stöffler, G. (1979), Nature,
     London, 280, 423-425.

THE INITIATION FACTORS

John W.B. Hershey

Department of Biological Chemistry, School of Medicine
University of California
Davis, CA 95616, U.S.A.

INTRODUCTION

Initiation of protein synthesis is the process whereby the ribosome binds mRNA and the first aminoacyl-tRNA to form an initiation complex which is capable of entering the elongation phase of protein synthesis. The pathway is complex and involves numerous steps: First, the 80S ribosome dissociates into 40S and 60S ribosomal subunits; the 40S subunit forms a preinitiation complex with methionyl-tRNA and mRNA; this is joined by the 60S subunit to complete formation of the 80S initiation complex. During these steps, two critical events occur: the ribosome selects for translation a specific mRNA from among numerous species of mRNAs; and the methionyl-tRNA interacts with a specific initiator site on the mRNA to assure proper translation in the correct phase. The reactions are promoted or catalyzed by a complex array of initiation factors and involve the hydrolysis of ATP and GTP.

A systematic study of initiation must address the following questions. What components are involved ? By what pathway do the components interact ? What kinds of molecular interactions occur and at what rates do the reactions proceed ? Finally, how is the initiation process controlled or modulated ? In this section we will focus specifically on the initiation factors from mammalian cells and will consider their identification, characterization and mechanism of action. Aspects of translational control and the interaction of factors with mRNAs are developed in later sections of this volume.

IDENTIFICATION OF INITIATION FACTORS

In order to identify and define the initiation factors, essen-

tially the only approach available is to isolate and purify those
proteins which stimulate *in vitro* assays for initiation. Another
approach, the analysis of mutant cell lines, has not been used because
mutations in initiation factor genes of eukaryotic organisms are not
generally available. Theo Staehelin and Max Schreier (1), in Basel,
and French Anderson and Bill Merrick (2), in Bethesda performed
pioneering studies on the purification of initiation factors from
rabbit reticulocytes. Their results showed that many factors are
involved in the eukaryotic initiation process. Subsequently Rob Benne
and I (3), in Davis, and Harry Voorma's group (4), in Utrecht, have
purified and characterized similar or identical factors from the same
source. Eight different factors, named eIF-1, eIF-2, eIF-3, eIF-4A,
eIF-4B, eIF-4C, eIF-4D and eIF-5, have been identified and implicated
in initiation. In addition, numerous other laboratories have contributed to our knowledge of these factors by purifying and studying
one or a few of them.

     The purification schemes employed by different laboratories are
quite similar and utilize classical techniques for purifying proteins.
Most workers begin by preparing a high salt wash of ribosomes and
then proceed to fractionate the proteins by ammonium sulfate precipitation, ion exchange column chromatography, and sizing steps such as
density gradient sedimentation and molecular sieve chromatography.
The resulting preparations of factors are usually greater than 60%
pure and some approach homogeneity.

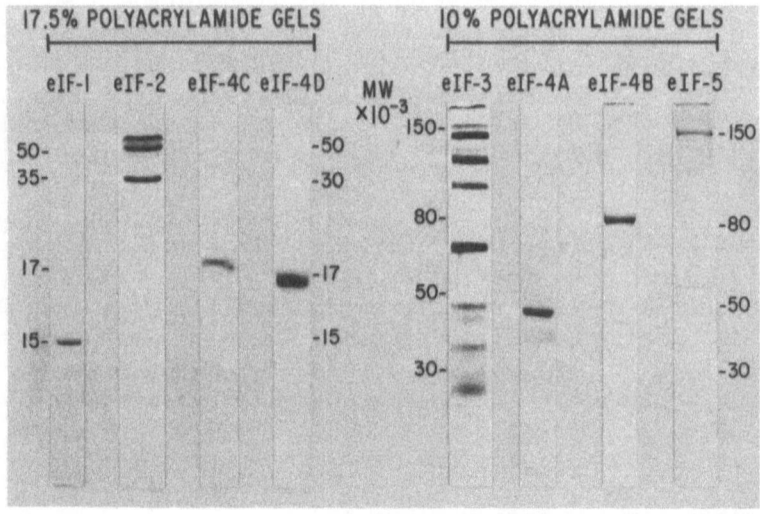

Fig. 1.   Analysis of preparations of initiation factors by SDS-poly-
          acrylamide gel electrophoresis. The method of Weber and
          Osborn (27) was used.

Preparations of the eight highly purified factors analyzed by SDS-polyacrylamide gel electrophoresis are shown in Fig.1. Six of the factors are comprised of single polypeptide chains, whereas two are multicomponent factors: eIF-2 consists of three protein subunits and eIF-3 is comprised of nine or more major protein components. The physical characteristics of these factors are described in detail in the next section below. We are interested now in asking whether all the factors have been identified and how one arrives at a definitive list of initiation factors.

An initiation factor is a non-ribosomal protein which participates uniquely in the initiation phase of protein synthesis. In practice, we use an operational definition, i.e., those proteins which stimulate cell-free *assays* of initiation. Staehelin and coworkers (1) introduced an assay for globin synthesis which contains washed ribosomes, globin mRNA and a post-ribosomal supernatant fraction; the assay is stimulated by seven factors (all except eIF-4D). The Bethesda group used an assay for methionyl-puromycin synthesis which does not require the post-ribosomal supernatant; all eight initiation factors stimulate this assay system to varying degrees (3, 5). Alternatively, eIF-2 activity may be assayed by its ability to form a ternary complex with Met-tRNA$_i^{Met}$ and GTP. Additional factors are implicated in the initiation process by using still other assays. A purified factor, Co-eIF-2A, affects the formation of the ternary complex of Met-tRNA$_i^{Met}$, GTP and eIF-2 (6), as do a number of less well purified or characterized preparations (7, 8). For a more detailed discussion of the co-factor, see chapter 16 of this volume. A protein, called the cap binding protein, can be crosslinked to the 7-methyl-guanosine portion of the cap structure of oxidized mRNA (9); the cap binding protein is found in preparations of purified eIF-3 and eIF-4B. Another protein, eIF-2A, promotes the binding of Met-tRNA$_i^{Met}$ to 40S ribosomal subunits in the presence of A-U-G (10).

The quantitative dependence on individual initiation factors in these assays has been determined by experiments in which factors are omitted one at a time. Results for globin and methionyl-puromycin synthesis are reported in Table 1.

Globin synthesis with crude pH 5 supernatant factors is stimulated five-fold or more by all factors except the small ones, eIF-1, eIF-4C and eIF-4D. Even greater dependence on the factors is seen when purified supernatant components are used in place of the pH 5 fraction (except for eIF-4D). When methionyl-puromycin is synthesized with A-U-G, only eIF-2, eIF-3, eIF-4C, eIF-4D and eIF-5 stimulate appreciably.

Methionyl-puromycin synthesis with globin mRNA requires all eight of the factors tested, although the stimulation by some is marginal. In general, stimulation by a factor is maximal when the molar amount of factor added equals the molar amount of ribosomes present in the

Table 1.   Dependence of Initiation Factors on Assays for Initiation

Factor Omitted	Globin synthesis		Methionyl-puromycin synthesis	
	pH 5 fraction	purified	A-U-G	globin mRNA
None	100	100	100	100
eIF-1	52	21	106	43
eIF-2	6	6	8	4
eIF-3	12	7	45	25
eIF-4A	20	5	103	85
eIF-4B	17	15	92	62
eIF-4C	38	29	69	66
eIF-4D	--	--	45	35
eIF-5	42	12	0	0

The results of single omissions experiments are reported as percent
of the controls (no omissions). The values for the globin synthesis
assays are taken from the work of Staehelin's laboratory (11); the
methionyl-puromycin synthesis results are from Benne and Hershey (12).
Control values (100%) are: for globin synthesis, with pH 5 enzymes,
1000 pmol amino acids incorporated; for globin synthesis with purified
aminoacyl-tRNAs, EF-1 and EF-2, 140 pmol amino acids incorporated; for
methionyl-puromycin synthesis with A-U-G and with globin mRNA, 3.42
and 1.25 pmol Met-puromycin, respectively.

assay. An exception is eIF-2, which often must be added in amounts
greater than the other factors. Since these assay systems are not
very active, the initiation factors do not necessarily recycle,
except for eIF-5 which is highly catalytic and maximally stimulates
these assays in very small amounts. Staehelin and coworkers (13)
compared the factor dependence for translations of globin and EMC
RNAs and found that EMC translation required all of the factors to
the same extent, except for eIF-4A and eIF-4B, which had to be added
at three-fold greater concentrations.

     Have all of the initiation factors been identified, or are there
still more factors to be discovered ? Essentially only one approach
is available for identifying components required for protein synthesis: The factors must be purified and used to construct a cell-free,
totally defined system for initiation which mimics the activity found
in vivo. The purified initiation factors stimulate the assay for
globin synthesis (1), which utilizes washed ribosomes and a pH 5
supernatant fraction, nearly as well as the crude ribosomal wash
fraction. Unfortunately, the reaction rates obtained *in vitro* are
much slower (ten-fold or more) than those in intact cells, especially
with the highly fractionated systems. This may be due to a number of
reasons: One or more components may be missing; the complex assay

system may not be optimized properly; the assay components may be more
dilute than in intact cells; and/or a number of the components may be
partially denatured or modified during their purification. The latter
point may be especially critical, as indicated by the identification
of a pressure-induced phosphorylation system in rabbit reticulocytes
(14). This inhibitory system, activated by high centrifugal pressure,
could explain why assays constructed simply by reconstituting pelleted
ribosomes and post-ribosomal supernatant are ten-fold less active than
the starting lysate. In any case, it is impossible to conclude rigorously that all of the components for initiation have been identified.

     Is each factor identified above *necessary* for protein synthesis
in vivo ? Although each of the factors shows stimulatory activity in
one or more of the assays devised, this alone does not constitute
proof that the factor actually functions during protein synthesis *in
vivo*. Because all cell-free assay systems, especially highly frac-
tioned ones, are innately artificial, stimulatory activities by some
components may be artifactual. The characterization of conditional
mutations in initiation factor genes is an elegant way to demonstrate
functional requirements. The lack of mutants obviates this approach
for mammalian cells, but the identification of mutations in yeast
genes for initiation factors (15) and the development of an active
cell-free system for protein synthesis (16) may allow such experiments
in this lower eukaryotic organism.

     A second approach is to use antibodies against a single initia-
tion factor and test whether protein synthesis is inhibited in crude
cell lysates. Naba Gupta has prepared chicken antisera against eIF-2
and Co-eIF-2A which inhibit the rabbit reticulocyte lysate (17).
Preliminary results of Larry Meyer in my laboratory indicate that goat
antibodies against eIF-3 and eIF-4B inhibit similarly. Thus only a
few of the initiation factors listed above are so far proved to be
necessary for protein synthesis.

PHYSICAL CHARACTERIZATION

     The initiation factors have been characterized by protein chemi-
cal techniques in a variety of laboratories, generally with results
in good agreement among the research groups. Molecular weight were
determined by SDS-polyacrylamide gel electrophoresis, and isoelectric
points, by focusing gels; amino acid compositions for many of the
factors are known. Values for the factors from rabbit reticulocytes
are given in Table 2, together with their major functions. The
factors range in size from 15 kd (eIF-1) to over 400 kd (eIF-3); the
largest polypeptide is about 200 kd and is a component of eIF-3. In
general, the initiation factors are acidic proteins, especially eIF-4C
and eIF-4A whose isoelectric points are 5.6 and 5.8, respectively.
The γ subunit of eIF-2 (pI = 8.9) is a striking exception. Although
the factors are acidic, most bind well to polyanionic resins. All of
the factors bind to phosphocellulose or heparin-Sepharose, except

Table 2. Initiation Factors from Rabbit Reticulocytes

Factor	Molecular weight	Number of poly-peptides	pI	Amino Acid Composition[1]	Functions
eIF-1	15,000	1	---	---	mRNA binding
eIF-2	122,000	3	6.4	18,19	ternary complex with Met-tRNA$_i$ and GTP
eIF-2A	65,000	1	---	10	binds Met-tRNA$_i$ to 40S
eIF-3	>400,000	>8	6.7	---	dissociation; promotes Met-tRNA$_i$ and mRNA binding
eIF-4A	49,000	1	5.8	---	mRNA binding
eIF-4B	80,000	1	6.3	---	mRNA binding
eIF-4C	17,000	1	5.6	20	promotes dissociation, Met-tRNA$_i$ binding
eIF-4D	16,000	1	6.1	20	stimulates Met-puromycin synthesis
eIF-5	160,000	1	6.4	21	required for 80S complex formation
cap binding	24,000	1	---	---	binds to cap of mRNA
Co-eIF-2A	19,000	1	---	---	stimulates ternary complex formation

[1] From references as indicated.

eIF-4A. This behavior may be due to specific sites on their surfaces which interact with RNA. Heparin or RNA affinity columns have been used to purify initiation factors (22-25).

Although in general specific factors isolated in different laboratories are very similar in physical characteristics and functions, some differences are seen. Proteolytic cleavage, either *in vivo* or during the isolation procedures, may be responsible for some of these differences. We have separated discrete molecular weight forms of eIF-5, ranging from 160 kd to 115 kd, all of which retain eIF-5 activity. The components of eIF-3 vary somewhat from preparation to preparation and proteolysis is thought to be a major factor in this variability (the structure of eIF-3 is discussed in greater detail below). Similarly, the γ subunit of eIF-2 may be partially digested to a minor form which is smaller by about 4 kd of protein. The relation of these minor forms to the putative parent protein was established by the peptide mapping technique of Cleveland et al. (26).

The structure of eIF-2, the factor which forms a ternary complex with Met-tRNA$_i^{Met}$ and GTP, is thought to be comprised of three non-identical subunits: α (32-38 kd), β (48-52 kd) and γ (50-57 kd). The α subunit is the substrate for the various highly specific eIF-2 kinases which are regulated by heme, double-stranded RNA or oxidized glutathione in reticulocytes and by the interferon-induced and double-stranded RNA dependent protein kinase (see chapter 18 of this volume). The γ subunit appears larger than the β subunit when analyzed by SDS-polyacrylamide gel electrophoresis according to Weber and Osborn (27), but the order of migration is reversed (28, 29) in gels prepared according to Laemmli (30). Analysis of separated polypeptides by sedimentation equilibrium gives molecular weights of: α, 32 kd; β, 35 kd; and γ, 55 kd (21). The unusual properties of the β subunit during electrophoresis in polyacrylamide gels may be due to a high charge density; the subunit is rich in basic amino acids and yet is an acidic protein (pI = 5.4). The α and γ subunits migrate normally in SDS gels and have isoelectric points of 5.1 and 8.9, respectively (19).

The stoichimetry of the subunits in the eIF-2 complex is 1:1:1, as determined by gel scans, amino acid composition and crosslinking (19). A molecular weight of 122,000, obtained by equilibrium sedimentation analysis is consistent with the stoichimetry. Recently, Harbitz and Hauge (31) and Stringer et al. (32) have reported the purification from pig liver and rabbit reticulocytes of active eIF-2 which contains only two subunits. We have purified eIF-2 from rabbit reticulocytes which either is deficient in or lacks the β subunit, as identified by peptide mapping. It is therefore presumably the β subunit which is lacking in the dimeric eIF-2 preparations. Curiously, all isolations of the dimeric (α, γ) form of eIF-2 have employed the protease inhibitor, phenylmethanesulfonyl fluoride (PMSF), during the purification procedures. The α subunit binds guanine nucleotides, whereas the β subunit binds RNAs (33). However, the physiological significance of the latter finding is difficult to evaluate.

eIF-3 is the most complex of the factors, containing numerous polypeptide components as shown in Fig. 1 (1, 34, 35). The complex sediments in sucrose gradients at 14 to 17S, suggesting a mass of 500,000 to 700,000 daltons. A molecular weight of 410,000 is calculated from equilibrium sedimentation data (36). Under non-denaturing conditions, the purified factor migrates as a single component on

electrophoresis in polyacrylamide gels or on sedimentation in density
gradients, and elutes from ion exchange columns without fractionation
of its components (34). This evidence suggests that eIF-3 is a homogeneous complex of many polypeptide components.

    In our earlier studies (34), we determined the molecular weights
and stoichiometries of the subunits and proposed a model of eIF-3
containing nine different proteins and a theoretical mass of 720 kd.
Recently we have compared the subunit polypeptides by peptide mapping
techniques and made the surprising observation that many of the components are related in primary structure. Partial digestion patterns
of iodinated eIF-3 polypeptides with Staphylococcal protease V8 are
shown in Fig. 2. It is immediately obvious that peptides d1 and d2,
and peptides c2 and c3, are related. Based on such data, we have
tentatively grouped the polypeptides into nine families of related
proteins (Fig. 2, left). Most of the largest polypeptides are grouped
into a single family, the a family. Comparisons of partial protease
fragmentation patterns of proteins differing greatly in mass are
ambiguous. Therefore more sophisitcated methods are required to prove
the relatedness of the a family which ranges in mass from 200 to 90 k .

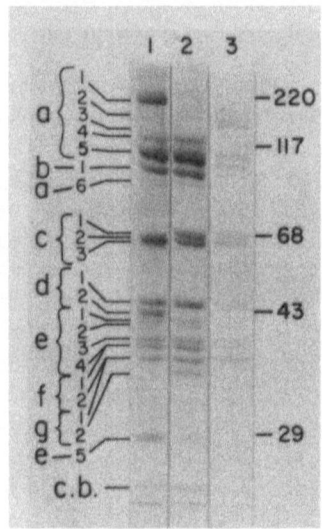 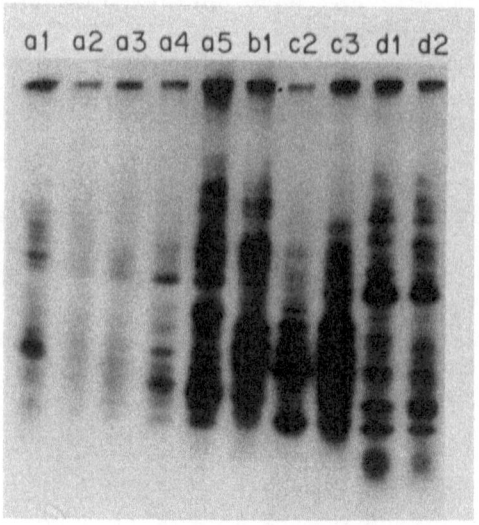

Fig. 2. Partial proteolytic digestion patterns of eIF-3 peptides.
         Bands containing major and minor components of iodinated
         eIF-3 from SDS gels (30), as identified on the left, were
         analyzed by the method of Cleveland *et al.* (26). An auto-
         radiogram of the digested components is shown on the right.

Similarly, the method does not allow us to determine whether the smaller protein components are derived from the largest members of the a family by proteolysis. It is not yet known precisely what the native structure of eIF-3 is in the intact cell. eIF-3 prepared in the presence of the protease inhibitor, PMSF, nevertheless contains multiple members of the various polypeptide families. It is conceivable that the eIF-3 complex is a dimer of the 200 kd protein which has undergone variable proteolytic digestion in the intact cell and/or during its isolation.

It is possible that some of the factors identified might be components in the complex factors eIF-2 or eIF-3, or are identical with other components of the protein synthesis system. This problem was addressed by examining the various factors in a two-dimentional polyacrylamide gel system which separates proteins by charge (acid urea) and by mass (SDS). None of the factor polypeptides co-migrate with each other or with ribosomal proteins, except that eIF-4A was found as a minor component in eIF-3 (3). Because of the problem of proteolysis, the failure to co-migrate does not prove that two proteins of different mass are not related in primary structure and possibly function. We have obtained partial protease fragmentation patterns of eIF-4A, eIF-4B and eIF-5 as well as the subunits of eIF-2 and eIF-3, and have compared these patterns. Except for eIF-4A and a component of eIF-3 (the d family), none of the factors is related to another. A third approach to establishing non-relatedness is by immunochemical methods. We have prepared antibodies against eIF-3, eIF-4A and eIF-4B and have tested the specificity of the antisera by double immunodiffusion techniques. Anti-eIF-4A and anti-eIF-4B form precipitin bands only with their cognate antigens; anti-eIF-3 cross-reacts weakly with eIF-4B, probably indicating impure antiserum.

Most of the studies on eukaryotic initiation factors have utilized rabbit reticulocytes as the cell source. In those cases where other mammalian cell sources were used, the factors purified have resembled closely those from rabbit reticulocytes. Trachsel et al. (37) purified seven initiation factors from Krebs II ascites cells and compared them with the rabbit factors. No differences in molecular weights, elution characteristics from ion exchange columns, or activities in in vitro assays were detected for the mouse factors. We have purified eIF-2, eIF-3, eIF-4A, eIF-4B and eIF-5 from HeLa cells, with similar results. Furthermore, the partial protease fragmentation patterns of corresponding polypeptides of the human and rabbit factors are indistinguishable. Antigenic determinants in rabbit eIF-3, eIF-4A and eIF-4B are shared in part by the corresponding human factors. It therefore appears that mammalian initiation factors are highly conserved in structure and function. eIF-2 and eIF-3 from Artemia salina can replace the corresponding reticulocyte factors in a reconstituted protein synthesizing system (A.J. Wahba, personal communication), indicating that considerable conservation extends even to lower animal forms.

Covalent modifications

With the discovery that the heme-regulated repressor in rabbit reticulocytes is a specific protein kinase which phosphorylates the α subunit of eIF-2, attention has been focused on the possible role of covalent modification of initiation factors in the regulation of protein synthesis. Best studied are the various eIF-2α kinases, which will be described in detail in later chapters of this volume. Following purification of the initiation factors, the proteins were tested *in vitro* as substrates for cAMP-regulated and non-regulated protein kinases (38, 39). A non-regulated kinase, called casein kinase II, phosphorylates eIF-2β, eIF-4B, a number of subunits of eIF-3, and possibly eIF-5. This kinase utilizes ATP or GTP as phosphate donor. The same pattern of phosphorylation is observed by examining partially purified factors isolated from cells incubated with (^{32}P) phosphate (28). However, when purified factors are phosphorylated preparatively and are compared with non-phosphorylated factors in cell-free assays for initiation, no change in specific activity has been detected. No functional role for the observed phosphorylations is known. A number of other protein kinases phosphorylate initiation factors and phosphoprotein phosphatases also act on the facts (40).

It is possible that modifications other than phosphorylation are important in regulating the activity of initiation factors. We have already described evidence for limited proteolysis of some of the factors, but no functional role is known. Traugh and coworkers have described a transacetylase which modifies a 45 kd component of eIF-3, as well as proteins in the 40S and 60S ribosomal subunits (41). Further work is required to determine whether acetylations or other possible covalent modifications are important for the function of these factors.

CELLULAR LEVELS AND BIOGENESIS

Knowledge of the cellular levels of initiation factors and the ratios of factors to ribosomes is important for understanding how the factors act during initiation and how translation may be regulated. Little information about cellular levels is yet available, however. Safer *et al.* (42) measured the size of the pool of eIF-2 in rabbit reticulocyte lysates by the principle of isotope dilution using (^{14}C) CH$_3$-eIF-2. A value of 20 to 30 pmol/ml of lysate was obtained, which corresponds to a eIF-2/ribosome molar ratio of about 0.1. The level of eIF-3 has been determined by two immunochemical methods (43): Quantitation by SDS-PAGE of eIF-3 protein in immunoprecipitates formed with reticulocyte lysates and a goat antibody against eIF-3, and rocket immunoelectrophoresis with anti-eIF-3. Both methods gave values of 0.3 to 0.5 eIF-3 molecules per ribosome. No other initiation factor has been measured directly, but values can be estimated on the basis of yields following purification. Estimates of factor: ribosome ratios fall in the range of 0.05 to 0.2, except that eIF-4A

may be greater and eIF-5 may be less. Thus mammalian initiation
factor:ribosome ratios appear to be similar to those determined in
bacteria (44).

Nothing is yet known about the biosynthesis of initiation factors
or the organization of their genes. The identification of mutations
for initiation factors in yeast creates the opportunity to study this
question in this lower eukaryote.

PATHWAY OF INITIATION

The pathway of initiation of protein synthesis in mammalian cells
is similar to that in bacteria. Ribosomes dissociate into subunits,
a preinitiation complex forms with the small ribosomal subunit, mRNA
and the initiator tRNA, and the complete initiation complex results
from junction of the large ribosomal subunit and the preinitiation
complex.

Because of the vastly greater complexity of mammalian initiation
factors, attention has been focused on when each acts in the pathway.
Two basic approaches have been utilized:

1)   intermediate complexes are assembled *in vitro* from purified
components; radioactive Met-tRNA$_i^{Met}$, mRNA and/or initiation factors
are used, and complex formation is usually monitored by sucrose density gradient centrifugation;

2)   intermediate complexes are isolated from cell lysates, often
following the addition of an antibiotic inhibitor of initiation.

Using these methods, a number of research groups have proposed
tentative schemes for the initiation pathway which are generally in
good agreement (12, 45, 46). One such model is shown in Fig. 3. We
shall examine each of the steps in detail, citing the major evidence
for each.

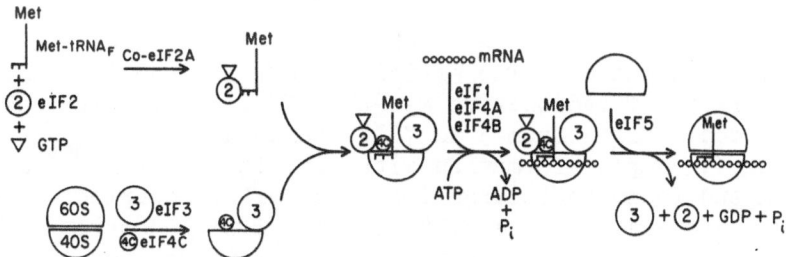

Fig. 3. Proposed pathway of initiation of protein synthesis.

Dissociation of Ribosomes Into Subunits

The dissociation of 80S ribosomes into 40S and 60S ribosomal subunits is the first step in the initiation pathway. Kaempfer (47)
showed that ribosomes active in protein synthesis transiently dissociate following termination. Isolated mammalian ribosomes are highly
associated at Mg^{++} ion concentrations above 1 mM and therefore must be
dissociated, presumably by the action of a dissociation (or anti-
association) factor. Two of the initiation factors are thought to
contribute to a shift in the equilibrium of rabbit reticulocyte ribosomes towards dissociation. eIF-3 binds stoichiometrically to the 40S
subunit (34) and increases the proportion of subunits versus 80S ribosomes (48). Analysis of eIF-3---40S complexes by SDS/PAGE indicates
that essentially all of the eIF-3 polypeptides bind to the ribosomal
subunit (34). Electron microscopic studies of native, unwashed subunits show eIF-3 bound to the same surface of the 40S subunit which
interfaces with the 60S subunit (49). It appears likely that eIF-3
acts as an anti-association factor, but a careful kinetic analysis of
mammalian ribosome dissociation has not yet been reported. eIF-4C
also binds to 40S subunits and somewhat affects ribosome dissociation
(50). It is possible that still other proteins are involved in ribosome dissociation in rabbit reticulocytes, but they have not yet been
identified. In the wheat germ system, a number of dissociation
factors have been identified, one of which binds to the 60S ribosomal
subunit (51).

Ternary Complex Formation

Met-tRNA$_i^{Met}$ forms a stable ternary complex with eIF-2 and GTP
(52-54). Formation of the complex with radiolabeled Met-tRNA$_i^{Met}$ is
assayed conveniently by filtration through nitrocellulose membranes,
which bind the complex but not free Met-tRNA$_i^{Met}$. GTP binding to
eIF-2 precedes Met-tRNA$_i^{Met}$ binding (55). The GTP is not hydrolyzed,
and non-hydrolyzable analogues of GTP can be substituted. eIF-2 only
binds the initiator, Met-tRNA$_i^{Met}$, but not the Met-tRNA species used
to insert methionine internally into protein (53). GDP is a powerful
competitive inhibitor of ternary complex formation (K_I = 3.4 x 10^{-7} M)
and binds to eIF-2 more tightly (K_D = 3.1 x 10^{-8} M) than does GTP
(K_D = 2.5 x 10^{-6} M) (56).

Most preparations of eIF-2 appear not to be fully active in
ternary complex formation. The efficiency of the reaction, expressed
as the molar ratio of radioactive Met-tRNA$_i^{Met}$ bound to nitrocellulose
over eIF-2 added, ranges from 5 to 70%. Benne and coworkers (57)
studied ternary, complex formation with both the Met-tRNA$_i^{Met}$ and eIF-2
radiolabeled and found that large losses of eIF-2 occurred. The ratio
of Met-tRNA$_i^{Met}$ to eIF-2 on the filters was 1:1 if an energy generating
system was included to assure that all of the nucleotides were in the
triphosphate form. Thus 100% efficiency of eIF-2 was observed when
the GTP/GDP ratio is high and protein losses are prevented. The

results also indicate that at physiological concentrations of eIF-2, the ternary complex is fully formed and apparently requires no further stabilization.

There are numerous reports of protein preparations which affect ternary complex formation. Some of these preparations may act by preventing the loss of eIF-2, and thus play a trivial role. Gupta and coworkers have purified a factor, Co-eIF-2 (22,000 daltons), which stabilizes the ternary complex(6). This point is extensively treated in chapter 16 of this volume. In their hands, the ternary complex formed with homogenous eIF-2 is unstable in the presence of Mg Further evidence that Co-eIF-2A plays a crucial role in protein synthesis was obtained by showing that antibody against Co-eIF-2A inhibits protein synthesis in reticulocyte lysates (17). Evaluation of the various conflicting results regarding the efficiency of complex formation, the sensitivity to Mg^{++} and the role of other factors is difficult. Most studies have utilized the nitrocellulose filter assay and the results reflect the stability of the complex. Data is lacking on rates of formation and dissociation of the complex, and how these kinetic parameters are altered by cofactors, energy charge, etc.

Ternary Complex Binding to 40S Subunits

The ternary complex is an obligatory intermediate for the binding of Met-tRNA$_i^{Met}$ to the 40S ribosomal subunit. The Met-tRNA$_i^{Met}$-40S complex forms in the absence of mRNA and is sufficiently stable that isolation by sucrose density gradient centrifugation is possible without fixation (13, 58, 59). eIF-3 and eIF-4C stimulate Met-tRNA$_i^{Met}$ binding, possibly by enhancing the stability of the ribosomal complex. None of the other initiation factors, with the possible exception of eIF-1 (13, 58), affects this reaction. The presence on the ribosome of all of the subunit polypeptides of eIF-2 and eIF-3, along with eIF-4C, has been demonstrated by using radioactive factors (12). The GTP carried by the ternary complex remains bound to the ribosome and is not hydrolyzed at this stage (13).

The formation *in vitro* of a stable Met-tRNA$_i^{Met}$-40S complex suggests that Met-tRNA$_i^{Met}$ binding may precede mRNA binding. The failure to bind mRNA to ribosomes in the absence of Met-tRNA$_i^{Met}$ is consistent with this view. Further evidence is obtained by examining glutaraldehyde-fixed native 40S complexes from reticulocyte lysates (60). Native 40S subunits are found complexed with about 900,000 daltons of protein (presumably eIF-3 and possibly other factors), some of which contain Met-tRNA$_i^{Met}$. However, no 40S subunits were found in association with mRNA. It is possible, but unlikely, that mRNA binding occurs first, followed by the rapid binding of Met-tRNA and the 60S subunit, and that Met-tRNA.40S complexes are non-functional. A rigorous kinetic analysis of Met-tRNA$_i^{Met}$ and mRNA binding is needed to establish with certainty the order of binding of these components to the 40S ribosome.

Binding of mRNA to 40S Subunits

Binding of mRNA to the Met-tRNA$_i^{Met}$. 40S complex complets formation of the 40S preinitiation complex. This step is in many ways the most crucial for initiation since it results in the selection of a particular mRNA for translation, and may be rate-limiting. The latter point is suggested by analyses of fixed native subunits: the presence of Met-tRNA$_i^{Met}$ in complexes with 40S subunits lacking mRNA and with 80S ribosomes containing mRNA is consistent with a kinetic model in which mRNA binding to 40S subunits is slow relative to both the Met-tRNA$_i^{Met}$ binding step and the subsequent junction step with the 60S subunit. It is therefore reasonable to postulate that the mRNA binding step may be subject to regulation. Studies with radiolabeled globin mRNA indicate that in addition to Met-tRNA$_i^{Met}$, eIF-2, eIF-3 and eIF-4C which are already present on the 40S ribosome, three additional initiation factors are required for maximal mRNA binding, namely eIF-4A and eIF-4B, and to a lesser extent, eIF-1 (12, 13). The hydrolysis of ATP is also required; non-hydrolyzable analogues of ATP and other nucleoside triphosphates can not substitute. The stoichiometric presence of Met-tRNA$_i^{Met}$, mRNA, eIF-2, eIF-3 and eIF-4C on the 40S subunit has been shown by using radioactive components (12, 50). The complex is sufficiently stable so that it can be isolated by sucrose density gradient centrifugation or molecular sieve chromatography without prior fixation.

The cap binding protein presumably is involved in the mRNA binding step also. However, there are no reports that this protein stimulates mRNA binding in highly fractionated systems, possibly because the 24 kd protein already is present in preparations of eIF-3 and eIF-4B. The presence of eIF-4A, eIF-4B and eIF-1 in the preinitiation complex has not yet been demonstrated unambiguously. If these proteins bind to 40S subunits, their affinities are not great enough to survive the analytical conditions of the sucrose gradients. Fixation with glutaraldehyde prior to analysis indicates binding of all three factors, but the physiological significance of this is difficult to assess. eIF-4A, eIF-4B and ATP hydrolysis are all needed for the binding and translation of mRNAs, but they are not required for the binding of A-U-G or for subsequent steps of initiation with this codon (12).

The structure of the mRNA plays an important role in mRNA binding to ribosomes. The cap structure at the 5'-terminus, m^{7}G^{5}ppp^{5}NmpN, increases the affinity of mRNAs for the 40S subunit, whereas the poly(A) tract at the 3'-terminus of many mRNAs does not seem to affect initiation. Marilyn Kozak (61) proposes that the 40S ribosome binds first to the cap structure, then scans the mRNA in order to locate the first A-U-G codon. Details of mRNA structure and the scanning hypothesis are given elsewhere in this volume. A further complication is that initiation events *in vivo* do not occur on "naked" mRNA as is generally the case in most assay systems. Either

the mRNA already is present in polysomes, or is highly complexed with
proteins (cytoplasmic free RNPs). RNPs, their structure and possible
role, are discussed elsewhere in this volume. Initiation events with
mRNA in polysomes appear not to differ from those with naked mRNA; the
requirements for eIF-3, eIF-4A and eIF-4B are indistinguishable (unpublished results).

It is important to distinguish whether mRNAs interact with initiation factors before binding to the 40S subunit. The best documented case for binding involves the cap binding protein, which interacts
with capped mRNAs and may be crosslinked to oxidized caps (9). eIF-2,
eIF-3, eIF-4B, eIF-5 and the cap binding protein all promote the
binding of radioactive mRNAs to nitrocellulose filters (62, 63). It
is difficult to assess the significance of these results, however.
eIF-2 and eIF-5 bind to most RNAs and therefore do not show specificity. The binding activity in eIF-3 and eIF-4B preparations, while
specific for mRNAs, may be due to the presence of the cap binding
protein. Nitrocellulose binding assays in general are subject to
artifacts and must be viewed with caution. It is desirable to study
these interactions by carefully determining stoichiometries and
kinetic parameters (rates of binding and release, and equilibrium
constants), preferably by a number of methods. Thus there is as yet
no compelling evidence that initiation factors interact with mRNAs
prior to binding to ribosomes.

Junction of the 60S Subunit and Formation of the
80S Initiation Complex

The last reaction in the proposed pathway is the junction of the
60S ribosomal subunit with the 40S preinitiation complex. Formation
of the 80S initiation complex proceeds only when the following conditions are met (12, 45). A complete 40S preinitiation complex is
required, i.e., the prior binding of mRNA is necessary. Another
initiation factor, eIF-5, is involved. Finally, hydrolysis of GTP,
presumably that originally brought to the ribosome in the ternary
complex, must occur; preinitiation complexes constructed with non-
hydrolyzable GTP analogues do not form 80S complexes. It is not yet
clear whether the ribosome or an initiation factor contribute the
active site for GTP hydrolysis, or what purpose the hydrolysis serves.

Associated with the junction reaction are the ejection of eIF-2
and eIF-3 from the ribosome (12). If other initiation factors are
present in the 40S preinitiation complex, it is likely that they are
ejected also. There is some controversy about whether release of
eIF-2 and eIF-3 occurs before or is concomitant with 60S junction.
When A-U-G replaces mRNA in 40S preinitiation complexes, eIF-5 in
the absence of 60S subunits can catalyze GTP hydrolysis and factor
release (64). With globin mRNA, however, no factor release occurs
when eIF-5 is added to 40S preinitiation complexes. The release of a
protein from ribosomes at a specific place in the pathway is an

important criterion for differentiating factors from structural ribosomal proteins. Released factors are presumably capable of recycling,
i.e., participating in another round of initiation, although this has
not yet been demonstrated unambiguously in highly fractionated
systems.

The 80S initiation complex contains Met-tRNA$_i^{Met}$ and mRNA, but
presumably no initiation factors. Such complexes react with puromycin
to form methionyl-puromycin. Therefore the Met-tRNA$_i^{Met}$ is bound in
the ribosomal P site and the 80S initiation complex is capable of
entering the elongation phase of protein synthesis. eIF-4D stimulates
the synthesis of methionyl-puromycin and appears to act on pre-formed
80S initiation complexes (12). However, this factor does not appear
to be required for formation of the first peptide bond with aminoacyl-
tRNA (65). It is therefore possible that the action of eIF-4D is
limited to reactions with puromycin. Substantial amounts of eIF-4D in
the post-ribosomal supernatant fraction have precluded its stimulation
of assays for protein synthesis with the pH 5 supernatant fraction.

MOLECULAR MECHANISM OF INITIATION

The goal of much research on protein synthesis is to explain the
pathway in terms of well understood chemical forces. Progress to
date has been limited primarily to defining the macromolecular elements involved, and in determining the temporal order of their interactions. Some information about molecular structure is available,
particularly concerning tRNAs and mRNAs. However, knowledge of the
three-dimensional structure of the eukaryotic ribosome is not very
extensive and is not adequate for making rigorous correlations of
structure and function. A similar situation exists in the study of
prokaryotic protein synthesis; much more information is available
about the primary structures of the ribosomal components and soluble
factors, but knowledge about detailed structure is not adequate for
elucidating the molecular mechanisms involved. In this section we
will discuss some of the major problems currently being addressed by
researchers and indicate the progress made.

mRNA-Ribosome Interactions

In bacteria, mRNA binding involves at least two RNA-RNA interactions: the initiator codon in the mRNA with the anti-codon of the
initiator tRNA, fMet-tRNA$_f^{Met}$; and the Shine-Dalgarno purine-rich
sequence in the mRNA on the 5'-side of the initiator codon, with the
pyrimidine-rich sequence at the 3'-terminus of the 16S ribosomal RNA.
Still other mRNA-rRNA interactions are possible, as are specific mRNA-
protein interactions, but these remain largely unknown. In eukaryotic
cells, the 3'-terminal sequence of the 18S ribosomal RNA is very
strongly conserved, suggesting involvement in a vital function, but
analysis of numerous mRNA sequences proximal to the initiator codon
do not reveal regions complementary to the 18S rRNA (61). It there-

fore appears unlikely that an interaction comparable to that proposed
by Shine and Dalgarno exists in eukaryotic protein synthesis. Only
two interactions are known for mammalian cells: The initiator codon-
anticodon interaction, and the binding of the 5'-terminus of mRNA to
the cap binding protein. These reactions are discussed in detail
later in this volume, and so will not be described further here.
Suffice it to say that much remains to be learned about mRNA binding.
In particular, we need to determine how the various interactions contribute to the binding kinetics; that is, do they influence the rate
of binding or the stability (rate of release) of bound mRNA ? Are
there a number of intermediate states between free mRNA and the short-
lived 40S preinitiation complex ? Do the proteins which are found in
mRNP particles and polysomes influence binding ? What features of
mRNA structure determine "strong" versus "weak" initiators ? The
answers to such questions on mechanism are vital for explaining the
putative translational control mechanisms which involve differential
mobilization of mRNPs during development and differentiation, and
the specific translation of viral versus cellular mRNAs.

Specific Factors for mRNA ?

The existence of specific proteins required for the translation
of a class of mRNAs is an attractive idea and has been postulated for
both prokaryotic and eukaryotic systems. The idea is controversial;
a caution view is that their existence is not yet proven. One of the
first examples was the purification of a protein thought to be specific for the translation of EMC RNA (66), but the factor was shown to
be identical to eIF-4A (13). Addition of eIF-4B to translation
systems influences the ratio of protein products where competition
between two mRNAs occurs (67, 68). It is proposed that eIF-4B binds
more avidly to one of the mRNAs, and when limiting it fails to promote
the translation of the less-avidly binding species. Heywood and coworkers (69, 70) report that proteins associated with eIF-3 from
muscle tissue specifically enhance the translation of myosin mRNA
vis-a-vis globin mRNA. In addition to protein factors, small RNAs,
called translational control RNAs (tcRNAs), may influence initiation,
either specifically or nonspecifically (71, 72). The concept of
masked mRNPs suggests that initiation on such complexes is blocked by
a component of the RNP, since the naked mRNA is capable of being
translated; the masking and unmasking of mRNAs may be sequence-specific. None of these examples are well characterized or understood
in detail. Some but not all of these observations may be explained
by the hypothesis of Lodish (73), which states that a non-specific
change in the rate of initiation at or before the mRNA binding step
can result in the differential translation of mRNAs with different
rate constants for initiation. For example, the translation of "weak"
mRNAs is more severely inhibited than that of "strong" mRNAs when the
overall rate of initiation is decreased. mRNA competition for initiation factor is further discussed in chapter 17 of this volume.

Ribosomal Sites for Initiation

Elucidation of the structure of the mammalian ribosome is not so advanced as that of bacterial ribosomes. 40S and 60S subunits resemble their prokaryotic counterparts when analyzed by electron microscopy (74, 75): the 40S subunit is divided into "head" and "body" regions by a constriction; the 60S subunit appears either rounder or more asymmetric. Analysis of native 43S particles which contain eIF-3 reveals that the factor binds on the surface of the 40S subunit which interfaces with the 60S subunit (49). Its site of binding is similar to the site of initiation for prokaryotic 30S sub-units. eIF-2 crosslinks to the 40S proteins S3, S3a and S15 (76); eIF-3 crosslinks to a larger number of proteins: S2, S3, S3a, S4, S6, S15 and S18 (77). Interestingly, the 3'-terminus of 18S rRNA can be crosslinked to S3a also (78). These results represent the beginning of an identification of the site of initiation on 40S subunits. Traut and coworkers have identified numerous protein-protein inter-actions in 40S subunits by crosslinking techniques and are now able to construct models showing the topography of the ribosomal proteins. Much more detailed information will be required before functional correlations can be made.

ACKNOWLEDGEMENTS

Work reported from this laboratory was supported by USPHS grant GM-22135. I thank Dr. Brian Safer for making available before publi-cation a review, entitled "The Regulation of Initiation of Mammalian Protein Synthesis" (79).

REFERENCES

1. Schreier, M.H., Erni, B., and Staehelin, T. (1977), J. Mol. Biol. 116, 727-753.
2. Merrick, W.C. (1979), Methods Enzymol. 60, 101-1C8.
3. Benne, R., Brown-Luedi, M.L., and Hershey, J.W.B. (1979), Methods Enzymol. 60, 15-35.
4. Voorma, H.O., Thomas, A., Goumans, H., Amesz, H., and van der Mast, C. (1979), Methods Enzymol. 60, 124-135.
5. Merrick, W.C. (1979), Methods Enzymol. 60, 108-123.
6. Dasgupta, A., Das, A., Roy, R., Ralston, R., Majumdar, A., and Gupta, N.K. (1978), J. Biol. Chem. 253, 6054-6059.
7. Das, A., and Gupta, N.K. (1977), Biochem. Biophys. Res. Commun. 77, 1307-1316.
8. deHaro, C., Datta, A., and Ochoa, S. (1978), Proc. Natl. Acad. Sci. U.S.A. 75, 243-247.
9. Sonenberg, N., Morgan, M.A., Merrick, W.C., and Shatkin, A.J. (1978), Proc. Natl. Acad. Sci. U.S.A. 75, 4843-4847.
10. Merrick, W.C., and Anderson, W.F. (1975), J. Biol. Chem. 250, 1197-1206.

11. Erni, B. (1976), Thesis, Swiss Federal Institute of Technology, Zurich.
12. Benne, R., and Hershey, J.W.B. (1978), J. Biol. Chem. 253, 3078-3087.
13. Staehelin, T., Trachsel, H., Erni, B., Boschetti, A., and Schreier, M.H. (1975), Proc. FEBS Meeting, 10th, 39, 309-323.
14. Henderson, A.B., Miller, A.H., and Hardesty, B. (1979), Proc. Natl. Acad. Sci. U.S.A. 76, 2605-2609.
15. Petersen, N., and McLaughlin, C.S. (1974), Molec. gen. Genet. 129, 189-200.
16. Gasior, E., Herrera, F., Sadnik, I., McLaughlin, C.S., and Moldave, K. (1979), J. Biol. Chem. 254, 3965-3969.
17. Ghosh-Dostider, P., Yaghmai, B., Das, A., Das, H.K., and Gupta, N.K. (1980), J. Biol. Chem. 255, 365-368.
18. Safer, B., Anderson, W.F., and Merrick, W.C. (1975), J. Biol. Chem. 250, 9067-9075.
19. Lloyd, M.A., Osborne, J.C., Safer, B., Powell, G., and Merrick, W.C. (1980), J. Biol. Chem. 255, 1189-1194.
20. Kemper, W.M., Berry, K.W., and Merrick, W.C. (1976), J. Biol. Chem. 251, 5551-5557.
21. Merrick, W.C., Kemper, W.M., and Anderson, W.F. (1975), J. Biol. Chem. 250, 5556-5562.
22. Waldman, A.A., Marx, G., and Goldstein, J. (1975), Proc. Natl. Acad. Sci. U.S.A. 72, 2352-2356.
23. Van der Mast, C., Thomas, A., Goumans, H., Amesz, H., and Voorma, H.O., (1977), Eur. J. Biochem. 75, 455-464.
24. Kaempfer, R. (1971), Methods Enzymol. 60, 247-255.
25. Moretti, S., Staehelin, T., Trachsel, H., and Gordon, J. (1979), Eur. J. Biochem. 97, 609-614.
26. Cleveland, D.W., Fischer, S.G., Kirschner, M.W., and Laemmli U.K. (1976), J. Biol. Chem. 252, 1102-1106.
27. Weber, K., and Osborn, M. (1969), J. Biol. Chem. 244, 4406-4412.
28. Benne, R., Edman, J., Traut, R.R., and Hershey, J.W.B. (1978), Proc. Natl. Acad. Sci. U.S.A. 75, 108-112.
29. Tahara, S.M., Traugh, J.A., Sharp, S.B., Lundak, T.S., Safer, B., and Merrick, W.C. (1978), Proc. Natl. Acad. Sci. U.S.A. 75, 789-793.
30. Laemmli, U.K. (1970), Nature, London, 227, 680-685.
31. Harbitz, I., and Hauge, J.G. (1979), Methods Enzymol. 60, 240-246.
32. Stringer, E.A., Chaudhuri, A., Valenzuela, D., and Maitra, U. (1980), Proc. Natl. Acad. Sci. U.S.A. 77; 3356-3359.
33. Barrieux, A., and Rosenfeld, M.G. (1977), J. Biol. Chem. 252, 3843-3847.
34. Benne, R., and Hershey, J.W.B. (1976), Proc. Natl. Acad. Sci. U.S.A. 73, 3005-3009.
35. Safer, B., Adams, S.L., Kemper, W.M., Berry, K.W., Lloyd, M., and Merrick, W.C. (1976), Proc. Natl. Acad. Sci. U.S.A. 73, 2584-2588.
36. Meyer, L.J. (1980), Thesis, University of California, Davis.

37. Trachsel, H., Erni, B., Schreier, M.H., Braun, L., and Staehelin, T. (1979), Biochim. Biophys. Acta 561, 484-490.
38. Traugh, J.A., Tahara, S.M., Sharp, S.B., Safer, B., and Merrick, W.C. (1976), Nature, London, 263, 163-165.
39. Issinger, O.-G., Benne, R., Hershey, J.W.B., and Traut, R.R. (1976), J. Biol. Chem. 251, 6471-6473.
40. Mumby, M., and Traugh, J.A. (1979), Methods Enzymol. 60, 522-534.
41. Traugh, J.A., and Sharp, S.B. (1979), Methods Enzymol. 60, 534-541.
42. Safer, B., Kemper, W., and Jagus, R. (1979), J. Biol. Chem. 254, 8091-8094.
43. Meyer, L., and Hershey, J.W.B., manuscript in preparation.
44. Howe, J.G., Yanov, J., Meyer, L., Johnston, K., and Hershey, J.W.B. (1978), Arch. Biochem. Biophys. 191, 813-820.
45. Trachsel, H., Erni, B., Schreier, M.H., and Staehelin, T. (1977), J. Mol. Biol. 116, 755-767.
46. Safer, B., and Anderson, W.F. (1978), Crit. Rev. Biochem. 5, 261-290.
47. Kaempfer, R. (1974), In Ribosomes (Nomura, M., Tissieres, A., and Lengyel, P., eds.) pp. 679-704, Cold Spring Harbor Laboratory, New York.
48. Thompson, H.A., Sadnik, I., Scheinbuks, J., and Moldave, K. (1977) Biochemistry 16, 2221-2230.
49. Emanuilov, I., Sabatini, D.D., Lake, J.A., and Freienstein, C. (1978), Proc. Natl. Acad. Sci. U.S.A. 75, 1389-1393.
50. Thomas, A., Goumans, H., Voorma, H.O., and Benne, R. (1980), Eur. J. Biochem. 107, 39-46.
51. Russell, D.W., and Spremulli, L.L. (1980), Arch. Biochem. Biophys. 201, 518-526.
52. Levin, D.H., and Kyner, D. (1971), Fed. Proc. 30, 1289.
53. Chen, Y.C., Woodley, C.L., Bose, K.K., and Gupta, N.K. (1972), Biochem. Biophys. Res. Commun. 48, 1-9.
54. Dettman, G.L., and Stanley, W.M. (1972), Biochim. Biophys. Acta 287, 124-133.
55. Safer, B., Adams, S.L., Anderson, W.F., and Merrick, W.C. (1975), J. Biol. Chem. 250, 9076-9082.
56. Walton, G.M., and Gill, G.N. (1976), Biochim. Biophys. Acta 418, 195-203.
57. Benne, R., Amesz, H., Hershey, J.W.B., and Voorma, H.O. (1979), J. Biol. Chem. 254, 3201-3205.
58. Benne, R., Wong, C., Luedi, M., and Hershey, J.W.B. (1976), J. Biol. Chem. 251, 7675-7681.
59. Peterson, D., Merrick, W.C., and Safer, B. (1979), J. Biol. Chem. 254, 2509-2516.
60. Hirsch, C.A., Cox, M.A., van Venrooij, W.J.W., and Henshaw, E.C. (1973), J. Biol. Chem. 248, 4377-4385.
61. Kozak, M. (1978), Cell 15, 1109-1123.
62. Shafritz, D.A., Weinstein, J.A., Safer, B., Merrick, W.C., Weber, L.A., Hickey, E.D., and Baglioni, C. (1976), Nature, London, 261, 291-294.

63. Brown-Luedi, M.L., Benne, R., Yau, P., and Hershey, J.W.B. (1978), Fed. Proc. 37, 1307.

64. Peterson, D.T., Safer, B., and Merrick, W.C. (1979), J. Biol. Chem. 254, 7730-7735.

65. Staehelin, T., Erni, B., and Schreier, M.H. (1979), Methods Enzymol. 60, 136-165.

66. Wigle, D.T., and Smith, A.E. (1973), Nature, London, New Biol. 242, 136-140.

67. Golini, F., Thach, S.S., Birge, C.H., Safer, B., Merrick, W.C., and Thach, R.E. (1976), Proc. Natl. Acad. Sci . U.S.A. 73, 3040-3044.

68. Kabat, D., and Chappell, M.R. (1977), J. Biol. Chem. 252, 2684-2690.

69. Heywood, S.M., and Kennedy, D.S. (1979), Arch. Biochem. Biophys. 192, 270-281.

70. Gette, W.R., and Heywood, S.M. (1979), J. Biol. Chem. 254, 9879-9885.

71. Heywood, S.M., Kennedy, D.S., and Bester, A.J. (1979), Methods Enzymol. 60, 541-549.

72. Pluskal, M.G., and Mukherjie, A. (1980), Fed. Proc. 39, 1868.

73. Bergmann, J.E., and Lodish, H.F. (1979), J. Biol. Chem. 254, 11927-11937.

74. Lutsch, G., Bielka, H., Wahn, K., and Stahl, J. (1972), Acta Biol. Med. Ger. 29, 851-876.

75. Lake, J.A., Sabatini, D.D., and Nomura, Y. (1974), In *Ribosomes* (Nomura, M., Tissieres, A., and Lengyel, P., eds.), pp. 543-557, Cold Spring Harbor Laboratory, New York.

76. Westermann, P., Heumann, W., Bommer, U.A., Bielka, H., Nygard, O., and Hultin, T. (1979), FEBS Letts. 97, 101-104.

77. Tolan, D., Hershey, J.W.B., and Traut, R.R., manuscript in preparation.

78. Svoboda, A.J., and McConkey, E.H. (1979), Biochem. Biophys. Res. Commun. 81, 1145-1152.

79. Jagus, R., Anderson, W.F., and Safer, B. (1980), Prog. Nucl. Acids Res. Mol. Biol., in press.

SECTION II:

ON THE IMPORTANCE OF BEING SPLICED

MESSENGER RNA STRUCTURE AND BIOSYNTHESIS

Robert P. Perry

Institute for Cancer Research
Fox Chase Cancer Center
Philadelphia, PA 19111, U.S.A.

INTRODUCTION

In principle the structure of an RNA molecule can be described
at three different levels: Primary, i.e., the sequence of nucleotides; secondary, i.e., the base pairing interactions between nucleotides; and tertiary, i.e., the three-dimensional conformation of the
entire molecule. For certain species of tRNA one can provide a
description of structure at all three levels. For messenger RNA one
can specify primary structure for several species, make some models
of energetically feasible secondary structures, and say essentially
nothing about tertiary structure.

The primary structure of eukaryotic mRNAs may be described in
terms of three parts: (1) a 5' untranslated region terminated by a
cap structure; (2) a coding region; and (3) a 3' untranslated region
usually, but not always, terminated by a poly A segment. Comparisons
of nucleotide sequences and experiments with modified mRNAs have
provided some insight into the functional significance of the untranslated regions, in particular the role of the 5' region in ribosome binding and translational initiation, and the role of the 3'
region in determining mRNA stability. Concern about mRNA secondary
structure is usually directed at the 5' and 3' untranslated regions,
the goal being to define topographical features that might explain
differences in translational efficiency amongst various species of
mRNA.

The biosynthesis of eukaryotic mRNA involves the production of a
high molecular weight precursor which must be cleaved and often
spliced to form the mature mRNA molecule. Formation of the 5'
terminal cap structure, methylation of internal adenylate residues

119

and construction of the 3' terminal poly A segment occur in the
nucleus during or shortly after transcription of the mRNA precursor.
For some mRNAs a further modification of the cap structure occurs
after the mRNA has been incorporated into the polyribosomes. Certain
features of the nucleotide sequences surrounding the splice junctions
have given clues about possible splicing mechanisms. According to
one current model the correct positioning of splice boundaries may be
achieved by specific base pairing with one or more small nuclear RNA
molecules.

    In the following presentation we shall consider messenger RNA
structure from both a static and a dynamic point of view:

    -As an example of a current experimental approach being used to
study mRNA structure I shall describe attempts at secondary structure
mapping of globin mRNA based on differential accessibility towards
nucleases (1).

    -As an example of the synthesis and processing of mRNAs I will
describe some of our work with immunoglobulin mRNAs.

    In addition to characteristics shared by most eukaryotic genes
the immunoglobulin genes have the special property of undergoing
rearrangement during the course of B cell development. By examining
the activity of the different genetic elements before and after rearrangement we can begin to make some distinctions as to what constitutes a fully functioning gene. Moreover, it can be shown that
alternative processing of the same genetic information leads to the
production of both secretory and membrane-associated immuno-globulins.

Determination of mRNA Secondary Structure

    One can usually predict the secondary structure of small RNA
molecules the size of tRNA with reasonably high success using only
the primary sequence data and a set of empirical rules (2) based on
maximizing the number of stable intramolecular base pairs. However,
for large RNA molecules the size of mRNA this approach usually yields
a family of alternative models with roughly equivalent stabilities.
It is, therefore, essential to apply some type of experimental
approach in order to distinguish between the various models.

    One such approach, which has recently been used for the analysis
of globin mRNA secondary structure (1), involves a discrimination
between unpaired and paired nucleotide sequences on the basis of the
latter's resistance to degradation by single stand specific nucleases.
The method consists of end labeling an RNA molecule at either the 5'
or 3' terminus, partially degrading it with alkali and a battery of
suitable enzymes, and analyzing the digestion products by polyacryl-
amide-gel electrophoresis under conditions suitable for sequence

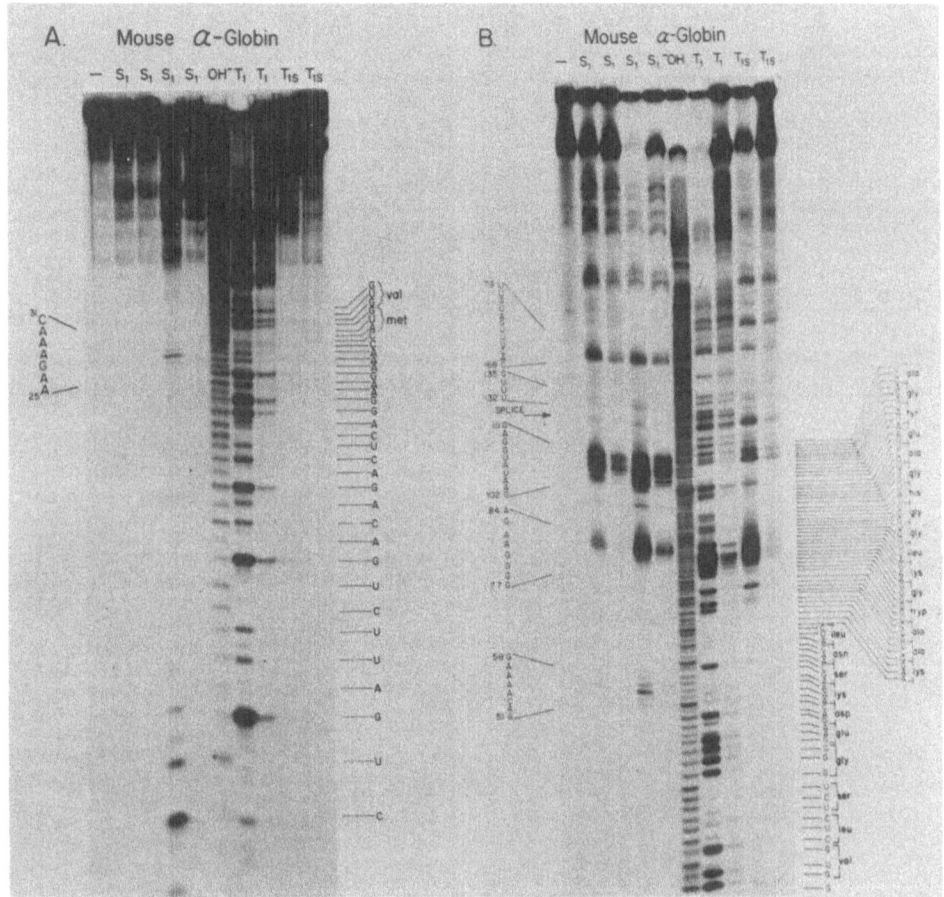

Fig. 1. Examples of sequence data used to map the secondary structure
         of mouse α- globin mRNA (from ref. 1). The data consist of
         autoradiograms of partial digests on 5' ^{32}P mouse α-mRNA
         electrophoresed on slab gels of 20%(A) and 15%(B) polyacryla-
         mide in 8.3 M urea. Partial digests contained the following
         amounts of enzyme per μg of RNA. From left to right: (-)
         enzyme; (S1) S1 nuclease, 0.006 U—10 min, 0.006 U—1 min ,
         0.06 U—10 min, 0.06 U—1 min; (-OH) partial alkaline dig st;
         (T1) T1 RNAse under denaturing conditions, 0.005 U, 0.0005 U;
         (T1S)T1 RNAase under nondenaturing conditions, 3 x 10^{-4}
         U/μg RNA, 3 x 10^{-5} U/μg RNA.

determination. The reliability of this method was tested using a
structurally well-characterized molecule tRNA^{Phe} and then applied to
α and β globin mRNAs.

An example of the data generated from such an experiment is illustrated in Fig. 1. The complete nucleotide sequence is obtained
from the alkali digest, whereas the nucleotides located in unpaired
regions appear as corresponding bands in Sl nuclease or Tl RNAse
digests. A pair of secondary structure models for α and β globin
mRNA consistent with these data are shown in Fig. 2. A comparison of
these structures and the nuclease accessibility data suggests that
the AUG initiator codon in β-globin mRNA may be at the apex of a
hairpin loop and relatively exposed whereas that in the α-globin mRNA
may be buried or relatively inaccessible. This structural difference
may be responsible for the more efficient translational efficiency of
β-globin mRNA relative to α-globin mRNA.

Messenger RNA Processing: Historical Background

The notion of mRNA processing began with the finding that heterogeneous nuclear RNA (hnRNA) and polyribosomal mRNA have strikingly

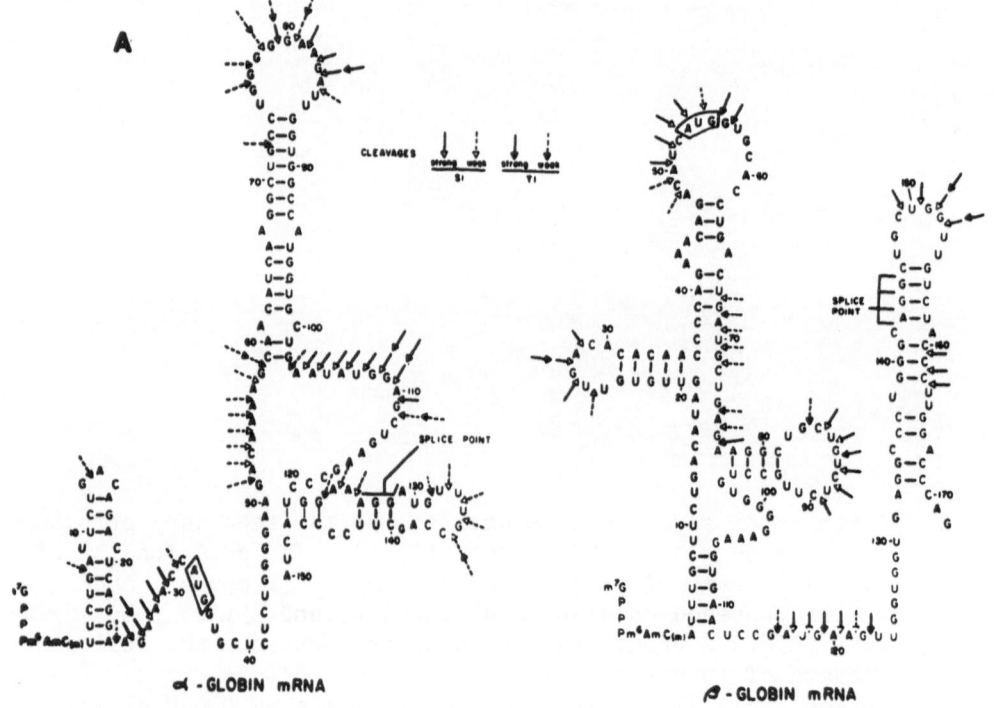

Fig. 2.   Computer generated secondary structure models of the 5'
          terminal regions of mouse α- and β- globin mRNAs using the
          data obtained from partial digests with Sl and Tl nucleases
          (see ref. 1).

different size distributions and yet very similar base compositions.
Before the development of recent techniques that have enabled us to
study the synthesis of individual species of mRNA, and investigator
wishing to probe the relationship between hnRNA and mRNA had to employ
methods that were suitable for complex mixtures of RNA sequences, and
to exploit, whenever possible, features such as poly A that are common
to a substantial fraction of the mRNA species. Nucleic acid hybridization is a technique that may readily be applied to complex mixtures
of RNA, and thus enjoyed wide popularity in the study of hnRNA and
mRNA beginning in the late 1960s and extending over more than a decade
(see 3-6 for refs.). The information gained from these studies,
together with concurrently acquired knowledge of the general properties of eukaryotic DNA sequences, e.g., the existence of single copy
and repetitive sequence elements (7), provided a new framework for
comparing the properties of hnRNA and mRNA. It was evident that
moderately repetitive as well as unique sequences are transcribed into
hnRNA and that at least a portion of these same sequences are processed into mRNA (8). However, the biological significance of these
repetitive sequence transcripts was not obvious, and, in fact, still
remains one of the challenging mysteries in our understanding of
eukaryotic gene expression.

Around 1970 two important discoveries helped accelerate progress
on the problem of mRNA processing. First, the finding that the majority of mRNA molecules and a significant fraction of hnRNA molecules
possess a 3' terminal poly A segment, 150 to 200 nucleotides long,
which is constructed posttranscriptionally (see 9 for refs.). The
poly A tail represented an interesting new aspect of processing, but,
even more importantly, it was rapidly exploited for purifying mRNA
away from the bulk of the cellular RNA (10-12). Second, the discovery
of reverse transcriptase (13, 14), which was later used to synthesize
DNA complementary to mRNA (cDNA), thus providing a valuable probe for
the study of mRNA frequency distributions and the homology relationships between mRNA and hnRNA (see 6 for refs.). Most cell types were
observed to have a very broad distribution of mRNA abundancies ranging
from a few species present at several thousand copies per cell to
thousands of species present in a few copies per cell. About 10-20%
of the hnRNA sequences are homologous to mRNA.

Modified Nucleotides-Cap Structures

In the mid-1970s it was discovered that the mRNAs of eukaryotic
cells and many types of viruses contain an unusual methylated "cap"
structure (Fig. 3) at their 5' terminus and one or more internal 6-
methyl adenine residues (see 15 for refs.). These modifications, like
poly A, are added posttranscriptionally to the mRNA precursors, and
then carried along through the rest of the processing stages. For
a long time it was thought that the capacity to be methylated was a
property confined to the structural RNAs, i.e., the RNAs that do not
encode proteins. This idea persisted because the level of methyl-

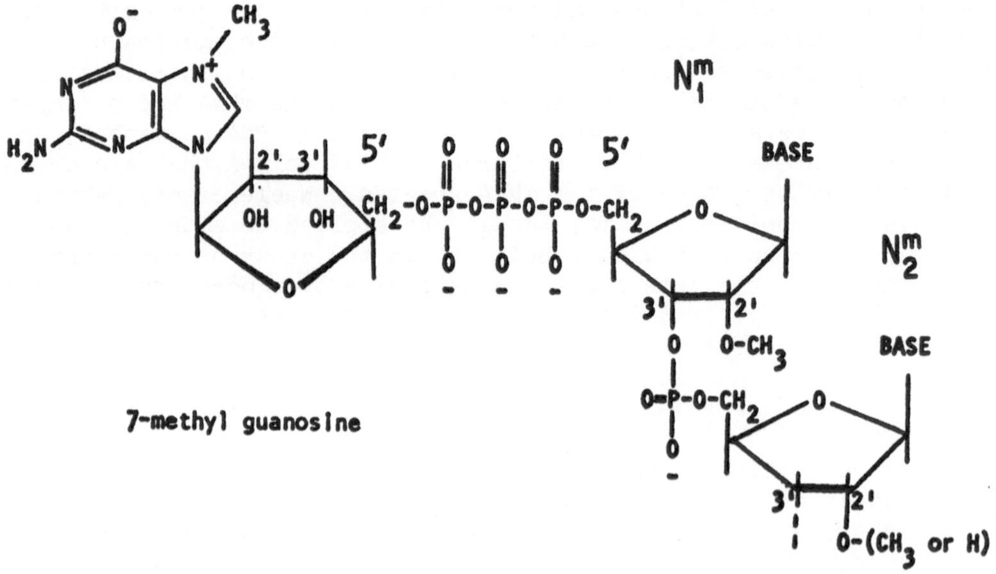

N_1^m

N_2^m

7-methyl guanosine

Fig.3.   The 5' terminal cap structure on eukaryotic messenger RNA.
         The 2'-O-methylation at position N_2 occurs in the cytoplasm
         on some, but not all mRNAs (22).

ation in mRNA is almost an order of magnitude lower than in rRNA, and
without a means for effectively separating these two RNA species the
mRNA methylation is entirely masked--especially if one doesn't know
that it is there in the first place. However, when methods for
isolating mRNA based on its unique poly A structure came into use, one
could obtain sufficiently pure preparations of mRNA so that an unambiguous identification of its methylated derivatives could be made
(16, 17). The parallel development of efficient cell-free systems
for the synthesis of certain viral mRNAs contributed similarly to the
characterization of their modified components, and, moreover, provided
an excellent means for studying the biochemistry of cap formation
(18, 19).

    The formation of a complete cap structure (Fig. 3) involves the
participation of four to six different enzymes (see 15 for refs.).
These are summarized by the following set of reactions:

1) $pppN_1pN_2p\ldots \rightarrow ppN_1pN_2p\ldots + P_i$

or 1') $pN_1pN_2p\ldots + ATP \rightarrow ppN_1pN_2p\ldots + ADP$

2) $pp\overset{\alpha}{p}G + \overset{\beta\alpha}{pp}N_1pN_2p\ldots \rightarrow G^{(5')}\overset{\alpha\beta\alpha}{ppp}(5')N_1pN_2p\ldots + pp_i$

3)§ AdoMet + $GpppN_1pN_2p\ldots \rightarrow {}^7mGpppN_1pN_2p\ldots$ + Ado-S-homocys

4)§ AdoMet + ${}^7mGpppN_1pN_2p\ldots \rightarrow {}^7mGpppN_1^mpN_2p\ldots$ + Ado-S-homocys

5)§ If $N_1^m = A^m$:

AdoMet + ${}^7mGpppA^mmN_2p\ldots \rightarrow {}^7mGppp{}^6mA^mpN_2p\ldots$ + Ado-S-homocys

6)§ AdoMet + ${}^7mGpppN_1^mpN_2p\ldots \rightarrow {}^7mGpppN_1^mpN_2^mp\ldots$ + Ado-S-homocys

The relevant enzymes being (1) RNA triphosphatase; (1') RNA 5' mono-
phosphate phosphokinase; (2) mRNA guanylyltransferase; (3) 7-methyl-
guanosine methyltransferase; (4) and (6) 2'-O-methyltransferase; and
(5) 6-methyl-(2'O-methyladenosine) methyltransferase. Reactions (1)
through (5) occur in the nucleus either during transcription of pre-
mRNA or soon after its completion (20, 21). Reaction (6) occurs in
the cytoplasm after the mRNA has been incorporated into polyribosomes
(22). The reaction leading to the synthesis of the cap structure of
reovirus mRNAs will be described in detail in Chapter 8 in relation
to the facilitating effect of the cap on translation.

There is presently some uncertainty about whether cap formation
occurs exclusively at the sites of transcriptional initiation (via
reaction (1)), or at internal cleavage sites as well (via reaction
(1') or by a variation in reaction (2), cf. (15)). Initially it was
believed that transcription could initiate only with purine nucleoside
triphosphates and that cap structures with pyrimidines in position N_1
(about 1/4 of the total mRNA in mammalian cells) must be formed at
internal cleavage sites. Indeed, studies of the 5' termini of hnRNA
seemed to confirm this idea (20). However, the recent demonstration
that transcription can also be initiated by pyrimidine nucleoside
triphosphates (23) suggests that this point should be re-examined
(cf. also (21)).

Sites of Transcriptional Initiation of mRNA

To date all of the known cap structures on mRNAs of defined
coding specificity have purines at position N_1, and for several of
these mRNAs there is evidence to indicate that N_1 is also the site
of transcriptional initiation. This evidence is sometimes of a
negative type, i.e., failure to detect any transcripts of sequences
that are located upstream from the cap site (cf. (24, 25)), so that
the possibility of extremely rapid processing of a 5' initiator
region cannot be rigorously excluded. However, in the case of
adenovirus mRNA synthesis there is strong positive evidence from both

§   The structure $G^{(5')}{}^{\alpha\beta\alpha}_{ppp}{}^{(5')}N_1pN_2p$ has been abbreviated $GpppN_1pN_2p$
       in reaction (3). The methylated derivatives in reactions
       (4), (5), and (6) are noted accordingly.

in vivo and *in vitro* studies to indicate that the mRNA cap sites and transcriptional initiation sites for polymerase II are, in fact, one and the same (26-28). Comparisons of nucleotide sequences in the 5' flanking regions of genes coding for several cellular and viral mRNAs (including the adenovirus mRNAs) has revealed the existence of a 7 base pair AT rich sequence about 28 nucleotides upstream from the cap site (29, 30). This sequence is similar, although not identical, to the so-called Pribnow box ($^{5'}$-TATAATG-$^{3'}$) which is universally part of the promoter regions of prokaryotic genes (31). Given an equivalence of initiation and cap sites (at least for some mRNAs) and the fact that the mRNA guanylyltransferase does not require a lengthly polynucleotide acceptor, it is reasonable to expect that cap formation will often occur on growing pre-mRNA chains (cf. (20, 21)).

It is conceivable that transcription of some genes can initiate at more than one site with different relative efficiencies, as happens with rRNA genes in *E. coli* (23). This might be an explanation for the heterogeneity of cap structures on certain SV40 and polyoma mRNAs (32, 33), and for the ability of certain SV40 mutants to survive deletions at a cap site (34). However, if such imprecise initiation ever occurs with cellular mRNAs it presumably is confined to relatively sparse mRNA species, since the various abundant mRNA species studied to date all seem to have a homogeneous 5' cap (15).

Splicing

Certainly one of the most surprising developments in the history of mRNA processing was the discovery of splicing. The initial observations were made in 1977 during investigations of adenovirus mRNA synthesis (35, 36), in which the relationship between viral mRNA and the DNA that encodes it were examined in the electron microscope using the powerful R-loop technique (37). The striking multiloop structures were correctly interpreted to mean that the mRNA was specified by several non contiguous genetic elements. In spite of its novelty, this interpretation was readily accepted because it explained the (then) puzzling observation that mRNAs made from distant portions of the adenovirus genome have the identical 5' terminal capped sequence (38, 39). Moreover, it also seemed to be a possible solution to the riddle of how a large hnRNA molecule with a cap structure on one end and a poly A tail on the other could be processed into a smaller mRNA molecule without losing either its cap or its poly A. Within a matter of months experiments employing restriction enzyme analysis with Southern's blotting technique (40) and R-loop or heteroduplex analyses of cloned gene fragments established the widespread occurrence of split genes and gave some idea of their organizational features. Studies of the organization of SV40 genes (41, 42) and of cellular genes like globin (43, 44), immunoglobulin (45), and ovalbumin (46, 47) indicated that the interruptions, termed intervening sequences or introns, can occur in the coding portions of the gene as well as in 5' untranslated leader sequences.

The expression of split genes always seems to involve production
of a composite RNA transcript and subsequent excision of the intron
sequences. This has been established first by showing that there are
large nuclear transcripts which are colinear with the complete gene
(48) and second by using kinetics and pulsechase experiments to demonstrate that the large transcripts are actually processed into mRNA
(49-52). The tendency of hnRNA to aggregate due to intermolecular
basepairing (53) makes it imperative to use rigorous denaturation
conditions in such studies; for example, fractionation of the hnRNA
by electrophoresis on methyl-mercury-agarose gels (54). This technique coupled with a blotting procedure by which the fractionated
 hnRNA is covalently attached to diazotized paper (55), enables one
to visualize precursors of any mRNA for which a pure sequence probe
is available. In fact by using an assortment of probes for structural
and intronic sequences one can in principle delineate the processing
pathway. An example of such an analysis for immunoglobulin light
chain mRNAs is shown in Figs. 4 and 5.

In a transcript containing multiple intronic sequences there may
be a preferential order of excision, but, in some cases the order is
not necessarily absolute (56). This is reminiscent of the alternative
temporal order previously observed in the processing of mammalian rRNA
(3). In certain viral systems like adenovirus and SV40 a given transcript can give rise to multiple mRNA species depending on the choice of
different splicing modes. In this case processing can have a role in
qualitatively determining which gene elements are utilized. Such
qualitative discrimination at the processing level could provide a
basis for certain types of cellular differentiation. An example of
this principle has been recently observed in early B lymphocytes in
which two distinct mRNA species encoding the membrane associated and
secreted forms of the μ-heavy chain are produced from a single set of
μ gene elements by variations in the modes of splicing (57-59). The
mRNAs are identical except for a region near the 3' end, which in one
case encodes the carboxy-terminal tailpiece of the secreted μ chain,
and in the other, a hydrophobic segment that apparently anchors the
membrane μ chain to the lipid bilayer (Fig. 6). During its ontogeny
the B-lymphocyte shifts from producing predominantly membrane μ chain
to predominantly secreted μ chain, presumably by shifting its major
mode of μ-mRNA processing.

Although the enzymes involved in mRNA splicing have not yet been
characterized, some clues concerning the splicing mechanism have come
from comparisons of the nucleotide sequences surrounding the splice
junctions and from studies of the consequences of perturbations in
gene organization. A compilation of a large number of junction
sequences of both cellular and viral pre-mRNAs has resulted in the
consensus sequence shown in Fig. 7a (60). The doubly underlined
nucleotides at the extreme ends of the intron are almost ubiquitous,
being present in over 95% of the sequences. Thus, it seems reasonable
to suppose that they are essential for the splicing reactions. An

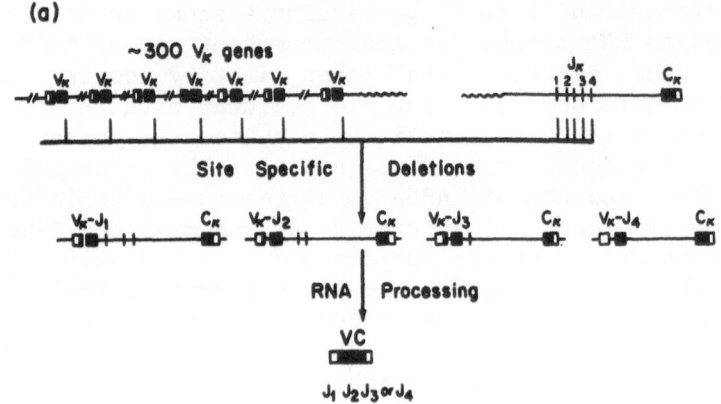

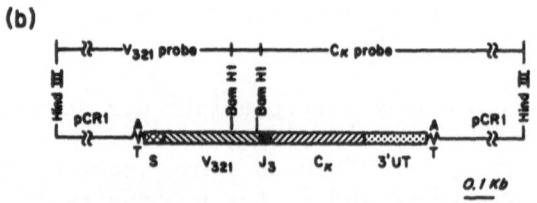

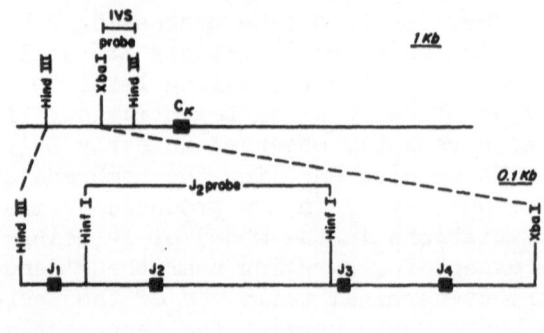

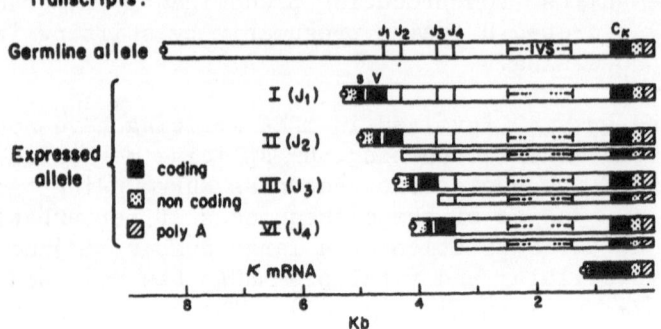

Fig. 4.   (see text opposite page).

interesting complementarity has been noted between the consensus
sequence and a 5' terminal sequence of one of the small nuclear RNAs
(snRNAs), termed U1 (61), suggesting a model in which U1-RNA helps
juxtapose the two splice junctions by appropriate base pairing interactions (Fig. 7b). A similar role has been invoked for the similarly
sized VA-RNA in the splicing of adenovirus mRNA (62). If this speculation turns out to be correct, the function of at least one small
nuclear RNA will have finally been elucidated. Although the snRNAs
were discovered more than 10 years ago, it has not been possible to

---

Fig. 4.   (a) Scheme for the formation and expression of a κ-chain
          immunoglobulin gene. In germline DNA the several hundred
          genes coding for the variable part of the κ chain (V_K genes)
          and the gene coding for the constant portion (C_K gene) are
          separated by an unknown distance. During B-cell differen-
          tiation, a site specific deletion occurs between one of the
          V_K genes and any one of four J_K segments located 2.4 to 3.9
          kb upstream from the C_K gene. This event creates a function-
          al κ-gene the size of which depends on the particular J_K
          segment being utilized. Such functional rearrangements are
          normally found on only one of the allelic pair of chromo-
          somes. The intervening sequence between J_K and C_K (J-C
          intron) is transcribed and the corresponding RNA sequence
          excised during RNA processing. The 3' untranslated sequence
          is contiguous to the C_K sequence. The 5' untranslated
          sequence and the sequence encoding the amino-terminal signal
          peptide are separated from the V_K gene by a small (∿0.1 kb)
          intron.

          (b) Four probes used in the analysis of κ-mRNA transcription
          and processing shown in Fig. 5. The V_K and C_K probes are
          obtained by restriction endonuclease digestion of a cloned
          DNA sequence corresponding to the κ-mRNA produced by MOPC 321
          myeloma cells. The J_2 and intervening sequence (IVS) probes
          are similarly obtained from a cloned fragment of germline DNA
          containing the J_K-C_K region (see ref. 75 for details).

          (c) Schematic representation of the transcripts produced by
          various myeloma cells. The unrearranged (germline) allele
          produces an 8.4 kb transcript which is not processed into
          any functional mRNA. The allele encoding the expressed κ
          chain is transcribed into a component the size of which
          varies according to the J segment being used (5.3, 5.0, 4.4
          and 4.1 kb, respectively, for J_1, J_2, J_3 and J_4 expressors).
          These precursors are processed into a common 1.2 kb κ-mRNA.
          In J_2, J_3 and J_4 expressors components are found which seem
          to arise by an asynchronous cleavage at the 5' boundary of
          the J-C intron (narrow bars).

assign them any specific cellular function. Fortunately, the uncertainty about their physiological significance did not deter studies
of their structural characteristics, and by the mid-1970s the complete
nucleotide sequence of two snRNA species was known (63). Indeed,
their novel highly modified 5' termini served as a model for sub-

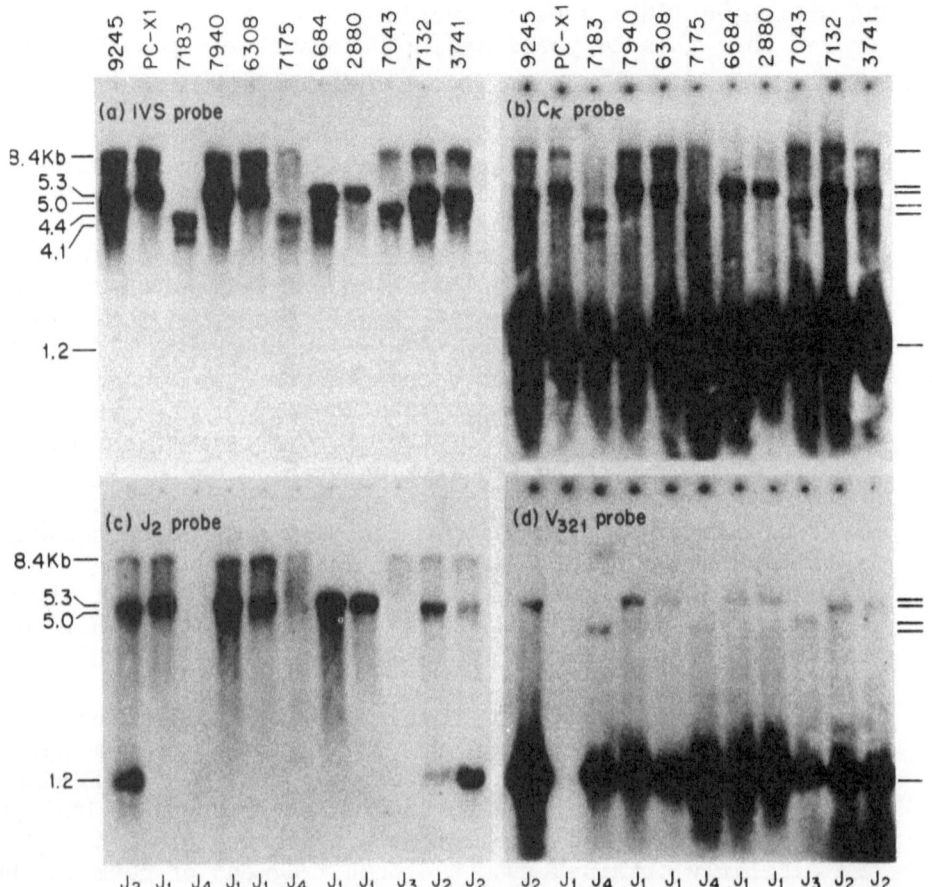

Fig. 5.   A "northern blot" of the poly A$^+$ nuclear RNA from eleven
          difference myelomas expressing distinctive κ-chains. The
          ply A$^+$ nuclear RNA was electrophoresed on methylmercury
          hydroxide gels and blotted onto diazotized paper. The im-
          mobilized RNA was annealed with the four probes described
          in Fig. 4b, and the nuclear components containing the cor-
          responding sequences were revealed by autoradiography. The
          tumor designation is shown at the top; the expressed J
          segment (from amino acid analysis of the κ-chain) is shown
          at the bottom. The size of the various bands (in kilobases)
          is at the left. The interpretation of these data is given
          in Fig. 4c (see ref. 75 for other details).

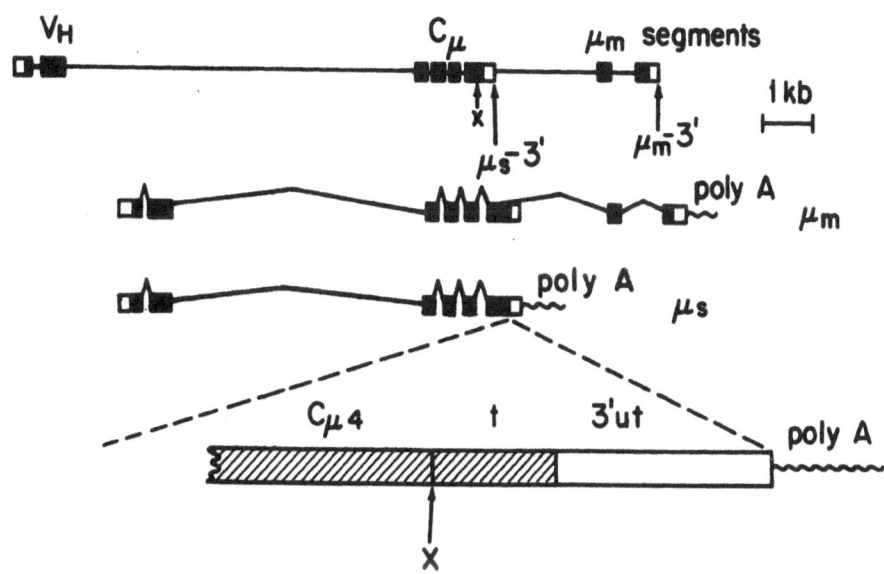

Fig. 6. Schematic diagram of the production of secreted (μ_s) and
         membrane-associated (μ_m)-mRNA from a common set of genetic
         elements. The production of a functional μ heavy chain gene
         occurs by DNA recombination analogous to the described in
         Fig. 4 for the κ light chain. The DNA sequences encoding
         the four domains of the μ constant region (Cμ) are separated
         by small introns. Adjacent to the Cμ_4 segment is a sequence
         specifying the 3' terminal tailpiece (t) and the untranslated
         region of μ_s, while about 2 kb further downstream are two
         regions specifying a hydrophobic anchor peptide and the 3'
         untranslated region of μ_m. RNA transcripts terminated at
         μ_m-3' (perhaps by cleavage of a growing chain and polyadenyl-
         ation) are processed into μ_m-mRNA by seven splicing events,
         one junction (X) being at the site separating the segments
         encoding Cμ_4 and t. RNA transcripts terminated at μ_s-3' are
         processed into μ_s-mRNA by five splicing events.

sequent elucidation of the cap structures of mRNA (64, 65). In the
snRNAs the ^{7}mG moiety is replaced by ^{2,2,7}m$_3$G; otherwise the structures are essentially the same. The snRNAs are evolutionarily
conserved, a fact which may be related to the apparent conservation
of mRNA processing systems (66-69).

Order of Processing Reactions

     Studies with total pre-mRNA populations (20, 21) and individual

a.

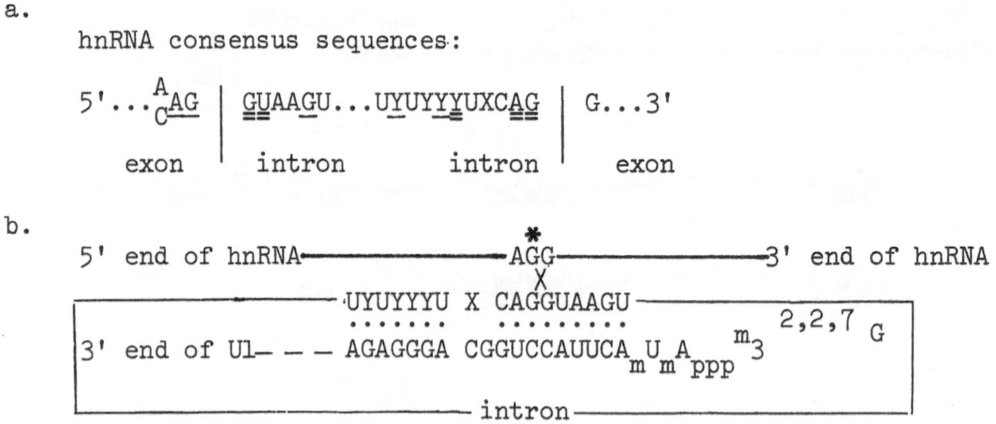

hnRNA consensus sequences:

$$5'...\overset{A}{\underset{C}{\underline{C}}}\underline{AG} \mid \underline{\underline{GU}}\underline{A}\underline{\underline{A}}\underline{GU}...\underline{U}\underline{Y}\underline{U}\underline{Y}\underline{\underline{Y}}\underline{Y}\underline{U}\underline{X}\underline{C}\underline{A}\underline{\underline{G}} \mid G...3'$$

     exon   intron   intron   exon

b.

Fig. 7. (a) A consensus sequence obtained by comparing 36 splice
junction sequences (60). To appear in this sequence a base
must be the most common in that position and occur with a
frequency of at least 45%; bases occurring in 75% of the
sequences are underlined; those present zith 95% or greater
frequency are underlined twice. Y indicates pyrimidines.
X marks the position of a single non conserved base in the
consensus sequence. Vertical lines mark the intron-exon
boundaries.

(b) A possible alignment of intron-exon boundaries by base
pairing between the 5' terminal portion of U1 RNA and
sequences at both ends of an intron. Processing would consist of cleavage of two G-G bonds and formation of a new
one (asterisk).

pre-mRNAs (52, 56, 70-73) indicate that cap formation, internal
methylation and polyadenylation usually *precede* the splicing out of
intronic sequences. Thus, in regard to the general order of processing reactions mRNA seems to resemble rRNA and tRNA, in that the
cleavages are directed at molecules which have already been subjected
to other types of post-transcriptional modification. Poly A formation
catalyzed by a terminal transferase enzyme, consists of the sequential addition of 150-200 adenylate residues to the 3'-end. The recognition signal for the terminal transferase seems to involve the hexa-
nucleotide AAUAAA since this sequence is approximately 20-25 nucleotides upstream from the 3' end of all poly A-containing mRNAs but
absent from poly A-lacking mRNAs. In some cases, e.g. adenovirus late
mRNAs, the poly A addition site may be formed by endonucleolytic

cleavage of the growing transcript rather than by termination of the RNA polymerase (74). The extent to which this applies to cellular mRNAs is presently unclear, although it is tempting to invoke such alternative termination in the differential processing of secreted and membrane-associated μ-mRNAs discussed earlier.

REFERENCES

1. Pavlakis, G.N., Lockard, R.E., Vamvakopoulos, N., Rieser, L., Raj Bhandary, U.L., and Vournakis, J.N. (1980), Cell. 19, 91-102.
2. Tinoco, I. Jr., Borer, P.N., Dengler, B., Levine, M., Uhlenbeck, O.C., Crothers, D.M. and Gralla, J. (1973), Nature, London, New Biol. 246, 40-41.
3. Perry, R.P. (1976), Annual Review of Biochemistry, E.E. Snell, ed., Annual Reviews, Inc., Palo Alto, CA, publ., Vol. 45, pp. 605-629.
4. Darnell, J. (1979), Prog. Nucl. Acid. Res. Mol. Biol. 22. 327-353.
5. Scherrer, K., Imaizumi-Scherrer, M.-T., Reynaud, C.-A., and Therwath, A. (1979), Mol. Biol. Rep. 5, 5-28.
6. Perry, R.P., Bard, E., Hames, B.D., Kelley, D.E. and Schibler, U. (1977), Prog. Nucl. Acid. Res. Mol. Biol. 19, 275-292.
7. Britten, R.J. and Kohne, D.E. (1968), Science 161, 529-540.
8. Greenberg, J.R. and Perry, R.P. (1971), J. Cell Biol. 50; 774-786.
9. Brawerman, G. (1980), CRC Reviews (in press).
10. Edmonds, M. and Caramela, M.G. (1969), J. Biol. Chem. 244, 1314-1324.
11. Lee, S.Y., Mendecki, J. and Brawerman, G. (1971), Proc. Natl. Acad. Sci. USA, 68, 1331-1335.
12. Sheldon, R., Jurale, C. and Kates, J. (1972), Proc. Natl. Acad. Sci. USA 69, 417-421.
13. Temin, H.M. and Mizutani, S. (1970), Nature, London, 226, 1211-1213.
14. Baltimore, D. (1970), Nature, London, 226, 1209-1211.
15. Banerjee, A.K. (1980), Microbiological Rev. 44, 175-205.
16. Perry, R.P. and Kelley, D.E. (1974), Cell 1, 37-42.
17. Desrosiers, R., Friderici, K. and Rottman, F. (1974), Proc. Natl. Acad. Sci. USA 47, 3971-3975.
18. Furuici, Y., Morgan, M., Muthukrishnan, S. and Shatkin, A.J. (1975), Proc. Natl. Acad. Sci. USA 72, 362-366.
19. Ensinger, M.J., Martin, S.A., Paoletti, E. and Moss, B. (1975), Proc. Natl. Acad. Sci. USA 72, 2525-2529.
20. Schibler, U. and Perry, R.P. (1977), Nucl. Acids Res. 4, 4133-4149.
21. Salditt-Georgieff, M., Harpold, M., Chen-Kiang, S. and Darnell, J.E. (1980), Cell 19, 69-78.
22. Perry, R.P. and Kelley, D.E. (1976), Cell 8, 433-442.
23. De Boer, H. and Nomura, M. (1979), J. Biol. Chem. 254, 5609-5612.

24. Tsuda, M., Ohshima, Y. and Suzuki, Y. (1979), Proc. Natl. Acad. Sci. USA 76, 4872-4876.
25. Roop, D.R., Tsai, K.-J. and O'Malley, B.W. (1980), Cell 19, 63-68
26. Ziff, E.B. and Evans, R.M, (1978), Cell 15, 1463-1475.
27. Manley, J.L., Sharp, P.A. and Gefler, M.L. (1979), Proc. Natl. Acad. Sci. USA 76, 160-164.
28. Weil, P.A., Luse, D.S., Segall, J. and Roeder, R.G. (1979), Cell 18, 469-484.
29. Konkel, D., Tilghman, S. and Leder, P. (1978), Cell 15, 1125-1132
30. Gannon, F., O'Hare, K., Perrin, F., Le Pennec, J.P., Benoist, C. Cachet, M., Breathnach, R., Royal, A., Garapin, A., Cami, B. and Chambon, P. (1979), Nature, London, 278, 428-434.
31. Pribnow, D. (1975), Proc. Natl. Acad. Sci. USA 72, 784-788.
32. Canaani, D., Kahana, C., Mukamel, A. and Groner, Y. (1979), Proc. Natl. Acad. Sci. USA 76, 3078-3082.
33. Flavell, A.J., Cowie, A., Legon, S. and Kamen, R. (1979), Cell 16, 357-371.
34. Villareal, L.P., White, R.T. and Berg, P. (1979), J. Virol. 29, 209-219.
35. Berget, S. M., Moore, C. and Sharp, P.A. (1977), Proc. Natl. Acad. Sci. USA 74, 3171-3175.
36. Chow, L.T., Gelinas, R.E., Broker, T.R. and Roberts, R.J. (1977), Cell 12, 1-8.
37. Thomas, M., White, R.L. and Davis, R.W. (1976), Proc. Natl. Acad. Sci. USA 73, 2294-2298.
38. Gelinas, R.E. and Roberts, R.J. (1977), Cell 11, 533-544.
39. Klessig, D.F. (1977), Cell 12, 9-21.
40. Southern, E.M. (1975), J. Mol. Biol. 98, 503-517.
41. Lavi, S. and Groner, Y. (1977), Proc. Natl. Acad. Sci. USA 74, 5323-5327.
42. Aloni, Y., Dhar, R., Laub, O., Horowitz, M. and Khoury, G. (1977) Proc. Natl. Acad. Sci. USA 74, 3686-3690.
43. Jeffreys, A.J. and Flavell, R.A. (1977), Cell 12, 1097-1108.
44. Tilghman, S.M., Tiemeier, D.C., Seidman, J.G., Peterlin, B.M., Sullivan, M., Maizel, J.V. and Leder, P. (1978), Proc. Natl. Acad. Sci. USA 75, 725-729.
45. Brack, C. and Tonegawa, S. (1977), Proc. Natl. Acad. Sci. USA 74, 5652-5656.
46. Breathnach, R., Mandel, J.L. and Chambon, P. (1977), Nature, London 270, 314-319.
47. Lai, E.C., Woo, S.L.C., Dugaiczyk, A., Catterall, J.F. and O'Malley, B. (1978), Proc. Natl. Acad. Sci. USA 75, 2205-2209.
48. Tilghman, S.M., Curtis, P.J., Tiemeier, D.C., Leder, P. and Weissman, C. (1978), Proc. Natl. Acad. Sci. USA 75, 1309-1313.
49. Ross, J. (1976), J. Mol. Biol. 106, 403-420.
50. Curtis, P.J., Mantei, N., Vanden Berg, J. and Weissman, C. (1977), Proc. Natl. Acad. Sci. USA 74, 3184-3188.
51. Gilmore-Hebert, M. and Wall, R. (1978), Proc. Natl. Acad. Sci. USA 75, 342-345.
52. Schibler, U., Marcu, K.B. and Perry, R.P. (1978), Cell 15, 1495-1509.

53. Fedoroff, N., Wellauer, P.K. and Wall, R. (1977), Cell 10, 597-610.

54. Bailey, J.M. and Davidson, N. (1976), Ann. Biochem. 70, 75-85.

55. Alwine, J.C., Kemp, D.J. and Stark, G.R. (1977), Proc. Natl. Acad. Sci. USA 74, 5350-5354.

56. Ryffel, G.U., Wyler, T., Muellener, D.B. and Weber, R. (1980), Cell 19, 53-61.

57. Perry, R.P. and Kelley, D.E. (1979), Cell 18, 1333-1339.

58. Rogers, J., Early, P., Carter, C., Calame, K., Bond, M., Hood, L. and Wall, R. (1980), Cell 20, 303-312.

59. Early, P., Rogers, J., Davis, M., Calame, K., Bond, M., Wall, R. and Hood, L. (1980), Cell 20, 313-319.

60. Lerner, M.R., Boyle, J.A., Mount, S.M., Wolin, S.L. and Steitz J.A. (1980), Nature, London, 283, 220-224.

61. Reddy, R., Ro-Choi, T.S., Henning, D. and Busch, H. (1974), J. Biol. Chem. 249, 6486-6494.

62. Murray, V. and Holliday, R. (1979), FEBS Letters 106, 5-7.

63. Busch, H. (1976), Perspectives in Biology and Medicine 19, 549-567.

64. Ro-Choi, T.S., Reddy, R., Choi, Y.C., Kaj, N.B. and Henning, D. (1974), Fed. Proc. 33, 1548.

65. Rottman, F., Shatkin, A.J. and Perry, R.P. (1974), Cell 3, 197-199.

66. De Robertis, E.M. and Olson, M.V. (1979), Nature, London, 278, 137-143.

67. Mantei, N., Boll, W. and Weissmann, C. (1979), Nature, London, 281, 40-46.

68. Hamer, D.H. and Leder, P. (1979), Nature, London, 281, 35-40.

69. Breathnach, R., Mantei, N. and Chambon, P. (1980), Proc. Natl. Acad. Sci. USA 77, 740-744.

70. Nevins, J.R. and Darnell, J.E. (1978), Cell 15, 1477-1493.

71. Bastos, R.N. and Aviv, H. (1977), Cell 11, 641-650.

72. Kinniburgh, A.J., Mertz, J.E., and Ross, J. (1978), Cell 14, 681-693.

73. Roop, D.R., Nordstrom, J.L., Tsai, S.Y., Tsai, M.-J. and O'Malley, B.W. (1978), Cell 15, 671-685.

74. Darnell, J.E. (1979), In *From Gene to Protein: Information Transfer in Normal and Abnormal Cells*, T.R. Russell, K. Brew, H. Faber and J. Schultz, eds., Academic Press, New York, publ. pp. 207-227.

75. Perry, R.P., Kelley, D.E., Coleclough, C., Seidman, J.G., Leder, P., Tonegawa, S., Matthyssens, G. and Weigert, M. (1980), Proc. Natl. Acad. Sci. USA 77, 1937-1941.

SV40 AS A MODEL SYSTEM FOR THE STUDY OF RNA TRANSCRIPTION

AND PROCESSING IN EUKARYOTIC CELLS

Yosef Aloni

Department of Genetics
Weizmann Institute of Science
Rehovot, Israel

INTRODUCTION

In the prokaryotic cell, translation of the mRNA by ribosomes
starts before transcription of the message from the DNA has been
completed. In a eukaryotic cell, by contrast, the process of transcription occurs in the nucleus while translation takes place mainly,
if not entirely, in the cytoplasm. The two operations are separated
by the nuclear membrane, providing an obvious opportunity for additional processing and maturation of the primary RNA product. The
various aspects of RNA processing are covered in comprehensive
reviews (1-12).

In the last three years it has become apparent that understanding the process of RNA transcript maturation is one of the great
frontiers in the field of eukaryotic cell physiology. The basis for
this belief has been the discovery of a novel maturation mechanism
termed "RNA splicing", identified first in the DNA and RNA tumor
viruses (13-28) and then extended to many eukaryotic organisms (29-
32). The principle of RNA splicing is that, during processing of a
primary RNA transcript, internal sequences in the molecule are excised, and the remaining pieces are ligated to give rise to a continuous, translatable RNA molecule. This mechanism implies that the
gene itself is split, in the sense that it contains sequences that
are ultimately translated, and sequences that are removed and not
translated. Gilbert (33) has referred to the intervening sequences,
those sequences on the DNA that do not end up in the translatable mRNA
as *introns*, the expressed sequences as *exons*.

The purpose of this review is to describe the use of SV40 as a
model system for the study of RNA transcription and processing in

137

eukaryotic cells, the discovery of RNA splicing and the first attempts
to determine the mechanism of this process. Reviews on split genes
and RNA splicing have been published (12, 33-35).

SV40 AS A MODEL SYSTEM

The use of animal viruses as model systems for probing the complexities of molecular control mechanisms has been particularly fruitful. It is generally felt that an understanding of genetic regulation
in viruses will provide insight into similar regulatory processes in
eukaryotic cells. The molecular biology of SV40 has been under intensive investigation for a number of years, and these studies have
provided considerable information regarding the regulation of gene
expression, in particular, transcriptional and posttranscriptional
processing of mRNA (36-39).

SV40 provides several unique advantages as a model system for
such studies. They include the following: 1) The viral genome is a
small circular molecule (M.W. 3.4×10^6) that contains genetic information for only five or six proteins. The DNA can be obtained in
large quantities, which is imperative for many experiments in molecular biology. 2) The same RNA polymerase (polymerase II) transcribes
both viral and cellular RNA. 3) The viral and cellular RNAs undergo
similar posttranscriptional modifications (r.g. polyadenylylation at
the 3' terminus, capping at the 5' terminus and internal methylation).
4) A number of mutants and hybrid viruses are available for study.
5) Transcriptional complexes are easy to obtain. 6) The entire nu-
cletide sequence of the genome of this virus has been determined.

The SV40 genome is comprised of early and late genes that are
localized in symmetrical halves of the viral DNA (40). The segment
between 0.67 and 0.17 on the map is transcribed in a <u>counterclockwise</u>
direction prior to the onset of viral DNA replication, and codes for
the *early* viral proteins. The second segment (from 0.67 to 0.17) is
transcribed in abundance (though not extensively) after initiation of
viral DNA replication in a <u>clockwise</u> direction. It enclodes the
information for the *late* proteins: VP_1, VP_2 and VP_3. The capsid
proteins have been mapped approximately between 0.95 and 0.16, 0.76
and 0.97, and 0.83 and 0.97 respectively (see Fig. 1) (36-38).
The late viral mRNAs are known as "16S" and "19S". The 16S RNA codes
for VP_1 while the 19S RNAs code for VP_2 and VP_3 (38). The viral DNA
segment between 0.67 and 0.76 is transcribed late in infection, but
the genetic information encoded in this region is unknown. Our
studies have been concerned with the expression of the *late* genes.
Reviews concerning the "early" genes have been published (36-39).

THE INITIATION OF TRANSCRIPTION OF SV40 DNA LATE AFTER INFECTION

The excitement that arose following the observation of RNA
splicing delayed the investigations of other mechanisms involved in

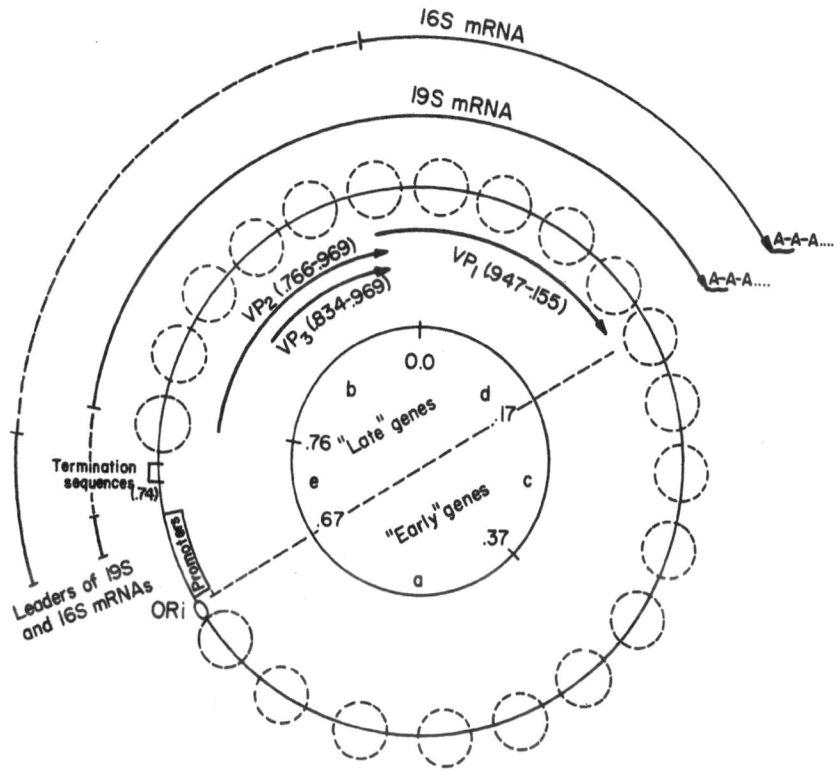

Fig. 1. Transcriptional map of SV40 indicating the five Eco RI;
         HpaI; and BgII restriction fragments of SV40 genome in the
         central circle. Arrows on RNAs indicate 3' termini, poly(A)
         tails, and direction of transcription. Decimal numbers
         represent map units of the SV40 genome. Dashed lines (---)
         indicate sequences spliced out of 16S and 19S mRNAs. The
         location of the coding regions are indicated by heavy lines.
         The small dashed circles denote the distribution of nucleo-
         somes about the SV40 minichromosome with an *exposed* region
         that contains the origin of replication (ORi) promoters for
         late transcription and transcription termination sequences
         (38). Only the major leaders of 19S and 16S mRNAs are
         represented; for more details see ref. 38.

the regulation of SV40 gene expression. Among them are 1) the controls that operate at the initial steps of transcription and determine
the specificity of transcriptional initiation, 2) the frequency of
completion of primary transcripts, and 3) the mechanism of strand
selection. I shall describe first our approaches to determine the
localization of the initiation sites of SV40 late transcription, and
the possible mechanisms involved in determining this specificity.

    The best approach to determine the initiation site for tran-

scription is by determining which nucleotides are at the 5' end of the
newly synthesized RNA. However, we and others (41) have failed to
detect any labeled 5' termini of SV40-specific RNAs. We therefore
used three independent approaches, which were undertaken in order to
localize the initiation site for transcription of SV40 DNA at late
time after infection (42). Two of these were based on the Dintzis
principles (43), while in the third we measured nascent RNA chains
attached to transcriptional complexes under the electron microscope.

The rationale of localization of the initiation site(s) based on
the Dintzis principles is that, after short pulses with radioactive
precursors, RNA molecules would contain some labeled sequences complementary to each region of the DNA, but the labeled RNA complementary to a fragment of DNA that includes the initiation site for
transcription would be in the shortest chains, while labeled RNA complementary to a DNA fragment far from the initiation site would be in
progressive longer chains.

RNA labeled *in vivo* was isolated from productively infected cells
and RNA labeled *in vitro* was isolated from transcriptional complexes
of SV40. The purified RNAs were denatured and fractionated by sedimentation through sucrose gradients. Labeled RNAs of various lengths
were hybridized with restriction fragments of SV40 DNA of a known
order. In both cases, the shortest RNAs hybridized with a fragment
that spans between 0.67 and 0.76 on the map (see Fig. 1). The hybridization with this fragment decreased with successively longer RNAs,
indicating that transcription initiates within this fragment or very
close to it. Similar enrichment for this fragment was obtained using
nascent RNA chains labeled *in vitro* with a short pulse. Electron
microscope observation of transcriptional complexes of SV40 has
revealed a substantial fraction with one short nascent RNA chain. The
initiation site of the nascent chains was placed at coordinate 0.67 $\pm$
0.02.

Based on these and other results in which we mapped the 5' end
of the nuclear viral RNA (24), we have concluded that late transcription initiates at alternative sites in a fragment of the genome that
spans 0.67 to 0.76. This conclusion is supported by the localization
on the map of the "caps" of the viral RNAs (44, 45) and by the localization of the 5' ends of poly(A)-containing viral RNA, using the
primer extension technique (38).

The next question was, what determines the specificity of initiation ? Is it only sequence-specific or does the structure of the
template contribute also to this specificity ?

Studies aimed at determining the DNA sequences required for
initiation of transcription were made possible by the development of
cell-free systems, where transcription was allowed to initiate
selectively and accurately (86-90). Based on these studies it has

been concluded that the TATAAAA sequences (or "Goldberg-Hogness" box),
located about 30 nucleotides upstream from the cap site (initiation
site) plays a fundamental role for promoting specific initiation by
polymerase B at least *in vitro* using naked DNA templates.

In more recent studies it has been found that *deletion* of the
"Goldberg-Hogness" box did *not* abolish transcription of the sea urchin
H_2A histon gene in *Xenopus laevis* oocytes (91), nor of the SV40 early
region in infected cells (97). Moreover, there is no "Goldberg-
Hogness" box upstream from the major initiation site of SV40 *late*
mRNA.

There is now clear evidence that sequences upstream from the
"Goldberg-Hogness" box play a positive or negative role in controlling
initiation of transcription by RNA polymerase B *in vivo*. The question
is still open as to whether these sequences are needed a) to bind the
RNA polymerase, b) to trigger the activation of some protein(s), or
c) to determine the structure of chromatin (91-92).

THE SV40 MINICHROMOSOME

SV40 DNA is found within infected cells in the form of a minichromosome (46), and a variety of methods have been developed for the
extraction of viral chromatin from the infected cells (47-49). The
SV40 minichromosome possesses a beaded structure composed of cellular
histones and supercoiled viral DNA in a molecular complex very similar
to that of cellular chromatin (50-52). The similarity between these
structures, together with the fact that the major viral functions
take place within cell nuclei via cellular machinery, have made SV40
an attractive system in which to study the organization and expression
of the more complex eukaryotic chromatin. Early studies on SV40
chromatin, attempting to determine the precise distribution of the
nucleosomes along the DNA, indicated that the nucleosomes were randomly distributed relative to the viral DNA sequences (53-56). Recent
studies have altered this picture somewhat by indicating that a region
close to the viral origin of replication is particularly sensitive to
nuclease digestion (57-60). The simplest interpretation of the
presence of a nuclease-sensitive region is that it contains a peculiar
arrangement of protein structures about this region.

We have analyzed SV40 minichromosomes in the electron microscope
in an attempt to determine whether alteration in the gross nucleopro-
tein structure of this region could be visualized. About 25% of the
SV40 minichromosomes observed contain a region of DNA between 0.67
and 0.75 on the map that is not organized into the typical nucleo-
some beaded structure (see Fig. 2) (61). This is the same region
where the initiation of late transcription occurs (42) (see Fig. 1).

It was therefore of interest to investigate whether there is
a correlation between the structure of the minichromosome and tran-

scriptional initiation. As a first approach, we transcribed the viral
minichromosome *in vitro* using *E. coli* RNA polymerase. The initiation
specificity of the bacterial RNA polymerase was highly altered by the
structure of the template, and initiation of transcription in the
region of the major late promoters of SV40 occurred very efficiently
on the viral minichromosome, while to only a minor extent, if at all,
on the naked viral DNA. Furthermore, the RNA synthesized on the minichromosome elongated along the *late* strand in a clockwise direction
in opposition to the *early* stand transcripts obtained from the naked
DNA (62). The *in vitro* transcription of viral minichromosomes with
E. coli RNA polymerase seemed to reproduce the situation found *in
vivo*, and transcription was limited to the *late* mRNAs. These observations suggested that the structure of the DNA-protein template may
be involved in determining transcriptional specificity.

   An additional regulatory mechanism that operates during the
initial steps of late transcription of SV40 DNA has been called *atten-*

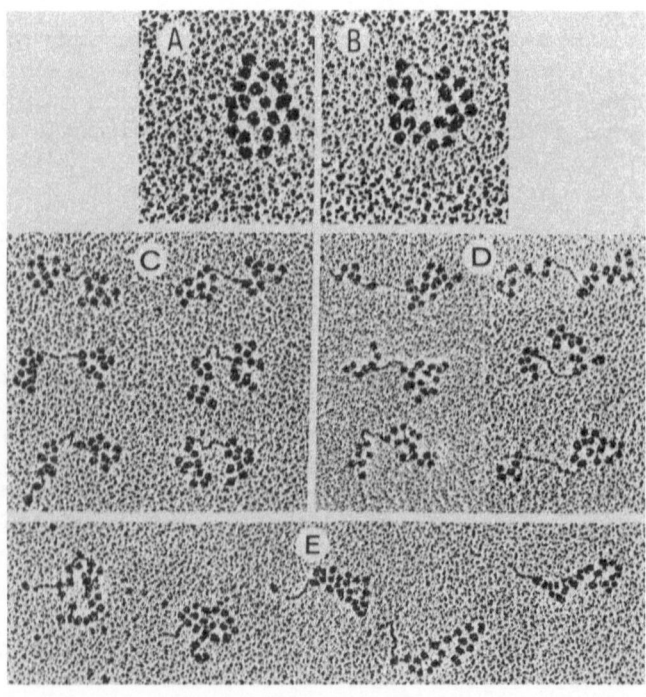

Fig. 2.   Electron microscopic visualization of SV40 minichromosomes.
         A. Minichromosome without a gap.   B. Minichromosome with a
         gap.   C, D, and E are minichromosomes with gaps cleaved with
         Bam HI (0.15 map units), Eco RI (0.0 map units) and BgII
         (0.67 map units), respectively.

uation. (42). According to this mechanism, there is controlled premature termination of a few hundred nucleotides beyond the initiation
sites (in Fig. 1). The premature termination is enhanced when the
infected cells were treated with the drug DRB (5,6-dichloro-1-β-ribo-
furanosylbenzimidazole) (63). It is interesting to note that in this
region of the genome there are sequences that signal transcriptional
termination (38). The site of attenuation is approximately at the 3'
end of the nucleosome-depleted region (61), and is also at the·site
of the 3' end of the leaders of late SV40 mRNAs.

SPLICING OF SV40 LATE mRNA

The transcripts of the late strand of SV40 DNA complementary to
the region between 0.67 to 0.76 on the map are abundantly represented
in the cytoplasmic RNA fraction of SV40 infected cells, and are also
retained on oligo(dT)-cellulose (64). Since the size of the RNA containing these sequences was not determined, it was not clear that
they contained a poly(A) stretch adjacent to the RNA sequence at 0.76.

To characterize the RNA transcribed from 0.67-0.76 of the genome,
we hybridized labeled poly(A)-containing RNA from infected cells to
this viral DNA fragment, eluted the RNA, and subjected it to two tests.
In the first test, the ^{32}P-labeled RNA was reannealed with the five
viral DNA fragments shown in Fig. 1 (central circle). In the second
test, the labeled RNA was sedimented through a sucrose gradient. In
a similar set of experiments, labeled poly(A)-containing RNA was
hybridized to an eluted from a fragment between 0.0-0.17 on the map,
a region of the genome known to encode the 3' portions of the 16S and
19S cytoplasmic RNAs (see Fig. 1). After elution, this RNA was also
analyzed both by reannealing to blots and by sucrose sedimentation.
Both of the eluted (^{32}P) RNAs rehybridized to the DNA fragments shown
in Fig. 1, and yielded a similar hybridization pattern: Hybridization
occurred with the b (0.76-0.0), d (0.0-0.17) and e (0.67-0.76) fragments. Analysis of the eluted (^{32}P) RNAs on sucrose gradients showed
that both of them contained the 16S and 19S species. Since the 16S
and 19S RNAs had been located previously between 0.95 and 0.17 and
between 0.77 and 0.17 respectively, their hybridization with fragment
e (0.67-0.76) was unexpected and prompted further investigations.

At this stage, we entertained two possibilities. The first was
that we were dealing with a new species of poly(A)-containing RNA
which is transcribed from 0.67-0.76 map units; the second was that
the RNA transcribed from this region of the genome was in some way
attached to the 16S and 19S viral RNAs. The approach we took to
distinguish between these two possibilities was to determine whether
the RNA transcribed from 0.67-0.76 was in fact polyadenylylated and
if so, at what site. To our surprise, we found that the abundant
sequences transcribed from 0.67-0.76 contained poly(A) at 3' ends
that were well *downstream* from 0.76, in fact at the same position

where the poly(A)s of the 16S and 19S RNAs were placed on the map.

The results we collected up to this stage showed that the sequences from 0.67 to 0.76 were found in high concentrations, that they sedimented in sucrose gradients together with the 16S and 19S SV40mRNA species, and that they were retained on oligo(dT)-cellulose by poly(A) tails attached to 3' ends mapping at sites not adjacent to 0.76 map units. A model that could accommodate all these results was that the abundant sequences located between 0.67 and 0.76 were covalently linked, *spliced*, to the 5' ends of the coding sequences of 16S and 19S viral RNAs.

To test this model, we performed experiments to map the 5' end of the 16S RNA, taking advantage of the occurrence of a methylated cap structure at this end (44, 45). Two sets of conditions were used. In the first experiment, the methyl-labeled poly(A)-containing RNA was hybridized in formamide at $37^\circ C$ to minimize thermal degradation. Under the second protocol, the labeled RNA was first fragmented to pieces 100-200 nucleotides in length and hybridized to restriction fragments at 68°, as shown in Fig. 1. We found radioactivity associated with fragments b, d and e in both conditions. However, the percent radioactivity associated with fragment e increased about two-fold and that associated with fragment b increased about 1.5-fold when the RNA was fragmented prior to hybridization. This indicated that the radioactivity associated with fragment d was in part due to nonhybridized RNA sequences covalently attached to hybridized sequences. This analysis, of course, did not discriminate between the label in cap structures and internal methyl-label.

The radiolabeled RNA associated with each of the fragments was then eluted from the nitrocellulose paper and digested with T_2, T_1 and pancreatic ribonucleases and the products were analyzed by electrophoresis on DEAE paper at pH 3.5. Radioactivity was found only in the cap structures and 6mA residues (17, 20, 44, 45). Furthermore, when intact RNA was analyzed, labeled caps were found associated with fragment b (43.7%), fragment d (46.7%) and fragment e (9.6%). However, analysis of fragmented RNA showed that more than 80% of the labeled caps annealed to fragment e, and none with fragment d (17). The most plausible interpretation of these results was that the 5' end of the coding region of the 16S poly(A)-containing RNA, previously mapped from 0.95 to 0.17 was covalently linked to sequences transcribed from fragment e (0.67-0.76).

We have used a different ensemble of fragments in order to locate more precisely within the EcoRI, HpaI and BgII fragment e the sequences ajacent to the cap, and have found that the caps of 16S and 19S poly(A)-containing SV40 RNAs are transcribed from a region between 0.67 and 0.73 on the map. Based on these results, we have suggested a model for the splicing of 16S viral RNA (17). According to this model, the 5'-terminal 200 ribonucleotides of late SV40 16S mRNA are

not transcribed immediately adjacent to their coding sequences. A
similar conclusion was drawn for the transcription of 19S mRNA.

MAPPING THE LEADER AND THE BODY OF THE VIRAL mRNAs BY ELECTRON MICROSCOPY

Confirmation for the biochemical approach, which indicated that
the leader sequences of the 16S mRNA were not transcribed adjacent to
the coding sequences, was obtained by electron microscopy. Two types
of analysis were used: In the first, RNA.DNA hybrids were generated
between the SV40 L-DNA strand, obtained by cleavage of form-I SV40 DNA
with EcoRl restriction endonuclease followed by strand separation, and
poly(A)-containing RNA purified from the cytoplasm of infected cells
(18). In the second analysis, the viral RNA was hybridized with
linear double-stranded SV40 DNA (23). Under appropriate conditions of
hybridization, individual RNA molecules can hybridize to double-
stranded DNA by displacing the part of the DNA strand identical to
this RNA and then hybridizing with the complementary DNA strand to
form a structure known as "R-loop" (65).

A. Analysis of DNA.RNA hybrids

The typical molecule in the first analysis is circular and has
two loops (Fig. 3). The small loop originates from the single-stranded DNA that lies between the body and the leader of the 16S mRNA. The
length of this loop corresponds to the distance between the 5' end of
the message and the 3' end of the leader. The large loop is partly
RNA.DNA duplex corresponding to the leader and the body of the 16S
mRNA, and partly single-stranded DNA corresponding to those DNA sequences that do not code for the 16S mRNA. Statistical analysis
performed on these molecules indicated that the small and large loops
represent about 18% and 82% of unit-length SV40 DNA, respectively.
The 5' end of the coding sequences of 16S RNA appears at about 0.94 on
the map; therefore, the 3' end of the leader should be 0.18 map units
from this site, at about 0.76. Measurements of the duplex regions
near the small loop suggested that the length of the leader corresponds to approximately 0.04 map units (about 200 nucleotides). This
confirmed the results obtained by the biochemical methods.

B. Analysis of R-loop Structures

R-loop structures were obtained by annealing double-stranded
linear SV40 DNA (generated by cleavage of SV40 form-I DNA with Bgl-I
restriction endonuclease, which cleaves at position 0.67) with poly-
(A)-containing SV40 RNA, purified from the cytoplasm of infected cells
late after infection. Two main groups of R loops were observed: one
representing 75% of the hybrid molecules, whose fractional length was
0.214 ± 0.010 (each value represents the mean ± S.D.), and the second
represents 15% of the hybrid molecules, whose fractional length was
0.399 ± 0.010. The remainder of the R-loops seemed to include a minor

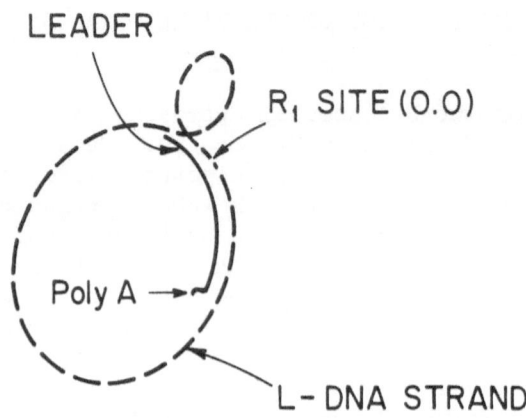

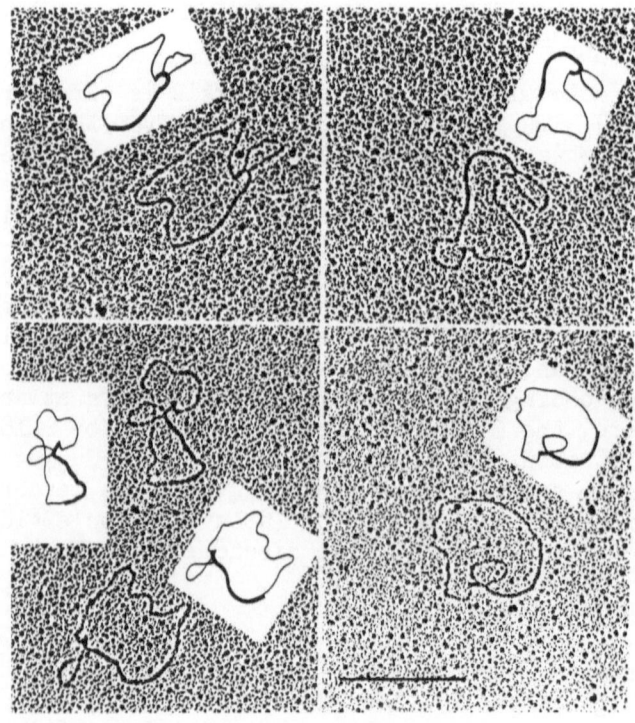

Fig. 3. Visualization of hybrids between L-DNA strands and 16S SV40
mRNA. In the schematic illustration the dashed line
represents the L-DNA strand, the solid line the 16S mRNA.
In the electromicrographs are some molecules scored with
their tracings. The bar represents 0.5 μm.

but specific size of loop. Figure 4A shows the appearance of the most abundant R-loop structures.

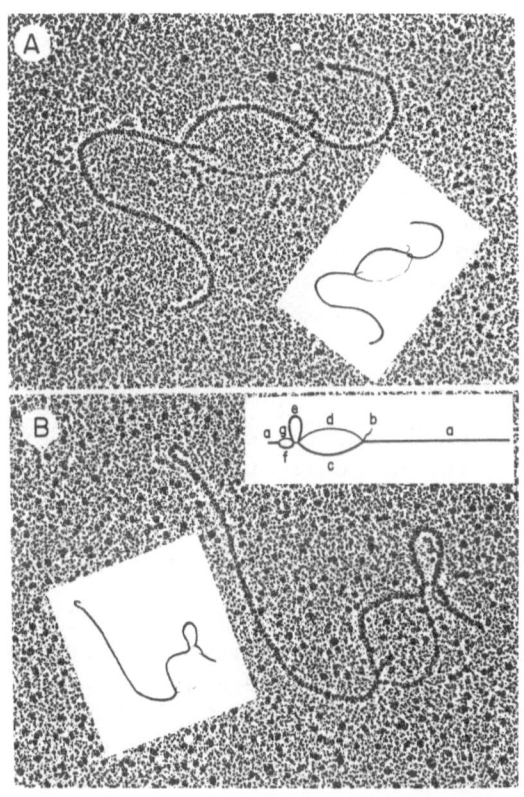

Fig. 4.   Electron micrographs of R-loop structures composed of linear
          SV40 DNA (Form I DNA cleaved at position 0.67 with Bgl I
          restriction endonuclease) and SV40 16S mRNA. In A, the
          leader is free; in B, the leader formed a small R-loop. The
          scheme in B shows; a) double-stranded DNA (0.50); b) free
          tail of poly(A) (not measured); c) a heteroduplex DNA.RNA
          that includes the body of the mRNA (0.22); d) displaced
          single-stranded DNA (not measured); e) double-stranded DNA
          intervening between the body and the leader of the mRNAs
          (see Fig. 1) (0.18); f) a heteroduplex DNA.RNA that includes
          the leader of the mRNA (0.04); g) displaced single-stranded
          DNA (not measured); a) double-stranded DNA (0.03). The
          figures in parentheses were obtained by measuring each seg-
          ment of the molecule as the fractional length of the whole
          DNA. The letters in the physical map of SV40 DNA are read
          in a counterclockwise direction from coordinate 0.67.

We computed the location of the termini of the R-loops with
referent to the Bgl-I cleavage site at 0.67 and found that the short
R-loops (Fig. 4A) span in a clockwise direction between 0.950 ± 0.007
and 0.168 ± 0.007 and the long R-loops (Fig. 5A) span in a clockwise
direction between 0.770 ± 0.009 and 0.167 ± 0.010. Based on previous
results, we concluded that the R-loops in Figures 4 and 5A are of the
16S and 19S late mRNAs, respectively. Furthermore, the forks proximal
to the short and long segments of DNA represent the map locations of
the 5' and 3' ends of the bodies of these viral mRNAs, respectively.
We suggested that the tails at the 3' ends represent non-hybridized
poly(A) sequences, and that the tails at the 5' ends represent the
leader sequences that are not coded immediately adjacent to the bodies
of the messages, and that are transcribed from an upstream segment in
the viral DNA. We found that the frequencies of recognizable tails at
the 5' and 3' ends were similar, which indicated that all or almost
all of the poly(A)-containing 16S and 19S late mRNAs have leader sequences at their 5' ends. We measured the lenghts of the leaders in
about 100 molecules of the 16S and 19S mRNAs and found them to be
0.043 ± 0.012 and 0.035 ± 0.013 map units respectively (23).

If the tails at the 5' end represent the leader, then in a
certain proportion of the molecules the sequences in the leader should
hybridize to DNA sequences upstream from the locations of the bodies
of the viral mRNAs and form a second R-loop with the intervening DNA
segment looping out. Figure 4B shows an example of the molecule with
a schematic tracing.

The R-loop analysis provided additional evidence that the leader
sequences at the 5' ends of SV40 late mRNAs are not coded at a site
immediately adjacent to the main portion of the mRNAs. Moreover, the
results obtained confirmed previous locations on the physical map of
SV40 DNA for the main portions of the late mRNAs and established the
lengths of the leaders and the map locations for their 5' and 3' ends
(23).

MODELS FOR THE JOINING OF THE LEADER TO THE CODING SEQUENCES

Several models for splicing of SV40 late mRNAs could be suggested. These include 1) intermolecular ligation of RNA, 2) deletion of
intervening DNA sequences, 3) looping out of intervening DNA sequences
so that the RNA polymerase could skip over short distances, and 4)
deletion of the appropriate intervening RNA sequences.

While analyzing the transcriptional complexes of SV40, we made
the following observations. First, as described above, there is one
major initiation site(s) for late transcription, which maps in a
fragment of the genome spanning from 0.67 to 0.76 on the map. There
was no initiation site where the 5' end of the body of 16S RNA maps
(42). This observation excluded the first possibility, which predicted *two* major initiation sites. The second finding was that the

template for transcription was mainly the supercoiled form of SV40 DNA (form I SV40 DNA) of unit length (42, 38). This observation excluded the second alternative. We were left, therefore, with the two most attractive alternatives, in which we had to distinguish between

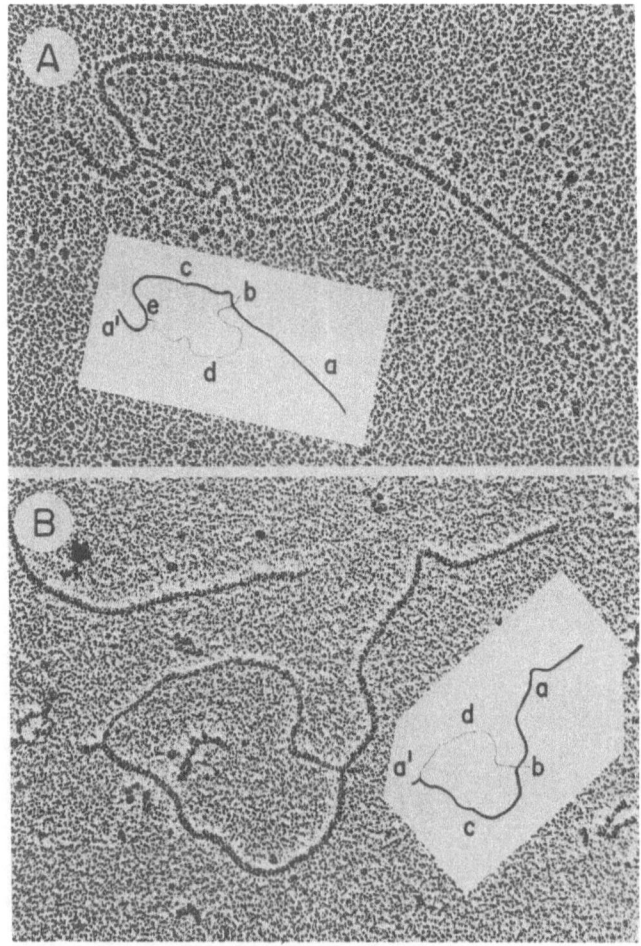

Fig. 5. Electron micrographs and their tracings of R-loops composed of linear SV40 DNA (Form I DNA cleaved at position 0.67 map units with Bgl I restriction endonuclease) and cytoplasmic 19S mRNA in A or nuclear, poly(A)-containing RNA in B. In the tracing the letters represent: a) double-stranded DNA; b) free tail of poly(A); c) a heteroduplex DNA.RNA that includes the body of the RNAs; d) displaced single-stranded DNA; e) free tail of a "leader"; a') double-stranded DNA.

whether splicing of the viral RNA occurred during transcription, or
at the post-transcriptional level, namely on the primary RNA transcript.

For this we performed R-loop analysis on nuclear viral RNA (24).
It was shown that the nuclear poly(A)-containing viral RNA accumulates
in the nucleus of the infected cell, sedimenting in sucrose gradient
as a 19S component (66).

R-loops were obtained by annealing nuclear or cytoplasmic 19S
poly(A)-containing RNA with double-stranded linear SV40 obtained by
cleavage of form-I DNA with Bgl I or Taq I restriction endonucleases.
(Taq I cleave form-I DNA at 0.58). Fig. 5 shows the appearance of
representative R-loop structures with their schematic tracings. There
are two striking differences in the appearance of the R-loops: 1) the
5' ends of the R-loops pertaining to the *nuclear* RNAs are closer to
the restriction enzyme cleavage sites than are the 5' ends of those
pertaining to the *cytoplasmic* RNAs; 2) free leaders are often seen at
the 5' ends of the R-loops pertaining to the cytoplasmic 19S RNA (in
about 70% of the molecules), whereas they are observed much less
frequently in the R-loops pertaining to the nuclear 19S viral RNA (in
about 10% of the molecules).

In an analysis of the map positions at the 5' ends of nuclear and
cytoplasmic 19S poly(A)-containing RNAs, the position of the 5'
termini of the *cytoplasmic* RNA species was represented by a resonably
sharp histogram with a mean at coordinate 0.77 ± 0.01 on the map. The
histogram for the location of the 5' termini of the *nuclear* viral RNAs
was skewed towards an upstream position and was more heterogeneous
(24).

The ability of the precursor nuclear poly(A)-containing viral RNA
to anneal to the entire length of the late portion of SV40 DNA (~ 0.67
to 0.17) including the corresponding interruptions between the leaders
and the bodies of the cytoplasmic mRNAs indicates that *splicing is a
post-transcriptional process that occurs following polyadenylation.*
Splicing of the precursor poly(A)-containing viral RNA may occur in
the nucleus before the RNA molecule is transported to the cytoplasm.
It is possible, however, that splicing occurs during the transport
from the nucleus to the cytoplasm, or immediately after the molecule
has reached the cytoplasm. In any case, splicing appears to be a
final exert in the process of maturation of mRNAs.

MODELS FOR SPLICING OF mRNA

At the moment, a mechanism that allows the accurate removal of
intervening sequences from mRNA precursor and the precise ligation
of the remaining sequences is unknown, and the ideas are largely
speculative. Though, an enzymic activity has been purified and
characterized (67), that is able to perform splicing of tRNA from

yeast. Nothing similar, however, has been so far identified for mRNAs.

Two main working hypotheses for splicing of pre-mRNA have been suggested. One is that the precursor RNA molecule itself contains the sequences and structure recognized by the enzymes involved in the splicing reaction. The second suggests that additional RNA molecules, either independently or as a part of the splicing enzymes, provide this specificity. In either case, it is possible that the proteins attached to pre-mRNA and/or to the RNA involved in this process also contribute to the enzymatic specificity.

In order to test the first possibility, several investigators have sequenced the boundaries between *introns* and *exons* (32, 38, 68, 69, see also the preceding chapter). High conservation of sequences spanning the splicing sites in RNA species from evolutionary distant organisms points both to a common general mechanism for processing, and to the possibility that, in a given organism, processing may be catalyzed by a single enzyme or enzyme complex.

All models for splicing predict that the ends of an intervening sequence must be brought close to each other *prior* to endonucleolytic cleavage, so that ligation can take place without physical separation of the 3' and 5' ends of the functional sequences that are to be spliced.

From sequencing these ends, the consensus sequence emerged:

$$5'\ldots\; {}^{A}_{C}\;AG\; \big|\; GUAAGU\ldots UYUYYYUXCAG\; \big|\; G\ldots 3'$$

$$\text{exon}\quad\big|\quad\text{intron}\qquad\qquad\text{intron}\quad\big|\quad\text{exon}$$

Y indicates pyrimidines. X marks the location of a single non-conserved base in the 3' consensus sequence. Vertical lines show the most likely location of the splice. In the position marked ${}^{A}_{C}$, A occurs 11 times and C 10 times in the 26 unique sequences examined (69). The dinucleotides G U --- A G were found at the 5' and 3' ends of the introns respectively in almost all poly(A)-containing RNAs studies. As sequences related to this prototype sequence occur elsewhere in exons, it is unlikely to be a sufficient signal to account for splicing specificity. The intervening sequences are, however, variable in length, and possess no marked homologies or symmetries of sequence. Attempts to draw the ends of an intervening sequence together by maximizing Watson-Crick base pairing, do not generate structures in which donor and acceptor sites would exhibit a reasonably constant configuration. Moreover, the signals for splicing seem to be restricted to a limited area around the junction between the

intron and exon. In SV40 early mRNAs, the maximum number of nucleotides required for splicing within the coding region is 11 nucleotides
on the 5'end of the G U donor site of small-t antigen mRNA precursor.
The minimum number of nucleotides within the intron of large-T antigen mRNA precursor is not more than 12 nucleotides beyond the donor
and the acceptor sites. Comparison of the nucleotide sequences of the
intron in the vicinity of the large-T and small-t antigen donor sites,
showed that there is no extensive homology. Yet both of these regions
are required for correct splicing (38, 70, 71). An alternative explanation is that the nucleotides flanking the splice site must be unpaired in order to provide a recognition mechanism through base
pairing with sequences remote from the splicing region, or with an
independent RNA (69, 72).

    Two main models have been suggested. The first predicts that the
information is built into the molecule itself, while the second
predicts the involvement of an additional small RNA molecule. It was
pointed out (72) that while the primary structure of the intervening
sequences are quite variable, they are rich in adenine and uridine.
The adenosine and uridine residues frequently occur as homopolymeric
runs of three to seven bases. On the basis of this observation and
data from studies on tertiary structures of tRNA and synthetic poly-
ribonucleotides, a model was suggested (72) in which the donor and
acceptor splicing sites are brought into juxtaposition by the formation of a triple-stranded structure between the poly(A) of the mRNA
precursor and regions from both the 5' and 3' ends of an intervening
sequence. Support for this model came from the observation that polyadenylylation precedes splicing (24, 28, 73), and that the histone
genes that were examined did not contain intervening sequences (74);
also, histone mRNAs are not polyadenylylated (75, 76). An exception
to this rule, interferon mRNAs does contain poly(A), though it is not
spliced.

    An independent class of small RNA that could be involved in RNA
splicing is the small RNAs found in the nuclei of eukaryotic cells
(77-81). Although several of these RNAs have been sequenced, their
function is so far unknown. They exist in cells as ribonucleoprotein
particles. One complex contains seven polypeptides with molecular
weights ranging from 12,000 and 35,000 (81). These complexes have
been highly conserved during evolution. The 5' terminal sequence in
one major small nuclear RNA shows extensive complementarity to the
consensus sequence (69). As can be seen in Fig. 7 of the preceding
chapter, nucleotides 3 to 8 from one small RNA known as U1 exactly
match that part of the consensus sequence found at the 5' end of the
intervening sequence, while nucleotides 9 to 11 exactly match the 3'
end. There are further possibilities for base pairing. Thus it has
been proposed that the 5'-terminal sequence of U1 RNA interacts with
the terminal sequences of the intervening regions and serves a template function allowing the correct alignement of the sequences to
be spliced. At the time this review is written, it is felt that,

whatever may be the final outcome of these predictions, they will provide an experimental framework for the study of the mechanism of RNA splicing in eukaryotic cells.

Recent studies with yeast mitochondrial RNAs point out that the normal sequence of nucleotides in an intron and its flanking sites is necessary but insufficient for the correct splice to occur. Other sequences quite remote from it are also important. The onset and progression of splicing most probably involves a temporal succession of specific three-dimensional structures that can be recognized by the enzymic machinery (82).

SPLICING INTERMEDIATES

Individual intron transcripts might be removed from a precursor RNA molecule as a one step process, or in a discrete number of steps. There is evidence that individual intron transcripts are not necessarily removed in one step in the processing of Adenovirus 2 (83), globin (84) and ovalbumin nuclear RNAs (85). If the GU-AG rule also applies to splicing for stepwise removal of a given intron transcript, it is clear that the sequence AGGT must be present in the transcript at the intermediate splicing point. Chambon and his colleagues (85) have shown that such a tetranucleotide was indeed found in introns A, C and E, and was absent in introns B and D of the ovalbumin gene. However, no evidence was provided to show that introns A, C and E are removed stepwise, while introns B and D are removed in one step.

CONCLUSIONS

It is evident that much more work will need to be carried out in order to comprehend the complex reactions involved in the proper regulation of viral and eukaryotic gene expression. The detailed studies presented here are concerned only with the overall organization of viral mRNA formation and regulation. This structuralistic analysis may provide guidelines to future studies in this exciting and fast-developing field.

ACKNOWLEDGMENTS

The author wishes to give special thanks to Drs. S. Bratosin, O. Laub, M. Horowitz and E.B. Jakobovits, among other colleagues, who made this contribution possible. Most of the author's work described in this review was supported by U.S. Public Health Service Research Grant CA 14995.

REFERENCES

1.   Darnell, J.E., Jelinek, W.R. and Molloy, G.R. (1973), Science, 181, 1215-1221.
2.   Weinberg, R.A. (1973), Ann. Rev. Biochem. 42, 329-354.
3.   Davidson, E.H. and Britten, R.J. (1973), Q. Rev. Biol. 48, 565-613.
4.   Brawerman, G. (1974), Ann. Rev. Biochem., 43, 621-642.
5.   Greenberg, J.R. (1975), J. Cell Biol. 64, 269-288.
6.   Lewin, B. (1975), Cell 4, 11-20.
7.   Lewin, B. (1975), Cell 4, 77-93.
8.   Perry, R.P. (1976), Ann. Rev. Biochem. 45, 605-629.
9.   Molloy, G. and Puckett, L. (1976), Prog. Biophys. Mol. Biol. 31, 1-38.
10. Chan, L., Harris, S.E., Rosen, J.M., Means, A.R. and O'Malley, B.W. (1977), Life Sci. 20, 1-16.
11. Revel, M. and Groner, Y. (1978), Ann. Rev. Biochem. 47, 1079-1126.
12. Darnell, J.E. (1979), Prog. Nuc. Ac. Res. and Mol. Biol. 22, 327-353.
13. Berget, S.M., Moore, C. and Sharp, P.A. (1977), Proc. Natl. Acad. Sci. U.S.A. 74, 3171-3175.
14. Gelinas, R.E. and Roberts, R.J. (1977), Cell 11, 533-544.
15. Chow, L., Gelinas, R.E., Broker, T. and Roberts, R.J. (1977), Cell 12, 1-8.
16. Klessig, D.F. (1977), Cell 12, 9-21.
17. Aloni, Y., Bratosin, S., Dhar, R., Laub, O., Horowitz, M. and Khoury, G. (1977), Proc. Natl. Acad. Sci. U.S.A. 74, 3686-3690.
18. Aloni, Y., Bratosin, S., Dhar, R., Laub, O., Horowitz, M. and Khoury, G. (1978), Cold Spring Harbor Symp. Quant. Biol. 43, 559-570.
19. Gelma, M., Dhar, R., Pan, J. and Weissmann, S. (1977), Nuc. Ac. Res. 4, 2549-2559.
20. Lavi, S. and Groner, Y. (1977), Proc. Natl. Acad. Sci. U.S.A. 74, 5323-5327.
21. Hsu. M.T. and Ford. S. (1979) Proc. Natl. Acad. Sci. U.S.A. 74 4982-4985.
22. Haegeman, G. and Fiers, W. (1978), Nature, London, 272, 70-73.
23. Bratosin, S., Horowitz, M., Laub, O. and Aloni, Y. (1978); Cell 13, 785-790.   .
24. Horowitz, M., Laub, O., Bratosin, S. and Aloni, Y. (1978), Nature London, 275, 558-559.
25. Horowitz, M., Bratosin, S. and Aloni, Y. (1978), Nuc. Ac. Res. 5, 4663-4675.
26. Rothenberg, E., Donoghue, D.J. and Baltimore, D. (1978), Cell 13, 435-452.
27. Panet, A., Gorecki, M., Bratosin, S. and Aloni, Y. (1978), Nuc. Ac. Res. 5, 3219-3230.
28. Tal, J., Ron, D., Tattersall, P., Bratosin, S. and Aloni, Y. (1979), Nature, London, 279, 649-651.

29. Jeffreys, A.J. and Flavell, R.A. (1977), Cell 12, 1097-1108.
30. Tilghman, S.M., Tiemeier, D.C., Seidman, J.G., Peterlin, B.M., Sullivan, M., Maizel, J.V. and Leder, P. (1978), Proc. Natl. Acad. Sci. U.S.A. 75, 725-729.
31. Brack, C. and Tonegawa, S. (1977), Proc. Natl. Acad. Sci. U.S.A. 74, 5652-5656.
32. Tonegawa, S., Maxam, A.M., Tizard, R., Bernhard, O. and Gilbert, W. (1978), Proc. Natl. Acad. Sci. U.S.A. 75, 1485-1490.
33. Gilbert, W. (1978), Nature, London, 271, 501.
34. Crick, F. (1979), Science 204, 264-271.
35. Flint, S.J. (1979), American Scientist 67, 300-311.
36. Kelly, T.J. and Nathans, D. (1977), Advances in Virus Research (Academic Press Inc. New York) 21, 85-173.
37. Acheson, N.H. (1976), Cell 8, 1-12.
38. Lebowitz, P. and Weissmann, S.M. (1979), Current Topics in Microbiology and Immunology, 87, 43-172.
39. Fareed, G. and Davoli, D. (1977), Ann. Rv. Biochem., 46, 471-552.
40. Khoury, G., Howley, P., Nathans, D. and Martin, M.A. (1975), J. Virol. 15, 433-437.
41. Ferdinand, F.J., Brown, M. and Khoury, G. (1977), Virology 78, 150-161.
42. Laub, O., Bratosin, S., Horowtiz, M. and Aloni, Y. (1979), Virology 92, 310-323.
43. Dintzis, H.M. (1961), Proc. Natl. Acad. Sci. U.S.A. 47, 247-261.
44. Canaani, D., Kahana, C., Mukamel, A. and Groner, Y. (1979), Proc. Natl. Acad. Sci. U.S.A. 76, 3078-3082.
45. Aloni, Y., Dhar, R. and Khoury, G. (1979), J. Virol. 32, 52-60.
46. Griffith, J.D. (1975), Science 187, 1202-1203.
47. White, M. and Eason, R. (1971), J. Virol. 8, 363-371.
48. Green, M.H., Miller, H.J. and Hendler, S. (1971), Proc. Natl. Acad. Sci. U.S.A. 68, 1032-1036.
49. Jakobovits, E.B. and Aloni, Y. (1980), Virology 102, 107-118.
50. Felsenfeld, G. (1978), Nature, London, 271, 115-122.
51. Korenberg, R.D. (1977), An. Rev. Biochem. 46, 931-954.
52. Chambon, P. (1977), Cold Spring Harbor Symp. Quant. Biol. 42, 1209-1234.
53. Cremisi, C., Pignatti, P.F., Crossant, O. and Yaniv, M. (1976), J. Virol. 17, 204-211.
54. Cremisi, C., Pignatti, P.F. and Yaniv, M. (1976), Biochem. Biophys. Res. Commun. 73, 548-554.
55. Ponder, B.A.J. and Crawford, L.V. (1977), Cell 11, 35-49.
56. Polisky, B. and McCarthy, B. (1975), Proc. Natl. Acad. Sci. U.S.A. 72, 2895-2899.
57. Varshavsky, A.J., Sundin, O.H. and Bohn, M.J. (1978), Nuc. Ac. Res. 5, 3469-3478.
58. Scott, W.A. and Wigmore, D.J. (1978), Cell 15, 1511-1518.
59. Sundin, O. and Warshavsky, P.J. (1979), J. Mol. Biol. 132, 535-546.
60. Waldeck, W., Fohring, B., Chowdhury, K., Gruss, P. and Sauer, G. (1978), Proc. Natl. Acad. Sci. U.S.A. 75, 5964-5968.

61. Jakobovits, E.B., Bratosin, S. and Aloni, Y. (1980), Nature,
     London, 285, 263-265.
62. Jakobovits, E.B., Sargosti, S., Yaniv, M. and Aloni, Y. (1980),
     Proc. Natl. Acad. Sci. U.S.A. in press.
63. Laub, O., Jakobovits, E.B. and Aloni, Y. (1980), Proc. Natl.
     Acad. Sci. U.S.A. in press.
64. Dhar, R., Zain, B.S., Weissmann, S., Pan, J. and Subramanian, K.
     N. (1974), Proc. Natl. Acad. Sci. U.S.A. 71, 371-375.
65. Thomas, M., White, R.L. and Davis, R.W. (1976), Proc. Natl. Acad.
     Sci. U.S.A. 73, 2069-2073.
66. Aloni, Y. (1974), Cold Spring Harbor Symp. Quant. Biol. 39,
     165-178.
67. Abelson, J. (1979), Ann. Rev. Biochem. 48, 1035-1069.
68. Breathnach, R., Benoist, C., O'Hare, K., Cannon, F. and Chambon,
     P. (1978), Proc. Natl. Acad. Sci. U.S.A. 75, 4853-4857.
69. Lerner, M.R., Boyle, J.A., Mount, S.M., Wolin, S.L. and Steitz,
     J.A. (1980), Nature, London, 283, 220-224.
70. Volckaert, G., Feunteun, J., Crawford, L.V., Berg, P. and Fiers,
     W. (1979), J. Virol. 30, 674-682.
71. Thimmappaya, B. and Shenk, T. (1979), J. Virol. 30, 668-673.
72. Bina, M., Feldman, R.J. and Deeley, R.G. (1980), Proc. Natl.
     Acad. Sci. U.S.A. 77, 1278-1282.
73. Lai, C-J., Dhar, R. and Khoury, G. (1978), Cell 14, 971-982.
74. Kedes, L.H. (1979), Ann. Rev. Biochem. 48, 837-870.
75. Adesnick, M. and Darnell, J. (1972), J. Mol. Biol. 64, 397-406.
76. Kedes, L.H. (1976), Cell 8, 321-331.
77. Weinberg, R. and Penman, S. (1968), J. Mol. Biol. 38, 289-304.
78. Zieve, G. and Penman, S. (1976), Cell 8, 19-31.
79. Jelinek, W. and Leinwand, L. (L978), Cell 15, 205-214.
80. Reddy, R., Ro-Choi, R.S., Henning, D. and Bush, H. (1974), J.
     Biol. Chem. 249, 6486-6494.
81. Lerner, M.R. and Steitz, J.A. (1979), Proc. Natl. Acad. Sci.
     U.S.A. 76, 5495-5499;
82. Halbreich, A., Pajot, P., Foucher, M., Grandchamp, C. and
     Slonimski, P. (1980), Cell 19, 321-329.
83. Chow, L. and Broker, T. (1978), Cell 15, 497-510.
84. Kinniburgh, A. and Ross, J. (1979), Cell 17, 915-921.
85. Benoist, C., O'Hare, K., Breathnach, R. and Chambon, P. (1980),
     Nuc. Ac. Res. 8, 127-142.
86. Weil, P.A., Luse, D.S., Segall, J. and Roeder, R.G. (1979),
     Cell 18, 469-484.
87. Manley, J.L., Fire, A., Cano, A., Sharp, P.A. and Gefter, M.L.
     (1980), Proc. Natl. Acad. Sci. U.S.A. 77, 3855-3859.
88. Luse, D.S. and Roder R.G. (1980), Cell 20, 691-699.
89. Wasylyk, B., Kendinger, C., Corden, J., Brinson, O. and Chambon,
     P. (1980), Nature, London, 285, 367-373.
90. Mathis, D. and Chambon, P. (1981), Nature, London, in press.
91. Grosscheal, R. and Birnsteil, M.L. (1980), Proc. Natl. Acad.
     Sci. U.S.A. 77, 1432-1436.
92. Benoist, C., and Chambon, P. (1981), Nature, London, in press.

MESSENGER RIBONUCLEOPROTEIN PARTICLES

John W.B. Hershey

Department of Biological Chemistry, School of Medicine
University of California
Davis, CA 95616 (U.S.A.)

INTRODUCTION

Messenger RNA, from the time of its synthesis in the nucleus
through its translation and degradation, does not exist as free RNA
in animal cells but rather always appear to be complexed with proteins. Free cytoplasmic ribonucleoprotein (RNP) particles containing
non-ribosomal RNA were discovered by Spirin and coworkers (1) in 1964
and were called *informosomes*. Shortly thereafter, similar particles
containing hnRNA were identified in cell nuclei (2). Both kinds of
mRNP particles are heterogeneous in size and are rich in protein, with
RNA/protein mass ratios of about 1:3. mRNPs have been found in all
eukaryotic organisms examined.

We can distinguish four kinds of informational (non-ribosomal)
RNPs: 1) nuclear RNPs, which form as the RNA is synthesized off
chromatin and may be involved in the processing of hnRNA and transport
of mRNA to the cytoplasm; 2) free cytoplasmic mRNPs, which appear to
be translationally *inactive* forms of mRNAs but are potentially capable
of being translated; 3) polysomal mRNPs, which are not actually
present in cells but are artificially derived from polysomes by treatment with EDTA or puromycin; and 4) viral RNPs, which are observed in
RNA virus-infected cells as intermediates in the assembly of new
virions. This brief review focuses primarily on the structure and
function of the cytoplasmic free and polysomal mRNPs, and their roles
in protein synthesis. Nuclear and viral RNPs will not be described
further. The review is neither comprehensive nor critical, but rather
attempts to identify the major conceptual problems and to illustrate
experimental approaches with selected examples from the literature.

BIOLOGICAL PROPERTIES

Before describing the isolation and detailed composition of free
mRNP particles, it is appropriate to discuss what is known about their
physiological role in the cell. Free mRNPs have been observed in a
large number of eukaryotic cells, and extensive studies have been
carried out with a wide variety of cell types. We shall examine some
aspects of a number of specific cell types from three general areas:
1) sea urchins during early development; 2) differentiating mammalian
cells; and 3) non-differentiating or terminally differentiated animal
cells.

Early Development in Sea Urchins

Several minutes after fertilization of the sea urchin egg there
is a dramatic increase in the rate of protein synthesis. By one hour
after fertilization the rate of protein synthesis is 15-30 times
greater than that of the unfertilized egg (3). High levels of ribosomes exist in unfertilized eggs, but less than 1% of the ribosomes
are present in polysomes. The percent of ribosomes in polysomes
increases within 30 minutes after fertilization to about 10% and
steadily increases throughout early development, reaching 20% at 2
hours and 60% at 15 hrs (4). The increased rate of protein synthesis,
which parallels the recruitment of ribosomes into polysomes, does not
depend on *de novo* RNA synthesis, but rather is dependent on the
utilization of maternal mRNA laid down during oogenesis. This has
been shown by monitoring protein synthesis in the absence of RNA
synthesis in enucleated eggs and in Actinomycin D-treated embryos
(5). The maternal mRNA in these cells is found in free mRNP particles
(inactive for protein synthesis) and in polysomes (active). Thus,
potentially functional translational machinery and mRNA are present in
unfertilized eggs and yet protein synthesis is suppressed.

The increase in the rate of protein synthesis could occur by one
or more of the following mechanisms: an increase in the initiation
rate, leading to a greater number of ribosomes per mRNA; an increase
in the elongation rate, i.e., a reduction in the ribosome transit time;
and an increase in the total number of polysomes, i.e. in the number of
mRNAs, being translated. Humphreys (4, 6) studied sea urchin eggs
before and after fertilization and found that neither average polysome
size nor average transit time changes significantly, whereas the
proportion of ribosomes in polysomes increases 30-fold, 2 hrs after
fertilization. These results suggest that mRNA is limiting in unfertilized eggs and that the subsequent increase in protein synthesis
is due to the mobilization of previously inactive mRNA into new polysomes. Others have found about a 2-fold increase in the elongation
rate following fertilization (7-9). Nevertheless, activation of
stored maternal mRNP appears to be the major contributor to the
increase in the rate of protein synthesis.

The apparent limitation of active mRNA in unfertilized eggs is
not due to the absence of translatable RNA: deproteinized free mRNP
from unfertilized eggs is fully translatable in a cell-free wheat germ
system (10). It is postulated that components in the RNPs *mask* the
mRNA and make it unavailable for translation. The sequences found in
the mRNP compartment are similar to those being translated in poly
somes. Brandhorst (11) analyzed sea urchin proteins synthesized
before and after fertilization by the two-dimensional gel electrophoresis technique of O'Farrell and found rather similar patterns
which differed primarily in overall intensity. Thus activation of
protein synthesis early in sea urchin development is due primarily to
a quantitative change in the proportion of all or most mRNAs in active
polysomes vs. free mRNPs. A similar mobilization of mRNAs from mRNPs
into polysōmes has been observed during early development in a large
variety of other biological systems including mammals, *Xenopus laevis*,
Artemia salina and plant seeds.

Differentiating Animal Cells

The myogenic system has been studied intensively as an example
of differentiating cells. Mononucleated myoblasts, which are committed to the formation of striated muscle, proliferate and then fuse
with other myoblasts to form multinucleated fibers or myotubes. The
fused cells no longer synthesize DNA or undergo nuclear division, but
instead begin the synthesis of myofibrillar proteins and the subsequent assembly of myofibrils (12). The system has many attractive
features. A number of well-characterized myofibrillar proteins are
expressed, providing a desirable level of complexity. The proteins
become abundant in the myotubes and their mRNAs are in high concentration and can be purified (13, 14). Myosin heavy chain mRNA, due to
its large size (26-28S), can be monitored directly in crude mixtures.
Recently it has been possible to clone sequences from the genes of
these proteins and thereby generate hybridization probes to measure
the levels of specific nuclear and cytoplasmic RNAs. In addition,
cell lines have been isolated which represent different stages of
muscle development and which will undergo relatively synchronous
differentiation into myotubes (15). The myogenic system is therefore highly suited for a study of how a large number of genes is
coordinately expressed at a precise developmental time.

Some early work has suggested that myosin heavy chain mRNA may
be synthesized and stored in the cytoplasm as free mRNP before it is
translated (16, 17, 18). The metabolic stability of the myosin RNA
(in mRNP) prior to myoblast cell fusion is reported to be low, but to
increase following fusion and the mobilization of the mRNA into
polysomes (17).   Free mRNPs containing principally myosin heavy
chain and actin mRNAs have been isolated from embryonic chick
muscle (19, 20).   The mRNA in these mRNPs is identical to the
corresponding mRNAs in polysomes with respect to size, sequences and
translability in cell-free protein synthesis systems. It is postu-

lated that myosin mRNAs in fused myoblasts initially may be stored as mRNPs, i.e., they are *masked* or inactive. Subsequently, the mRNPs are thought to be unmasked and the myosin mRNAs mobilized into polysomes. This view is not universally accepted, however. As in the case of sea urchin development, the problem is to demonstrate rigorously for specific types of mRNA a precursor-product relationship, and to elu⎯ cidate the molecular basis for masking and unmasking of free mRNPs. It is expected that the use of cloned probes for specific mRNAs will allow the precise determination of RNA levels in the different cyto- plasmic compartments as a function of differentiation and will help answer, unambiguously, the question whether there is a significant delay between the transport of a mRNA into the cytoplasm and its entry into polysomes.

Non-Differentiating or Terminally Differentiated Mammalian Cells

A large proportion of the cytoplasmic non-ribosomal RNA of mam- malian cells *in vivo* or cultured *in vitro* is found as free mRNP. From 20 to 60% of the mRNA mass is mRNP, depending on the physiological state of the cells. The role of these mRNP particles is not known, but the storage model used to explain developmental systems seems less appropriately applied to cells growing under steady-state conditions. Spohr, Scherrer and coworkers (21, 22) studied the kinetics of label- ing of HeLa cell mRNAs and found that both polysomal and RNP compart- ments become labeled, but that most of the label in RNPs cannot be chased into polysomes. Presumably the bulk of the mRNP is not simply mRNA in transit from nucleus to polysome. The fact that HeLa RNA in the mRNP compartment turns over more rapidly than the mRNA in polysome (half lives of 2 h and greater than 8 h, respectively), argues that the two mRNA compartments are distinct.

Martin and coworkers (23, 24) studied the poly(A^+) mRNAs of mouse Taper ascites cells, in which 25% of the RNA is found as free mRNPs. They characterized the mRNA sequences from polysomes and RNP particles by measuring their complexities by mRNA-cDNA hybridization. Half of the mass of the mRNAs in RNPs consists of about 15 species, the other half of about 400 different sequences. The polysomal poly(A^+) mRNA consists of three abundance classes of 25, 500 and 8,500 different mRNA sequences. From analysis of *in vitro* translational products by two-dimensional gel electrophoresis, they showed that most of the mRNP sequences are a subset of the polysomal sequences. Thus the high and middle abundant species are found in both polysomes and RNPs, but the highly complex, low abundant species may be present only in poly- somes.

Avian erythroblasts are a particularly good system to study because these cells synthesize primarily globin proteins and yet retain their nuclei throughout their lifetime. Scherrer and coworkers (25) have shown that duck erythroblast polysomes are composed of approximately 240 different messengers, whereas the mRNP population in contrast to the ascites cells above is of greater complexity, containing 2000 species. The globin mRNAs are primarily in the polysome compartment although from 10 to 40% of these mRNAs are in free RNPs, depending on the stage of erythroblast maturity. Rabbit reticulocytes, which lack a nucleus, also contain mRNPs. The dominant globin mRNA in the RNP fraction is α-globin mRNA, suggesting that weaker initiating mRNA species may be enriched in RNPs relative to polysomes (26).

The quantitative distribution of mRNA molecules between RNPs and polysomes changes when the rate of protein synthesis is altered. A shift of mRNA species from polysomes to RNPs or vice versa has been observed in a number of cases. The following are a few examples of such phenomena. The number of polysomes decreases and the RNP pool increases when mouse sarcoma 180 cells are starved for amino acids; the process is reversed upon feeding (27). A similar reversible shift is seen when cultured mouse fibroblasts pass from exponential to stationary phase (28). Shafritz and coworkers (29) used cDNA probes to monitor albumin mRNA levels in rat liver; upon feeding, all of the albumin mRNA was found in polysomes, but when the rats were subjected to a short fast, 60% of the mRNA appeared in the mRNP compartment. The presence or absence of iron affects the distribution of ferritin mRNA between polysomes and mRNP particles (30). In all of these cases, mRNA appears to be able to move from polysomes into RNPs or from RNPs into polysomes. It is not entirely clear whether a change in the distribution of mRNA between polysomes and mRNPs is the result of a change in the rate of initiation of protein synthesis, or whether the masking of mRNA itself causes the change in the rate of protein synthesis.

In summary, the studies sketched above indicate that there may be a number of different kinds of free mRNPs in the cytoplasm. 1) RNPs may be mRNAs in transit from nucleus to polysomes. This is conceptually compelling, but such RNPs appear to constitute a minor fraction of the total. 2) They may in part represent mRNAs on the way to being degraded. Again, lifetime measurements argue that only a small proportion can be of this type. 3) The RNPs could be storage forms of mRNAs, which are masked and not capable of entry into polysomes. 4) RNPs may be potentially active mRNAs which have not yet undergone initiation of protein synthesis. Most results indicate that categories 3 and 4 are the dominant ones, but methods for distinguishing unambiguously between these categories are not generally available and quantitative knowledge on this point is lacking. Also not known is how the various types of RNPs differ in structure and how they interact dynamically in the cell.

ISOLATION AND COMPOSITION

Free cytoplasmic mRNPs are usually isolated from cell extracts by differential centrifugation and sedimentation in sucrose gradients (1). Most of the mRNP particles have sedimentation coefficients of

20 to 80S and are thereby separated from the bulk of the cellular
protein. However, ribosomes and ribosomal subunits contaminate a
substantial fraction of such mRNPs. Poly(A$^+$) mRNAs may be separated
from ribosomal subunits by affinity chromatography with oligo(dT)
cellulose (31). Thermal elution from oligo(dT) cellulose columns can
be used to separate free cytoplasmic RNPs from polysomal RNPs (32).
Alternatively, density gradient centrifugation with metrizamide (33)
or sucrose gradient electrophoresis (34) also have been used successfully. The buoyant density of the free cytoplasmic mRNP particles is
1.40-1.42 g/cm^3, which corresponds to a protein content of about 75%.
The major difficulty in judging the effectiveness of an isolation
procedure is the lack of criteria for defining the *in vivo* form of
RNPs. This problem is compounded by the likelihood that there are a
variety of types of free mRNPs which themselves differ in composition
and stability. Since there are no adequate assays for RNPs, we cannot
evaluate whether the components found associated with the mRNAs are
contaminants, or whether the isolation conditions have already removed
important constituents of the particles.

The protein composition of these particles has been examined in
a variety of laboratories by one-dimensional SDS-polyacrylamide gel
electrophoresis. Free globin RNPs from duck erythrocytes contain a
number of proteins in the 20,000-30,000 dalton range and a 50,000
protein (35). A rather similar pattern is reported for rabbit reticulocytes (26). Striking is the lack of the 78,000 dalton protein
(see below). Bulk free mRNPs isolated from embryonic chick muscle
contain six major proteins with masses of 44, 52, 58, 64, 78 and 95
kilodaltons, and a number of minor components in the mass range 35-95
kilodaltons (20). A similarly complex, but apparently not identical,
array of proteins is obtained from free mRNPs from Ehrlich ascites
tumor cells (36, 37) and HeLa cells (38). Free cytoplasmic mRNPs for
specific mRNAs have been purified and their proteins characterized
(19, 20). These mRNPs also contain numerous proteins. It is difficult to evaluate whether the proteins found in the particles described above are normally associated with mRNPs *in vivo* and play a
functional role there, or whether some are contaminants in the preparations.

Polysomal mRNPs are less complex, containing fewer proteins.
These RNP particles are derived from polysomes by treatment with
either EDTA or puromycin and are separated from ribosomes by sucrose
density gradient centrifugation. Most polysomal mRNP particles
contain two major proteins of about 52 and 78 kd, and a large number
of minor components (20, 25, 26, 36, 38, 39). The 52 and 78 kd
proteins are throught to be present in one copy each per mRNA (36) and
to correspond to two of the components of the free mRNPs. However
many of the proteins in the free mRNP particles appear to leave the
mRNA as polysomes form.

The 78 kd protein binds to the poly(A) portion of the mRNA; the

52 kd protein presumably binds elsewhere. Rose et al. (40) suggest that the 78 kd protein is the poly(A) polymerase; it may also play a role in the transport of poly(A^+) mRNPs from the nucleus to the cytoplasm (41). It binds very tightly to poly(A) and is not removed in CsCl gradients (36). On the other hand, this protein is found in preparations of histone polysomal mRNPs (42); since these mRNAs lack poly(A) tails, this result suggests that the 78 kd protein may interact elsewhere on the RNA. When free poly(A^+) mRNPs are digested with T_2 RNase, a pattern of fragments is obtained that are multiples of about 27 nucleotides (43); by assaying for the ability to produce this pattern in reconstituted particles, Baer and Kornberg purified a 75 kd protein, presumably the mRNP component discussed here (personal communication). The precise role of the 78 kd protein, or any of the other components found in mRNPs, is not yet known.

TRANSLATION OF mRNPs

Essentially all of the polysomal mRNPs studied are active in cell-free systems for protein synthesis and stimulate such assays as well as fully deproteinized mRNAs. Furthermore, no difference in the dependence on eIF-3, eIF-4A or eIF-4B for translation of globin mRNA and polysomal mRNP could be detected (unpublished results). Whether the 52 and 78 kd proteins facilitate protein synthesis in any way is not known.

Since free mRNPs may not be active in protein synthesis *in vivo*, it is expected that they would not be readily translated in cell-free assays. However, many isolated free mRNPs promote protein synthesis very well. Recently there have been reports of the purification of mRNPs which are inactive for protein synthesis unless they are washed with high salt buffers (44-46). Since many of the procedures used for isolating free mRNPs utilize high salt, active particles may have lost the element responsible for their masking. Precisely what is removed by high salt has not yet been determined. The free globin mRNPs (20S) isolated from duck erythroblasts do not support translation in cell-free systems (35). When washed with 0.5 M KCl buffer, the 20S particles are converted into 19S, 16S and 13S forms which nevertheless are inactive. Thus the element which masks these mRNAs is resistant to high salt. Fully deproteinized mRNA is fully competent for protein synthesis, however.

Heywood and coworkers (47) have isolated small (2-3S) RNAs which inhibit the translation of mRNAs. These so-called translational control RNAs (tcRNAs) contain an oligo(U)-rich region and Heywood proposes that the tcRNA hybridizes to the poly(A) tail of the mRNP, thereby masking its activity for protein synthesis (48). Other laboratories have reported the presence of similar RNA molecules which affect protein synthesis (49, 50). Recently Bag and Sells (51) and Sarkar and coworkers (52) have identified a 4S RNA in embryonic chick muscle which is a potent inhibitor of protein synthesis. This RNA does not hybridize to poly(A) or poly(U), and therefore appears to be distinct from that reported by Heywood. The 4S iRNA is found free in the cytoplasm as a 10S RNP particle which is also an active inhibitor (52). It is possible that some preparations of free mRNPs contain this inhibitory RNA and thereby fail to support protein synthesis.

SUMMARY AND CONCLUSIONS

     Free mRNPs are found in essentially all eukaryotic cells. These mRNPs appear to consist of a number of different forms of mRNAs: transit; masked; or active mRNA in equilibrium with polysomes. The precise mechanisms which provide for masking and which lead to acti- vation of mRNPs are not known. Additional work is needed to prove that specific mRNAs are stored as mRNPs and then mobilized into poly- somes. We need to understand why some mRNAs are substantially or predominately in the form of mRNP and what purpose this serves. Iden- tification of the putative masking element(s) and how its presence is controlled is required. The roles played by the protein (and small RNA ?) components must be determined. Whereas much has been learned, nevertheless many of the results described in this review are contra- dictory and the interpretations may be invalid. The difficulty in answering the questions above and in defining the precise structure and function of mRNPs derives in large part from there being no ade- quate *in vitro* assay for these particles. It is therefore not sur- prising that this fascinating and important area of molecular biology remains controversial and confusing.

ACKNOWLEDGEMENT

     I thank Drs. M.E. Buckingham, M.B. Dworkin, A.A. Infante, S. Sarkar and M.W. Winkler for many helpful discussions.

REFERENCES

1.   Sprin, A.S., Belitsina, N.V., and Ajtkhozhin, M.A. (1964), Zh. Obshch. Biol. 25, 321. (English translation in Fed. Proc. 24, T907, 1965).
2.   Samarina, O.P., Krichevskaya, A.A., and Georgiev, G.P. (1966), Nature, London, 210, 1319-1322.
3.   Regier, J.C., and Kafatos, F.C. (1977),Develop. Biol. 57, 270-283
4.   Humphreys, T. (1971), Develop. Biol. 26, 201-208.
5.   Davidson, E.H. (1976), *Gene Activity in Early Development*, Academic Press, New York.
6.   Humphreys, T. (1969), Develop. Biol. 20, 435-458.
7.   Brandis, S.W., and Raff, R.A. (1978), Develop. Biol. 67, 99-113.
8.   Hille, M.B., and Albers, A.A. (1979), Nature, London, 278, 469- 471.
9.   Brandis, S.W., and Raff, R.A. (1979), Nature, London, 278, 467- 469.

10. Gross, K.W., Jacobs-Lorena, M., Baglioni, C., and Gross, P.R. (1973), Proc. Natl. Acad. Sci. U.S.A. 70, 2614-2618.

11. Brandhorst, B.P. (1976), Develop. Biol. 52, 310-317.

12. Emerson, C.P., Jr., and Beckner, S.K. (1975), J. Mol. Biol. 93, 431-447.

13. Heywood, S.M., and Nwagwu, M. (1969), Biochemistry 8, 3839-3845.

14. Mondel, H., Sutton, A., Chen, V.-J., and Sarkar, S. (1974), Biochem. Biophys. Res. Commun. 56, 988-996.

15. Martin, G.R. (1975), Cell 5, 229-243.

16. Heywood, S.M., and Rich, A. (1968), Proc. Natl. Acad. Sci. U.S.A. 59, 590-597.

17. Buckingham, M.E., Caput, D., Cohen, A., Whalen, R.G., and Gros, F. (1974), Proc. Natl. Acad. Sci. U.S.A. 71, 1466-1470.

18. Robbins, J., and Heywood, S.M. (1978), Eur. J. Biochem. 82, 601-608.

19. Bag, J., and Sarkar, S. (1975), Biochemistry 14, 3800-3807.

20. Jain, S.K., and Sarkar, S. (1979), Biochemistry 18, 745-753.

21. Spohr, G., Granboulan, N., Morel, C., and Scherrer, K. (1970), Eur. J. Biochem. 17, 296-318.

22. Mauron, A., and Spohr, G. (1978), Eur.J. Biochem. 82, 619-625.

23. McMullen, M.D., Shaw, P.H., and Martin, T.E. (1979), J. Mol. Biol. 132, 679-694.

24. Kinneburgh, A.J., McMullen, M.D., and Martin, T.E. (1979), J. Mol. Biol. 132, 695-708.

25. Maudrell, K., Maxwell, E.S., Civelli, O., Vincent, A., Goldenberg S., Buri, J.-F., Imaigumi-Scherrer, M.-T., and Scherrer, K. (1979), Molec. Biol. Reports 5, 43-51.

26. Princen, H.M.G., van Eekelen, C.A.G., Asselbergs, F.A.M., and van Venrooij, W.J. (1979), Molec. Biol. Reports 5, 59-64.

27. Sonensheim, G.E., and Brawerman, G. (1977), Eur. J. Biochem. 73, 307-312.

28. Lee, G.T-Y., and Engelhardt, D.L. (1978), J. Cell. Biol. 79, 85-96.

29. Yap, S.H., Strair, R.K., and Shafritz, D.A. (1978), J. Biol. Chem. 253, 4944-4950.

30. Zahringer, J., Baliga, B.S., and Munro, H.N. (1976), Proc. Natl. Acad. Sci. U.S.A. 73, 857-861.

31. Lindberg, V. and Sundquist, B. (1974), J. Mol. Biol. 86, 451-468.

32. Jain, S.K., Pluskal, M.G., and Sarkar, S. (1979), FEBS Letts. 97, 84-90.

33. Buckingham, M.E., and Gros, F. (1975), FEBS Letts. 53, 355-359.

34. Liautard, J.-P., and Kohler, K. (1976), Biochimie 58, 317-323.

35. Civelli, O., Vincent, A., Maundrell, K., Buri, J.-F., and Scherrer, K. (1980), Eur. J. Biochem. 107, 577-585.

36. Barrieux, A., Ingraham, H.A., Nystul, S., and Rosenfeld, M.G. (1976), Biochemistry 15, 3523-3528.

37. Van Venrooij, W.V., van Eekelen, C.A.G., Jansen, R.T.P., and Princen, H.M.G. (1977), Nature, London, 270, 189-191.

38. Liautard, J.-P., Setyono, B., Spindler, E., and Kohler, K. (1976) Biochim. Biophys. Acta 425, 373-383.

39. Morel, C., Gander, E.S., Herzberg, M., Dubochet, J., and Scherrer
     K. (1973), Eur. J. Biochem. 36, 455-464.
40. Rose, K.M., Jacob, S.T., and Kumar, A. (1979), Nature, London,
     279, 260-262.
41. Schwartz, H., and Darnell, J.E. (1976), J. Mol. Biol. 104, 833-
     851.
42. Liautard, J.-P., and Jenteur, Ph. (1979), Nucl. Acid Res. 7,
     135-150.
43. Baer, B.W., and Kornberg, R.D. (1980), Proc. Natl. Acad. Sci.
     U.S.A. 77, 1890-1892.
44. Jenkins, N.A., Kaumeyer, J.F., Young, E.M., and Raff, R.A.
     (1978), Develop. Biol. 63, 279-298.
45. Geoghegan, T., Cereghini, S., and Brawerman, G. (1979), Proc.
     Natl. Acad. Sci. U.S.A. 76, 5587-5591.
46. Liautard, J.-P., and Egly, J.M. (1980), Nucl. Acids Res. 8,
     1793-1804.
47. Bester, A.J., Kennedy, D.S., and Heywood, S.M. (1975), Proc.
     Natl. Acad. Sci. U.S.A. 72, 1523-1527.
48. Heywood, S.M., Kennedy, D.S., and Bester, A.J. (1979), Methods
     Enzymol. 60, 541-549.
49. Lee-Huang, S., Sierra, J.M., Naranjo, R., Filipowicz, W., and
     Ochoa, S. (1977), Arch. Biochem. Biophys. 180, 276-287.
50. Slegers, H., Mettrie, R., and Kondo, M. (1977), FEBS Letts. 80,
     390-394.
51. Bag, J., Hubley, M., and Sells, B. (1980), J. Biol. Chem. 255,
     7055-7058.
52. Pluskal, M.G., and Mukherjee, A. (1980), Fed. Proc. 39, 1868.

SECTION III:

ON SELECTING THE RIGHT MESSENGER

RECOGNITION OF INITIATION SITES

IN EUKARYOTIC MESSENGER RNAs

Marilyn Kozak

Department of Biological Sciences
University of Pittsburgh
Pittsburgh, PA 15260 (U.S.A.)

INTRODUCTION

The mechanism by which ribosomes select the correct AUG codon for initiation of protein synthesis has been the subject of considerable speculation and experimentation. In the case of prokaryotic messages and ribosomes, there is convincing evidence supporting Shine and Dalgarno's proposal (1) that base pairing occurs between the pyrimidine-rich 3'-end of 16S ribosomal RNA and a purine-rich sequence located approximately 10 nucleotides to the left of the AUG or GUG initiator codon (2-4). This interaction plays a central role in recognition of initiation sites by *E. coli* ribosomes, although additional features in the message also influence the efficiency of ribosome binding (5-8). Eukaryotic translational systems display certain peculiarities which argue against the notion that the mechanism of initiation in eukaryotes is more-or-less an extension of the accepted prokaryotic mechanism. For example, the mono-cistronic character of eukaryotic messenger RNAs (9-11) and the facilitating effect of the 5'-terminal m^{7}G cap (12) are features, not observed in prokaryotes, which must be assimilated into any model for translational initiation in eukaryotes.

CHARACTERISTICS OF INITIATION REGIONS IN EUKARYOTIC MESSENGER RNAs

Definition of the "ribosome binding site" in a eukaryotic message is complicated by several observations: (a) The region of the message protected by a 40S ribosomal subunit (with associated initiation factors) can extend up to 65 nucleotides (13). This is considerably bigger than the 25 to 30 nucleotide fragment protected by an 80S ribosome, under initiation conditions. It is not clear whether the "extra" sequences protected by the 40S ribosomal sub-

Table 1. Characteristics of Initiation Regions in Eukaryotic Messenger RNAs.

Messenger RNA	Proximity of initiator codon to 5'-terminus	Translation begins at first AUG	Initiator AUG identified by...	Splicing within 5' noncoding region	Sequence flanking initiator codon $\underset{AUG}{}$	Reference
Chicken ovalbumin	65	yes	aa sequence	+	ACC...GGC	18,19
Chicken conalbumin	77	yes	"	-	AAC...AAG	20
Rabbit α-globin	37	yes	"	-	ACC...GUG	21,22
Human α-globin	38	yes	"	-	ACC...GUG	23-25
Mouse α-globin	33	yes	"	-	ACC...GUG	26,27
Rabbit β-globin	54	yes	"	-	AGA...GUG	28-30
Human β-globin	51	yes	"	-	ACC...GUG	23,25
Mouse β-globin (major)	53	yes	"	-	AUC...GUG	26,27,31
Mouse β-globin (minor)	53	yes	"	-	AUC...GUG	26,31
Chicken β-globin	∿52	yes	"	-	GCC...GUG	32
Human γ-globin	54	yes	"	-	GCC...GGU	33
Silkworm fibroin	25	yes	open frame	-	AAG...AGA	34
Bovine preproparathyroid hormone	∿103	yes[a]	aa sequence		AAU...AUG	35
Rat preproinsulin I	∿58	yes	aa sequence[d]	+	AAC...GCC	36,37
Human preproinsulin	∿60[b]	yes	aa sequence[d]	+	GCC...GCC	38
Chicken preproinsulin	∿55[b]	no	aa sequence[d]	+	AUC...GCU	39
Mouse dihydrofolate reductase	∿83	yes		-[c]	AUC...GUU	40
Glyceraldehyde-3-phosphate dehydrog.	50 to 100	yes	aa sequence		AAA...GUU	41
S. purpuratus histone H1	∿39	yes	--	-	AAG...GCU	42,43
" histone H2A	∿71	yes	aa sequence	-	AUC...UCU	42,43
" histone H2B	∿79'	yes	open frame	-	AUC...GCU	42,43
" histone H3	∿56	yes	aa sequence	-	ACU...GCA	42,43
" histone H4	∿68	yes	--	-	AUC...UCA	42,43
P. miliaris (clone h22) histone H1	∿39	yes	open frame	-	AAG...ACU	44-46
" " histone H2A	∿75	yes	aa sequence	-	AUC...UCU	44-46
" " histone H2B	∿79	yes	--	-	ACC...	44-46
" " histone H3	∿55	yes	aa sequence	-	ACC...GCA	44-46
" " histone H4	∿62	yes	aa sequence	-	AUA...UCA	44-46
Human fibroblast interferon	∿73	yes	aa sequence[d]	-	AAC...ACC	47,48
Human leukocyte interferon	∿57	yes	aa sequence[d]	-	ACG...GCC	49
Dictyostelium actin, clone 5	∿43	yes	aa sequence[e]	-	AAA...GAC	50[f]
" clone 6	∿40	yes	"	-	AAA...GAU	50[f]
" clone 8	∿42	yes	"	-	AAA...GAC	50[f]
Dictyostelium discoidin-I (WR7)	∿72	yes	open frame	-	AAA...UCU	51[f]
" (I4)	∿71	yes	open frame	-	AAA...UCU	51[f]
Human chorionic gonadotropin (α)	∿51	yes	aa sequence		GCC...GAU	223
Rat prolactin	∿52	yes	aa sequence		ACC...AAG	224
Chicken lysozyme	∿30	yes	aa sequence	-	AAC...AGG	225
Mouse α-amylase (pancreatic)	18	yes	open frame	-	AAA...AAG	226
Mouse α-amylase (salivary)	94	no	open frame	+	AAA...AAA	226
Satellite tobacco necrosis virus	30	yes	aa seq; protect	-	AAC...GCA	52,53
Turnip yellow mosaic virus - coat	20	yes	aa sequence	-	AAC...GAA	54
Tobacco mosaic virus genome	69	yes	protection	-	ACA...GCA	55,56
Tobacco mosaic virus coat protein	10	yes	aa sequence	-	AAU...UCU	57
Brome mosaic virus RNA-4 (coat)	10	yes	aa sequence	-	AUA...UCG	58
Brome mosaic virus RNA-3	92	yes	open frame	-	CCG...UCU	59
Alfalfa mosaic virus RNA-4 (coat)	37	yes	aa seq; protect	-	AUC...AGU	60
Reovirus σ2 protein (s46)	19	yes	protection	-	GUU...GCU	61,62
" σNS protein (s45)	28	yes	"	-	ACU...GCU	61,62
" σ3 protein (s54)	32	yes	"	-	GCA...GAG	61,62
" μ1 protein (m52)	30	yes	"	-	AAG...GGG	61,62
" μNS protein (m44)	19	yes	"	-	GUC...GCU	61,62
" μ2 protein (m36)	14	yes	"	-	GUC...GCU	61,62
" large message	14	yes	"	-	AGG...AAG	unpub.[g]
Vesicular stomatitis virus NS prot.	11	yes	"	-	AUC...AAG	63,64
" " L protein	11	yes	"	-	AUC...GAA	63,64
" " M protein	42	yes	"	-	AUC...AGU	63,64
" " G protein	30	yes	"	-	ACU...AAG	63,64
" " (Indiana) N prot.	14	yes	"	-	AAA...UCU	63,64
" "(New Jersey) N prot.	14	yes	"	-	AAC...GCU	65
Adenovirus hexon protein	240	yes	aa sequence (66a)	+	AAG...GCU	66
Adenovirus fiber protein	203[h]	yes	aa sequence "	+	AAG...AAA	67
Adenovirus (late) polypeptide IX	25	yes	aa sequence "	-	GCC...AGC	68
Adenovirus (early) E1a mRNAs	62	yes		-	AAA...AGA	69,70
SV40 early mRNAs	74	yes	aa sequence	-	AAG...GAU	71,72
SV40 late 16S mRNA (VP1)	239[i]	no	aa sequence	+	CUU...AAG	73
SV40 late 19S mRNA (VP2)	115[i]	no	open frame	+	UCC...GGU	74
Polyoma late 16S mRNA (VP1)	124,181,238[j]	yes	aa sequence(84)	+	AAG...GCC	75-78
Polyoma late 19S mRNA (VP2)	93,150,207[j]	yes	aa sequence(84)	+/-	AAA...GGA	75-78
Polyoma late 18S mRNA (VP3)	126,183,240[j]	yes	aa sequence(84)	+	AAU...GCG	75-78
Influenza (A/Udorn/72) HA mRNA	39-44	yes	aa sequence	+[k]	ACC...AAG	79
Influenza virus matrix protein	39	yes	aa sequence	+[k]	AAG...AGU	227-228
Adenovirus region E1b 15K protein	13	yes	aa sequence (66a)	-	CUC...GAG	229
Rous sarcoma virus genome	372	no	aa sequence	-	AGC...	230

Footnotes to Table 1

   In calculating the proximity of the AUG codon to the 5'-terminus,
numbering begins with the nucleotide adjacent to the m^{7}G cap and
includes the A of the initiator codon. In some genes the start site
of transcription is staggered, resulting in microheterogeneity at the
5'-end of the message. In such cases the data shown in the table
generally refer to the longest form detected for a given message.
In many instances the 5'-proximal sequence of a message was deduced
indirectly by analysis of cDNA, and I have indicated the *approximate*
cap-to-AUG distance for these messages; the possibility that a cDNA
copy may not extend all the way to the 5'-end of the message must be
remembered. In cases where S1 nuclease was used to map the 5'-termi-
nus of a message, there might be a small error due to imprecise
trimming by the nuclease (46). Sequences derived solely from analysis of genomic DNA are not included in the table unless there is
supplementary evidence identifying the (approximate) start site for
transcription, and demonstrating colinearity of the 5'-noncoding
region of the message with the gene. I have also omitted a few
messages for which sequence information is available, but in which
the functional initiator codon has not been identified. In the
fourth column, the notation "aa sequence" means that the initiator
AUG was identified by comparison with the known N-terminal amino
acid sequence of the protein. The notation "protection" means that
a ribosome-protection assay was used to pinpoint the AUG initiator
codon. Existence of an open reading frame is, by itself, a relatively weak criterion for identifying the functional initiator codon.
However, in cases where the molecular weight of the protein is known
and the next AUG is a considerable distance downstream from the
first, the argument for initiation at the indicated AUG is reasonably strong.

   [a]In the sequence reported by Kronenberg *et al.* (35) for preproparathyroid hormone cDNA, the region corresponding to the 5'-terminal
∿50 nucleotides of the message apparently became inverted during
cloning. The correct sequence has recently been determined by C.A.
Weaver, D.F. Gordon and B. Kemper (pers. commun.) and it contains no
AUG triplets upstream of the functional initiator codon.

   [b]In the case of human and chicken preproinsulin mRNAs, the 5'-
terminus and the end points of the intervening sequence were tentatively located by comparison with the rat sequence.

   [c]Contrary to an earlier report, the 5'-noncoding region of the
mouse dihydrofolate reductase gene does not appear to be spliced
(C. Simonsen and R. Schimke, personal communication).

   [d]The AUG initiator codon was identified based on the known N-
terminal amino acid sequence of the mature protein, and the expectation that the nascent form of the protein carries an additional

unit are important for the initiation process. (b) Neither 80S-
protected nor 40S-protected mRNA fragments can rebind efficiently to
ribosomes unless the m^{7}G cap is included on the fragment (14).   Thus,

---

Footnotes to Table 1 (cont.)

"signal peptide" at the amino terminus.

^{e}The initiator codon was identified by comparing the nucleotide
sequence of these genes with the N-terminal amino acid sequence of
Physarum actin.

^{f}Unpublished sequence data on these messages and information
regarding the start site of transcription were kindly provided by
Dr. Richard Firtel.

^{g}Unpublished data, M.K.

^{h}There is a second (translatable) form of the adenovirus fiber
message, in which the 5'-untranslated region is 384 nucleotides long
(67).

^{i}There is extensive heterogeneity in the 5'-proximal portion of
SV40 late mRNAs (74, 81) and the data in the table refer to only the
most abundant form of 16S and 19S mRNA. In the case of 16S mRNA,
the AUG in position 239-241 was identified as the initiator codon in
vitro (experiments by A. Smith *et al.* cited in ref. 82). It is not
known whether translation begins at position 239 or 245 *in vivo*. The
second site corresponds to the N-terminal amino acid sequence of the
mature VP1 protein (83).

^{j}Multiple species of polyoma virus mRNA are apparently derived
by repetition (within the leader region of the message) of the
sequence extending from position 216 to 272 on the genome (75, 78,
80). There may be mRNA species with leader segments even longer
than those listed.

^{k}The 5'-noncoding region of influenza virus mRNA is not spliced
(in the sense of removing an "intron"), but the region is interrupted
due to transfer of the 5'-terminal 10 to 15 nucleotides from a
cellular message to the nascent viral mRNA (85). A variety of
cellular messages can serve as donor, thus generating heterogeneous
5'-termini in the viral mRNA population (79).

^{l}The genomic RNA of Rous sarcoma virus is believed to be the
message for Pr76^{gag}. A portion of the 5'-sequence is given in ref.
230; the remainder was determined by R. Swanstrom and J.M. Bishop
(personal communication).

the 5'-terminal cap--which is not protected by the ribosome if the
cap-to-AUG distance is greater than about 50 nucleotides--neverthe-
less must be considered part of the initiation region. There is
abundant evidence from other types of experiments implicating the
m⁷G cap in initiation (12, 15). (c) Circumstantial evidence suggests
that an entry site for ribosomes exists upstream of the AUG initiator
codon. The current working hypothesis is that a 40S ribosomal subunit binds initially at or near the 5'-end of a message and subsequently migrates along the RNA chain, stopping when it encounters the
first AUG codon. In view of this mechanism, for which evidence is
presented in Section 3 of this Chapter, it seems reasonable to define
the "initiation region" as the sequence extending from the 5'-
terminus up to (and perhaps slightly beyond) the AUG initiator codon.
This definition is obviously too broad, since portions of the 5'-
noncoding region have been deleted from some messages without impair-
ing translation (16, 17). Nevertheless, this seems at the moment
to be the most practical definition.

Table 1 summarizes some characteristics of initiation regions
from a large number of viral and cellular messages. I have included
only those mRNAs for which the entire sequence is known from the
5'-terminus up to the AUG initiator codon. The length of the 5'-
noncoding region varies tremendously--from 10 nucleotides (58) to
over 300 nucleotides (67, 78). The extremely short and extremely
long 5'-noncoding segments are found on viral messages. Initiation
regions of cellular messages are somewhat more uniform; most fall
within the range of 40 to 80 nucleotides, as indicated in the upper
half of Table 1. This difference between viral and cellular messages
might be related to the higher translational efficiency observed with
many viral messages, although (as explained in Section 4) there is no
simple correlation between translational efficiency and length of
the 5'-noncoding region. The 5'-untranslated regions of eukaryotic
messages also show enormous variation in composition. For example,
the A + U composition varies from 38% for the long leader on adenovirus hexon mRNA (66) to 77% in the 5'-untranslated region of tobacco
mosaic virus genome RNA (55), and greater than 90% in *Dictyostelium*
actin messages (50). In most of the messages listed in Table 1, the
G content of the 5'-noncoding region is lower than the 25% that would
be expected, on a random basis.

In contrast with the variability in length and overall composition of 5'-untranslated segments, the sequences immediately flanking
the AUG initiator codon show rather striking uniformity. Table 2
(part A) shows that, in 98% of eukaryotic messages, a purine (most
often A) occurs 3 residues before the AUG codon. And in 85% of
eukaryotic messages the residue immediately following the initiator
AUG is also a purine--most often G. As a control of sorts, I
surveyed the sequences surrounding AUG's which are *not* involved in
translation: part B of Table 2 shows a much more random distribution
of nucleotides flanking AUG triplets in the introns of eukaryotic

Table 2.

A. Sequences Flanking the Initiator AUG in Eukaryotic Messages

	Position 1	Position 2	Position 3	Position 7	Position 8	Position 9
U	1%	27%	12%	14%	15%	45%
A	83%	38%	24%	20%	22%	16%
C	1%	31%	48%	1%	48%	15%
G	15%	4%	16%	65%	15%	24%

B. Sequences Flanking AUG Codons Which Occur in Introns of Eukaryotic Genes:

	Position 1	Position 2	Position 3	Position 7	Position 8	Position 9
U	36%	26%	37%	36%	25%	22%
A	39%	27%	29%	29%	36%	35%
C	13%	22%	19%	19%	16%	26%
G	12%	25%	15%	16%	23%	17%

C. Sequences Flanking the Initiator Codon in Prokaryotic Messages

	Position 1	Position 2	Position 3	Position 7	Position 8	Position 9
U	17%	36%	30%	14%	16%	34%
A	60%	29%	32%	42%	28%	46%
C	18%	24%	28%	11%	41%	11%
G	5%	11%	10%	33%	15%	9%

1 2 3 4 5 6 7 8 9
        Nucleotides flanking the AUG codon are numbered X-X-X-A-U-G-X-X-X. Data
used in part A were taken from Table 1 and from references 86 to 94. The initiation site for SV40 VP3 was included (ref. 72) but VP1 was omitted since it
is not known whether translation begins at position 239 or 245. A total of 75
messages are represented in part A. The intervening sequences surveyed in
part B are from the chick ovalbumin gene (95), silkworm fibroin (34), rat preproinsulin (37), rabbit β-globin (96), mouse α- and β-globin (97) and a mouse
immunoglobulin light chain gene (86). For part C, the seventy-two prokaryotic
initiation sites in references 98 and 99 were used.

genes. Interestingly, the sequences flanking the AUG or GUG
initiator codon in prokaryotic messages (Table 2, part C) show a
bias similar to--although not quite as extreme as--that seen in
eukaryotic initiation sites. While it is tempting to attribute
functional significance (see Section 4) to the conserved sequences
flanking eukaryotic initiator codons, presence of a purine in positions 1 and 7 is clearly not an absolute requirement, since a few
messages deviate from the pattern (59, 100). Comparison of the 5'-
proximal sequence of the N-message in two strains of vesicular
stomatitis virus (65) reveals only three differences--one of which
is the residue immediately following the initiator codon.

        Because prokaryotic ribosome binding sites nearly always contain
a termination codon--either UAA or UGA--closely preceding the initiator codon, Atkins (98) has speculated that termination codons and
release factors might be involved in the initiation process in
prokaryotes. Inspection of the 5'-noncoding sequences in eukaryotic
messages reveals that the frequency of terminator codons (UAA, UGA
or UAG) is approximately as expected on a random basis--one terminator
in any frame per seven codons. The terminator codons occur at varying distances from the AUG triplet, and there is no reason to suspect
their involvement in the initiation process in eukaryotes.

        The most striking characteristic of eukaryotic initiation
regions is absence of AUG triplets from the 5'-noncoding portion of
the mRNA. In 69 of the messages listed in Table 1, the AUG codon
located closest to the 5'-end is the one used to initiate translation.
The few exceptions to this rule are discussed in Section 3. (In
addition to the exceptions noted in Table 1, very recent studies have
revealed a few more messages in which translation does *not* begin at
the first AUG. Although some of these sequences were not available
in time to be included in the table, they are mentioned in Section
3.) Splicing has been detected within the 5'-noncoding region of
many viral, and a few cellular messages--including those of ovalbumin
(101), preproinsulin (37-39) and mouse α-amylase (U. Schibler, R.A
Young and O. Hagenbuchle, personal communication). It is curious
that in each of the preproinsulin genes, the rather short intron that
interrupts the 5'-noncoding segment contains one or more AUG codons.
Thus, one might argue that the intervening sequence *must* be removed--
if absence of AUG codons upstream of the initiation site is important
for mRNA function, as proposed below. Although GUG triplets occur
in the 5'-proximal portion of many eukaryotic messages, there is no
evidence that GUG can function as an initiation codon in eukaryotes
(102, 103). This restriction might be due to the hypermodified t^6A
adjacent to the anticodon in eukaryotic initiator Met-tRNA (104 and
see Chapter 1 of this volume).

MECHANISMS WHICH HAVE BEEN PROPOSED TO EXPLAIN SELECTION OF
INITIATION SITES BY EUKARYOTIC RIBOSOMES

Figure 1 summarizes some of the mechanisms by which eukaryotic
ribosomes might recognize the unique initiator codon in a message.
The scanning model (82), listed first, postulates an entry site for
40S ribosomal subunits which is distinct from the AUG-containing
site. The hypothesis is that a 40S subunit (with associated
initiation factors and Met-tRNA) binds initially at or near the 5'-
terminus of the message and subsequently migrates toward the interior
of the mRNA, stopping when it encounters the first AUG codon. According to this mechanism, the initiator codon is defined simply by
its position: the AUG closest to the 5'-terminus is the one used
to initiate protein synthesis.

Whereas the scanning mechanism postulates sequential interactions, model 2 postulates that the ribosome interacts simultaneously
with the 5'-terminus and the AUG triplet. Since the distance between
the cap and the AUG varies among eukaryotic messages from 10 to nearly
400 nucleotides, mechanism 2 requires that secondary/tertiary folding
within the 5'-region of the message brings the cap and the AUG close
together.

The third model, like the first, proposes a two-step mechanism
for initiation. The 40S ribosome would enter at the 5'-terminus of
the message and then re-position itself at the AUG-containing site
without traversing the entire 5'-noncoding sequence. Since there is
no evidence for initiation at multiple sites, even when the 5'-
segment of a message contains several nearby AUG's, the mechanism by
which the ribosome is transferred from the 5'-terminus must be
precise. Thus, model 3 requires a second signal to tell the ribosome
which AUG to aim for when it jumps. The recognition feature could be
a hairpin loop preceeding the AUG, as suggested by Lomedico *et al.*
(37), but such structures are found in the 5'-noncoding region of
only a few eukaryotic messages. The most popular alternative suggestion is that recognition of the initiator AUG is mediated by base
pairing between the 3'-end of 18S ribosomal RNA and a complementary
sequence located to the left (105) or right (34, 106) of the AUG
triplet. Some eukaryotic messages do have a sequence complementary
to the purine-rich tract 3'-UAGGAAGGCGU-5' which is present near the
3'-end of 18S rRNA, but occurrence of the required sequence within
the 5'-proximal region of eukaryotic mRNAs does not appear to be
statistically significant (107). (Noting the absence of a common
sequence upstream of the AUG triplet, Baralle and Brownlee (26)
proposed that a base pairing mechanism might operate in eukaryotes
similar to that which occurs in prokaryotes, except that it is the
AUG codon itself that pairs with ribosomal RNA. Such an interaction
would not help to explain how ribosomes distinguish the unique
initiator AUG from all others, however.) In recent studies a
psoralen derivative was used to crosslink apposed regions of mRNA and

(1) SCANNING MODEL:

Step 1: AUG-independent binding at 5'-terminus of message

Step 2: 40S subunit SLIDES from 5'-terminus to first AUG codon.

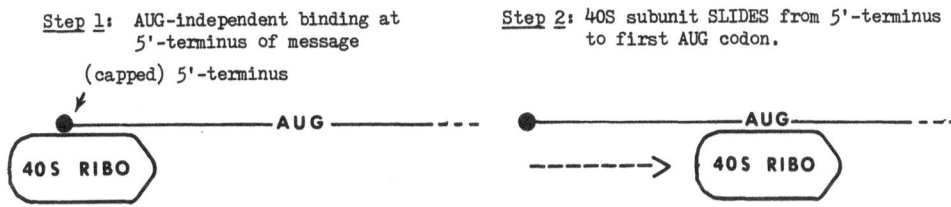

(2) SIMULTANEOUS INTERACTION WITH 5'-TERMINUS AND AUG CODON (which are oriented by secondary/tertiary structure of the message).

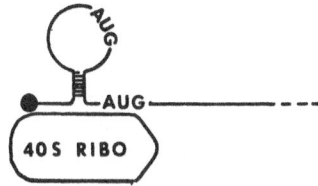

(3) Ribosome BINDS AT 5'-END then JUMPS TO AUG (which is recognized by flanking primary or secondary structural features).

(4) DIRECT ATTACHMENT AT AUG-CONTAINING SITE (which is recognized by flanking structural features). No important role for 5'-end.

(5) Direct attachment at AUG-initiator codon DUE TO ALL OTHER AUGs BEING SEQUESTERED by secondary/tertiary structure of the message.

Fig. 1.   Mechanisms by which eukaryotic ribosomes might select the correct AUG initiator codon. See text for explanation and evaluation of each model.

ribosomal RNA in 40S and 80S initiation complexes (108). The results suggest that the 3'-terminal segment of 18S rRNA is positioned near the 5'-end of reovirus mRNA, although further experiments are needed to verify this interpretation. The same type of crosslinked complex could be formed with capped poly(U), which has little potential to base pair with rRNA. Thus, although these experiments demonstrate that messenger RNA and 18S rRNA are closely apposed in initiation complexes, additional work is needed to determine if the two RNA components are actually base paired, precisely which portions of each RNA are involved, and whether the interaction has functional significance.

While the potential for base pairing with ribosomal RNA appears marginal in most eukaryotic messages, a special case might be made for adenovirus early (109) and late mRNAs (110), globin mRNAs (33), polyoma late 16S mRNA (111, 112), and the messages for interferon (47), ovalbumin (19), and the *Dictyostelium* M4 gene (A. Kimmel and R. Firtel, personal communication). Each of these messages shows an impressive 6 to 9 nucleotide complementarity with a purine-rich segment near the 3'-end of 18S rRNA. The problem is that the putative "Shine/Dalgarno sequence" in these messages does not occur near the

AUG codon. Thus, if that sequence has functional significance, it
seems more likely that interaction with rRNA might enhance the
efficiency of translation--perhaps by stabilizing a "pre-initiation
complex"-- than that it would help to identify the correct initiator
codon. In summary, although 5'-terminal sequences are now known for
over seventy eukaryotic messages, it is difficult to identify a
common structural feature in the vicinity of the AUG which might be
recognized by the ribosome, as necessitated by mechanism 3. There is
a similar problem with respect to the fourth mechanism.

The fifth mechanism, however, does not require special recognition of the initiator AUG. Rather, the hypothesis is that the
functional initiator codon is chosen by default: it is the only AUG
not sequestered by the secondary/tertiary structure of the message.
Although the model predicts that the functional AUG should always be
located in an exposed region, there are several examples of eukaryotic
messages in which the initiator codon is likely to be involved in a
stable base paired structure (27, 34, 59, 113). Model 5 also predicts
that denaturation of a message should activate new initiation sites.
Contrary to this prediction, however, recent studies showed that 80S
ribosomes continue to bind exclusively at the 5'-terminal region of
unfolded reovirus mRNA (114, 115). This constitutes strong evidence
against mechanism 5, and also contradicts mechanism 2, which predicts
that disruption of the secondary/tertiary structure should seriously
impair initiation. (The observed effect of abolishing mRNA secondary
structure was *enhanced* binding to ribosomes (114).)

Although mechanisms 2 through 5 have not been rigorously dis-
proven, they are contradicted to some extent by the experiments
alluded to above. On the other hand, a variety of experimental
approaches have yielded evidence (albeit circumstantial) supporting
the scanning model.

EVALUATION OF THE SCANNING MECHANISM FOR INITIATION

A Summary of the Evidence

Various types of experiments, carried out in many different
laboratories, have provided evidence consistent with the scanning
model.

a)   The mechanism rationalizes the facilitating effect of the m^{7}G
cap (12), which is at the putative entry site for 40S ribosomes.
The importance of the methylated cap was shown by comparing the
translational activity of capped and uncapped forms of a given
message (116-118), and by demonstrating that *in vitro* translation
of capped mRNAs is inhibited by addition of cap analogues (119) or by
incubation with antibodies specific for 7-methylguanosine (120). It
is noteworthy that the cap enhances translation even in messages in
which the initiator codon is hundreds of nucleotides downstream of

the 5'-terminus (121, 122). In attempting to deduce how the cap
exerts its effect, one point that must be remembered is that ribosomes initiate at the *same* 5'-proximal AUG codon irrespective of the
presence or absence of the terminal m'G (14, 123). Thus, although
a methylated cap greatly increases the efficiency of initiation, the
cap does not function as a recognition signal telling the ribosome
where to bind. It seems to be the 5'-end of an RNA molecule as such,
not the m'G cap, that comprises the primary recognition site for
ribosome attachment. It follows that one need not postulate different
initiation mechanisms for capped versus uncapped messages.

b)   Ribosome binding is prevented by circularization of the messsge
(124). Linear forms of various (uncapped) synthetic polymers were
shown to bind to wheat germ and reticulocyte ribosomes *in vitro*, but
the same polymers failed to bind after they were circularized by RNA
ligase. This again emphasizes that--even without the m G cap--the
5'-terminus of a message plays an essential role in ribosome attachment.

c)   Fragmentation of a message allows ribosomes to bind to sequences
derived from the interior of the RNA molecule. (These internal sites
are not accessible to ribosomes in intact mRNA, even after extensive
denaturation (114).) Ribosome binding to spurious internal sites has
been observed upon fragmentation of a variety of messages including
cowpea mosaic virus RNA (125), foot-and-mouth disease virus mRNA (17),
ovalbumin mRNA (113), reovirus mRNA (126), adenovirus mRNA (127) and
polyoma cRNA (112). The experiments suggest that presence of an
exposed 5'-end on an RNA molecule is necessary *and sufficient* to
permit ribosome attachment.

d)   The scanning model predicts that 40S ribosomal subunits should
be able to attach to RNA molecules which lack an AUG codon. This has,
indeed, been observed with a wide variety of synthetic ribopolymers
(128).

e)   The model predicts that when ribosomes are incubated *in vitro*
with a heterogeneous set of RNA fragments, 40S subunits should interact with (and protect from nuclease digestion) a greater variety of
sequences than 80S ribosomes. That is, a 40S ribosome should be able
to attach to almost any RNA fragment, whereas a 60S subunit should
join only if the sequence includes an AUG codon. Consistent with
this prediction, two-dimensional fingerprints of protected mRNA
fragments recovered from 40S initiation complexes were far more
complex than fingerprints of 80S-protected fragments (112, 129). (In
view of the somewhat contradictory result described in the following
paragraph, systematic studies are needed--using RNA templates of
known sequence--to confirm that an AUG codon is needed for 60S joining.)

f)   Messages with a long 5'-noncoding region permit more than one

ribosome to bind in the presence of an inhibitor of elongation, such
as sparsomycin. Binding of several ribosomes to the long 5'-leader
segment of tobacco mosaic virus genome RNA (130), brome mosaic virus
(BMV) RNA-3 (59) and polyoma virus late mRNAs (112) is consistent
with the notion that there is an entry site for ribosomes upstream of
the AUG codon. (The long 5'-noncoding region in BMV-3 RNA actually
permitted binding of a second 80S ribosome--an interesting result
not predicted by the scanning model. It is not known, however, if the
80S complex that forms upstream of the AUG codon is a functional
intermediate or an abortive complex.)

g)   A key postulate of the scanning mechanism is that 40S subunits
must be able to migrate. This has been demonstrated in a variety of
ways. When the interaction between Met-tRNA and the AUG codon was
perturbed (by inhibitors such as edeine (13) or pactamycin (131), or
by using ITP-substituted mRNA as template (114)), 40S subunits migrated *beyond* the 5'-proximal AUG. Under these circumstances, messenger RNA accumulated in complexes containing ten to twelve 40S
subunits--each subunit having attached apparently at the 5'-end of
the message and then migrated into the interior. (The fact that
denaturation of reovirus mRNA permits 40S subunits to advance beyond
the 5'-proximal AUG (114) raises the possibility that the secondary/
tertiary structure in *native* mRNA might serve to *slow* migration of
40S ribosomes, thereby facilitating recognition between the AUG codon
and the complementary anticodon in Met-tRNA.)

Recent studies have revealed that migration of wheat germ 40S
ribosomal subunits requires ATP hydrolysis (132). When migration was
inhibited by depleting ATP reserves, a single 40S ribosome was trapped
near the 5'-end of the message--upstream of the AUG initiator codon.
This experiment confirms a key step in the scanning mechanism.

h)   Examination of the 5'-proximal nucleotide sequences of over
seventy eukaryotic messages has revealed that in most cases translation begins at the AUG closest to the 5'-terminus. The data are
summarized in Table 1, and the few exceptions are discussed in the
following section. Absence of AUG codons from the 5'-untranslated
region of most eukaryotic messenger RNAs--including a number of
messages in which the 5'-noncoding segment is hundreds of nucleotides long--provides striking support for the scanning mechanism.
A further prediction of the model is that if a new AUG triplet were
to be introduced upstream of the natural initiator codon, the adven-
titious AUG should supplant the natural AUG codon. In order to test
this prediction, sodium bisulfite was used to convert an ACG sequence
(fortuitously located in the 5'-noncoding region of two reovirus
messages) to AUG. Subsequent sequence analysis of the ribosome-
protected mRNA fragments recovered from sparsomycin-blocked 80S
initiation complexes revealed that a high percentage of the ribosomes
were indeed situated at the "unnatural" 5'-proximal AUG created by
the bisulfite treatment (115).

In an elegant genetic analysis of yeast iso-1-cytochrome c
mutants, Sherman and his colleagues showed that when the normal initiator codon was inactivated by mutation, introduction of a new AUG
codon anywhere within a 37-nucleotide region restored translation
(100). The N-terminal region of the yeast iso-1-cytochrome c gene
has the sequence

    AAA UUA AUA *AUG* ACU GAA UUC AAG GCC GGU UCU GCU AAG.

The AUG shown in italics is the initiator codon in the wild type gene,
and the underlined sequences indicate positions where an AUG initiator
codon has been introduced in various revertants. Once the start site
for transcription has been ascertained for this gene, the genetic
(100) and nucleotide sequence data (89) should provide definitive
evidence for a model in which ribosomes initiate at the 5'-proximal
AUG codon irrespective of the flanking sequences. (Comparison of the
N-terminal amino acid sequences of wild type yeast iso-1- and iso-2-
cytochrome c (90) suggests, amusingly, that Nature may have mimicked
Sherman's experiment in relocating the AUG initiator codon in one of
those genes.)

i) The scanning mechanism rationalizes the (functionally) monocistronic character of eukaryotic messages. Even in yeast, where
there are coordinately regulated genes for various biochemical pathways (superficially reminiscent of bacterial operons), the messenger
RNAs have been shown to be monocistronic (133). There are numerous
examples of viral messages (reviewed in references 11 and 82) which
are structurally polycistronic--that is, the transcript encodes two
or more polypeptides. In every instance, however, translation is
limited to the 5'-proximal cistron. This peculiar restriction on
initiation in eukaryotes actually reflects, not the structure of the
mRNA, but the properties of the translational machinery. Thus, even
with a polycistronic mRNA from bacteriophage lamba, only the first
cistron was translated by wheat germ ribosomes (134). In the case
of the polycistronic genomic message of RNA phages, it seems clear
that authentic phage polypeptides--including coat protein (located
in the center of the genome), synthetase (located near the 3'-end)
and lysis peptide (overlapping the coat and synthetase cistrons)--
can be synthesized in eukaryotic cell-free extracts (135-137). However, recent studies with R17 RNA using a wheat germ extract indicated that cleavage of the phage RNA was a prerequisite for initiation at the internal cistrons (126).

j) A final prediction of the scanning mechanism is that initiation
should be relatively insensitive to major perturbations in the 5'-
proximal region of the message. Translation was remarkably unaffected by *deletion* of 5'-noncoding sequences in SV40 mRNA (16, 138, 139),
polyoma early mRNA (140), foot-and-mouth disease virus RNA (17) and
rabbit β-globin mRNA (141). (Although each of these shortened
messages clearly directed synthesis of normal polypeptide(s), in most
cases the efficiency of translation of the altered message could not

be determined.) Ribosome binding was not prevented by a large
insertion upstream of the AUG codon in adenovirus mRNA (142), or by
extensive *alteration* of the primary sequence flanking the AUG codon
in reovirus (115) and in yeast cytochrome c mRNA (100). Thus, it is
not surprising that the 5'-untranslated regions of closely related
messages display considerable evolutionary divergence--for example,
among the actin genes in slime molds (50); between yeast iso-1- and
iso-2-cytochrome c (90); between isocoding messages in two different
subsets of histone genes (clones h19 and h22) in a single sea urchin
species (44); and between the late mRNAs of SV40 and polyoma virus
(75). Although the 5'-noncoding sequences of human and rabbit β-
globin mRNAs are quite similar (143), there is little evidence of
conservation when the comparison is extended to the chicken β-globin
message (32). All this variation suggests that ribosomes do not
recognize or require a precise sequence in the 5'-untranslated
portion of messenger RNA. Finally, as noted above, irreversible
denaturation of reovirus mRNA did not diminish its capacity to interact with ribosomes (114). Thus, one is led to postulate a mechanism
for initiation in eukaryotes in which neither the primary nor the
secondary structure of the message plays an obligatory role in directing ribosome binding.

Variations on the Theme

A number of variations on the scanning model have been suggested. One possibility is that the 40S ribosomal subunit bypasses AUG
codons which are sequestered in base-paired structures, and that
translation begins at the first *exposed* AUG. This mechanism was
suggested to account for failure to initiate at the first AUG codon
in late SV40 mRNAs (73, 74). The idea seems to be contradicted,
however, by several other messages in which the functional AUG codon
apparently is located in a stable base-paired region of the message
(27, 34, 59, 113). In fact, as noted above, the best way to induce
a 40S ribosome to bypass an AUG triplet is not to sequester it, but
to abolish the secondary structure of the region (114) !

Another proposed variation on the scanning model is that a 60S
subunit might couple with the 40S subunit at or near the 5'-terminal
entry site, and it is the 80S ribosome which then scans the 5'-
proximal region of the message and stops at the first AUG. This idea
is compatible with the observed binding of a second 80S ribosome
(rather than just a 40S subunit) within the long 5'-noncoding region
that precedes the AUG initiator codon in brome mosaic virus RNA-3
(59). It has not been shown, however, (and indeed it seems unlikely)
that an 80S complex which assembles upstream of the AUG codon can
subsequently migrate, in the presence of sparsomycin. On the other
hand, analysis of the 40S ribosome-protected regions of vesicular
stomatitis virus (64), polyoma virus (112), satellite tobacco
necrosis virus (53) and reovirus mRNAs (61) clearly indicates that,
in each of these messages, the 40S ribosome forms a stable complex

centered around the AUG codon. Thus, the 40S ribosomal subunit--in the absence of 60S subunits--seems capable of both migrating and recognizing the AUG "stop-signal".

How Can the Exceptions Be Explained

Two experimental approaches have revealed apparent exceptions to the rule that translation is confined to the first AUG in a message. One approach involves analysis of the products of *in vitro* protein synthesis. There are a number of reports in which a single messenger RNA species directs synthesis of two or more proteins (of which one usually predominates), which appear to be independently initiated. This has been shown most convincingly with the genome RNA of polio-virus (144, 145), and a possible explanation for the poliovirus phenomenon is mentioned in the following section. Initiation at a second site within the interior of the message has also been claimed for Semliki Forest virus 42S RNA (146), tobacco rattle virus RNA-2 (147), tobacco necrosis virus (148), Southern bean mosaic virus (149), carnation mottle virus (150), alfalfa mosaic virus RNA-3 (151) and cowpea mosaic virus genome RNA (152). In most of these cases (with the notable exception of poliovirus), the possibility that a second initiation site was activated by *cleavage* of the message must be considered. Lawrence (127) has emphasized the ease with which this occurs during incubation in cell-free extracts. Although formation of disomes in the absence of protein chain elongation would seem to constitute good evidence that ribosomes can bind to a second, internal site (151-153), this criterion is meaningless if the message contains a long 5'-noncoding region, which may allow a queue of ribosomes to bind upstream of the *single*, 5'-proximal initiator codon (59, 130).

I have omitted from the above discussion some very preliminary reports in which the possibility of internal initiation is raised (154, 155), as well as a few reports in which early experiments suggesting internal initiation have been contradicted by more recent data. One of these concerns the 14S mRNA for calf lens α-crystallin, which appeared from early experiments to be a bicistronic message (156). More recent data indicate, however, that each polypeptide is translated from a different species of mRNA (157). Similarly, the 13S mRNA from maize that directs synthesis of two zein components(158) has been shown to consist of two separable mRNAs (159). Early reports that 42S flavivirus RNA directs synthesis of several inde-pendently initiated proteins (160) must also be re-evaluated in the light of recent evidence for a single initiation site in the flavi-virus genome (161).

The second approach which has unequivocally revealed exceptions to the first-AUG-rule involves determination of nucleotide sequences. In two messages, *chicken preproinsulin* (39) and *Semliki Forest virus genome* RNA[§] (162)--nonfunctional AUG codons occur very close to the cap. Since ribosomes might be expected to have a "blind spot" between

the cap-binding site and the AUG-recognition site, AUG codons occurring within the first 10 or so residues adjacent to the cap might not
be recognized. (Brome mosaic virus is probably the best available
yardstick: in that message the functional AUG occurs in position 10
to 12.   In other messages in which the 5'-penultimate and subpenulti-
mate nucleotides are methylated--or in which the nucleotide composition of the cap-adjacent region differs--the point at which the
ribosome begins "scanning" might be shifted by a couple of residues.)

Other messages in which the first AUG (which occurs at a considerable distance from the cap) is not the functional initiator
codon pose more of a problem.   Translation begins at the second AUG
in *mouse* α-*amylase* mRNA (226) and in a *Dictyostelium* message designated M4 (A. Kimmel and R. Firtel, personal communication). (In the
case of the M4 message, identification of the initiator codon is
based on the presence of an open reading frame. Since a protein
encoded by this region has not been demonstrated, however, I have not
included it in Table 1). The messages encoding *Rous sarcoma virus*
(RSV) *gag, src,* and *env* proteins share a leader segment over 350
nucleotides long in which the first three AUG triplets are followed
by in-phase terminator codons (R. Swanstrom and J.M. Bishop, personal
communication).   Translation of the RSV *gag* protein begins at the
fourth AUG. The exact point at which the leader is spliced to the
body has not yet been determined for the *env* and *src* messages, but it
seems clear that at least three AUG triplets occur upstream of the
functional initiator codon. (The suggestion that the first GUG
might initiate translation in type C retroviruses (163) seems unlikely, since GUG is not a functional initiator codon in any other
eukaryotic message that has been characterized, nor was GUG found to
initiate polypeptide synthesis when synthetic ribopolymers were tested
at low magnesium concentration (102). Moreover, if translation were
to begin at the GUG that is included in the ribosome-protected region
of RSV RNA (164), the N-terminal amino acid sequence of the resulting
polypeptide would not correspond to the known N-terminal sequence of
Pr76^{gag}.) SV40 *late messenger* RNAs provide another clear violation
of the rule that translation begins at the first AUG. The first AUG
in late 19S mRNA (74) and the first two AUG's in 16S mRNA (73) cannot
initiate translation of the viral structural proteins, due to in-phase
termination signals. In the case of late 19S mRNA, it appears as if
the first AUG is nonfunctional, the second AUG is the initiator codon
for VP2, and the third AUG might function to initiate synthesis of
VP3--since only trace amounts of a smaller message (retaining the VP3
but not the VP2 initiation site) have been detected in infected cells
(74, 165). (This interpretation applies only to SV40; in the case

_____

§ *Semliki Forest virus* RNA is not included in Table 1 because the
sequence around the functional initiator codon is not known. The conclusion that translation does not begin at the 5'-proximal AUG is
based on the presence of an in-phase termination codon short distance
downstream from the first AUG.

of polyoma virus it is clear that VP2 and VP3 are translated from
separate messages (166).) In summary, *there are at least nine
eukaryotic messages in which translation does not begin at (or is not
limited to) the first AUG.*

        The question of how ribosomes select the correct internal AUG in
these exceptional messages is intriguing. It might be noted that
there is no evidence that ribosomes initiate *exclusively* at the
correct AUG (preceding the coding region for the viral structural
protein) in SV40 or RSV messages, since neither N-terminal dipeptides
nor ribosome-protected initiation sites have been analyzed in detail.
Indeed, the only study in which a ribosome-protected fragment of RSV
genomic RNA was recovered in amounts sufficient for analysis suggested that the ribosome was positioned at or near the *first* AUG (164).
One can envision a number of mechanisms by which ribosomes might
initiate at a downstream AUG--in addition to, or instead of, the first
AUG:

a)   Due to absence of secondary structure or to some other peculiarity of the leader region, 40S subunits might migrate more rapidly than
normal. This might, in turn, reduce the efficiency of recognition of
AUG codons--resulting in some ribosomes bypassing the first and
initiating instead at the second or third AUG triplet. This possibility is supported to some extent by experiments carried out with
denatured reovirus RNA (114).

b)   40S subunits might scan the 5'-noncoding region as usual, but
fail to stop at the first AUG due to absence of an additional signal
that is needed for initiation. The nonfunctional upstream AUG
triplets in these messages, however, resemble functional initiator
codons in having a purine in at least one of the two conserved
positions noted in Table 2. If an additional signal is needed it has
not been identified.

c)   It is possible that migration of 40S ribosomes is discontinuous;
i.e., ribosomes might skip over portions of the leader involved in
stable base-paired structures. In the case of SV40, Piatak *et al.*
(165) have proposed that nonfunctional AUG codons in the leader
region might be sequestered by base pairing. However, the fact that
in some messages the *functional* AUG appears to be present in the stem
of a hairpin structure (27, 34, 59, 113) argues that conformational
constraints do not effectively exclude ribosomes. This is, admit-
tedly, a weak rebuttal since the available information about mRNA
secondary structure is meager.

d)   Translation might begin at the first AUG, terminate after a short
distance, and then *re-initiate* at the "proper" AUG. This mechanism
is tenable for SV40 16S mRNA, RSV messages, mouse α-amylase mRNA
and *Dictyostelium* M4 mRNA--since the first AUG in each of these
messages is followed by an in-phase terminator codon, upstream

of the "correct" AUG. In the case of SV40 late 19S mRNA, however,
a ribosome initiating at the first AUG would not encounter a nonsense
codon until after the ribosome had entered the coding region of the
VP2 gene, making it difficult for re-initiation to occur at the VP2
initiation site. Although re-initiation at a downstream AUG is
theoretically possible for the other examples listed, it is difficult
to understand how eukaryotic ribosomes could re-initiate internally
in just these few instances, when they appear unable to do so in the
majority of messages.

       All of the mechanisms described above might be considered
modified forms of the scanning model. The most radical alternative
is that ribosomes enter *directly* at the internal AUG. Consider the
hypothesis that, after binding at the 5'-terminus of SV40 16S mRNA,
the ribosome is transferred directly to the AUG initiator codon at
position 239-241. This would necessitate a signal identifying that
AUG as the target site. Substitution of the VP1 coding sequence
(and five nucleotides preceding the AUG) by the coding and most of
the 5'-noncoding sequence of rabbit β-globin mRNA did not prevent
ribosomes from finding the correct internal initiator codon in 16S
mRNA--in this case, that of β-globin (167). This implies that the
putative signal for internal binding does not reside in the sequences
immediately flanking the VP1 initiator codon. This conclusion is
reinforced by a recent experiment in which introduction of heterologous sequences (namely, a portion of the mouse β-globin gene, which
includes several AUG codons) near the 5'-end of SV40 16S mRNA *prevented* translation of VP1 (231). In this case, the normal VP1 coding
sequence and some 200 nucleotides immediately upstream of the
initiator codon were retained--but ribosomes failed to initiate at
the VP1 site. One might also argue that if the sequences flanking
the AUG initiator codon for VP1 included a signal that permitted
direct attachment of ribosomes, then VP1 should be translated from
19S as well as from 16S mRNA. But only 16S mRNA appears to direct
synthesis of VP1, *in vitro*. Even in these unusual messages (SV40,
RSV, etc.) in which translation does not begin at the first AUG, the
requirement for a nearby 5'-terminus persists. Although there is as
yet no answer to the question of how--in a handful of messages--
eukaryotic ribosomes are able to initiate at the second, third, or
fourth AUG, the possibility that the scanning mechanism operates in
a "relaxed" manner seems somewhat more likely than the possibility
that ribosomes attach directly at an internal initiation site in
these messages.

QUESTIONS AND SPECULATIONS

An Economical Message Might Initiate at the First and the Second AUG

       One consequence of the scanning mechanism is that the second AUG
in a message might be viewed as a potential "secondary ribosome
binding site". If a small percentage of 40S subunits were to bypass

the normal initiation site and stop instead at the second AUG, two polypeptides might be synthesized from a single message. Rose has presented evidence for low level binding of ribosomes at the second AUG during *in vitro* translation of one of the vesicular stomatitis virus messages (63), although there is no evidence that this occurs *in vivo*. The interesting phenomenon that Ehrenfeld has described (144) in which ribosomes initiate at either or both of two sites in poliovirus mRNA--synthesizing two quite unrelated polypeptides--might have its explanation in the ability of ribosomes to sometimes bypass the first, and initiate at the second AUG. Knowledge of the 5'-proximal sequence of poliovirus mRNA should help to unravel this phenomenon.

From a different perspective, it would seem that one way to ensure exclusive and efficient translation of the "correct" polypeptide might be to place a second AUG very close to, and in the same reading frame as, the first. This sort of redundancy is noticeable in a few eukaryotic messages, including bovine preproparathyroid hormone mRNA (35), SV40 VP1 mRNA (73), rat preproinsulin mRNA (36, 37) and adenovirus hexon mRNA (66). Obviously, it is more likely that the second AUG in each of these messages reflects a need for methionine in the corresponding position of the polypeptide, than that the second AUG enhances the fidelity of initiation. But the possibility exists. In the case of an immunoglobulin light chain mRNA which has the sequence AUGxxxAUG, initiation at both AUG's has been demonstrated *in vitro* (168), consistent with the notion that a second in-phase initiator codon can serve as a back-up to the first. On the other hand, adenovirus type 2 hexon protein, synthesized in vitro, has a unique amino-terminal sequence (66a); if the AUG triplets in positions 5 and 6 (downstream of the normal initiator codon) function as alternate initiation sites, they do so with very low efficiency. The fact that the intiator codon is *not* reiterated (either in tandem or at two nearby sites) in most eukaryotic messages is further evidence that a single AUG triplet generally functions quite efficiently to stop migration of the 40S subunit and to initiate peptide bond formation.

Role of 5'-Terminal Methylated Residues

The mechanism by which the methylated cap promotes initiation is obscure. Denaturation of mRNA obviates the requirement for m^{7}G (114), raising the interesting possibility that the cap might mediate unfolding of the 5'-terminus of the native message, thereby facilitating ribosome entry. The observation that messenger RNAs are less dependent on the m^{7}G cap at low ionic strength (169-171), which increases the conformational flexibility of mRNA, might be taken as support for this hypothesis.

The significance of methylation of the 5'-penultimate and sub-penultimate nucleotides also remains to be explained. The extent of methylation of these residues varies from one cellular message to another (27, 172), and some viral messages lack 2'-O-methylated residues adjacent to the cap(173-175). *In vivo* experiments using inhibitors of methylation have shown that 2'-O-methylation of the penultimate nucleotide is not absolutely required for translation (176, 177). Although 2'-O-methylation of the cap-adjacent residue was reported to enhance binding of synthetic polymers to ribosomes in certain cell-free extracts (178), the effect has not been observed consistently (179). Since 2'-O-methylation causes the 5'-terminus of the RNA molecule to have a more extended structure (180), its role may be to augment the m⁷G cap in making the 5'-end of the RNA chain more accessible to ribosomes.

Determinants of Messenger Efficiency

The question of what features in mRNA modulate translational efficiency has just begun to be explored. The existence of messenger-discriminatory initiation factors (181, 182) is part of the answer--but what features in messenger RNAs underlie their differential response to initiation factors ? Although I emphasized above that drastic alteration of the 5'-untranslated segment of a message does not impair the *fidelity* of initiation, one might expect (even though experimental evidence is lacking) that the 5'-noncoding portion of a message would modulate translational *efficiency*. Thus, the very interesting observation (U. Schibler, R.A. Young and O. Hagenbuchle, personal communication) that the leader sequence on mouse α-amylase mRNA varies, depending on the tissue of origin, might provide a basis for regulating expression of that enzyme in various tissues.

Translational efficiency is not simply related to length of the 5'-noncoding region, if one considers a large spectrum of eukaryotic messages (169). On the other hand, in a number of studies in which a small set of related viral messages were compared, there was an inverse relationship between translational efficiency and length of the 5'-noncoding segment. Examples include brome mosaic virus mRNAs (183, 184), vesicular stomatitis virus mRNAs (185), turnip yellow mosaic virus RNAs (186) and the adenovirus message for polypeptide IX versus other late adenovirus messages (187). But other studies contradict the simple conclusion that the most efficient messages are those in which the AUG is closest to the cap. For example, β-globin mRNA--which has a somewhat longer 5'-noncoding region than α-globin mRNA--is more efficiently initiated both *in vitro* and *in vivo* (188). Moreover, SV40 late 16S mRNA, which has an extraordinarily long 5'-leader segment, is more resistant than bulk host mRNA to hypertonic conditions, which presumably discriminate against less efficient messages (189, 190). Thus, *if* a long leader tends to reduce translational efficiency, there must be ways to circumvent the effect of length. It may be that conformational folding within a long leader region, rather than length *per se*, lowers translational efficiency.

A negative effect of secondary structure has been postulated to explain the difference in efficiency between α- and β-globin mRNAs (21, 27), and systematic studies with base-substituted reovirus mRNAs have indeed revealed an inverse correlation between translational efficiency and the stability of the secondary structure (114). Obviously, much additional work is needed to define which specific region(s) of the message are important in this reagard. Enhancement of translation following treatment with methylmercury hydroxide (191) is consistent with the notion that mRNA secondary structure has a general negative effect on translational efficiency.

Although I have argued above that eukaryotic ribosomes tolerate extensive variation in the 5'-proximal sequence of messenger RNAs, in fact, as shown in Table 2, the nucleotides flanking the AUG initiator codon are not random. It seems more than coincidental that most eukaryotic initiation sites have the sequence AUGG (complementary to the sequence 3'-UACC-5' in the anticodon loop of eukaryotic initiator Met-tRNA), whereas the most common sequence in prokaryotic initiation sites is AUGA (complementary to the sequence 3'-UACU-5' in prokaryotic fMet-tRNA). In the case of one prokaryotic message, there is evidence that the flanking A residue enhances initiation (8). In view of the great number of spurious "initiation sites" that can be activated by fragmentation of reovirus messenger RNA (126) or by mutational relocation of the AUG codon in the yeast cytochrome c gene (100), presence of a particular set of flanking sequences cannot be an *absolute* requirement for initiation in eukaryotes. Sherman (100) has noted, on the other hand, that the efficiency of translation in certain cytochrome c revertants is influenced by the sequence preceding the AUG codon. It might be that interaction of Met-tRNA with the AUG codon is facilitated by the presence of purines in positions 1 and 7, in somewhat the same way as the sequences flanking terminator codons have been shown to modulate the efficiency of binding of suppressor tRNAs (192-194).

Finally, one must consider the ability of distant regions of the message to directly or indirectly influence initiation. Deletion of 3'-noncoding sequences from rabbit β-globin mRNA apparently did not impair translation *in vitro* (141). And no functional difference has been discerned among the multiple forms of mouse dihydrofolate reductase mRNA, in which the 3'-untranslated region varies from 80 to over 900 nucleotides (D. Setzer, M. McGrogan, J. Nunberg and R. Schimke, personal communication). On the other hand, the 3'-noncoding region does seem to influence the *in vivo* stability of some messages. When the normally untranslated 3'-region of α-globin mRNA is translated (as in production of Hb Constant Spring), the stability of α-globin mRNA is drastically reduced (195). Similarly, although the poly(A) segment commonly found on eukaryotic messages is not an absolute requirement for *in vitro* translation (196), in long term experiments

presence of a 3'-terminal poly(A) tract was shown to enhance re-
initiation (196, 197). Whether the poly(A) tail merely stabilizes
mRNA against degradation by nucleases or directly promotes re-initiation of translation remains to be determined.

Translation of Viral Messages

Viral messenger RNAs have been observed to differ from host mRNAs
in their sensitivity to interferon (198, 199), high salt (189, 200-
202), and certain antibiotics (203). The implied difference in the
structure of (most) viral versus (most) cellular messages makes it
possible for viruses to usurp the translational machinery of the
host cell. In the case of picornaviruses, it has been known for some
time that the viral mRNA is uncapped (204). Thus, the observation
that poliovirus inhibits translation of capped, cellular messages by
inactivating a cap-binding protein (205) came as no surprise--although
the experimental proof was not easily obtained ! A more subtle
mechanism of host shut-off is required for vesicular stomatitis virus,
herpes, adenovirus, vaccinia, alphaviruses, and other agents whose
messenger RNAs (like those of the host cell) rely on the m^7G cap.
Studies are underway to identify the altered translational compo-
nent(s) in vaccinia virus infected cells (206, 207). Although adenovirus causes a striking inhibition of host mRNA transport (208),
interference at the level of translation might also be needed to
preclude binding of ribosomes to stable host mRNAs pre-existing in
the cytoplasm. The viral-coded 100K protein associated with messenger
ribonucleoprotein particles in adenovirus-infected cells might be
involved in translational regulation (209). In the case of vesicular
stomatitis virus, the molecular mechanism effecting shut off of host
protein synthesis is not yet understood, but analysis of mutants
indicates that, under some circumstances, translation of viral as
well as host mRNA can be inhibited (210, 211). Moreover, abrogation
of the viral function that inhibits host protein synthesis seems to
be a key step in establishing persistent infections (210). One can
anticipate that some interesting regulatory mechanisms will be uncovered when the molecular bases of these phenomena are understood.

A final intriguing question about RNA viruses concerns the
mechanism(s) by which ribosomes are excluded from the population of
viral "plus strands" which must participate in other functions--such
as replication and assembly of progeny virions. Noncovalent circularization of the RNA genome of Sindbis virus (212) may accomplish
this objective. Another possibility is covalent attachment of a
protein to the 5'-terminus of virion RNA--as occurs with poliovirus
(213), calicivirus (214), cowpea mosaic virus (215, 216), tobacco
ringspot virus (217) and Southern bean mosaic virus RNA (218).
Although our current understanding of the translation initiation
mechanism in eukaryotes predicts that the 5'-linked protein should
effectively prevent ribosome binding, direct proof of this has been
difficult to obtain due to the ubiquity in cell-free extracts of an

enzyme that cleaves the protein/RNA linkage (219). This activity seems to be lower in reticulocyte lysates than in other cell-free extracts, and Golini *et al.* (220) have exploited this finding to show that poliovirus RNA with covalently-linked protein *can* bind to reticulocyte ribosomes. This interesting and unexpected result will be easier to evaluate when the exact RNA sequence selected by ribosomes (in the presence and absence of 5'-terminal protein) has been identified. Since *in vitro* translation of protein-linked RNA is at variance with the earlier observation that all of the poliovirus mRNA extracted from polysomes *in vivo* has a free 5'-terminus (221), additional studies are needed to clarify the situation. On the other hand, it is clear that the 5'-linked protein is not *required* for translation, since prior treatment with protease did not impair *in vitro* translation of mengovirus (222), foot-and-mouth disease virus RNA (17), or cowpea mosaic virus RNA (216).

ACKNOWLEDGEMENT/NOTES

I am indebted to Dr. Aaron Shatkin for helpful conversations on many of the topics discussed above, and for critical reading of the manuscript. I am also grateful to many colleagues who sent preprints, and who permitted me to cite their unpublished data. Work from my laboratory was supported by a research grant and a Career Development Award from the National Institute of Allergy and Infectious Diseases.

The literature search for this review extends through June, 1980.

REFERENCES

1. Shine, J. and Dalgarno, L. (1974), Proc. Nat. Acad. Sci. USA, 71, 1342-1346.
2. Steitz, J.A. (1979), "Genetic Signals and Nucleotide Sequences in Messenger RNA," pp. 349-399. In *Biological Regulation and Development*, Vol. I, Gene Expression (R.F. Goldberger, ed.), Plenum Press, New York and London.
3. Steitz, J.A. and Jakes, K. (1975), Proc. Nat. Acad. Sci. USA, 72, 4734-4738.
4. Dunn, J.J., Buzash-Pollert, E. and Studier, F.W. (1978), Proc. Nat. Acad. Sci. USA, 75, 2741-2745.
5. Lodish, H.F. (1970), J. Mol. Biol., 50, 689-702.
6. Iserentant, D. and Fiers, W. (1980), Gene 9, 1-12.
7. Borisova, G.P., Volkova, T.M., Berzin, V., Rosenthal, G. and Gren, E.J. (1979), Nucleic Acids Res., 6, 1761-1774.
8. Taniguchi, T. and Weissmann, C. (1978), J. Mol. Biol., 118, 533-565.
9. Jacobson, M.F. and Baltimore, D. (1968), Proc. Nat. Acad. Sci. USA, 61, 77-84.

10. Petersen, N.S. and McLaughlin, C.S. (1973), J. Mol. Biol., 81,
     33-45.
11. Smith, A.E. (1977), "Cryptic initiation sites in eukaryotic virus
     mRNAs." pp. 37-46. In *Gene Expression* (B.F.C. Clark, et al.,
     eds), Federation of European Biological Societies Symposium,
     Vol. 43, Pergamon Press, Oxford.
12. Shatkin, A.J. (1976), Cell, 9, 645-653.
13. Kozak, M. and Shatkin, A.J. (1978), J. Biol. Chem., 253, 6568-
     6577.
14. Kozak, M. and Shatkin, A.J. (1978), Cell, 13, 201-212.
15. Filipowicz, W. (1978), FEBS Letters, 96, 1-11.
16. Villarreal, L.P., White, R.T. and Berg, P. (1979), J. Virol.,
     29, 209-219.
17. Sangar, D.V., Black, D.N., Rowlands, D.J., Harris, T.J.R. and
     Brown, F. (1980), J. Virol., 33, 59-68.
18. McReynolds, L., O'Malley, B.W., Nisbet, A.D., Fothergill, J.E.
     Givol, D., Fields, S., Robertson , M. and Brownlee, G.G. (1978),
     Nature, London, 273, 723-728.
19. Kuebbing, D. and Liarakos, C.D. (1978), Nucleic Acids Res., 5,
     2253-2266.
20. Cochet, M., Gannon, F., Hen, R., Maroteaux, L., Perrin, F. and
     Chambon, P. (1979), Nature, London, 282, 567-574.
21. Baralle, F.E. (1977), Nature, London, 267, 279-281.
22. Heindell, H.C., Liu, A., Paddock, G.V., Studnicka, G.M., and
     Salser, W.A. (1978), Cell, 15, 43-54.
23. Baralle, F.E. (1977), Cell, 12, 1085-1095.
24. Wilson, J.T., Wilson, L.B., Reddy, V.B., Cavallesco, C., Ghosh,
     P.K., deRiel, J.K., Forget, B.G., and Weissman, S.M. (1980),
     J. Biol. Chem., 255, 2807-2815.
25. Chang, J.C., Temple, G.F., Poon, R., Neumann, K.H., and WaiKan,
     Y. (1977), Proc. Nat. Acad. Sci. USA, 74, 5145-5149.
26. Baralle, F.E. and Brownlee, G.G. (1978), Nature, London, 274,
     84-87.
27. Pavlakis, G.N., Lockard, R.E., Vamvakopoulos, N., Rieser, L.,
     RajBhandary, U.L. and Vournakis, J.N. (1980), Cell, 19, 91-102.
28. Efstratiadis, A., Kafatos, F.C. and Maniatis, T. (1977), Cell,
     10, 571-585.
29. Lockard, R.E and RajBhandary, U.L. (1976), Cell, 9, 747-760.
30. Baralle, F.E. (1977), Cell, 10, 549-558.
31. Konkel, D.A., Maizel, J.V. and Leder, P. (1979), Cell, 18, 865-
     873.
32. Richards, R.I., Shine, J., Ullrich, A., Wells, J.R.E. and
     Goodman, H.M. (1979), Nucleic Acids Res., 7, 1137-1146.
33. Chang, J.C., Poon, R., Neumann, K.H. and Wai Kan, Y. (1978),
     Nucleic Acids Res., 5, 3515-3522.
34. Tsujimoto, Y. and Suzuki, Y. (1979), Cell, 18, 591-600.
35. Kronenberg, H.M., McDevitt, B.E., Majzoub, J.A., Nathans, J.,
     Sharp, P.A., Potts, J.T., JR. and Rich, A. (1979), Proc. Nat.
     Acad. Sci. USA, 76, 4981-4985.

36. Cordell, B., Bell, G., Tischer, E., DeNoto, F.M., Ullrich, A., Pictet, R., Rutter, W.J. and Goodman, H.M. (1979), Cell, 18, 533-543.

37. Lomedico, P., Rosenthal, N., Efstratiadis, A., Gilbert, W., Kolodner, R. and Tizard, R. (1979), Cell, 18, 545-558.

38. Bell, G.I., Pictet, R.L., Rutter, W.J., Cordell, B., Tischer, E. and Goodman, H.M. (1980), Nature, London, 284, 26-32.

39. Perler, F., Efstratiadis, A., Lomedico, P., Gilbert, W., Kolodner, R. and Dodgson, J. (1980), Cell, 20, 555-566.

40. Nunberg, J.H., Kaufman, R.J., Chang, A., Cohen, S.N. and Schimke, R.T. (1980), Cell, 19, 355-364.

41. Holland, J.P. and Holland, M.J. (1979), J. Biol. Chem., 254, 9839-9845.

42. Sures, I., Levy, S. and Kedes, L.H. (1980), Proc. Nat. Acad. Sci. USA, 77, 1265-1269.

43. Sures, I., Lowry, J. and Kedes, L.H. (1978), Cell. 15, 1033-1044;

44. Busslinger, M., Portmann, R., Irminger, J.C. and Birnstiel, M.L. (1980), Nucleic Acids Res., 8, 957-977.

45. Schaffner, W., Kunz, G., Daetwyler, H., Telford, J., Smith, H.O., and Birnstiel, M.L. (1978), Cell, 14, 655-671.

46. Hentschel, C., Irminger, J-C., Bucher, P. and Birnstiel, M.L. (1980), Nature, London, 285, 147-151.

47. Houghton, M., Stewart, A.G., Doel, S.M., Emtage, J.S., Eaton, M.A.W., Smith, J.C., Patel, T.P., Lewis, H.M., Porter, A.G., Birch, J.R., Cartwright, T. and Carey, N.H. (1980), Nucleic Acids Res., 8, 1913-1930.

48. Derynck, R., Content, J., DeClercq, E., Volckaert, G., Tavernier, J., Devos, R. and Fiers, W. (1980), Nature, London, 285, 542-547.

49. Mantei, N., Schwarzstein, M., Streuli, M., Panem, S., Nagata, S. and Weissmann, C. (1980), Gene. 10, 1-10.

50. Firtel, R.A., Timm, R., Kimmel, A.R. and McKeown, M. (1979), Proc. Nat. Acad. Sci. USA, 76, 6206-6210.

51. Rowekamp, W., Poole, S. and Firtel, R.A. (1980), Cell, 20, 495-505.

52. Leung, D.W., Browning, K.S., Heckman, J.E., RajBhandary, U.L. and Clark, J.M., Jr. (1979), Biochemistry, 18, 1361-1366.

53. Browning, K.S., Leung, D.W, and Clark, J.M., Jr. (1980), Biochemistry, 19, 2276-2283.

54. Guilley, H. and Briand, J.P. (1978), Cell, 15, 113-122.

55. Richards, K., Guilley, H., Jonard, G. and Hirth, L. (1978), Eur. J. Biochem., 84, 513-519.

56. Jonard, G., Richards, K., Mohier, E. and Gerlinger, P. (1978), Eur. J. Biochem., 84, 521-531.

57. Guilley, H., Jonard, G., Kukla, B. and Richards, K.E. (1979), Nucleic Acids Res., 6, 1287-1308.

58. Dasgupta, R., Shih, D.S., Saris, C. and Kaesberg, P. (1975), Nature, London, 256, 624-628.

59. Ahlquist, P., Dasgupta, R., Shih, D.S., Zimmern, D. and Kaesberg, P. (1979), Nature, London, 281, 277-282.

60. Koper-Zwarthoff, E.C., Lockard, R.E., Alzner-deWeerd, B.,
    RajBhandary, U.L. and Bol, J.F. (1977), Proc. Nat. Acad. Sci.
    USA, 74, 5504-5508.
61. Kozak, M. (1977), Nature, London, 269, 390-394.
62. Darzynkiewicz, E. and Shatkin, A.J. (1980), Nucleic Acids Res.,
    8, 337-350.
63. Rose, J.K. (1980), Cell, 19, 415-421.
64. Rose, J.K. (1978), Cell, 14, 345-353.
65. Rowlands, D.J., (1979), Proc. Nat. Acad. Sci. USA, 76, 4793-4797.
66. Akusjarvi, G. and Pettersson, U. (1979), Cell, 16, 841-850.
66a. Anderson, C.W. and Lewis, J.B. (1980), Virology, 104, 27-41.
67. Zain, S., Sambrook, J., Roberts, R.J., Keller, W., Fried, M. and
    Dunn, A.R. (1979), Cell, 16, 851-861.
68. Alestrom, P., Akusjarvi, G., Perricaudet, M., Mathews, M.B.,
    Klessig, D.F. and Pettersson, U. (1980), Cell, 19, 671-681.
69. Perricaudet, M., Akusjarvi, G., Virtanen, A. and Pettersson, U.
    (1979), Nature, London, 281, 694-696.
70. Van Ormondt, H., Maat, J., DeWaard, A. and Van der Eb, A.J.
    (1978), Gene, 4, 309-328.
71. Reddy, V.B., Ghosh, P.K., Lebowitz, P., Piatak, M. and Weissman,
    S.M. (1979), J. Virol., 30, 279-296.
72. Reddy, V.B., Thimmappaya, B., Dhar, R., Subramanian, K.N., Zain,
    B.S., Pan, J., Ghosh, P.K., Celma, M.L. and Weissman, S.M. (1978)
    Science, 200, 494-502.
73. Ghosh, P.K., Reddy, V.B., Swinscoe, J., Choudary, P.V., Lebowitz,
    P. and Weissman, S.M. (1978), J. Biol. Chem., 253, 3643-3647.
74. Ghosh, P.K., Reddy, V.B., Swinscoe, J., Lebowitz, P. and Weissman
    S.M. (1980), J. Mol. Biol., 126, 813-846.
75. Arrand, J.R., Soeda, E., Walsh, J.E., Smolar, N. and Griffin,
    B.E. (1978), J. Virol., 33, 606-618.
76. Kamen, R., Favaloro, J. and Parker, J. (1980), J. Virol., 33,
    637-651.
77. Flavell, A.J., Cowie, A., Arrand, J.R. and Kamen, R. (1980),
    J. Virol., 33, 902-908.
78. Zuckermann, M., Manor, H., Parker, J. and Kamen, R. (1980),
    Nucleic Acids Res., 8, 1505-1519.
79. Dhar, R., Chanock, R. and Lai, C-J. (1980), Cell, 21, 495-500.
80. Legon, S., Flavell, A.J., Cowie, A. and Kamen, R. (1979), Cell,
    16, 373-388.
81. Reddy, V.B., Ghosh, P.K., Lebowitz, P. and Weissman, S.M. (1978),
    Nucleic Acids Res., 5, 4195-4213.
82. Kozak, M. (1978), Cell, 15, 1109-1123.
83. Van de Voorde, A., Contreras, R., Rogiers, R. and Fiers, W.
    (1976), Cell, 9, 117-120.
84. Hewick, R.M., Mellor, A., Smith, A.E. and Waterfield, M.D. (1980)
    J. Virol., 33, 631-636.
85. Krug, R.M., Broni, B.A. and Bouloy, M. (1979), Cell, 18, 329-334.
86. Bernard, O., Hozumi, N. and Tonegawa, S. (1978), Cell, 15, 1133-
    1144.
87. Fiddes, J.C. and Goodman, H.M. (1979), Nature, London, 281, 351-356.
    351-356.

88. Valenzuela, P., Gray, P., Quiroga, M., Zaldivar, J., Goodman, H.M. and Rutter, W.J. (1979), Nature, London, 280, 815-819.
89. Smith, M., Leung, D.W., Gillam, S., Astell, C.R., Montgomery, D.L. and Hall, B.D. (1979), Cell, 16, 753-761.
90. Montgomery, D.L., Leung, D.W., Smith, M., Shalit, P., Faye, G. and Hall, B.D. (1980), Proc. Nat. Acad. Sci. USA, 77, 541-545.
91. Both, G.W. and Air, G.M. (1979), Eur. J. Biochem., 96, 363-372.
92. Jenkins, J.R. (1979), Nature, London, 279, 809-811.
93. Roskam, W.G. and Rougeon, F. (1979), Nucleic Acids Res., 7, 305-320.
94. Friedmann, T., LaPorte, P. and Esty, A. (1978), J. Biol. Chem., 253, 6561-6567.
95. Robertson, M.A., Staden, R., Tanaka, Y., Catterall, J.F., O'Malley, B.W. and Brownlee, G.G. (1979), Nature, London, 278, 370-372.
96. Van Ooyen, A., Van den Berg, J., Mantei, N. and Weissmann, C. (1979), Science, 206, 337-344.
97. Konkel, D.A., Tilghman, S.M. and Leder, P. (1978), Cell, 15, 1125-1132.
98. Atkins, J.F. (1979), Nucleic Acids Res., 7, 1035-1041.
99. Post, L.E. and Nomura, M. (1980), J. Biol. Chem., 255, 4660-4666.
100. Sherman, F., Stewart, J.W. and Schweingruber, A.M. (1980), Cell, 20, 215-222.
101. Catterall, J.F., O'Malley, B.W., Robertson, M.A., Staden, R., Tanaka, Y. and Brownlee, G.G. (1978), Nature, London, 257, 510-513.
102. Brown, J.C. and Smith, A.E. (1970), Nature, London, 226, 610-612.
103. Stewart, J.W., Sherman, F., Shipman, N.A. and Jackson, M. (1971), J. Biol. Chem., 246, 7429-7445.
104. Sherman, F., McKnight, G. and Stewart, J.W. (1980), Biochim. Biophys. Acta, 609, 343-346.
105. Hagenbuchle, O., Santer, M., Steitz, J. and Mans, R.J, (1978), Cell, 13, 551-563.
106. Both, G.W. (1979), FEBS Letters, 101, 220-224.
107. DeWachter, R. (1979), Nucleic Acids Res., 7, 2045-2054.
108. Nakashima, K., Darzynkiewicz, E. and Shatkin, A.J. (1980), Nature, London, 286, 226-230.
109. Baker, C.C. Herisse, J., Courtois, G., Galibert, F. and Ziff, E. (1979), Cell, 18, 569-580.
110. Ziff, E.B. and Evans, R.M. (1978), Cell, 15, 1463-1475.
111. Soeda, E., Arrand, J.R. and Griffin, B.E. (1980), J. Virol., 33, 619-630.
112. Legon, S. (1979), J. Mol. Biol., 134, 219-240.
113. Schroeder, H.W., Liarakos, C.D., Gupta, R.C., Randerath, K. and O'Malley, B.W. (1979), Biochemistry, 18, 5798-5808.
114. Kozak, M. (1980), Cell, 19, 79-90.
115. Kozak, M. (1980), J. Mol. Biol., 144, 291-304.
116. Both, G.W., Banerjee, A.K. and Shatkin, A.J. (1975), Proc. Nat. Acad. Sci. USA, 72, 1189-1193.

117. Muthukrishnan, S., Moss, B., Cooper, J.A. and Maxwell, E.S. (1978), J. Biol. Chem., 253, 1710-1715.
118. Lockard, R.E. and Lane, C. (1978), Nucleic Acids Res., 5, 3237-3247.
119. Hickey, E.D., Weber, L.A. and Baglioni, C. (1976), Proc. Nat. Acad. Sci. USA, 73, 19-23.
120. Munns, T.W., Morrow, C.S., Hunsley, J.R., Oberst, R.J. and Liszewski, M.K. (1979), Biochemistry, 18, 3804-3810.
121. Beemon, K. and Hunter, T. (1977), Proc. Nat. Acad. Sci. USA, 74, 3302-3306.
122. Canaani, D., Revel, M. and Groner, Y. (1976), FEBS Letters, 64, 326-331.
123. Shih, D.S., Dasgupta, R. and Kaesberg, P. (1976), J. Virol., 19, 637-642.
124. Kozak, M. (1979), Nature, London, 280, 82-85.
125. Pelham, H. (1979), FEBS Letters, 100, 195-199.
126. Kozak, M. (1980), J. Virol., 35, 748-756.
127. Lawrence, C.B. (1980), Nucleic Acids Res., 8, 1307-1317.
128. Both, G.W., Furuichi, Y., Muthukrishnan, S. and Shatkin, A.J. (1976), J. Mol. Biol., 104, 637-658.
129. Legon, S., Model, P. and Robertson, H.D. (1977), Proc. Nat. Acad. Sci. USA, 74, 2692-2696.
130. Filipowicz, W. and Haenni, A-L. (1979), Proc. Nat. Acad. Sci. USA, 76, 3111-3115.
131. Kozak, M. (1979), J. Biol. Chem., 254, 4731-4738.
132. Kozak, M. (1980), Cell, 22, 459-467.
133. Hopper, J.E. and Rowe, L.B. (1978), J. Biol. Chem., 253, 7566-7569.
134. Rosenberg, M. and Paterson, B.M. (1979), Nature, London, 279, 696-701.
135. Atkins, J.F., Steitz, J.A., Anderson, C.W. and Model, P. (1979), Cell, 18, 247-256.
136. Davies, J.W. and Kaesberg, P. (1973), J. Virol., 12, 1434-1441.
137. Morrison, T.G. and Lodish, H.F. (1973), Proc. Nat. Acad. Sci. USA, 70, 315-319.
138. Haegeman, G., Iserentant, D., Gheysen, D. and Fiers, W. (1979), Nucleic Acids Res., 7, 1799-1814.
139. Subramanian, K.N. (1979), Proc. Nat. Acad. Sci. USA, 76, 2556-2560.
140. Bendig, M.M. and Folk, W.R. (1979), J. Virol., 32, 530-535.
141. Kronenberg, H.M., Roberts, B.E. and Efstratiadis, A. (1979), Nucleic Acids Res., 6, 153-166.
142. Dunn, A.R., Mathews, M.B., Chow, L.T., Sambrook, J. and Keller, W. (1978), Cell, 15, 511-526.
143. Kafatos, F.C., Efstratiadis, A., Forget, B.G. and Weissman, S.M. (1977), Proc. Nat. Acad. Sci. USA, 74, 5618-5622.
144. Knauert, F. and Ehrenfeld, E. (1979), Virology, 93, 537-546.
145. Celma, M.L. and Ehrenfeld, E. (1975), J. Mol. Biol., 98, 761-780.
146. Van Steeg, H., Pranger, M.H., Van der Zeijst, B., Benne, R. and Voorma, H.O. (1979), FEBS Letters, 108, 292-298.

147. Fritsch, C., Mayo, M.A. and Hirth, L. (1977), Virology, 77, 722-732.
148. Salvato, M.S. and Fraenkel-Conrat, H. (1977), Proc. Nat. Acad. Sci. USA, 74, 2288-2292.
149. Salerno-Rife, T., Rutgers, T. and Kaesberg, P. (1980), J. Virol., 34, 51-58.
150. Salomon, R., Bar-Joseph, M., Soreq, H., Gozes, I. and Littauer, U.Z. (1978), Virology, 90, 288-298.
151. Pinck, L., Franck, A. and Fritsch, C. (1979), Nucleic Acids Res., 7, 151-166.
152. Pelham, H.R.B. (1979), Virology, 96, 463-477.
153. Neeleman, L. and Van Vloten-Doting, L. (1979), pp. 410-417, In *Methods in Enzymology*, Vol. 60 (K. Moldave and L. Grossman, eds.) Academic Press, New York.
154. Dolja, V.V., Sokolova, N.A., Tiulkina, L.G. and Atabekov, J.G. (1979), Molec. Gen. Genet., 175, 93-97.
155. Somogyi, P. and Dobos, P. (1980), J. Virol., 33, 129-139.
156. Chen, J.H. and Spector, A. (1977), Proc. Nat. Acad. Sci. USA, 74, 5448-5452.
157. Cohen, L.H., Westerhuis, L.W., de Jong, W.W. and Bloemendal, H. (1978), Eur. J. Biochem., 89, 259-266.
158. Larkins, B.A., Jones, R.A. and Tsai, C.Y. (1976), Biochemistry, 15, 5506-5511.
159. Wienand, U. and Feix, G. (1978), Eur. J. Biochem., 92, 605-611.
160. Westaway, E.G. (1977), Virology, 80, 320-335.
161. Wengler, G., Beato, M. and Wengler, G. (1979), Virology, 96, 516-529.
162. Wengler, G., Wengler, G. and Gross, H.J. (1979), Nature, London, 282, 754-756.
163. Lovinger, G.G. and Schochetman, G. (1980), Cell. 20, 441-449.
164. Darlix, J-L., Spahr, P-F., Bromley, P.A. and Jaton J-C. (1979), J. Virol., 29, 597-611.
165. Piatak, M., Ghosh, P.K., Reddy, V.B., Lebowitz, P. and Weissman, S.M. (1979), pp. 199-215, In *Extrachromosomal DNA (ICN/UCLA SYMPOSIUM)*, Academic Press, New York.
166. Siddell, S.G. and Smith, A.E. (1978), J. Virol., 27, 427-431.
167. Mulligan, R.C., Howard, B.H. and Berg, P. (1979), Nature, London, 277, 108-114.
168. Zemell, R., Burstein, Y. and Schechter, I. (1978), Eur. J. Biochem., 89, 187-193.
169. Herson, D., Schmidt, A., Seal, S., Marcus, A. and van Vloten-Doting, L. (1979), J. Biol. Chem., 254, 8245-8249.
170. Weber, L.A., Hickey, E.D., Nuss, D.L. and Baglioni, C. (1977), Proc. Nat. Acad. Sci. USA, 74, 3254-3258.
171. Weber, L.A., Hickey, E.D. and Baglioni, C. (1978), J. Biol. Chem., 253, 178-183.
172. Lockard, R.E (1978), Nature, London, 275, 153-154.
173. Colonno, R.J. and Stone, H.O. (1976), Nature, London, 261, 611-614.

174. Hefti, E., Bishop, D.H.L., Dubin, D.T. and Stollar, V. (1976), J. Virol., 17, 149-159.

175. Cleaves, G.R. and Dubin, D.T. (1979), Virology, 96, 159-165.

176. Kaehler, M., Coward, J. and Rottman, F. (1979), Nucleic Acids Res., 6, 1161-1175.

177. Dimock, K. and Stoltzfus, C.M. (1979), J. Biol. Chem., 254, 5591-5594.

178. Muthukrishnan, S., Morgan, M., Banerjee, A. and Shatkin, A.J. (1976), Biochemistry, 15, 5761-5768.

179. Keith, J.M., Muthukrishnan, S. and Moss, B. (1978), J. Biol. Chem., 253, 5039-5041.

180. Kim, C.H. and Sarma, R.H. (1978), J. Amer. Chem. Soc., 100, 1571-1590.

181. Golini, F., Thach, S.S., Birge, C.H., Safer, B., Merrick, W.C. and Thach, R.E. (1976), Proc. Nat. Acad. Sci. USA, 73, 3040-3044.

182. Gette, W.R. and Heywood, S.M. (1979), J. Biol. Chem., 254, 9879-9885.

183. Shih, D.S. and Kaesberg, P. (1976), J. Mol. Biol., 103, 77-88.

184. Chroboczek, J., Puchkova, L. and Zagorski, W. (1980), J. Virol., 34, 330-335.

185. Lodish, H.F. and Froshauer, S. (1977), J. Biol. Chem., 252, 8804-8811.

186. Benicourt, C. and Haenni, A-L. (1978), Biochem. Biophys. Res. Commun., 84, 831-839.

187. Cherney, C.S. and Wilhelm, J.M. (1979), J. Virol., 30, 533-542.

188. Lodish, H.F. (1971), J. Biol. Chem., 246, 7131-7138.

189. England, J.M., Howett, M.K. and Tan, K.B. (1975), J. Virol., 16, 1101-1107.

190. Wolgemuth, D.J., Yu, H-Y. and Hsu, M-T. (1980), Virology, 101, 363-375.

191. Payvar, F. and Schimke, R.T. (1979), J. Biol. Chem., 254, 7636-7642.

192. Feinstein, S.I. and Altman, S. (1978), Genetics, 88, 201-219.

193. Akaboshi, E., Inouye, M. and Tsugita, A. (1976), Molec. Gen. Genet., 149, 1-4.

194. Colby, D.S., Schedl, P. and Guthrie, C. (1976), Cell, 9, 449-463.

195. Weatherall, D.J. and Clegg, J.B. (1979), Cell, 16, 467-479.

196. Soreq, H., Nudel, U., Salomon, R., Revel, M. and Littauer, U.Z., (1974), J. Mol. Biol., 88, 233-245.

197. Doel, M.T. and Carey, N.H. (1976), Cell 8, 51-58.

198. Yakobson, E., Prives, C., Hartman, J.R., Winocour, E. and Revel, M. (1977), Cell, 12, 73-81.

199. Metz, D.H. (1975), Cell, 6, 429-439.

200. Garry, R.F., Bishop, J.M., Parker, S., Westbrook, K., Lewis, G. and Waite, M. (1979), Virology, 96, 108-120.

201. Garry, R.F., Westbrook, K. and Waite, M. (1979), Virology, 99, 179-182.

202. Nuss, D.L., Oppermann, H. and Koch, G. (1975), Proc. Nat. Acad. Sci. USA, 72, 1258-1262.

203. Ramabhadran, T.V. and Thach, R.E. (1980), J. Virol., 34, 293-296.
204. Hewlett, M.J., Rose, J.K. and Baltimore, D. (1976), Proc. Nat. Acad. Sci. USA, 73, 327-330.
205. Trachsel, H., Sonenberg, N., Shatkin, A.J., Rose, J.K., Leong, K., Bergmann, J.E., Gordon, J. and Baltimore, D. (1980), Proc. Nat. Acad. Sci. USA, 77, 770-774.
206. Beaud, G. and Dru, A. (1980), Virology, 100, 10-21.
207. Schrom, M. and Bablanian, R. (1979), Virology, 99, 319-328.
208. Beltz, G.A. and Flint, S.J. (1979), J. Mol. Biol., 131, 353-373.
209. Tasseron-DeJong, J., Brouwer, J., Rietveld, K., Zoetemelk, C. and Bosch, L. (1979), Eur. J. Biochem., 100, 271-283.
210. Stanners, C.P., Francoeur, A.M. and Lam, T. (1977), Cell, 11, 273-281.
211. Davis, N.L. and Wertz, G.W. (1980), Virology, 103, 21-36.
212. Frey, T.K., Gard, D.L. and Strauss, J.H. (1979), J. Mol. Biol., 132, 1-18.
213. Lee, Y., Nomoto, A., Detjen, B. and Wimmer, E. (1977), Proc. Nat. Acad. Sci. USA, 74, 59-63.
214. Schaffer, F.L., Ehresmann, D.W., Fretz, M.K. and Soergel, M.E. (1980), J. Gen. Virol., 47, 215-220.
215. Daubert, S.D., Bruening, G. and Najarian, R.C. (1978), Eur. J. Biochem., 92, 45-51.
216. Stanley, J., Rottier, P., Davies, J.W., Zabel, P. and Van Kammen, A. (1978), Nucleic Acids Res., 5, 4505-4522.
217. Mayo, M.A., Barker, H. and Harrison, B.D. (1979), J. Gen. Virol., 43, 735-740.
218. Ghosh, A., Dasgupta, R., Salerno-Rife, T., Rutgers, T. and Kaesberg, P. (1979), Nucleic Acids Res., 7, 2137-2146.
219. Ambros, V., Pettersson, R.F. and Baltimore, D. (1978), Cell, 15, 1439-1446.
220. Golini, F., Semler, B.L., Dorner, A.J. and Wimmer, E. (1980), Nature, London, 287, 600-603.
221. Nomoto, A., Kitamura, N., Golini, F. and Wimmer, E. (1977), Proc. Nat. Acad. Sci. USA, 74, 5345-5349.
222. Perez Bercoff, R. and Gander, M. (1978), FEBS Letters, 96, 306-312.
223. Fiddes, J.C. and Goodman, H.M. (1979), Nature, London, 281, 351-356.
224. Cooke, N.E., Coit, D., Weiner, R.I., Baxter, J.D. and Martial, J.A. (1980), J. Biol. Chem., 255, 6502-6510.
225. Jung, A., Sippel, A., Grez, M., and Schütz, G. (1980), Proc. Nat. Acad. Sci. USA, 77, 5759-5763.
226. Hagenbüchle, O., Bovey, R., and Young, R.A. (1980), Cell, 21, 179-187.
227. Both, G.W. and Air, G.M. (1979), Eur. J. Biochem., 96, 363-372.
228. Caton, A.J. and Robertson, J.S. (1980), Nucleic Acids Res., 8, 2591-2603.
229. Maat, J. and Van Ormondt, H. (1979), Gene, 6, 75-90.
230. Shine, J., Czernilofsky, A.P., Friedrich, R., Bishop, J.M. and Goodman, H.M. (1977), Proc. Nat. Acad. Sci. USA, 74, 1473-1477.
231. Gruss, P. and Khoury, G. (1980), Nature, London, 286, 634-637.

A CLOSER LOOK AT THE 5'-END OF mRNA IN RELATION TO INITIATION

Aaron J. Shatkin

Roche Institute of Molecular Biology

Nutley, New Jersey 07110 (U.S.A.)

INTRODUCTION

Eukaryotic mRNAs are generally monocistronic (1, 2). Ribosomes attach at or near the 5'-end of the mRNA, and translation is limited to the 5'-proximal cistron (3, see preceding chapter of this volume). By contrast, on prokaryotes ribosomes initiate protein synthesis on some mRNAs at multiple internal sites, resulting in polycistronic translation (4). Two kinds of mRNA structural features may account at least in part for this functional difference between the messages of higher and lower organisms. One is the 5'-terminal cap structure, $m^7G(5')ppp(5')N$, which is unique to eukaryotic cellular and most viral mRNAs (5). A variety of experimental findings indicate that the presence of a cap promotes translation by facilitating stable binding of ribosomes (6, 7). Implied from these results is a recognition of the cap by putative cap binding proteins during initiation of protein synthesis. Messengers in bacteria apparently are not methylated or otherwise modified and nascent transcripts are read directly by a process of coupled transcription/translation (8). The mRNAs of prokaryotes include a purine-rich sequence, located ∿10 nucleotide upstream from initiator triplets, that helps to stabilize initiation complexes by base-pairing with the 3'-terminal end of 16S ribosomal RNA (4, 9). A corresponding complement to the 3'-region of 18S ribosomal RNA is not commonly found in the 5'-terminal leaders of eukaryotic mRNAs (10-13). However, some viral (14) and cellular messengers (15) do contain such a sequence which could affect initiation complex formation.

In an effort to explore whether proteins and/or ribosomal RNA interact with the 5'-terminal region of mRNA during eukaryotic initiation, two experimental approaches were used. In the first, initiation

complexes and preparations of partially purified initiation factors
were tested for cap binding proteins, i.e. polypeptides that recognize, bind to, and can be specifically cross-linked to the capped end
of mRNA. On the basis of this approach we isolated by m'GDP-Sepharose
affinity chromatography cap binding proteins that stimulate capped
mRNA translation *in vitro*. In a second line of investigation, initiation complexes were photoreacted with an RNA-RNA cross-linking agent,
4'-aminomethyl-4,5',8-trimethylpsoralen (AMT). The results indicate
that the mRNA 5'-proximal initiation site, defined as the region
protected by ribosomes against RNase digestion, is closely apposed to
18S ribosomal RNA in 80S as well as 40S initiation complexes. In
this report these findings are reviewed and their possible functional
significance for initiation discussed.

FACILITATING EFFECT OF THE CAP ON mRNA TRANSLATION
AT THE LEVEL OF RIBOSOME BINDING

The 5'-terminal cap structure characteristic of eukaryotic mRNAs
consists of 7-methylguanosine linked by the 5'-position through a triphosphate bridge to the penultimate residue of the polynucleotide
chain as shown in Fig. 1 and described in detail previously in this
volume (chapter 5; see also ref. 5).

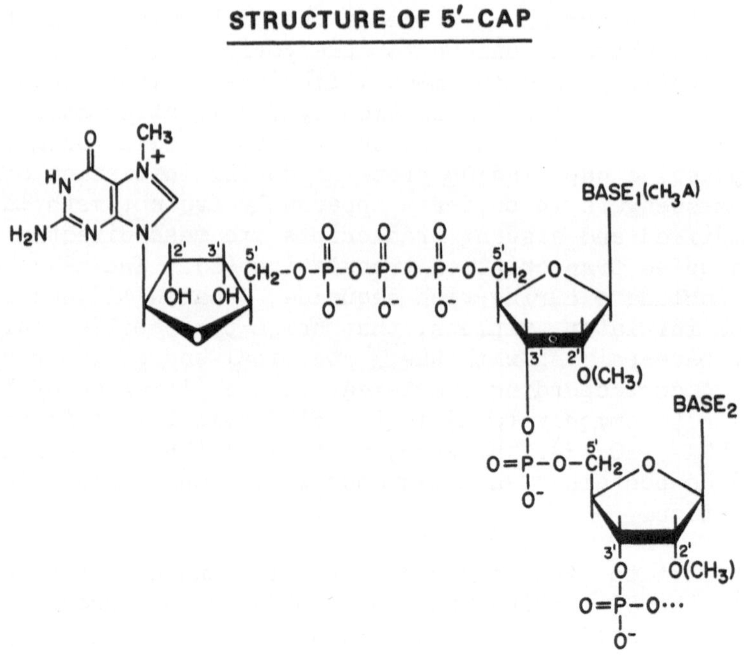

STRUCTURE OF 5'-CAP

Fig. 1. Structure of 5'-cap.

Caps are formed at an early stage of mRNA formation and conserved during processing of nuclear transcripts to cytoplasmic messengers (16, 17). The mechanism of formation of caps in its general aspects has been discussed by R.P. Perry (chapter 5), and I shall describe here this process as it was first worked out by using purified reovirions. This was possible thanks to the fact that reovirus contain enzyme activities which synthesize capped mRNAs *in vitro* (18). Reovirus mRNA caps were formed at the initiation stage of transcription, i.e. short, initiator oligonucleotides synthesized by reovirus cores in incomplete transcriptase mixtures were already capped. The series of capforming reactions catalyzed by five viral core-associated activities in the presence of two ribonucleoside triphosphates and S-adenosylmethionine (SAM) can be summarized as follows:

1) $\quad$ pppG + pppC $\xrightarrow{\text{RNA polymerase}}$ pppGpC + PP$_i$

2) $\quad$ pppGpC $\xrightarrow{\text{Nucleotide phosphohydrolase}}$ ppGpC + P$_i$

3) $\quad$ pppG + ppGpC $\xrightleftharpoons{\text{Guanylyltransferase}}$ G$^{(5')}$ppp$^{(5')}$GpC + PP$_i$

4)§ $\quad$ GpppGpC + SAM $\xrightarrow{\text{Methyltransferase 1}}$ m^{7}GpppGpC + SAH

5)§ $\quad$ m^{7}GpppGpC + SAM $\xrightarrow{\text{Methyltransferase 2}}$ m^{7}GpppG^{m}pC + SAH

A similar series has been established for the viral mRNAs made on the genome DNA template of vaccinia virus (19). Furthermore, enzymes that catalyze the same reactions have been purified from HeLa cell nuclei (20, 21) indicating that cellular transcripts are probably capped by the same general mechanism.

One of the first functional effects of the cap was observed when capped and uncapped viral mRNAs were compared as template for translation in wheat germ protein synthesizing extracts (22). For both reovirus and vesicular stomatitis virus (VSV) mRNAs, the presence of a m^{7}G cap increased translation many-fold. It was also found that the extract contained capping activities that converted unmethylated mRNA 5'-termini to m^{7}G caps, resulting in a parallel increase in messenger activity (23). Subsequent studies from many laboratories confirmed that m^{7}G capped mRNAs were more effective in directing protein synthesis *in vitro* than the corresponding RNAs lacking a 5'-terminal m^{7}G (5, 6, 24). The latter included molecules synthesized in the absence of a methyl donor and containing unblocked (5'-pppN) or blocked, unmethylated (5'-GpppN) termini as well as mRNAs that had been decapped chemically by oxidation and β-elimination (5'-pppN$^{(m)}$) or enzymatically with pyrophosphatases purified from tobacco (25) or potato (26) (5'-pN$^{(m)}$). The extent of cap-dependence *in vitro* varied

---

§ In reactions (4) and (5) the structure G$^{(5')}$ppp$^{(5')}$GpC has been abbreviated GpppGpC.

with the source of protein synthesizing extract, e.g. wheat germ was
more stringent than reticulocyte lysate for the presence of a 5'-
terminal m⁷G (24, 27), and with the conditions of incubation such as
ionic strength (28) and initiation factor source and concentration
(29). In all cases naturally capped mRNAs became less effective
without the m⁷G. Under certain conditions, namely decapped mRNA incubated in wheat germ extract or microinjected into *Xenopus* oocytes,
the functional loss could have been explained by mRNA degradation
since decapping was accompanied by destabilization of the message (25,
30). However, unmethylated RNA (5'-GpppN) which was not destabilized
was also untranslated, consistent with m⁷G having a positive effect
on translation (30).

Although many *in vitro* results support the functional importance
of the mRNA cap for protein synthesis, it has been difficult to test
its role directly *in vivo*. Mammalian tissue culture cells treated
with S-tubercidinylhomocysteine, an inhibitor of RNA methylation,
synthesized mRNA that was deficient in internally located m^6A and in
2'-O-methyl groups on the 5'-penultimate residue (31). Similar findings were obtained for avian sarcoma virus RNA produced in transformed
chick embryo cells treated with another methylation inhibitor, cyclo-
leucine (32). Disappointingly, with both inhibitors N-7 methylation
of the mRNA 5'-terminal G was relatively drug-resistant, and it was
not possible to inhibit the formation of m⁷G without also decreasing
RNA synthesis. Thus assessment of the m⁷G requirement for protein
synthesis was difficult. However, on the basis of other studies of
cycloleucine-treated, VSV-infected CHO cells it was suggested that mRNA
methylation is required for translation (33). Similarly, in a single
study of herpesvirus-infected cells treated with still another inhibitor, S-isobutyryl-adenosine, results consistent with prerequisite
5'-terminal methylation of mRNA for viral protein were also obtained
(34). In VSV-infected BHK cells, the viral mRNAs in polysomes were
capped while the small fraction of viral mRNAs not associated with
ribosomes contained unblocked 5'-termini (35). The latter result was
in agreement with a variety of *in vitro* experiments showing that the
facilitating effect of the cap on protein synthesis was at the level
of initiation, promoting stable initiation complex formation (36, 37).

An example of mRNA dependence on the m⁷G cap for ribosome binding
is shown in Fig. 2. Capped reovirus mRNA incubated in wheat germ
extract under conditions of translation initiation, but with sparsomycin present to inhibit polypeptide chain elongation, formed 80S
initiation complexes. After treatment with potato pyrophosphates
which removed the m⁷G and one or two cap phosphates (26), the decapped
mRNA failed to form 80S complexes and was extensively degraded (panel
A). Reovirus mRNA that was synthesized in the presence of S-adenosyl-
homocysteine (SAH) to prevent methylation also failed to bind ribosomes but was not degraded, presumably because it contained a blocked
5'-end (30) (panel B).

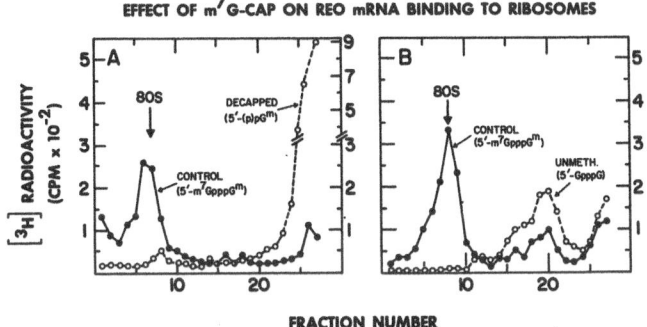

EFFECT OF m⁷G-CAP ON REO mRNA BINDING TO RIBOSOMES

Fig. 2. Positive effect of m⁷G on ribosome binding.

A. 5'-³H-methyl-labeled reovirus mRNA was incubated in wheat germ extract under conditions of initiation before and after cleavage of the 5'-terminal cap by digestion with potato nucleotide pyrophosphatase (26).

B. Ribosome binding of ³H-uridine-labeled reovirus mRNAs that were synthesized in the presence of SAM (5'-m⁷GpppGᵐ) or SAH (5'-GpppG) (18, 30).

Other diverse observations provide strong support for the suggestion that recognition of the capped 5'-end of mRNA occurs during initiation: (i) Caps in some mRNAs form part of the ribosome binding site as defined by protection against RNase digestion in initiation complexes (38, 72). (ii) In the majority of eukaryotic mRNAs the A-U-G closest to the 5'-end is utilized for initiation (as reviewed in this volume by M. Kozak). (iii) In different cell-free protein synthesizing systems translation and ribosome binding are inhibited by cap analogs such as m⁷G⁵p and m⁷G⁵pp (39-41). The m⁷G nucleoside and 3'-nucleotide are not active. Alteration of the 5'-nucleotide, for example by methyl esterification of the phosphate (E. Darzynkiewicz et al. unpublished results) or by ring-opening of the imidazole portion of the base with loss of the extra plus charge (42) (see Fig. 1), essentially abolished the inhibitory activity of m⁷G⁵p. Other changes such as replacement of the N-7 methyl group by an ethyl substituent had little or no effect (43). These findings indicate that specific features of the cap analog, and by inference the cap in intact mRNA, are recognized during ribosome attachment.

It should be noted that the translation of satellite tobacco necrosis virus RNA, a naturally uncapped messenger (44) that directs protein synthesis by a cap-independent mechanism, was found in some studies to be sensitive to m⁷G⁵p (45). *In vitro* translation of decapped VSV messenger RNAs was also sensitive to cap analogs under certain conditions (46). These observations suggested that an element

of translation that recognized the capped 5'-end of mRNA, for example
an initiation factor complex (47), also might be involved in uncapped
mRNA translation. In an effort to detect proteins with the ability to
bind to the capped end of mRNA, we developed a cross-linking assay
based on the inverted 5'-5' linkage of the m[7]G in caps.

DETECTION OF CAP BINDING PROTEINS BY CHEMICAL CROSS-LINKING
TO mRNA 5'-END

    Initial experiments demonstrated that the high salt wash of brine
shrimp *Artemia salina* ribosomes contains cap binding protein(s) as
measured by retention of ^{3}H-m[7]GpppG[(m)]pC on nitrocellulose filters
(48). The activity was labile, making purification difficult. Preparations of reticulocyte initiation factors eIF-2 (49) and eIF-4B
(50) also were found to have capped mRNA binding activity by a similar
filter binding technique. However, this indirect method did not allow
assignment of the cap-binding activity to a specific constituent in
preparations that contained more than a single polypeptide. A more

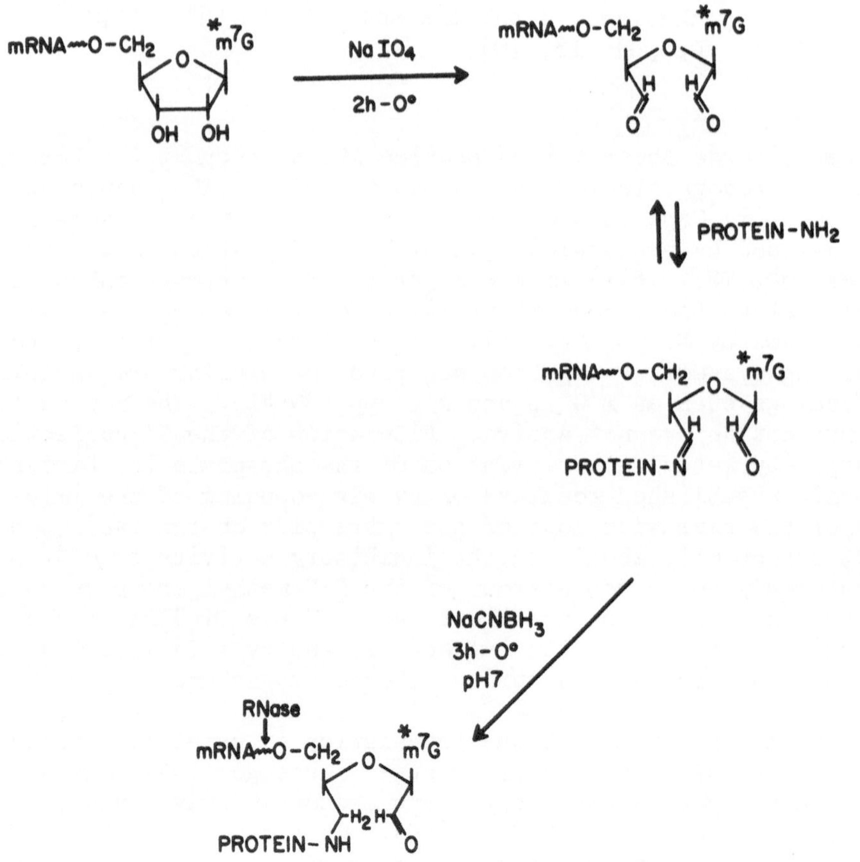

Fig. 3.   Scheme for cross-linking cap binding proteins to mRNA 5' ends.

direct test for identifying proteins that interact specifically with
the cap is outlined in Fig .3 (51). In this scheme we took advantage
of two structural features of the cap: (i) the presence of the 7-
methyl group which could be radiolabeled by synthesizing reovirus mRNA
in incubation mixtures containing ^{3}H-methyl-SAM and (ii) the *cis*-diol
in the 5'-linked m^{7}G which could be converted to the reactive dial-
dehyde form by periodate oxidation. Proteins, for example in initiation complexes or factor preparations, that contained accessible amino
groups and bound at or near the oxidized mRNA 5'-cap were covalently
joined to the m^{7}G by Schiff base formation. The linkage was stabilized by subsequent reduction with sodium cyanoborohydride. Cap-binding proteins that became radiolabeled after cross-linking to ^{3}H-
methyl-labeled mRNA were analyzed by gel electrophoresis and fluoro-
graphy after hydrolyzing the mRNA chains by exhaustive digestion with
RNases T$_1$ and A. Since any protein with free amino groups could
potentially form Schiff bases with the cap of oxidized mRNA, specificity was assessed by using cap analogs as inhibitors of cross-
linking.

Cap-Binding Activity in Cell-Free Extracts

The cross-linking method was applied to wheat germ protein synthesizing extracts incubated under initiation conditions with a
variety of different oxidized, 5'-^{3}H-methyl-labeled mRNAs. In each
case radiolabeled putative cap binding proteins of apparent molecular
weights ∿135,000, 93,000 and 26,000 were detected (51, 52). Cross-
linking was dependent on initiation complex formation as shown by
parallel inhibition of both processes by cap analogs and the non-
hydrolyzable ATP analog, AMPP(NH)P. The cap in ribosome-bound messengers, but not in unbound mRNA, was protected against cleavage by
nucleotide pyrophosphatase, presumably by associated cap binding
proteins.

Similar experiments with reovirus mRNA and rabbit reticulocyte
lysates yielded by gel electrophoresis a cap-specific polypeptide of
molecular weight ∿1600 and a lower intensity 35,000 band, both cross-
linked in lysates only under conditions of initiation (52). Analyses
of isolated reticulocyte 40S initiation complexes revealed two additional cap-specific polypeptides of ∿130,000 and 60,000 daltons by
the cross-linking method. Reconstitution experiments with fractionated initiation components indicated that these polypeptides were
derived from the ribosomal high salt wash. The intriguing possibility that one or more polypeptides in initiation factor preparations
might have cap binding activity was also suggested from the finding
that ^{3}H-methyl-labeled reovirus mRNA cross-linked to partially puri-
fied(0.5 M KCl washed) mouse ascites factor eIF-3 yielded labeled
subunits corresponding to molecular weights of ∿160,000, 130,000 and
60,000, i.e. similar to those obtained in reticulocyte 40S complexes.
Taken together, the results suggested that in mammalian initiation
complexes the 5'-end of mRNA may be positioned close to eIF-3, a

factor which is known to promote 40S ribosomal subunit binding to
natural mRNAs (see chapter 3 of this volume).

Cap Affinity of Initiation Factors

    To test whether any of the initiation factors in high salt wash
of reticulocyte ribosomes can bind directly to the capped end of mRNA
in the absence of ribosomes, factor mixtures and individual partially
purified factors were analyzed by the cross-linking scheme (53). In
the factor mixture a single radio labeled band with apparently high
cap affinity was obtained following binding at 0°C (Fig. 4, left).
At 30° a doublet at ∿50,000 dalton was also present in the gel
fluorogram. Labeling of the smaller polypeptides was blocked by the
cap analog, m^{7}GDP, but only by GDP, consistent with cap-specificity
of the 24,000 dalton band. The ∿50,000 dalton doublet was identified
as elongation factor 1 which binds GTP, explaining the inhibition of
its cross-linking by GDP but not by the cap analog. Among the many
individual factors tested the 24,000 dalton cap-specific polypeptide
was detected exclusively in preparations of eIF-3 and eIF-4B (Fig. 4).

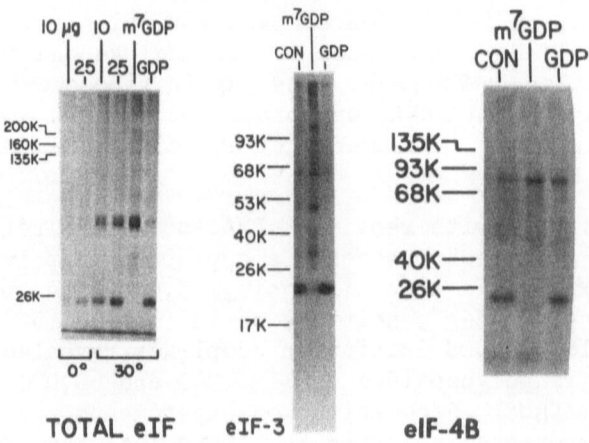

Fig. 4.   Gel electrophoresis patterns of reticulocyte initiation
          factors radiolabeled by cross-linking to capped end of oxi-
          dized, ^{3}H-methyl-labeled reovirus mRNA (53). *Left panel*:
          Mixture of factors partially purified by DEAE cellulose
          chromatography; 15 µg protein used for assays containing
          1 mM m^{7}GDP and GDP. *Center panel*: 15 µg eIF-3, 0.4 mM m^{7}GDP,
          GDP. *Right panel*: 2 µg eIF-4B, 0.4 mM G nucleotide. CON =
          control without nucleoside diphosphate.

Table 1. Differential Stimulation of Capped mRNA Translation by
          Protein Synthesis Factors

| | FOLD-STIMULATION | | RATIO |
FACTOR	+Sindbis RNA	+EMC RNA	Capped/ Uncapped
Expt. 1			
eIF-2 (3.5 µg)	2.2	1.1	2.0
eIF-3 (5µg of 0.5 M KCl washed)	0.7	1.4	0.5
eIF-4A (6 µg)	5.6	2.2	2.5
eIF-4B (1.5 µg)	13.6	1.5	9.1
eIF-4C+D (2 µg)	0.8	0.5	1.6
eIF-5 (0.5 µg)	0.6	1.6	0.4
EF-2 (2 µg)	1.0	1.8	0.6
Expt. 2			
EF-1 (1 µg)	1.2	0.9	1.3
eIF-3 (4 µg of 0.1 M KCl washed)	12.1	0.9	12.3

HeLa cell-free extracts were incubated for 60 min. at $37^{o}C$ with 0.5 µg
mRNA/25 µL and ^{35}S-methionine (24 µCi, spec. act. = 893 Ci/mmol) as
radioactive precursor. For expts. 1 and 2 the levels of synthesis
without added factors were 1,754 and 7,880 cpm with sindbis RNA and
8,830 and 87,652 cpm with EMC RNA, respectively (54).

From stained gel profiles it appeared to be present in both factors
at less than stoichiometric levels. These same two factors differentially stimulated translation of capped mRNA in HeLa cell-free
extracts, suggesting a functional role of the factor-associated cap
binding protein (54, Table 1). When eIF-3 was washed with 0.5 M KCl,
a process which removed the 24, 000 dalton protein, the differential

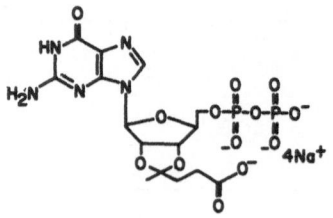

LEVULINIC ACID - (7-METHYL
GUANOSINE-O⁵'-DIPHOSPHATE-
O²', O³'-ACETAL)

LEVULINIC ACID- (GUANOSINE-O⁵'-
DIPHOSPHATE-O²', O³'-
ACETAL)

Fig. 5.   Structures of compounds that were coupled to AH-Sepharose 4B
          for affinity chromatography.

stimulatory activity (but not eIF-3 activity assayed in a reconsti-
tuded *in vitro* system) was also lost (Table 1).   These and other
observations encouraged us to purify cap binding proteins by affinity
chromatography for further functional studies.

FUNCTIONAL CAP BINDING PROTEINS PURIFIED BY m⁷GDP-SEPHAROSE
AFFINITY CHROMATOGRAPHY

     Affinity resins were prepared from the levulinic acid acetals
of m⁷GDP and GDP (55) (Fig. 5).   The nucleotide derivatives were
coupled to AH-Sepharose 4B by carbodiimide catalyzed formation of
peptide bonds between their carboxyl groups and the amino groups
exposed at the ends of the six-carbon spacer chains on the resin.
High salt wash of reticulocyte ribosomes was fractionated by 0-40%
ammonium sulfate precipitation, and the precipitate was further
purified by sedimentation in a glycerol gradient.   The slow-sediment-
ing material (<10S) was applied to m⁷GDP-Sepharose in 0.1 M KCl and
developed with m⁷GDP to elute putative cap binding proteins.   As
shown in the stained gel profiles of the eluted fractions (Fig. 6),
highly purified cap binding protein was obtained by this procedure.

     The affinity purified cap binding protein possessed functional
activity, i.e. addition of small amounts to HeLa cell extracts effectively stimulated translation of the capped mRNA of Sindbis virus
but not the naturally uncapped Encephalomyocarditis Virus (EMC) RNA
(54, 55) (Fig. 7).   One possible mechanism by which capped mRNA
function is enhanced by cap binding protein is suggested in Fig. 8.
In this model the protein binds to the capped 5'-end of mRNA in a
pre-initiation complex.   The cap binding protein has an affinity for
eIF-3 (Fig. 4) and consequently the mRNA complex is directed to

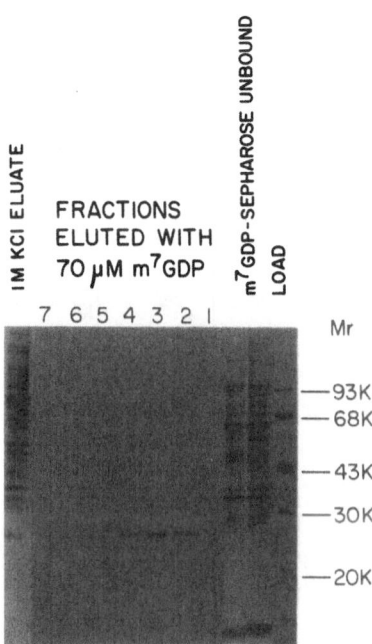

Fig. 6. Stained polyacrylamide gel profiles of cap binding protein
         purified by binding to m⁷GDP-Sepharose affinity resin and
         elution by cap analog. For experimental details see (55).

active, eIF-3 containing 40S subunits. After stable complex formation
(also involving other interactions such as initiator met-tRNA anticodon pairing with the A-U-G), the cap binding protein is released
for recycling, and its interaction with the mRNA cap replaced by the
160,000, 135,000 or 60,000 molecular weight subunits of eIF-3 (52).

Another mechanism by which capped mRNA initiation might be promoted by cap binding proteins is suggested from observations made with
native *versus* denatured mRNAs. Native capped mRNAs formed 80S initiation complexes when incubated in wheat germ extract in the presence
of the inhibitor of elongation, sparsomycin, but IMP-substituted mRNAs
also yielded heavier complexes (56, 57) (Fig. 9A). The loss of secondary structure resulting from replacement of G by I apparently
allowed ribosomes to *scan* through the 5'-proximal initiation site (3),
additional ribosomes attaching at the 5' entry site (56). Diminished
secondary structure also resulted in a decreased cap dependence for
ribosome binding; 1 mM m⁷GMP inhibited I-substituted mRNA binding by
∿30% as compared to ∿80% with native G-containing mRNAs (Fig. 9B).
The ATP requirement for initiation was also strikingly lower for
denatured RNA as also observed by Kozak (58). AMPP(NH)P decreased

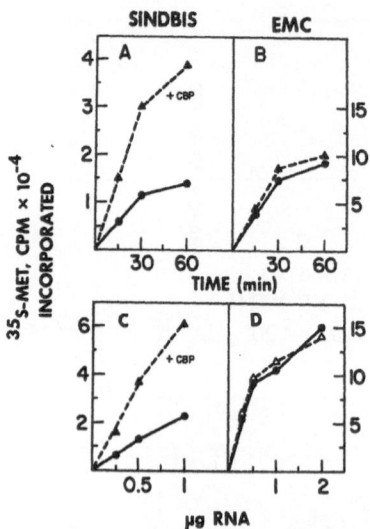

Fig. 7. Differential stimulation of capped mRNA translation by affinity purified cap binding protein. Capped Sindbis virus
mRNA and uncapped EMC virus RNA were translated in extracts
of HeLa cells with or without addition of ~30 µg of the
24,000 dalton cap binding protein per 25 µl incubation as
described in detail (54).

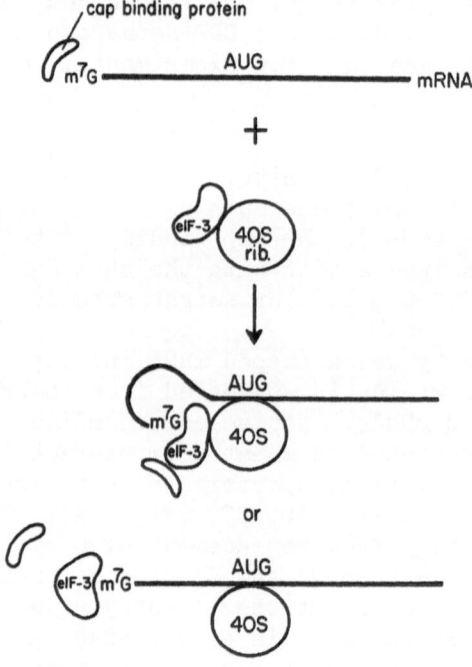

Fig. 8. Model of cap interactions with proteins during initiation of
translation.

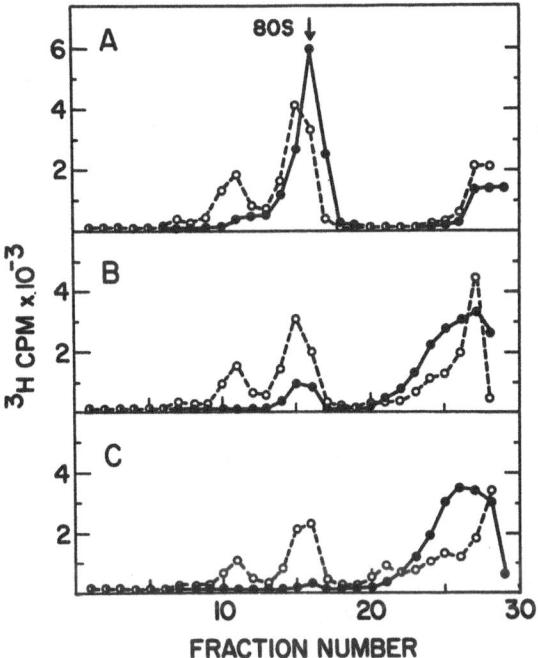

Fig. 9.   Diminished dependence on ATP and the m[7]G cap for initiation
          by I-substituted mRNAs. Reovirus mRNA was synthesized in
          the presence of ^{3}H-methyl-SAM and ITP in place of GTP for
          I-substituted mRNA (57). The modified, denatured mRNAs
          (O---O) and native G-containing mRNAs(●---●) were incubated
          in wheat germ extracts under initiation conditions and
          analyzed in separate glycerol gradients. A. Control.
          B. + 1 mM m[7]GMP. C. ATP replaced with 1 mM AMPP(NH)P.

initiation by only 30% with I-substituted mRNA but inhibited initiation with G-containing mRNA by >95% (Fig. 9C). These and other
similar findings lead to speculation that the cap binding protein may
denature the capped mRNA 5' leader sequence by an ATP dependent un-
winding process, resulting in increased accessibility for attachment
of ribosomes. In this regard, ATP-dependence of some cap binding
proteins has been detected recently by the cross-linking assay (59).

    In poliovirus-infected HeLa cells the proteins made 2-3 hours
after infection are exclusively viral specific (60). Host mRNAs
apparently remain intact but become incapable of initiating protein
synthesis (60-62). This *in vivo* situation was confirmed with infected
cell extracts that translated the naturally uncapped viral RNA but not
capped globin or Vesicular Stomatitis Virus (VSV) mRNAs (63). These
results suggested that the host shut-off phenomenon might involve inactivation of cap binding proteins togehter with translation of uncapped viral RNA by a cap-independent mechanism. Similar events may

occur at late times in reovirus type 3 infected mouse L cells (64).
Consistent with a functional loss of cap binding proteins in infected
cells, addition of the 24,000 dalton cap binding protein, purified
by affinity chromatography or by convential procedures, restored the
ability of extracts of poliovirus-infected HeLa cells to translate
capped mRNAs (65). However, the purified restoring activity was
highly unstable in comparison to cruder fractions of cap binding
protein.

To test the possibility that additional components necessary for
stable restoring activity were being removed or degraded during preparation, a modified cap binding protein purification procedure including the use of protease inhibitors was developed by Dr. S. Tahara
(66). Reticulocyte ribosome high salt wash was precipitated with 70%
ammonium sulfate and fractionated by sucrose gradient centrifugation
into material sedimenting at <5S (fraction I) and in the 10-16S region
of the gradient (fraction II). Each pool purified by m^{7}GDP-Sepharose
affinity chromatography contained the 24,000 cap specific protein as
determined by cross-linking. Stained gel analysis of fraction II also
revealed polypeptide bands at ~47,000 and ~210,000 daltons in amounts
similar to the 24,000 dalton polypeptide. Fractions I and II both
stimulated viral capsid synthesis in uninfected HeLa cell extracts
directed by Sindbis capped mRNA. However, only fraction II had restoring activity in poliovirus-infected cell extracts and it was
stable after storage and freeze-thawing. The translational role of
cap binding proteins, in particular the stabilizing effect of the additional polypeptides in fraction II on restoring activity requires
further study. The possibility that cap binding proteins are involved
in other cellular processes, such as mRNA transport from nucleus to
cytoplasm also remains to be explored.

mRNA 5'-REGION PROXIMITY TO 18S RIBOSOMAL RNA IN INITIATION COMPLEXES

The 4'-substituted psoralens, including 4'-aminomethyl-4,5',8-
trimethylpsoralen (AMT), are recently synthesized, water soluble
furocoumarins that bind effectively to RNA (67). Upon irradiation of
RNA-psoralen complexes with long wavelength ultraviolet light, pyrimidine adducts form (Fig. 10). Because AMT is an intercalating agent
containing a double-bond at each end of its planar structure, it can
form cross-links in RNA by photoaddition to pyramidines in alternating
purine-pyrimidine base pairs. AMT photoreaction has already proven
useful as a method for probing ribosomal RNA conformation (68) and
virion genome structures. For example, the double-stranded genome RNA
of reovirus, either after extraction from virions or *in situ*, can be
extensively cross-linked by 4'-substituted psoralens (67, 69). It was
further shown by this approach that the RNA polymerase and not the
other cap-forming enzyme activities associated with purified reovirus
is template dependent (69). Recently we have used AMT photoreaction
in an attempt to determine if mRNA interacts with ribosomal RNA in
protein synthesis initiation complexes (70).

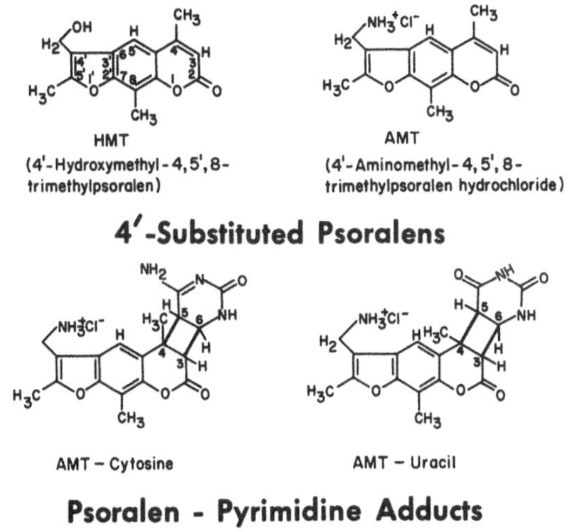

4'-Substituted Psoralens

HMT
(4'-Hydroxymethyl-4,5',8-trimethylpsoralen)

AMT
(4'-Aminomethyl-4,5',8-trimethylpsoralen hydrochloride)

AMT – Cytosine

AMT – Uracil

Psoralen - Pyrimidine Adducts

Fig. 10. Structures of 4'-substituted psoralens and pyrimidine adducts formed by photoaddition of AMT.

VSV mRNA containing ^{3}H-methyl labeled, capped 5'-ends was nicked with alkali to ∿200 nucleotide lengths and used to form 80S initiation complexes in reticulocyte lysate. The isolated complexes were irradiated in the presence of AMT, and the RNAs extracted and analyzed by glycerol gradient sedimentation. As shown in Fig. 11, approximately half of the ribosome-bound radioactivity from complexes irradiated with AMT was cross-linked to and sedimented with the 18S RNA. Similar results were obtained with intact mRNA or when gradient analyses were done under denaturing conditions, but in these conditions the mRNA and 18S rRNA were less well resolved. The use of other mRNAs and initiating systems did not change the basic findings. For example, Fig. 12 shows results obtained with reovirus mRNA fragments labeled with ^{3}H-methyl at the 5'-end and internally with ^{32}P-UMP and incubated in wheat germ extract under conditions of 80S complex formation. As observed for wheat germ 40S initiation complexes, photoreaction of 80S complexes with AMT yielded a large fraction of the mRNA fragments cross-linked to 18S rRNA. Although there was little cross-linking to the larger rRNA, it was also accessible to psoralen in the 80S complexes. Both rRNA species formed ^{3}H-labeled AMT adducts at average molar ratios of 0.8-1.5 under conditions that resulted in cross-linking of ^{32}P-labeled mRNA mainly to the 18S rRNA (Fig. 12, panel C).

With either intact mRNA or fragments, interaction with 18S rRNA was dependent on initiation complex formation. No cross-linked hybrid molecules were obtained by photoreacting mixture of mRNA and purified ribosomal RNAs with AMT. Additional evidence for initiation dependence was obtained by AMT photoreaction of total wheat germ initiation mixtures after incubation in the presence or absence of ATP which is required for mRNA binding to ribosomes. Without ATP addition there

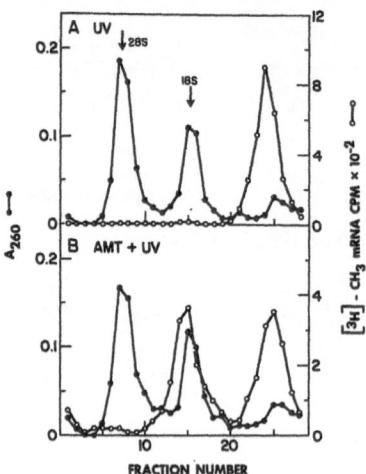

Fig. 11. Cross-linking of VSV mRNA to 18S rRNA by AMT photoreaction
          of reticulocyte initiation complexes. 80S Complexes con-
          training ^{3}H-methyl labeled, alkali fragmented viral mRNAs
          (5'-m^{m}GpppA^m) were irradiated at λ_{352nm} for 20 min. at 0°C
          in the absence (panel A) or presence (panel B) of 2.8 x 10^{-4}
          M AMT. The complexes were digested with protease, and the
          phenol extracted RNAs were analyzed in glycerol gradients as
          described (70).

was a parallel loss of initiation (56, 57) and of AMT cross-linking
of labeled mRNA to rRNA (Fig. 13).

        Polynucleotide phosphorylase has been used previously to syn-
thesize different ribopolymers containing a 5'-terminal reovirus mRNA
cap, ^{3}H-methyl-m^{m}GpppG^{m}pC (37). Such capped ribopolymers have been
useful for studies on the functional effects of the cap on transcription of influenza (71) and on RNA binding to ribosomes (37). In
earlier experiments it was shown that in wheat germ extract capped
poly(U) formed only 40S complexes. Capped poly(A,U) on the other
hand also yielded 80S complexes (Fig. 14A). As in the case of initiation complexes made with natural mRNAs, AMT photoreaction of the
40S-poly(U) complexes and poly(A,U)-40S and -80S complexes resulted
in cross-linking of both polymers to 18S rRNA (Figs. 14B and C).

        Of particular interest in all these cross-linked mRNA-rRNA
"hybrid" molecules is the location of the AMT cross-links. In attempting to position them in the rRNA, cross-linked complexes were digested
with RNase T$_1$, labeled at free 3'-termini with ^{32}P-pCp and analyzed
by gel sequencing after photoreversal of the AMT cross-links. The
results indicated that at least some 18S rRNA 3'-termini were present
in the cross-linked "hybrids" (70). This finding probably does not
indicate the presence of eukaryotic *Shine and Dalgarno* structures

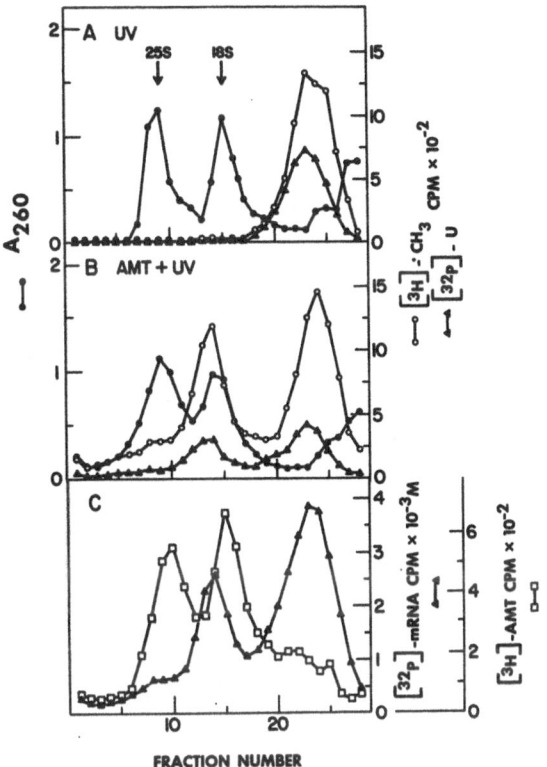

Fig. 12. Reovirus mRNA cross-linked to wheat germ 18S rRNA and covalent binding of ^{3}H-AMT to both rRNA species in 80S
initiation complexes. Complexes containing ^{3}H-methyl- and
^{32}P-UMP-labeled reovirus mRNA fragments were irradiated and
analyzed as in Fig. 11. For panel C, ^{32}P-labeled mRNA and
1.4×10^{-4} M ^{3}H-labeled AMT (kindly provided by Dr. J.
Hearst) were used (70).

(4, 9) because most messengers including VSV (72) and reovirus (73)
mRNAs (not to mention polyU) do not contain within the 5' leader a
sequence of sufficient complementarity to allow stable base-pairing
with 18S rRNA 3'-termini. Despite this caveat, other experiments
described below indicate that the 5'-proximal region of mRNA, more
specifically the initiation site, is in close proximity to 18S in
initiation complexes.

Wheat germ initiation complexes were formed with a mixture of
^{32}P-UMP-labeled reovirus mRNAs; The complexes were digested with
RNase T1 under conditions that yield the initiation sites as 80S
ribosome-protected oligonucleotide fragments of chain lengths 30 to
54 (74). The initiation complexes containing the protected ^{32}P-la-

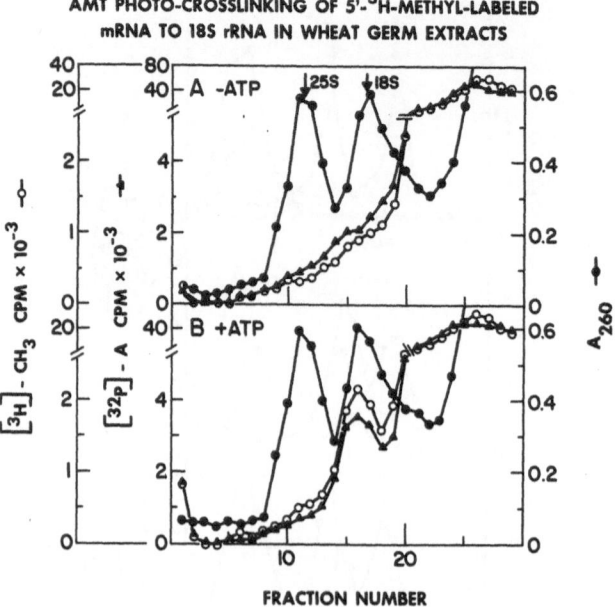

Fig. 13. Cross-linking dependence on initiation complex formation.
Reovirus mRNA labeled at the 5'-end with ^{3}H-methyl and
internally with ^{32}P was incubated in wheat germ extract
without ATP (panel A) or in the presence of ATP to obtain
80S initiation complexes (panel B). After incubation for
10 min. at 25°C, mixtures (0.1) were chilled, diluted with
ten-fold with 6.6% glycerol containing 0.5 mM Mg acetate,
48 mM KCl and 7 mM Tris-HCl, pH 7.4, and photoreacted for
20 min. with 2.8 x 10[-4] AMT. RNA was then extracted and
analyzed by gradient sedimentation.

beled mRNA fragments were isolated and photoreacted with AMT. RNA
extracted from the complexes was analyzed by electrophoresis in polyacrylamide sequencing gels. Consistent with AMT cross-linking to
rRNA, most of the ^{32}P-labeled mRNA fragments remained at the gel
origin (Fig. 15, lane 1). After photoreversal of the AMT cross-links
by irradiation at $\lambda_{254\ nm}$ (75) the ^{32}P-labeled fragments were released
and migrated slower than xylene cyanol in the position expected for
reovirus mRNA T_1 initiation sequences (73, 74). The control sample
of RNase T_1-protected fragments derived from initiation complexes
that were irradiated in the absence of AMT yielded a similar gel
patterns (lane 3). Furthermore, the profile obtained after exposure
to photoreversal conditions was identical to that of the photore-
versed, AMT reacted sample (compare lane 4 with lane 1). These
results indicate that the 5'-proximal initiation sites in reovirus
mRNAs are positioned close to (and possibly are base-paired with) 18S
rRNA in wheat germ initiation complexes.

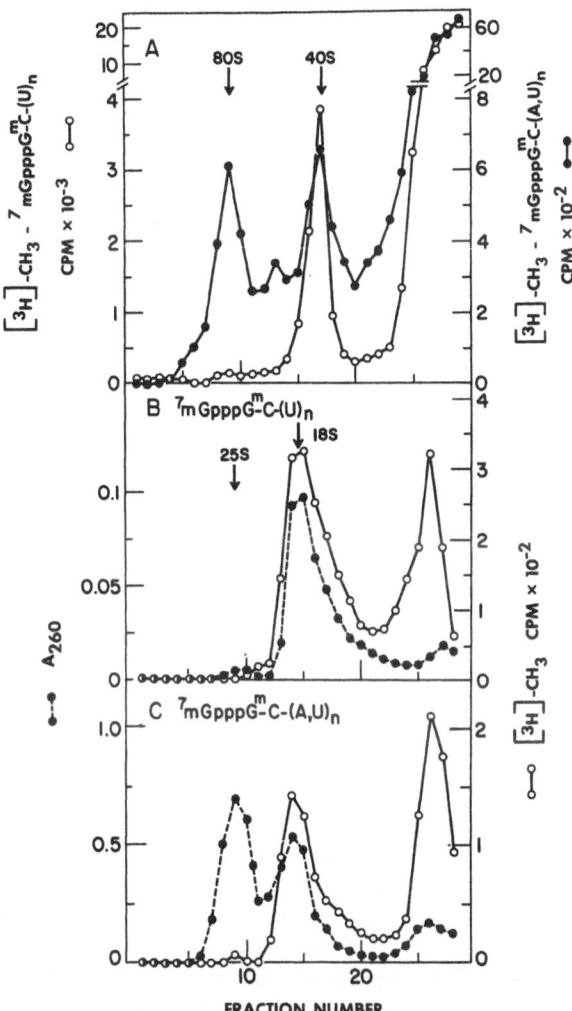

Fig. 14. Formation of wheat germ initiation complexes by capped ribopolymers and cross-linking to 18S rRNA. ^{3}H-methyl-labeled
m[7]GpppG[m]pCp(U)n and m[7]GpppG[m]pCp(A,U)n of chain length ∿100
nucleotides were synthesized with *M. luteus* polynucleotide
phosphorylase under primer-dependent conditions and incubated
in wheat germ extracts under conditions of initiation complex
formation (37). Complexes were analyzed in separate glycerol
gradients, and the results are shown in panel A. The 40S
complexes obtained with m[7]GpppG[m]pCp(U)n (fractions 15-19)
and the 40S plus 80S complexes with m[7]GpppG[m]pCp-(A,U)n
(fraction 7-19) were pooled separately and irradiated for
20 min. at λ$_{352 nm}$ in 2.8 x 10[4] M AMT. The samples were
deproteinized, and the extracted RNAs analyzed by glycerol
gradient sedimentation (panels B and C).

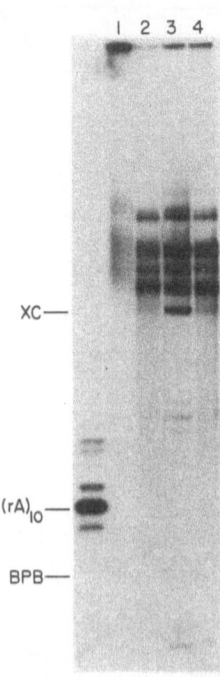

Fig. 15. Cross-linking of the ribosome-protected, 5'-proximal initia-
         tion sites of mRNA to 18S rRNA in wheat germ initiation
         complexes. Reovirus mRNA labeled with ^{32}P-UMP (specific
         activity ~10^6 cpm/μg) was incubated in wheat germ extract
         (20 μg mRNA/0.4 ml incubation mixture) under conditions of
         initiation in the presence of 0.2 mM sparsomycin. After
         10 min. at 23°C, 150 units of RNase T_1 were added and the
         incubation was continued for 15 min. Initiation complexes
         were then separated by gradient sedimentation. The pooled
         fractions were divided, and the two samples irradiated at
         $\lambda_{354\ nm}$ ± AMT as in Fig. 11. RNA was extracted and purified
         in another gradient. Radioactivity sedimenting with rRNA
         was collected and analyzed in a 20% polyacrilamide gel
         before (lanes 1 and 3) and after (lanes 2 and 4) photo-
         reversal. ^{32}P-labeled marker oligo(A) was kindly provided
         by Dr. Y. Furuichi. XC and BPB = xylene cyanol and bromo-
         phenol blue dyes.

     It should be possible using different mRNAs and radioactive AMT
to localize the exact sites of mRNA-rRNA proximity. Clearly, further
studies are also needed to evaluate whether putative interactions
between 18S rRNA and ribosome binding sites in mRNA have functional
importance in the initiation process of eukaryotic protein synthesis.

ACKNOWLEDGMENTS

The contributions of many coworkers and collaborators who carried out these studies during the past several years are gratefully acknowledged.

REFERENCES

1. Oberg, B.F. and Shatkin, A.J. (1972), Proc. Nat. Acad. Sci. U.S.A. <u>69</u>, 3589-3593.
2. Glanville, N., Ranki, M., Morser, J., Kääriäinen, L. and Smith, A.E. (1976), Proc. Nat. Acad. Sci. U.S.A. <u>73</u>, 3059-3063.
3. Kozak, M. (1978), Cell <u>15</u>, 1109-1123.
4. Steitz, J.A. (1979), In *Biological Regulation and Development*, (Goldberger, R.F., ed.)Plenum Press, New York and London, pp. 349-399.
5. Shatkin, A.J. (1976), Cell <u>9</u>, 645-653.
6. Filipowicz, W. (1978), FEBS Lett. <u>96</u>, 1-11.
7. Banerjee, A.K. (1980), Microbiol. Revs. <u>44</u>, 175-205.
8. Oxender, D.L., Zurawski, G. and Yanofsky, C. (1979), Proc. Nat. Acad. Sci. U.S.A. <u>76</u>, 5524-5528.
9. Shine, J. and Dalgarno, L. (1974), Proc. Nat. Acad. Sci. U.S.A. <u>71</u>, 1342-1346.
10. Hägenbuchle, O., Santer, M., Steitz, J.A. and Mans, R.J. (1978), Cell <u>13</u>, 551-563.
11. Baralle, F.E. and Brownlee, G.G. (1978), Nature, London, <u>274</u>, 84-87.
12. Both, G.W. (1979), FEBS Lett. <u>101</u>, 220-224.
13. De Wachter, R. (1979), Nucleic Acids Res. <u>7</u>, 2045-2054.
14. Ziff, E.B. and Evans, R.M. (1978), Cell <u>15</u>, 1463-1475.
15. Schroeder, Jr., H.W., Liarakos, C.D., Gupta, R.C., Randerath, K. and O'Malley, B.W. (1979), Biochemistry <u>18</u>, 5798-5808.
16. Rottman, F.M. (1978), In *Biochemistry of Nucleic Acids* II, (Clark, B.F.C., ed.) University Park Press, Baltimore Vol. 17, pp. 47-73.
17. Darnell, Jr., J.E. (1979), Prog. Nuc. Acid. Res. Mol. Biol. <u>22</u>, 327-353.
18. Furuichi, Y., Muthukrishnan, S., Tomasz, J. and Shatkin, A.J. (1976), J. Biol. Chem. <u>251</u>, 5043-5053.
19. Martin, S.A. and Moss, B. (1976), J. Biol. Chem., <u>251</u>, 7313-7321.
20. Venkatesan, S. and Moss, B. (1980), J. Biol. Chem. <u>255</u>, 2835-2842.
21. Venkatesan, S., Gershowitz, A. and Moss, B. (1980), J. Biol. Chem. <u>255</u>, 903-908.
22. Both, G.W., Banerjee, A.K. and Shatkin, A.J. (1975), Proc. Nat. Acad. Sci. U.S.A. <u>72</u>, 1189-1193.
23. Muthukrishnan, S., Both, G.W., Furuichi, Y. and Shatkin, A.J. (1975), Nature, London, <u>255</u>, 33-37.
24. Rose, J.K. and Lodish, H.F. (1976), Nature, London <u>262</u>, 32-37.

25. Shimotohno, K., Kodama, Y., Hashimoto, J. and Miura, K.I. (1977),
     Proc. Nat. Acad. Sci. U.S.A. 74, 2734-2738.
26. Zan-Kowalczewska, M., Bretner, M., Sierakowska, H., Szczesna, E.
     Filipowicz, W. and Shatkin, A.J. (1977), Nucleic Acids Res. 4,
     3065-3081.
27. Muthukrishnan, S., Morgan, M., Banerjee, A.K. and Shatkin, A.J.
     (1976), Biochemistry, 15, 5761-5768.
28. Weber, L.A., Hickey, E.D., Nuss, D.L. and Baglioni, C. (1977),
     Proc. Nat. Acad. Sci. U.S.A., 74, 3254-3258.
29. Held, W.A., West, K. and Gallagher, J.F. (1977), J. Biol. Chem.
     252, 8489-8497.
30. Furuichi, Y., Lafiandra, A. and Shatkin, A.J. (1977), Nature,
     London, 266, 235-239.
31. Kaehler, M., Coward, J. and Rottman, F. (1977), Biochemistry 16,
     5770-5775.
32. Dimock, K. and Stoltzfus, C.M. (1978), Biochemistry 17, 3627-
     3632.
33. Caboche, M. and La Bonnardiere, C. (1979), Virology, 93, 547-557.
34. Jacquemont, B. and Huppert, J. (1977), J. Virol. 22, 160-167.
35. Rose, J.K. (1975), J. Biol. Chem. 250, 8098-8104.
36. Both, G.W., Furuichi, Y., Muthukrishnan, S. and Shatkin, A.J.
     (1975), Cell 6, 185-195.
37. Both, G.W., Furuichi, Y., Muthukrishnan, S. and Shatkin, A.J.
     (1976), J. Mol. Biol. 104, 637-658.
38. Kozak, M. and Shatkin, A.J. (1976), J. Biol. Chem. 251, 4259-
     4266.
39. Roman, R.J.D., Booker, S.N. and Marcus, A. (1976), Nature,
     London, 260, 359-360.
40. Hickey, E.D., Weber, L.A. and Baglioni, C. (1976), Proc. Nat.
     Acad. Sci. U.S.A. 73, 19-23.
41. Canaani, D., Revel, M. and Groner, Y. (1976), FEBS Lett. 64,
     326-331.
42. Adams, B.L., Morgan, M., Muthukrishnan, S., Hecht, S.M. and
     Shatkin, A.J. (1978), J. Biol. Chem. 253, 2589-2595.
43. Furuichi, Y., Morgan, M. and Shatkin, A.J. (1979), J. Biol. Chem.
     254, 6732-6738.
44. Wimmer, E., Chang, A.Y., Clark, Jr., J.M. and Reichmann, M.E.
     (1968), J. Mol. Biol. 38, 59-73.
45. Seal, S.N., Schmidt, A., Tomaszewski, M. and Marcus, A. (1978),
     Biochem. Biophys. Res. Commun. 82, 553-559.
46. Bergmann, J.E. and Lodish, H.F. (1979), J. Biol. Chem. 254,
     459-468.
47. Bergmann, J.E., Trachsel, H., Sonenberg, N., Shatkin, A.J. and
     Lodish, H.F. (1979), J. Biol. Chem. 254, 1440-1443.
48. Filipowicz, W., Furuichi, Y., Sierra, J.M., Muthukrishnan, S.,
     Shatkin, A.J. and Ochoa, S. (1976), Proc. Nat. Acad. Sci. U.S.A.
     73, 1559-1563.
49. Kaempfer, R., Rosen, H. and Israeli, R. (1978), Proc. Nat. Acad.
     Sci. U.S.A. 75, 650-654.

50. Shafritz, D.A., Weinstein, J.A., Safer, B., Merrick, W.C., Weber, L.A., Hickey, E.D. and Baglioni, C. (1976), Nature, London, 261, 291-294.

51. Sonenberg, N. and Shatkin, A.J. (1977), Proc. Nat. Acad. Sci. U.S.A. 74, 4288-4292.

52. Sonenberg, N., Morgan, M.A., Testa, D., Colonno, R.J. and Shatkin, A.J. (1979), Nucleic Acids Res. 7, 15-29.

53. Sonenberg, N., Morgan, M.A., Merrick, W.C. and Shatkin, A.J. (1978), Proc. Nat. Acad. Sci. U.S.A. 75, 4843-4847.

54. Sonenberg, N., Trachsel, H., Hecht, S. and Shatkin, A.J. (1980), Nature, London, 285, 331-333.

55. Sonenberg, N., Rupprecht, K.M., Hecht, S.M. and Shatkin, A.J. (1979), Proc. Nat. Acad. Sci. U.S.A. 76, 4345-4349.

56. Kozak, M. (1980), Cell 19, 79-90.

57. Morgan, M.A. and Shatkin, A.J., Biochemistry, in press.

58. Kozak, M., Cell, in press.

59. Sonenberg, N., submitted.

60. Willems, M. and Penman, S. (1966), Virology 30, 355-367.

61. Ehrenfeld, E. and Lund, H. (1977), Virology 80, 297-308.

62. Fernandez-Munoz, R. and Darnell, J.E. (1976), J, Virol. 126, 719-726.

63. Rose, J.K., Trachsel, H., Leong, K. and Baltimore, D. (1978), Proc. Nat. Acad. Sci. U.S.A. 75, 2732-2736.

64. Skup, D. and Millward, S. (1980), Proc. Nat. Acad. Sci. U.S.A. 77, 152-156.

65. Trachsel, H., Sonenberg, N., Shatkin, A.J., Rose, J.K., Leong, K., Bergman, J.E., Gordon, J. and Baltimore, D. (1980), Proc. Nat. Acad. Sci. U.S.A. 77, 770-774.

66. Shatkin, A.J., Darzynkiewicz, E., Nakashima, K., Sonenberg, N. and Tahara, S. (1980), Proceedings of Juselius Symposium, Helsinki, Finland, Academic Press, in press.

67. Isaacs, S.T., Shen, C.J., Hearst, J.E. and Rapoport, H. (1977), Biochemistry, 16, 1058-1064.

68. Cantor, C.R. (1980), Annals N.Y. Acad. Sci. 346, 379-385.

69. Nakashima, K., Lafiandra, A.J. and Shatkin, A.J. (1979), J. Biol. Chem. 254, 8007-8014.

70. Nakashima, K., Darzynkiewicz, E. and Shatkin, A.J., Nature, London, in press.

71. Krug, R.M., Broni, B.A., Lafiandra, A.J., Morgan, M.A. and Shatkin, A.J., Proc. Nat. Acad. Sci. U.S.A. in press.

72. Rose, J.K. (1978), Cell 14, 345-353.

73. Kozak, M. (1977), Nature, London, 269, 390-394.

74. Kozak, M. and Shatkin, A.J. (1979), Methods Enzymol. 60, 360-375.

75. Rabin, D. and Crothers, D.M. (1979), Nucleic Acids Res. 7, 689-703.

INITIATION FACTOR/mRNA INTERACTIONS AND mRNA RECOGNITION

Raymond Kaempfer

The Hebrew University
Hadassah Medical School
Jerusalem, Israel

GENERAL ASPECTS OF mRNA RECOGNITION

The recognition of mRNA and its binding to ribosomes are generally considered to be of key importance to translational control. That
is, the decision whether a given mRNA species will be translated
during cell growth, differentiation or virus infection is largely
determined at initiation. Once bound to ribosomes, the mRNA will
generally be translated completely, as polypeptide chain extension and
completion ordinarily do not limit protein synthesis.

Translation of different species of mRNA may be initiated, in
the same cell or cell-free system, with frequences that can differ
as much as fifty-fold, and perhaps more. Which features of the mRNA
molecules themselves, and of the protein synthetic machinery of the
cell contribute to such regulation ? We should consider:

a) <u>Structure and conformation of the mRNA</u>. What determines the
efficiency of a given mRNA in translation is not yet understood, but
the structure and conformation of the 5'-leader sequence and the
accessibility of the AUG initiation codon seem to be crucial.

b) <u>Met-tRNA$_f$/mRNA interaction</u>. Since binding of Met-tRNA$_f$ to 40S
ribosomal subunit, directed by eIF-2, must occur before mRNA can be
bound, this interaction is absolutely essential.

c) <u>rRNA/mRNA interaction</u>. Recent experiments point to a proximity
of these RNA species in initiation complexes (1), but it is not yet
clear if there is an interaction comparable in importance to that
seen in prokaryotes (2). While many eukaryotic mRNA species do
contain sequences complementary to the conserved 3' end of 18S rRNA,

UGCGGAAGGAU (3) in their 5'- leader regions, others clearly do not(1).
In this respect, the difference between prokaryotes and eukaryotes is
well illustrated by the properties of Satellite Tobacco Necrosis
Virus (STNV) RNA. The 5'-terminal sequence of this RNA is shown in
Fig. 1. Although there is no complementarity to the 3' end of 18S
rRNA in the nucleotides preceeding the AUG initiation codon at positions 30-32, this RNA is a highly efficient messenger both *in vivo*
and in reticulocyte lysates. On the other hand, although eukaryotic
mRNA species generally fail to be translated in bacterial cell-free
systems and are not recognized at the correct initiation sites by
bacterial ribosomes (4), STNV RNA is translated correctly, and efficiently, in *E. coli* extracts (5). Most likely this is because of
the presence of an AGGA sequence in positions 10-14 of the leader
sequence, located in a loop (Fig. 1) that can base-pair with the
3'-terminus of the bacterial 16S rRNA, CCUCCUUA.

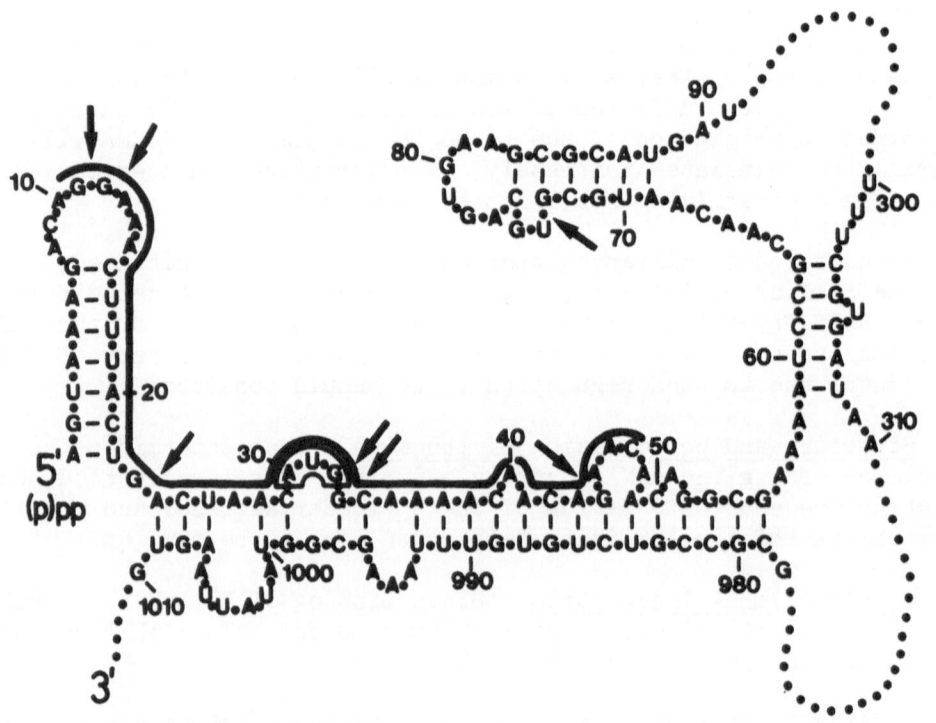

Fig. 1.   Secondary structure model for the 5' end of STNV RNA. The
          model (35) depicts stable secondary interactions. The line
          indicates nucleotides protected by 40S ribosomal subunits
          against nucleases (28). Arrows denote prominent sites of
          ribonuclease (T$_1$) cleavage (25).

d)   Initiation factor/mRNA interactions. These interactions are the
main topic of the chapter, and they will be considered in relationship to items (1) and (2) cited above.

APPROACHES TO THE STUDY OF mRNA/INITIATION FACTOR INTERACTIONS

Work done on this aspect may be analyzed according to the different techniques used:

a)   Sucrose gradient centrifugation. Here, binding of mRNA to 40S
ribosomal subunits or into 80S initiation complexes is studied with
purified initiation factors, and the effect of single omissions on the
binding is analyzed. In essence, one asks here which factors are
required to form complexes stable enough to survive in a centrifugal
field, and measures the presence of initiation factors in the 40S/mRNA
complex, the key intermediate in the mRNA-binding step of protein
synthesis. Binding of globin mRNA in 40S or 80S initiation complexes
is almost completely abolished when eIF-2 or eIF-3 are omitted, but
omission of eIF-4A or eIF-4B causes only a *marginal* reduction, even
though the latter two factors do stimulate the binding of mRNA (6-8).
Moreover, all of the polypeptide components of labeled eIF-2 and
eIF-3 are present in the mRNA-containing 40S initiation complex, but
binding of none of the other factors could be detected, indicating
that they do not bind to the 40S complex at all, or else their binding is too weak to survive centrifugation (8).

eIF-3 is thought to be the first initiation factor that interacts
with the 40S ribosomal subunit upon its release from mRNA, and stabilizes it by preventing association with 60S ribosomal subunits (10),
much as does bacterial IF-3 (11). eIF-2 forms a ternary complex with
Met-tRNA$_f$ and GTP (12-14) and directs the binding of Met-tRNA$_f$ to the
40S ribosomal subunit, a necessary prerequisite to the subsequent
binding of mRNA (6, 15).

b)   Crosslinking. A protein of 24,000 daltons that binds specifically to the cap in mRNA was identified by the fact that cap analogs
inhibit the crosslinking. The identification of the cap-binding
protein has been described in detail by A.J. Shatkin in the precedent
chapter, and will not be discussed further. The protein has a tendency to stick to eIF-3, and perhaps some of the nine polypeptides
that compose eIF-3 (9) may themselves crosslink with the cap (16).
It was suggested that the cap-binding protein could, by interacting
with both the cap and IF-3, contribute to the binding of mRNA to 40S
ribosomal subunits (16).

c)   Filter binding and translation. A combination of studies
measuring, on one hand initiation factor-dependent retention of mRNA
on nitrocellulose filter membranes (17), and on the other hand, translation in a complete cell-free system have revealed a functionally
significant recognition of mRNA by eIF-2, as well as eIF-4B.

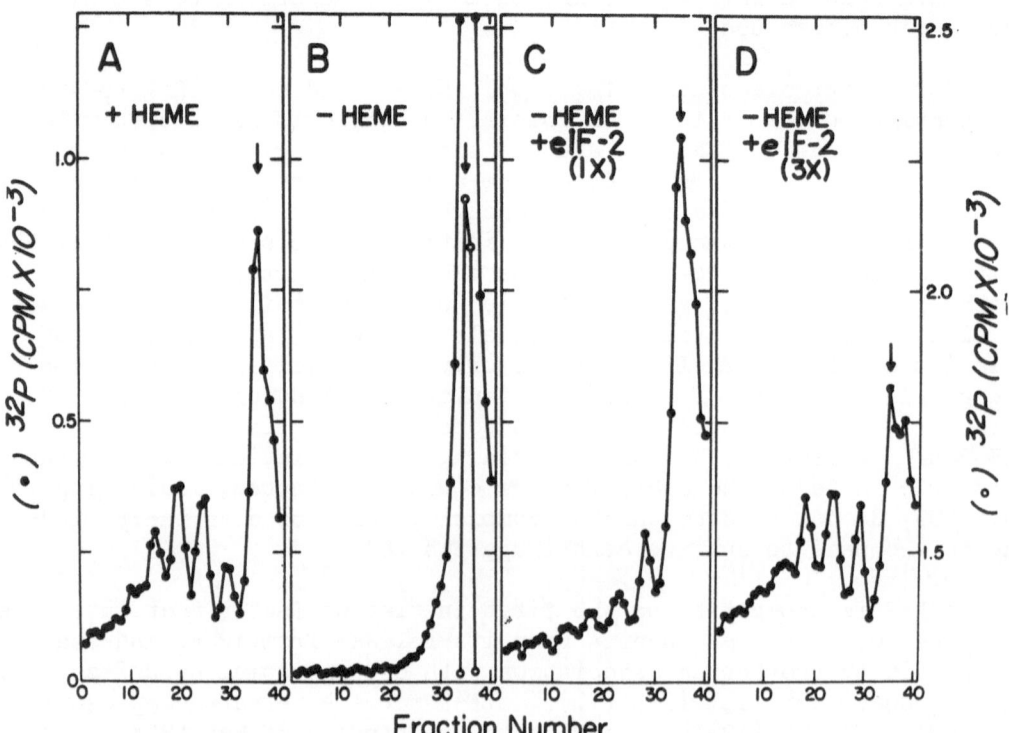

Fig. 2.   Continued recycling of reticulocyte ribosomes over mRNA in
          the absence of heme: requirement for eIF-2. Rabbit reti-
          culocyte lysate, containing ribosomes labeled *in vivo* by
          three intravenous injections of 10 mCi of ^{32}P at 36,24 and
          12 hr before bleeding, was incubated for 20 min in con-
          ditions for translation (33), with or without 50 µM hemin
          as indicated, and analyzed on sucrose gradients.   Arrow
          denotes position of single ribosomes.   A, the dominant poly-
          some species are tri- to pentasomes, characteristic for the
          size of globin mRNA. B, polysome run-off occurs in the
          absence of added hemin.   C and D, the effect of eIF-2, in
          increasing amounts, is to maintain recycling of ribosomes.

d)   Nucleotide sequence analysis. Such analysis has revealed specific binding of eIF-2 to STNV RNA at a 5'-terminal sequence comprising the ribosome binding site (25) (see Fig. 1).

RECOGNITION OF mRNA BY eIF-2

eIF-2 was initially discovered as a protein that overcomes the
block in initiation of translation observed in the absence of heme
(Fig. 2) or in the presence of dsRNA (18). The interferon-induced
block of translation observed in $vitro$ also is overcome by eIF-2 (44).
Because eIF-2 interacts with Met-tRNA$_f$ and GTP with extremely high
specificity, and because binding of Met-tRNA$_f$ to the 40S ribosomal
subunit is absolutely required before mRNA can be bound (15), any
studies of mRNA binding to ribosomes must always be conducted in the
presence of eIF-2. The well-defined task of eIF-2 to bind Met-tRNA$_f$
at first caused more attention to be focused on proteins whose only
detectable activity was to stimulate binding of mRNA to 40S subunits.
However, as we shall see, eIF-2 itself recognizes mRNA in a highly
specific manner.

eIF-2 Binds to mRNA

Binding of eIF-2 to mRNA has been observed in several laboratories (18, 23). That the interaction is specific was shown by the
finding that all mRNA species tested possess a high-affinity binding
site for eIF-2, including mRNA species lacking the 5'-terminal cap
or 3'-terminal poly(A) moieties (Fig. 3), while RNA species not
serving as mRNA, such as negative strand RNA (Fig. 3) (22), tRNA (24)
and rRNA (ref. 20 and Fig. 7) bind much more weakly, if at all.

The evidence that eIF-2 itself binds mRNA, rather than a contaminant, rests on several experiments. The mRNA-binding and Met-
tRNA$_f$-binding properties copurify completely and are present in eIF-2
preparations that are about 98% pure (21, 24). A crucial observation
is that the binding of globin mRNA to eIF-2 is completely sensitive
to competitive inhibition by Met-tRNA$_f$, but not by uncharged tRNA,
provided GTP is present (24). This shows that the only mRNA-binding
component in these preparations is eIF-2.

Equilibrium binding analyses show that globin mRNA forms an
equimolar complex with eIF-2 (23), as is already evident from the
first-order binding observed in Fig. 3. Binding is tight, as evi-
denced by an apparent K diss = 5×10^{-10} M at physiological salt
concentration (23).

The 3'-Untranslated Portion of mRNA and Poly(A) Are Not Recognized
by eIF-2

Removal of poly(A) or the 3'-untranslated sequence from globin
mRNA does not diminish its affinity for eIF-2, indicating that these

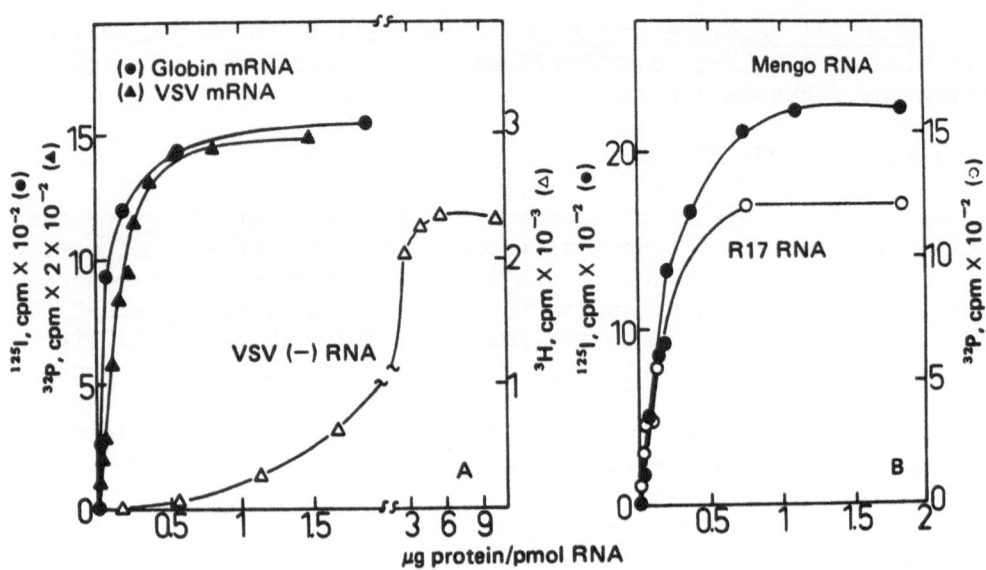

Fig. 3.   Relative affinities of mRNA and VSV negative-strand RNA for
          eIF-2. ^{125}I-labeled purified globin mRNA (0.031 pmol;
          2,300 cpm) (17), ^{32}P-labeled VSV mRNA (3.6 pmol, assuming
          average molecular weight 0.55 x 10[6]; 800 cpm), ^{3}H-labeled
          RNA extracted from VSV virions (0.21 pmol; 2,800 cpm),
          ^{125}I-labeled Mengovirus RNA (0.0023 pmol; 2,630 cpm), and
          ^{32}P-labeled R17 RNA (2.25 pmol; 1,660 cpm) were incubated
          with increasing amounts of purified eIF-2, and RNA binding
          was assayed (17). Background without protein was subtracted;
          for all RNA species, this background was less than 3%. Note
          scale change in curve of VSV negative-strand RNA.

regions are not required for the interaction (23). In addition, free
poly(A) does not compete the binding of eIF-2 to mRNA (20). Thus,
eIF-2 recognizes other features in globin mRNA.

Role of 5'-Terminal Cap and Internal mRNA Sequences in Binding
of eIF-2

     Non-capped mRNA species, such as Mengovirus RNA (Fig. 3) or
STNV RNA (25), bind extremely well to eIF-2, indeed more tightly than
globin mRNA. This must mean that the cap is not essential for recognition of these mRNA species by eIF-2. Nevertheless, cap analogs
such as m[7]GMP or m[7]GpppGm inhibit both binding of Met-tRNA$_f$ and of
mRNA to eIF-2, while m[7]G or GMP do not (26). While it is possible to

invoke a non-specific effect of the analogs on filter-binding assays
(16), this could not explain the findings that eIF-2 relieves the cap-
analog induced inhibition of globin mRNA translation, and that both
the translation of Mengovirus RNA and binding of this RNA to eIF-2
are thirty to forty-fold more resistant to inhibition by cap analog
(26). The inhibition of globin mRNA binding to eIF-2 requires cap
analog in very large molar excess (10^2-fold) over globin mRNA, and is
reversed readily by the addition of excess globin mRNA (26). This,
and the afore-mentioned high affinity of non-capped mRNA species for
eIF-2, both suggest that eIF-2 binds primarily to a nucleotide sequence in mRNA, and secondarily may recognize the cap.

Specific Binding of eIF-2 to a 5'-Terminal Sequence Comprising the Ribosome Binding Site

The mRNA-binding property of eIF-2 was further examined by
studying its interaction with STNV RNA (25). This RNA has an unmodified 5'-end that makes it readily suitable for 5'-end-labeling
with polynucleotide kinase. As seen in Fig. 4c (inset), the labeled
viral RNA preparation was composed of a collection of fragments
heterogeneous in size, and contained only a minority of labeled,
intact STNV RNA molecules; the latter are more refractory to the
dephosphorylation-kination reaction. The labeled STNV RNA molecules
bound by limiting amounts of eIF-2 were isolated and digested with
pancreatic and T_1 ribonucleases. The fingerprints in Fig. 4a and 5a
show that the STNV RNA offered to eIF-2 was quite heterogeneous, containing at least 30 different 5' ends. By contrast, the RNA selected
by eIF-2 possessed predominantly one 5' end (Figs. 4b and 5b). The
major spot observed in Figs. 4b and 5b migrate exactly as the single
spot observed in fingerprints of isolated, intact STNV RNA (see Fig.
1c), and could be identified as pApGpUp and pApGp, respectively (not
shown), indeed, STNV RNA starts with the sequence 5'-pApGpUp... (27).

Binding analysis of individual 5'-terminal fragments generated
from isolated, intact STNV RNA by partial digestion with ribonuclease
T_1 showed that eIF-2 does not bind detectably to the 32-nucleotide
fragment ending with the initiation codong AUG (Fig. 6, No. 1), or
shorter ones, but does bind to the 44-nucleotide fragment (No. 2) that
contains the ribosome binding site (Fig. 1).

The Role of RNA Conformation

In addition to the structural features localized at the 5'-end
of STNV RNA, eIF-2 appears to recognize a conformation found only in
larger molecules, for intact RNA and large 5'-terminal fragments are
bound preferentially over smaller ones (see Fig. 6). However, binding of short 5'-terminal STNV RNA fragments is specific, as judged by
competition with STNV and ribosomal RNA (Fig. 7). Although the need
for greater amounts of eIF-2 to bind the smaller fragments makes
quantitative comparison with intact RNA difficult, it is clear that

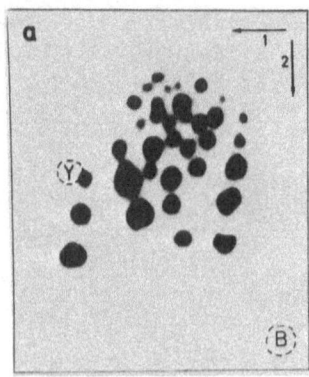 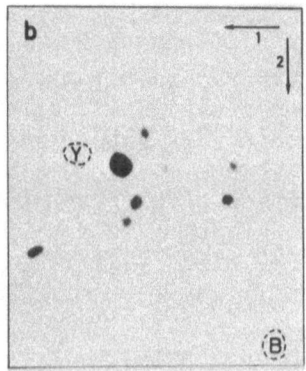 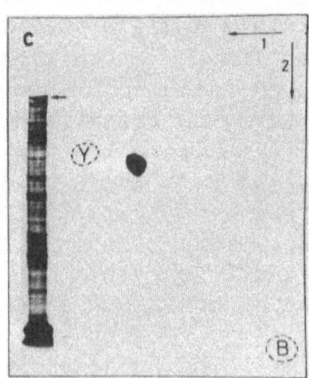

Fig. 4.   Fingerprint analysis of pancreatic ribonuclease digests of
          total 5' end-labeled STNV RNA, eIF-2-selected RNA, and
          purified, intact RNA. Pancreatic ribonuclease digests were
          analyzed by electrophoresis at pH 3.5 in the first dimension
          and homochromatography on a DEAE thin layer plate in the
          second dimension (see arrows), followed by autoradiography.
          (a) Total 5' end-labeled STNV RNA. (b) eIF-2-selected RNA.
          Total 5' end-labeled STNV RNA was incubated in conditions
          for RNA binding (17) with increasing amounts of purified
          eIF-2, and bound RNA was isolated by retention on 0.45 μm
          nitrocellulose filters and eluted with 0.1% NaDodSO$_4$, follow-
          ed by ethanol precipitation. At saturating amounts of
          eIF-2, 70% of the labeled RNA could be bound; in the sample
          analyzed, 7.2% of the input label was bound, (c) Intact
          5' end-labeled STNV RNA purified from the total RNA mixture
          by electrophoresis in a 4% polyacrylamide gel containing
          7 M urea (inset; arrow points to intact RNA). Marker dyes
          were methyl orange (Y) and xylene cyanol FF (B).

the differential competition by STNV and rRNA is preserved. Thus, the
structural features leading to recognition by eIF-2 may not only be
localized in the 5'-terminal region of STNV RNA, but may also be
determined, or strengthened, by the overall conformation of the RNA.
The observation that capping of intact STNV RNA does not enhance
initiation complex formation, while capping of 5'-terminal fragments
does have a stimulatory effect (29), and the finding that 5'-terminal
fragments bind to eIF-2 more weakly than does intact RNA may both be
related to the same phenomenon, i.e., a contribution of the overall
conformation to the affinity properties of these molecules.

Alteration of 5'-Proximal RNA Conformation Induced by Binding of eIF-2

     Binding of eIF-2 to STNV RNA facilitates the action of nucleases
at sites close to the 5' end of this RNA (25). This follows from

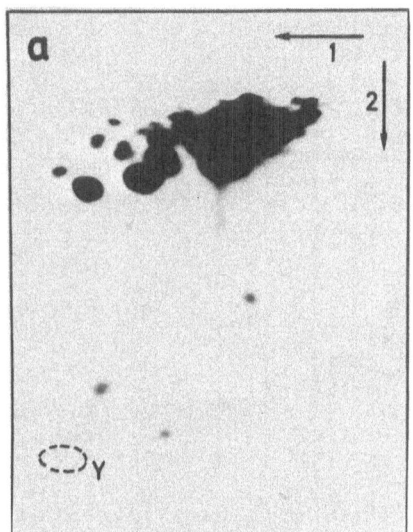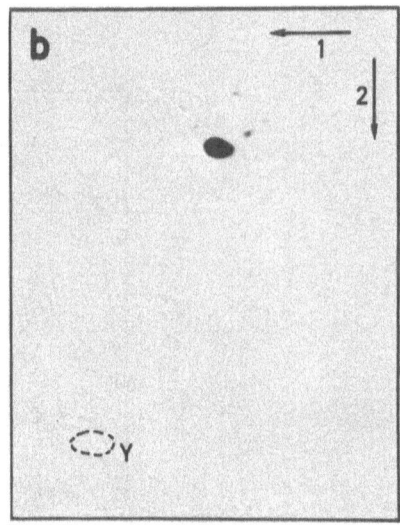

Fig. 5.   Fingerprint analysis of ribonuclease T_1 digests of total and
          eIF-2-selected 5' end-labeled STNV RNA. Ribonuclease T_1
          digests of the samples used for Fig. 4 were analyzed by
          electrophoresis at pH 3.5 in the first dimension and by homo-
          chromatography on a polyethleneimine thin layer plate in the
          second dimension (arrows). (a) Total 5' end-labeled RNA.
          (b) eIF-2-selected RNA. The intact RNA gave one spot co-
          migrating with the major one in (b) (not shown). B, xylene
          cyanol FF.

experiments in which intact 5'end-labeled RNA was partially digested
with ribonuclease T_1 to yield a mixture of products still containing
intact RNA as the predominant species. When eIF-2 was added shortly
before the end of the digestion period, and RNA bound to eIF-2 was
isolated, it consisted almost entirely of the 5'-terminal 44-nucleo-
tide fragment, while intact RNA was absent. Since eIF-2 binds more
readily to intact RNA than to the 44-nucleotide fragment (Fig. 6), and
since intact RNA was present in excess over smaller fragments, this
result must mean that binding of eIF-2 to STNV RNA leads to local unfolding of the RNA that allows cleavage by ribonuclease T_1 down to
position 44. Similar results were obtained with P1 and carboxyl-
methylated pancreatic ribonucleases (25).

     Binding of eIF-2 to STNV RNA, therefore, leads to an opening up
of the RNA structure. Since the viral RNA is most likely folded
compactly (e.g. to resist nuclease attack or for encapsidation), unfolding of the 5'-terminal region would be expected to occur during
protein synthesis and is also supported by the results of ribosome
binding studies (see below).

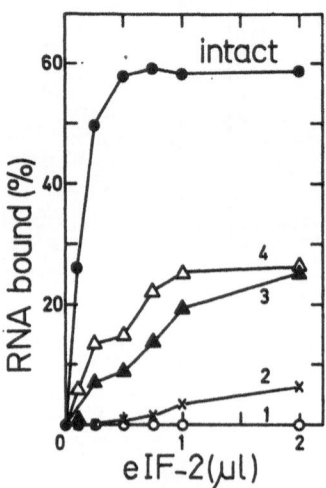

Fig. 6.   eIF-2 saturation binding curves of intact STNV RNA and
           individual 5'-terminal fragments. Intact STNV RNA (1,018
           cpm), and 32-nucleotide (no. 1) (293 cpm), 44-nucleotide
           (no. 2) (283 cpm), 73-nucleotide (no. 3) (483 cpm) and 115-
           nucleotide (no. 4) (201 cpm) fragments isolated from a T_1
           partial digest of intact, 5' end-labeled STNV RNA were
           incubated in the presence of the indicated amounts of eIF-2
           (0.15 µg/µl) and the percentage of radioactivity retained on
           nitrocellulose filters was determined (17).

Relationship Between Binding of eIF-2 and Binding of the Ribosome

        In studying the binding of ribosomes in radioiodinated STNV RNA
in conditions for initation of translation, Browning *et al*. (28)
observed that 80S ribosomes fully protect against nuclease attack the
region 19-44 nucleotides from the 5' end, and partially protect the
sequence from position 19 to 52, while 40S ribosomal subunits fully
protect the region 10-47 nucleotides, and partially protect the region
3-52 nucleotides from the 5' end. They concluded that the 5'-terminal
stem and loop conformation (see Fig. 1) is opened on binding of the
ribosome. The experiments (25) with STNV RNA carrying label at the
5' end allow determination of the 3'-proximal boundary of the eIF-2
binding site at or near the G residue at position 44. The fact that
such a fragment is selectively recovered bound to eIF-2 in conditions
of T_1 ribonuclease digestion that cause extensive degradation of
intact STNV RNA (see previous section), means that the normally accessible residues at positions 11, 12, 23, 32 and 33 (see Fig. 1) are
now shielded by the factor. It remains to be established whether the
interaction also involves the physical 5' terminus.

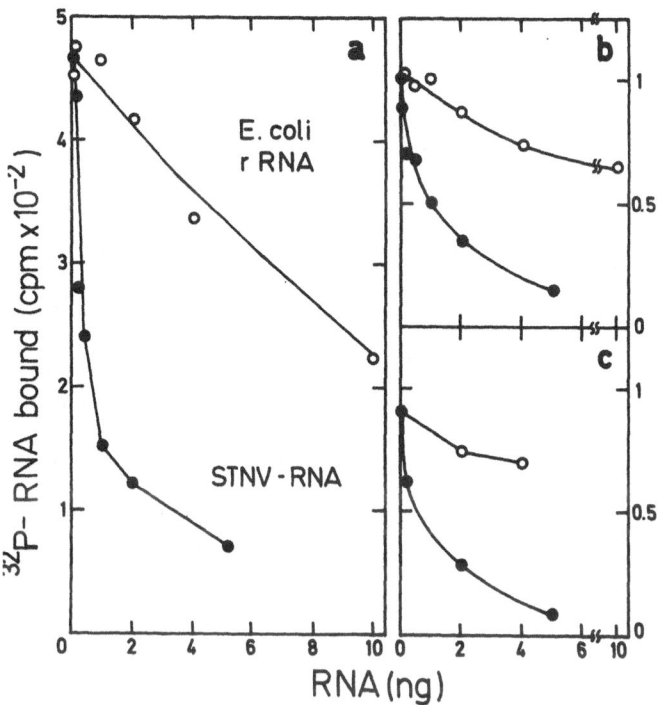

Fig. 7.   Competition binding behavior of intact STNV RNA and the 73-
          and 115-nucleotide 5'-terminal fragments. Purified, intact
          STNV RNA (1,290 cpm) (a) and 115-nucleotide (250 cpm) (b) and
          73-nucleotide (303 cpm) (c) fragments (see Fig. 6) were
          incubated with a limiting amount of eIF-2 (7.8 ng for a, 78
          ng for b, and 155 ng for c) and the indicated amounts of
          over 90% intact, unlabeled STNV RNA or *E. coli* rRNA. The
          amount of radioactivity retained on nitrocellulose filters
          was determined (17). Binding of the 44-nucleotide fragment
          was similarly analyzed; 58% of the binding was competed by
          5 ng of intact, unlabeled STNV RNA, and 11% by 5 ng of *E.
          coli* rRNA (not shown).

     There exists, therefore, a striking resemblance between the
binding site for eIF-2 and that for 40S ribosomal subunits in STNV
RNA. Considering that the interaction between eIF-2 and the RNA was
studied in the absence of Met-tRNA$_f$, ribosomes and other components
for protein synthesis, this raises the interesting possibility that
binding of a ribosome to mRNA during initiation of translation is
guided to an important extent by eIF-2. Additional strong evidence
for this concept is provided by the finding that in the 7500-base
long Mengovirus RNA, eIF-2 specifically protects against nuclease

attack three unique, large T_1 oligonucleotides that are identical
with three out of the four large oligonucleotides protected in both
40S and 80S initiation complexes (45).

mRNA Competition for eIF-2 during Translation

As will be detailed elsewhere in this Volume (30), Mengovirus RNA
competes thirty to forty-fold more effectively than globin mRNA in
translation. The competition can be relieved by addition of eIF-2.
Mengo RNA binds to eIF-2 some thirty-fold more tightly than globin
mRNA.   Likewise, eIF-2 relieves translational competition between α-
and β-globin mRNA, and binds more tightly to the latter (31).   eIF-2
also relieves the salt-induced inhibition of mRNA binding to 40S ribosomal subunits (31).   Moreover, there is a direct correlation between
the salt sensitivity of globin mRNA translation on one hand, and of
globin mRNA binding to eIF-2 on the other (31).   Finally, the prokaryotic phage R17 RNA binds to eIF-2 ten to fifteen-fold more weakly
than globin mRNA (23) and is translated in a eukaryotic cell-free
system at ten to fifteen-fold lower efficiency (32).

Interaction Between eIF-2 and Double-Stranded RNA

Double-stranded RNA (dsRNA) is a powerful inhibitor of initiation
of protein synthesis in rabbit reticulocyte lysates. The inhibition
is known to involve the inactivation of eIF-2 (18, 33), but the
mechanism of inactivation is not yet understood. An unexpected
involvement of mRNA in the action of dsRNA was revealed by the finding
that translation of Mengo- or Coxsackie virus RNA in a template-
dependent reticulocyte lysate is resistant to inhibition by dsRNA in
conditions where translation of globin or ascites tumor cell mRNAs is
inhibited (Fig. 8). The unabated translation of Mengo RNA in the
presence of dsRNA is dependent on continued initiation, for translation of Mengo RNA is as sensitive as that of globin mRNA to the
specific inhibitors of initiation, pactamycin and aurin tricarboxy-
late. The resistance of Mengo RNA translation to dsRNA is not caused
by a lesser dependence on initiation factors, but by a failure of
dsRNA to establish inhibition when Mengo RNA is used as messenger
(34).

It must be noted here that eIF-2 can bind directly to dsRNA (18).
There is, however, a basic difference between the interactions of
mRNA and of dsRNA with eIF-2. ·While binding of mRNA to eIF-2 involves a specific site or sequence, binding of dsRNA does not, but
instead seems to be determined primarily by the conformation of dsRNA
(22). This is illustrated by the experiment of Fig. 9, which shows
that binding of phage R17 RNA to eIF-2, even at greater than saturating amounts of eIF-2 (arrow), does not lead to protection of a
significant per cent of sequences to pancreatic ribonuclease digestion
in low salt conditions, where the complex is almost completely stable;

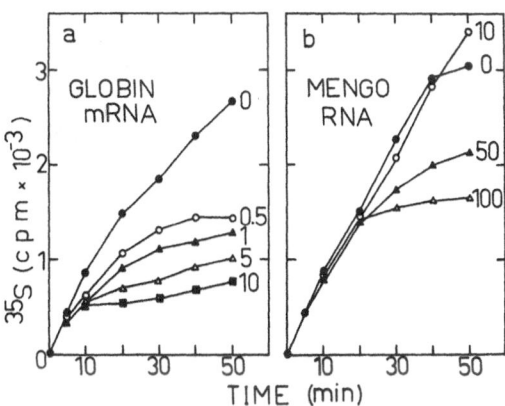

Fig. 8.   Differential sensitivity of Mengovirus RNA and globin mRNA
          translation to inhibition by dsRNA.   A rabbit reticulocyte
          lysate treated with micrococcal nuclease was incubated with
          globin mRNA (1.5 μg) (a) or Mengovirus RNA (1.5 μg) (b).
          DsRNA from *Penicillium chrysogenum* was present at the
          indicated concentrations (in ng/ml).   Translation was at
          30°C in 50-μl reaction mixtures containing 30 μl lysate,
          6 μg mouse liver tRNA 95 mM of added KCl and 0.6 mM of
          added Mg^{2+}.   This Mg^{2+} concentration was determined as
          optimal for both mRNA species.   Aliquots of 5μl were sampled
          at time intervals and hot CCl_3COOH-precipitable ^{35}S-methio-
          nine was determined.   Background without mRNA (300 cpm) was
          subtracted.

by contrast, binding of dsRNA to eIF-2 leads to almost complete
protection of its sequences.

   Direct RNA-binding competition studies reveal that dsRNA
competes with mRNA for eIF-2, binding this factor more strongly than
globin mRNA, but more weakly than Mengovirus RNA (Fig. 10).   The
progressively greater affinities of globin mRNA, dsRNA, and Mengovirus
RNA for eIF-2 are also reflected by the ability of these RNA species
to competitively inhibit the binding of methionyl-tRNA$_f$ to eIF-2 (34).

The correlation between the results demonstrating mRNA-speci-
ficity in the inhibitory action of dsRNA on translation, and those of
the RNA binding competition experiments suggests that the affinity of
a given mRNA species for eIF-2 is essential in determining the sensitivity of its translation to dsRNA (34), and again exmphasizes the
importance of the interaction of mRNA with eIF-2 for translational
control.

Mutually exclusive Binding of mRNA and Met-tRNA$_f$ to eIF-2

How can the property of eIF-2 to bind Met-tRNA$_f$ and GTP in a

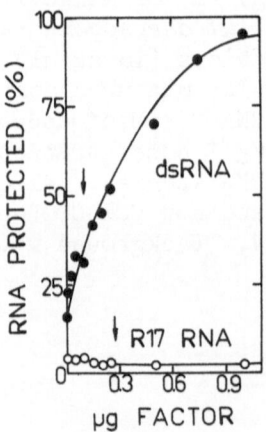

Fig. 9.   Ribonuclease sensitivity of complexes between eIF-2 and
          mRNA or dsRNA. ^{32}P-labeled φ6 dsRNA (0.63 pmol; 2,730 cpm)
          and ^{32}P-labeled R17 RNA (1.7 pmol; 9,600 cpm) were incubated
          for 10 min at 37° in the RNA-binding assay (17) with
          increasing amounts of purified eIF-2. Pancreatic ribo-
          nuclease was then added to a concentration of 5.4 µg/ml, and
          incubation at 37° was continued for another 30 min. The KCl
          concentration during binding and digestion was 10 mM.
          After cooling, cold CCl$_3$COOH-precipitable radioactivity was
          determined. No background was subtracted.

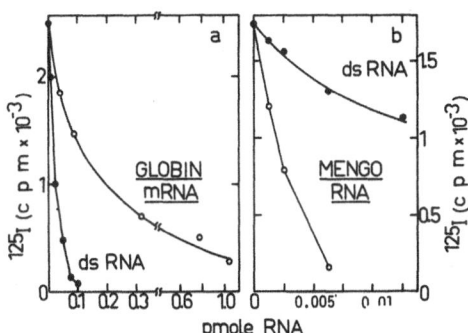

Fig. 10. Competition of dsRNA, globin mRNA and Mengovirus RNA for
eIF-2. RNA-binding assays in a contained ^{125}I-labeled
globin mRNA (0.025 pmole; 9,500 cpm) and 10.6 ng eIF-2.
In b, they contained ^{125}I-labeled Mengovirus RNA (0.002
pmole; 4,470 cpm) (17) and 0.28 ng eIF-2. Unlabeled dsRNA,
globin mRNA or Mengovirus RNA were present in the indicated
amounts. Control without eIF-2 (110 cpm in a, 70 cpm in b)
was subtracted.

ternary complex which subsequently binds to the 40S ribosomal subunit,
be reconciled with its property to recognize and bind mRNA ? An
important clue is provided by the observation that Met-tRNA$_f$ and mRNA
are mutually exclusive in their binding to eIF-2 (20, 22, 24). That
is, mRNA competes with labeled Met-tRNA$_f$ for binding of eIF-2 (20,
22), and in this respect Mengovirus RNA competes on a molar basis
thirty to forty-fold more effectively than globin mRNA (34). Conversely, binding of labeled mRNA to eIF-2 is totally sensitive to
competition by Met-tRNA$_f$, provided GTP is present, but not by uncharged tRNA (Fig. 11).

Upon binding of Met-tRNA$_f$ and GTP to purified eIF-2, the three
subunits of the factor remain associated; by contrast, the binding of
mRNA to free eIF-2 has been reported to induce dissociation, most
likely of the small subunit, from the ribonucleoprotein complex (21).
Both mRNA and Met-tRNA$_f$ bind to one of the large subunits of eIF-2
(20).

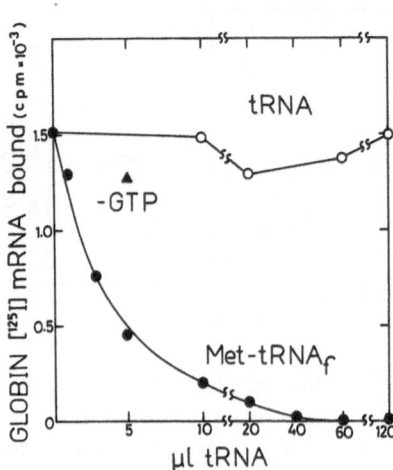

Fig. 11. Globin mRNA binding to eIF-2:   competitive inhibition by
         Met-tRNA$_f$.   Reaction mixtures (150 µl) contained 150 mM
         KCl, 20 mM Tris-HCl (pH 7.4), 0.1 mM GTP, 0.1 mM MgCl$_2$, 10
         mM 2-mercaptoethanol, ^{125}I-labeled globin mRNA (4.8 x 10^5
         cpm/µg; input, 2,700 cpm), the indicated amounts of a 0.1
         mg/ml solution of mouse liver tRNA (charged in the presence
         or absence of unlabeled L-methionine with *E. coli* synthe-
         tase; charging was monitored in a parallel sample contain-
         ing ^{35}S-labeled methionine) and purified eIF-2 (6 ng).
         After incubation for 7.5 min at 30°C, samples were passed
         through nitrocellulose filters as in the Met-tRNA$_f$-binding
         assay (17).   Background without protein (120 cpm) was sub-
         tracted.

eIF-2 and Initiation of Translation

        Whether during protein synthesis a molecule of eIF-2 will form
a complex with Met-tRNA$_f$ and GTP, or bind to mRNA, will depend on the
equilibrium conditions.   If the molar concentrations of Met-tRNA$_f$ and
GTP in the cytosol greatly exceed that of mRNA, as seems likely
during active translation, free eIF-2 molecules will tend to form
ternary complexes rather than bind to mRNA directly.   In this context,
it should be noted that at physiological salt concentrations the rate
of exchange of mRNA in complexes with eIF-2 far exceeds that of Met-
tRNA$_f$ with the ternary complex, eIF-2/Met-tRNA$_f$/GTP   (26).

        Studies of the effect of salt on translation support the con-
tention that mRNA interacts with eIF-2 molecules located in 40S/Met-

tRNA$_f$ complexes. eIF-2 relieves the inhibition of total protein
synthesis, and the sharpened competition between α- and β-globin mRNA,
occurring at elevated KCl concentrations (30, 31). Since high concentrations of Cl$^-$ion inhibit initiation of translation by preventing
primarily the binding of mRNA to 40S initiation complexes, while
binding of Met-tRNA$_f$ to 40S subunits is only slightly inhibited,(46)it
seems likely that the ability of eIF-2 to relieve the inhibitory
effect of KCl is based on its action at the mRNA-binding step rather
than at the step involving 40S/Met-tRNA$_f$ complex formation. Indeed
KCl directly inhibits the binding of mRNA to eIF-2 at concentrations
that hardly affect the formation of ternary complexes between eIF-2,
Met-tRNA$_f$, and GTP (31, 26).

It is most plausible, therefore, to suggest that the interaction
of a molecule of mRNA with eIF-2 takes place only after the ternary
complex is bound to the 40S ribosomal subunit. That is, both binding
of mRNA to eIF-2 and mRNA competition for eIF-2 would involve eIF-2
molecules located on 40S ribosomal subunits. At this point, the
interaction of a molecule of mRNA with eIF-2 may well displace the
bound Met-tRNA$_f$ from this factor. Thus, during initiation of protein
synthesis three processes may occur in step: binding of mRNA to
eIF-2, displacement of Met-tRNA$_f$ from eIF-2, and base-pairing between
mRNA and Met-tRNA$_f$ at the AUG intiation codon which, as we have seen
for STNV RNA, is located in the region recognized by eIF-2. Accordingly, the displaced Met-tRNA$_f$ will now be ready to form the first
peptide bond, and most likely would be stabilized at this point in the
ribosomal structure. How ribosome components and/or other initiation
factors participate in this process is not known.

BINDING OF OTHER INITIATION FACTORS TO mRNA

Binding of eIF-3 to poly(A)-containing mRNA can be observed, but
this binding does not appear to be specific as judged by competition
with other RNA species (36).

A protein consisting of two subunits of 56,000 and 61,000
daltons, respectively, has been isolated from rabbit reticulocytes;
it binds preferentially to mRNA (37). This protein does not appear
to be related to any of the known initiation factors, and its role
in protein synthesis, if any, is not known.

Maximal binding of mRNA to 40S ribosomal subunits may involve,
in addition to eIF-2 and eIF-3, the initiation factors eIF-1, eIF-4A,
eIF-4B and eIF-4C (8, 9), and the cap-binding protein (38). Of these,
it is difficult to visualize a major role for eIF-1 and eIF-4C in
mRNA recognition, as their omission lowers mRNA binding only 15 to
50% (8). This is also true for eIF-4B; indeed, the requirement for
this factor is most noticeable by its stimulatory effect on overall
translation of globin mRNA (8, 9), and this could be due in part to
contaminating cap-binding protein (38). Omission of eIF-4A leads to
a marginal reduction of mRNA binding (50-75%), and as mentioned

earlier, neither eIF-4A nor eIF-4B are recovered bound to the 40S
initiation complex (8).   eIF-4A does not bind detectably to mRNA,
but eIF-4B is able to retain labeled mRNA on membrane filters.

An involvement of eIF-4A and eIF-4B in mRNA recognition was first
suggested by the ability of these proteins to affect mRNA competition
in an overall translation system (39, 40).   Indeed, eIF-4B possesses
properties similar to eIF-2 in that it relieves translational competition between host mRNA and picorna virus RNA (40), and binds the
latter RNA more tightly than globin mRNA (41).   These properties will
be dealt with more extensively elsewhere in this volume (30).   At
this time, these studies are difficult to interpret because they were
conducted with initiation factor preparations that were less well
characterized than those available later, and indeed eIF-4B was found
to contain cap-binding protein (16, 38).   Thus, the individual roles
of eIF-4A and eIF-4B are not clear at present.

How the cap binding protein, eIF-2, eIF-4A and eIF-4B might
function together in the binding of mRNA remains to be elucidated.
The studies reviewed above strongly support a central role for eIF-2
in mRNA binding and recognition.   There could be a time-ordered or a
concerted interaction.   For example, initial binding of mRNA might
involve interaction with cap-binding protein (16) at the cap, and with
eIF-2 at the 5'-terminal initiation region.   It is conceivable that
eIF-2 undergoes a secondary interaction with the cap thereafter.
eIF-4A and/or eIF-4B may interact with the mRNA, perhaps in ways
differing from eIF-2 and cap-binding protein, to stabilize its binding
to the 40S ribosomal subunit.   Evidence for such mechanisms should now
be sought.

CONCLUSION

A role for eIF-2 in recognition and binding of mRNA was overlooked initially, largely because of its well-defined function in the
binding of Met-tRNA$_f$.   Now, a variety of experiments support a direct,
specific interaction between eIF-2 and mRNA during initiation.   What
share eIF-2, eIF-3, eIF-4B and the cap-binding protein take in the
initial binding of mRNA is not yet clear, but eIF-2 serves as a target
for mRNA competition, and both translation and binding of mRNA to
eIF-2 are similarly affected by a variety of agents.   The data
indicate that during initiation of protein synthesis, binding of mRNA
to eIF-2, displacement of Met-tRNA$_f$ from eIF-2, and base-pairing
between mRNA and Met-tRNA$_f$ may occur in step.   The studies with STNV
RNA show that eIF-2 recognizes a 5'-terminal region strongly resembling that protected in that 40S ribosomal initiation complex against
nuclease attack.   In binding to this region, eIF-2 causes local unfolding of STNV RNA, perhaps to allow accommodation of the 40S ribosomal subunit during protein synthesis.

As regards the possible molecular basis for the specific interaction between eIF-2 and the 5' end of mRNA, the results with STNV
RNA lead one to consider a *free* AUG (e.g., not involved in secondary
or tertiary interactions), a 5' end, and a favorable overall conformation. The presence of a 5' end is not sufficient for recognition by eIF-2, as demonstrated by the results of Figs. 4 and 5 which
show that eIF-2 is able to select a *specific* mRNA fragment from among
a collection of fragments bearing different 5' ends. Clearly, the
presence of pyrophosphate linkages at the 5' end is not an absolute
requirement, as witnessed by the specific binding of eIF-2 to 5'-mono-
phosphorylated STNV RNA. That the AUG initiation codon in STNV RNA
is indeed free follows from the observation that partial digestion of
intact STNV RNA with ribonuclease T_1 yields the 5'-terminal 32-nucleo-
tide fragment ending in this codon (see Fig. 1) as one of the major
digestion products. Moreover, the AUG initiation codon in mouse or
rabbit β-globin mRNA also is readily accessible to ribonuclease T_1,
while by contrast in the less efficient α-globin mRNA it apparently is
not (42). This fits well with the finding that β-globin mRNA binds
to eIF-2 with higher affinity than does α-globin mRNA (31), and tends
to underline the contribution of a free AUG codon. The importance of
overall RNA conformation in the 5'-terminal region is observed not
only for binding of eIF-2 (see Fig. 6) (25), but also for the binding
of 40S ribosomal subunits during initiation: they fail to select the
correct initiation site in extensively denatured reovirus mRNA (43).

ACKNOWLEDGEMENTS

    The authors's work reported in this review was partly supported
by GSF (München), the Israel Academy of Sciences, the Centrum voor
Phytovirologie of the IWONL and the Geconcerteerde Akties (Belgium),
as well as by EMBO Long-Term and FEBS Fellowships.

REFERENCES

1.   Nakashima, K., Darzynkiewicz, E. and Shatkin, A.J. (1980),
     Nature, London, 286, 226-230.
2.   Shine, J. and Dalgarno, L. (1974), Proc. Nat. Acad. Sci. U.S.A.
     71, 1342-1346.
3.   Hagenbuchle, O., Santer, M., Steitz, J.A. and Mans, R.J. (1978),
     Cell 13, 551-563.
4.   Kozak, M. (1978), Cell 15, 1109-1123.
5.   Klein, W.M., Nolan, C., Lazar, J.M., and Clark, J.M. (1972),
     Biochemistry 11, 2009-2014.
6.   Trachsel, H., Erni, B., Schreier, M. and Staehelin, T. (1977),
     J. Mol. Biol. 116, 755-767.
8.   Benne, R. and Hershey, J. (1978), J. Biol. Chem. 253, 3078-3087.
9.   Benne, R., Brown-Luedi, M. and Hershey, J. (1979), *Methods in
     Enzymol.* 60, 15-35.

10. Thompson, H.A., Sadnik, I., Scheinbuks, J. and Moldave, K. (1977), Biochemistry 16, 2221-2230.
11. Kaempfer, R. (1972), J. Mol. Biol. 71, 583-598.
12. Dettman, G.L., and Stanley, W.M., Jr. (1972), Biochim. Biophys. Acta 287, 124-133.
13. Levin, D.M., Kyner, D. and Acs, G. (1973), Proc. Nat. Acad. Sci. U.S.A. 70, 41-45.
14. Schreier, M.H. and Staehelin, T. (1973), Nature, London, New Biol. 242, 35-38.
15. Darnbrough, C.M., Legon, S., Hunt, T. and Jackson, R.J. (1973), J. Mol. Biol. 76, 379-403.
16. Sonenberg, N., Morgan, M.A., Testa, D., Colonno, R.J. and Shatkin, A.J. (1979), Nuc. Acids Res. 7, 15-29.
17. Kaempfer, R. (1979), *Methods in Enzymology* 60, 380-392.
18. Kaempfer, R. (1974), Biochim. Biophys. Res. Commun. 61, 591-597.
19. Hellerman, J.G. and Shafritz, D.A. (1975), Proc. Nat. Acad. Sci. U.S.A. 72, 1021-1025.
20. Barrieux, A. and Rosenfeld, M.G. (1977), J. Biol. Chem. 252, 3843-3847.
21. Barrieux, A. and Rosenfeld, M.G. (1978), J. Biol. Chem. 253, 6311-6315.
22. Kaempfer, R., Hollender, R., Abrams, W.R. and Israeli, R. (1978), Proc. Nat. Acad. Sci. U.S.A. 75, 209-213.
23. Kaempfer, R., Hollender, R., Soreq, H. and Nudel, U. (1979), Eur. J. Biochem. 94, 591-600.
24. Rosen, H. and Kaempfer, R. (1979), Biochem. Biophys. Res. Commun. 91, 449-455.
25. Kaempfer, R., Van Emmelo, J. and Fiers, W. (1981), Proc. Nat. Acad. Sci. U.S.A., in press.
26. Kaempfer, R., Rosen, H. and Israeli, R. (1978), Proc. Nat. Acad. Sci. U.S.A. 75, 650-654.
27. Lesnaw, A. and Reichmann, M.E. (1968), J. Mol. Biol. 38, 59-73.
28. Browning, K.S., Leung, D.W. and Clark, J.M., Jr. (1980), Biochemistry 19, 2276-2282.
29. Smith, R.E. and Clark, J.M., Jr. (1979), Biochemistry 18, 1366-1371.
30. Kaempfer, R. (1981), in *Protein Biosynthesis in Eukaryotes* (R. Perez-Bercoff, Ed.), Plenum Press, New York and London.
31. Di Segni, G., Rosen, H. and Kaempfer, R. (1979), Biochemistry 18, 2847-2854.
32. Schreier, M., Staehelin, T., Gesteland, R. and Spahr,,P. (1973), J. Mol. Biol. 75, 575-578.
33. Kaempfer, R. and Kaufman, J. (1973), Proc. Nat. Acad. Sci. U.S.A. 70, 1222-1226.
34. Rosen, H., Knoller, S. and Kaempfer, R. (1980), submitted.
35. Ysebaert, M., Van Emmelo, J. and Fiers, W. (1980), J. Mol. Biol. 143, 273-287.
36. Rosenfeld, M.G. and Barrieux, A. (1979), *Methods in Enzymology* 60, 392-401.

37. Rosenfeld, M.G. and Barrieux, A. (1977), Biochemistry 16, 514-518.
38. Sonenberg, N., Rupprecht, K.M., Hecht, S.M., and Shatkin, A.J. (1979), Proc. Nat. Acad. Sci. U.S.A. 76, 4345-4349.
39. Kabat, D. and Chappell, M. (1977), J. Biol. Chem. 252, 2684-2690.
40. Golini, F., Thach, S.S., Birge, C.H., Safer, B., Merrick, W.C. and Thach, R.E. (1976), Proc. Nat. Acad. Sci. U.S.A. 73, 3040-3044.
41. Baglioni, C., Simili, M. and Shafritz, D.A. (1978), Nature, London, 275, 240-243.
42. Pavlakis, G.N., Lockard, R.E., Vanvakopoulos, N., Rieser, L., Raj-Bhandary, U.L., Vourmakis, J.M. (1980), Cell 19, 91-102.
43. Kozak, M. (1980), Cell 19, 79-90.
44. Kaempfer, R., Israeli, R., Rosen, H., Knoller, S., Zilberstein, A., Schmidt, A. and Revel, M. (1979), Virology 99, 170-173.
45. Perez-Bercoff, R. and Kempfer, R. (1981), submitted.
46. Weber,L.A.,Hickey,E.D.,Maroney,P.A. and Baglioni,C. (1977),J. Biol.Chem.252,4007-4010

BUT IS THE 5' END OF MESSENGER RNA

ALWAYS INVOLVED IN INITIATION ?

Raul Pérez-Bercoff

The Institute for Virology
University of Rome
Viale di Porta Tiburtina 28
00185-ROME (Italy)

INTRODUCTION

A preeminent feature of eukaryotic messenger RNAs is the presence of a peculiar structure of methylated nucleotides "capping" the 5' end of the molecule. The cap structure seems to facilitate recognition of the messenger by eukaryotic ribosomes during the early steps of translation (1). The need for so direct an involvement of the 5' end of the molecule in the process of initiation (conceptually in line with the generally accepted notion that all mRNA species are functionally *monocistronic* and, hence must contain only one initiation site for translation) appears far less evident when we come to consider the case of naturally un-capped mRNAs. Indeed, if the high affinity of the eukaryotic ribosome and its associated initiation factors for the 5'-terminal cap secures the proper positioning of the initiation complex at the 5' end of the messenger, it becomes questionable why, in the absence of a cap structure, should still be the 5' end the only possible entry site for ribosomes. And why should internal sequences be forbidden in messengers that have bypassed the need of a 5'-terminal cap.

This seems to be precisely the case of the genomic RNA of picornaviruses, a naturally non-capped RNA of unusual efficiency in translation. Curiously, the suspicion that mRNAs *lacking a cap* might initiate translation in a different way arose retroactively, after a series of studies on the mechanism of translation of picornavirus RNA lend strong support to the concept that, in contrast to other eukaryotic mRNA species, these mRNAs were able to express *two* cistrons, at least during their *in vitro* translation.

It appeared, therefore, proper and fitting to the purpose of this volume to review here the presently available evidence for the involvement of *internal* (rather than 5'-terminal) sequences of mRNA in the process initiation of translation. Several reasons (not the last one the personal preferences of this author...) suggested to concentrate on the case of picornavirus RNA, with the explicit proviso that a similar situation may still be found in other non-capped messengers.

THE GENOMIC RNA OF PICORNAVIRUS

The genome of picornaviruses is a single-stranded RNA molecule, about 7500 nucleotides in length (2) which can serve directly as messenger for translation of viral proteins *in vivo* and *in vitro*. It contains a 3'-terminal poly(A) stretch (3) and, in the case of cardio- and aphto-viruses, there is a characteristic poly(C) region of variable lenght (70-150 bases), close to (3) but not at (4) the 5' end.

The 5' end of picornavirus RNA is exceptional in more than one way:

i)   it lacks a cap structure (2), and

ii) a small protein is covalently linked to the 5'-terminal residue (4,5).

These unusual features of the 5' end, however, do not hinder the unmatched translational ability of picornavirus RNA, which ranks among the most efficient messengers so far known in eukaryotic systems. Moreover, this un-capped RNA has been shown to outcompete cellular and viral mRNAs during translation (6-11). The molecular basis for such an efficiency are not yet known. In the absence of a recognizable cap at the 5' end other mechanisms should be operative. Some years ago, we proved that the small protein covalently linked to the 5'-terminal pUp of Mengovirus RNA was not needed to secure its efficient *in vitro* translation (4). Similarly, Wimmer and collaborators have recently shown that the presence of VPg does not hamper the translation ability of poliovirus RNA (12).

EVIDENCE FOR MORE-THAN-ONE INITIATION SITE IN PICORNAVIRUS RNA

Twelve to fifteen peptides were identified as the products of translation of picornavirus RNA (13). Since their combined molecular weights exceeded the total coding capacity of the genomic RNA, the paradox was solved by postulating that the viral peptides were generated by cleavage of a large, unique precursor (13-16). The most obvious implication of such a mechanism of translation is that all viral peptides must be present in *equimolar* amounts. However, two independent reports pointed out that during the late stages of

the infectoius cycle of polio- (17) or Mengo-virus (18) the molar
yields of some viral proteins did not fit into the predicted value,
suggesting a sort of "asymmetrical" translation of picornavirus RNA
(17). The experimental evidence available at the time, however, did
not consent a univocal interpretation, and the reported "asymmetry"
could still be looked at as arising from premature termination of
translation, or else by preferential degradation of portions of the
product.

The concept of picornavirus RNA being able to initiate translation at *two* different sites (or alternatively, being translated from
a single site in two different reading frames) received a more direct
support when shortly thereafter E.Ehrenfeld identified two different
N-termini in the tryptic fragments of the peptides synthesized *in
vitro* in a coupled system for transcription/translation derived from
poliovirus-infected HeLa cells (19). Further analyses by polyacrylamide gel electrophoresis confirmed that these two N-terminal ends
actually belonged to two virus-coded peptides of quite different
sizes (20).

The relevance of these findings was questioned (21), the main
objection raised being that the relative proportion of the two
N-termini varied according to the concentration of Mg^{++} present in
the reaction: By lowering the Mg^{++} concentration -the argument went-
the 40S ribosomal subunit is simply induced to skip over the first
AUG codon, a situation already observed as main effect of the antibiotic edeine (21-23). However, later studies on the mechanism of
in vitro translation of Mengovirus RNA showed that the relative
proportion of the two N-terminal tryptic fragments was constant over
a wide range of Mg^{++} concentrations, as far as they were compatible
with translation (24).

In conclusion, it appears in our view that the evidence for the
ability of picornavirus RNA to express two cistrons (at least under
the conditions of *in vitro* translation) cannot be dismissed as a
mere technical artifact. Two cistrons, however, did not imply
necessarily the existence of an *internal* initiation site: The
possibility had to be entertained that limited fragmentation of the
RNA into subgenomic pieces might generate new 5' ends.   That this
is *not* the case will be discussed in the next section.

INVOLVEMENT OF INTERNAL REGIONS OF PICORNAVIRUS RNA IN INITIATION

In all picornavirus/host cell systems so far studied, infection
of susceptible cells results in a rapid disaggregation of cellular
polysomes, and their substitution with viral ones (25-27). The
latter are characteristically bigger (average 350S) and rather homogenous. Although they have been extensively studied for over twenty
years, to the best of our knowledge there is heretofore no report

of picornavirus polysomes of subgenomic size. Obviously, this kind
of negative evidence does not imply that a closer scrutiny (especially into the slowly moving ones) will fail to reveal fragments of
viral RNA actually associated with polysomes. But there is little
doubt that the sort of massive fragmentation needed to generate an
homogeneous population of sub-genomic polysomes, if it happens to
occur, has passed so far unnoticed. Moreover, all the presently
available evidence is consistent with the opposite contention, i.e.:
that the viral RNA serving as messenger in the polysomes *does* conserve the original 5' end. Fingerprint analysis of poliovirus RNA
extracted from virions and polysomes proved that both contained the
same 5' terminal oligonucleotide, the only difference being the no-
torious absence of VPg in polysomal RNA (28,29). Yet, ribosomes
appear to start translation of poliovirus RNA at position 741:
Wimmer and collaborators have recently determined the full sequence
of poliovirus RNA (30). A direct correlation was then established
between the nucleotide and aminoacid sequences of viral RNA and
proteins, respectively. By means of such a direct analysis it was
shown that translation of poliovirus RNA starts at the AUG codon
extending from positions 741 to 743, that is: far removed from the
un-capped 5' end. (Incidentally, a closer inspection of the sequence of poliovirus RNA reveals that seven presumably silent AUG
triplets rank in the area between the 5' end and the actual initiation of translation. This makes very difficult (though not impossible) to visualize a mechanism of initiation involving the compulsory
entry site of ribosomes at the 5' end and the subsequent "scanning"
of the leader sequence: The 40S ribosomal subunit would have to
"ovelook" a number of potential initiation sites before selecting the
the right one.)

    Further evidence for the involvement of *internal* (rather than
5'-terminal) regions of picornavirus RNA during initiation of translation comes from experiments of *in vitro* translation of defined-
size fragments of Foot-and-Mouth Disease (FAMD) virus RNA (31). As
already mentioned, the RNA of aphtoviruses contains a long poly(C)
tract located some 400 bases distal from the 5' end. By hybridizing
this tract to oligo(dG), and subsequently digesting the complex with
RNase H (a nuclease that attacks the ribonucleic moiety of RNA/DNA
hybrids) it was possible to prepare fragments of the viral RNA
extending from the region immediately adjacent to the poly(C) tract
up to the poly(A) stretch at the 3' end of the molecule. The long
fragments were efficiently translated *in vitro*. A comparison of the
(^{35}S)-formyl-methionine-labeled products with the similarly labeled
peptides produced *in vivo*, in virus-infected cells, proved that both
were identical. The conclusion drawn from these studies was that an
authentic initiation site for translation of FAMD virus RNA must be
3'-*distal* with respect to the poly(C) tract (31).

    This view is further reinforced by recent findings suggesting
that eukaryotic ribosomes do *not* "scan" the un-capped RNA of pico-

rnaviruses, but are able to enter directly at the initiation site:
In fact, migration of the 40S ribosomal subunit from the 5' end to
the initiation codon requires ATP hydrolysis (21,32). Yet (as reviewed by R.J.Jackson elsewhere in this volume) the RNA of encephalomyocarditis (EMC) virus can form initiation complexes very efficiently in the absence of ATP or in the presence of non-hydrolyzable ATP
analogs. Moreover, were the entry site for the 40S ribosomal subunit
at the 5' end of the molecule, one would expect that blocking translocation would lead to the accumulation of several 40S subunits
"queuing" in the long sequence (500-700 nucleotides) preceding the
initiation site. However, this was not observed (R.J.Jackson, chapter
15 of this book).

In summary, the combined evidence of RNA sequencing and translation analysis points towards the involvement of *internal* (rather
than 5'-terminal) regions of picornavirus RNA during initiation of
protein synthesis.

STUDIES ON THE RIBOSOME-BINDING SITES OF MENGOVIRUS RNA

A more direct demonstration of the ability of eukaryotic ribosomes to recognize internal sequences of picornavirus RNA at initiation of translation was recently obtained in experiments aimed at
isolating the ribosome-binding sites of Mengovirus RNA (33). Advantage was taken of the known ability of ribosomes to interact with
specific areas of mRNA and, under proper conditions, to protect them
against nuclease attack. A similar approach had been originally used
to isolate the ribosome-binding sites of several phage RNAs (34,35).
In an attempt to study the binding of eukaryotic ribosomes to (^{32}P)-
labeled Mengovirus RNA, lysates of mouse L-929 or Krebs tumor-ascites
cells were treated with diphtheria toxin and NAD to block translocation (see N. Brot, chapter 11), and subsequently with sparsomycin,
to prevent peptide bond formation. After incubation to allow the
formation of the initiation complex, the RNA was trimmed with nuclease, and the (^{32}P)-labeled fragments protected in either the 40S or
the 80S initiation complexes were fingerprinted: The material protected in either the 40S or the 80S initiation complexes yielded
four unique large T_1-oligonucleotides. A physical map of the large
T_1-oligonucleotides had been constructed, and the four oligonucleotides characteristic of the fragments protected in either the 40S
or the 80S initiation complexes were found to map internally, within
the region between the poly(C) tract and the 3' end. (Incidentally,
the same four oligonucleotides were recovered from the RNA present
in 80S initiation complexes formed in lysates in which unlabeled
Mengovirus RNA had been translated extensively, indicating that recognition of the initiation sites by ribosomes was not modulated detectably by a viral translational product.)

In a series of parallel experiments, the recognition of intact
(^{32}P)-labeled Mengovirus RNA by the eukaryotic initiation factor 2

(eIF-2) was examined by direct complex formation. Fingerprint
analysis of the RNA protected by eIF-2 against nuclease digestion
yielded three T_1-oligonucleotides that were identical to three out
of the four oligonucleotides protected in either 40S or 80S complexes.

More relevant to our point, however, was the fact that the four
oligonucleotides characteristic of the fragment protected by either
ribosomes or eIF-2 could not be arranged in a continuous sequence,
but must constitute a minimum of two widely separated domains.

The conclusions drawn were that during initiation of translation
eukaryotic ribosomes recognize and bind to *more than a single sequence* of Mengovirus RNA located, here again, at *internal* sites that
are far removed from the un-capped 5' end of the molecule. The specific binding of eIF-2 to *internal* sequences of Mengovirus RNA (which
overlaps almost entirely with those protected in both 40S and 80S
initiation complexes) suggests that a free 5' end is *not* indispensable for this interaction.

The absence of 5'-terminal sequences in both eIF-2- and ribosome-protected segments of Mengovirus RNA supports the contention
that the RNA of picornaviruses may have evolved a highly efficient
mechanism of initiation that bypasses the need for either a 5' end
or a 5'-terminal cap. In line with this concept it should be
recalled here that studies on the interaction of eIF-2 with the RNA
of satellite tobacco necrosis virus (STNV), a highly efficient uncapped messenger RNA, had recently shown that eIF-2 and ribosome-
protected sites do also overlap to a considerable extent, and lie
close to the 5'end (36).

" IN VITRO VERITAS "

A preliminary consideration is due before drawing some final
conclusions: Most of the studies reviewed here have been conducted
in vitro, using cell lysates that bear a pale resemblance with the
physiological conditions of the intact cell: The question can be
raised as to how relevant, how meaningful these results are. While
this author would agree in principle with such objection, there is
reason to believe that the general conclusions reached at during
these studies can be safely extrapolated to the intact cell. Indeed,
the basic mechanisms involved in the process of initiation of translation are known to be operative in cell lysates which, from this
viewpoint, do not differ excesively from the intact cell: *"In vitro
veritas "*.

Bearing in mind such qualification, it appears that the general
principle emerging from these studies is that a suitable structure
of the initiation site may be the primary requirement for high efficiency of translation.

The proximity of a 5' end and, possibly the presence of a capping group provide additional signals for the recognition of the messenger and, perhaps for the proper positioning of ribosomes during initiation.

In the case of picornavirus RNA, the first condition seems to be amply fulfilled and suffices to secure a most efficient translation. STNV RNA, on the other hand, appears to satisfy the first two requirements, and all three might be needed in capped eukaryotic mRNAs. All available evidence suggests that in the latter the presence of a cap structure plays so prominent a role in the process of recognition of the messenger as to eclipse all other structural requirements. But the case of picornavirus RNA should serve to recall us that under proper conditions eukaryotic ribosomes (like their prokaryotic counterpart) have conserved the ability to initiate translation at internal sites of the messenger.

REFERENCES

1.   Shatkin, A.J. (1976), Cell, 9, 645-653
2.   Fellner, P. (1979), in: *"The Molecular Biology of Picornaviruses"* (R.Pérez-Bercoff, ed.) pp.25-47,Plenum Press, New York and London.
3.   Pérez-Bercoff, R. and Gander, M. (1977), Virology, 80, 426-429.
4.   Pérez-Bercoff, R. and Gander, M. (1978), FEBS Letters, 96, 306-312.
5.   Wimmer, E. (1979), in: *"The Molecular Biology of Picornaviruses"* (R. Pérez-Bercoff, ed.) pp.175-190, Plenum Press, New York and London.
6.   Golini, F., Thach, S.S., Birge, C.H., Safer, B., Merrick, W.C., and Thach, R.E. (1976), Proc.Natl.Acad.Sci.USA, 73,3040-3044.
7.   Lawrence, C. and Thach, R.E. (1974), J.Virol.14, 598-610.
8.   Rose, J.K., Trachsel, H.H., Leong, K., and Baltimore, D.(1978), Proc.Natl.Acad.Sci.USA, 75, 2732-2736.
9.   Hackett, P.B., Egberst, E. and Traub, P. (1978), Eur.J.Biochem. 83, 341-352.
10. Egberts, E.,Hackett, P.B. and Traub, P. (1977), Hoppe-Seyler's Z.Physiol.Chem. 358, 463-474
11. Hackett, P.B., Egberts, E. and Traub, P. (1978), Eur.J.Biochem. 83, 353-361.
12. Golini, F., Semler, B.L., Dorner, A.J. and Wimmer, E. (1980) Nature (London), 287, 600-603.
13. Summers, D.F., Maizel, J.V.jr. and Darnell, J.E. (1965), Proc. Natl.Acad.Sci.USA, 54, 505-513.
14. Summers, D.F. and Maizel, J.V.jr. (1968), Proc.Natl.Acad.Sci. USA, 59, 966-971.
15. Jacobson, M.F. and Baltimore, D. (1968), Proc.Natl.Acad.Sci.USA 61, 77-84.
16. Lucas-Lenard, J. (1979), in: *"The Molecular Biology of Picornaviruses"* (R.Pérez-Bercoff, ed.) pp.127-147, Plenum Press,

New York and London.

17. Paucha, E., Seehafer, J. and Colter, J.S. (1974), Virology, 61, 315-326.
18. Lucas-Lenard, J. (1974), J.Virol. 14, 261-269.
19. Celma, E. and Ehrenfeld, E. (1975), J.Mol.Biol. 98, 761-780.
20. Jense, H., Knauert, F. and Ehrenfeld, E. (1978), J.Virol. 28, 387-394.
21. Kozak, M. (1978), Cell, 15, 1109-1123.
22. Kozak, M. (1979), J.Biol.Chem. 254, 4731-4738.
23. Kozak, M. and Shatkin, A.J. (1978), J.Biol.Chem. 253, 6568-6577.
24. Pérez-Bercoff, R. and Degener, A.M. (1981), submitted.
25. Penman, S., Scherrer, K., Becker, Y. and Darnell, J.E. (1963), Proc.Natl.Acad.Sci.USA, 49, 654-661.
26. Dalgarno, L., Cox, R.A. and Martin, E.M. (1967), Biochim.Biophys. Acta, 138, 316-328.
27. Lucas-Lenard, J. (1979), in: *"The Molecular Biology of Picorna-viruses"* (R. Pérez-Bercoff, ed.) pp.73-99,Plenum Press, New York and London
28. Nomoto, A., Kitamura, N., Golini, F. and Wimmer, E. (1977), Proc. Natl.Acad.Sci.USA, 74, 5345-5349.
29. Petterson, R.F., Flanegan, J.B., Rose, J.K. and Baltimore, D. (1977), Nature (London), 268, 270-272.
30. Kitamura, N., Semler, B.L., Rusberg, P.G., Larsen, G.R., Adler, C.G., Dorner, A.J., Enini, E.D., Hanecak, R., Lee, J.J., Van den Werf, S., Anderson, C.W. and Wimmer, E. (1981), Nature (London) in press.
31. Sangar, D.V.,Black, D.M., Rowlands, D.J., Harris, T.J.R. and Brown, F. (1980), J.Virol. 33, 59-68.
32. Kozak, M. (1980), Cell, 22, 459-467.
33. Pérez-Bercoff, R. and Kaempfer, R. (1981), submitted.
34. Steitz, J.A.(1969), Nature (London), 224,957-964.
35. Hindley, J. and Staples, D.H. (1969), Nature (London), 224, 964-967.
36. Kaempfer, R., Van Emmelo, J. and Fiers, W. (1981), Proc.Natl. Acad.Sci.USA, in press

SECTION IV:

SYNTHESIS AND PROCESSING OF PROTEINS

PEPTIDE CHAIN ELONGATION

AND TERMINATION IN EUKARYOTES

Nathan Brot

Roche Institute of Molecular Biology

Nutley, N.J. 07110 (U.S.A.)

INTRODUCTION

The steps of protein synthesis in both prokaryotes and eukaryotes involve similar reactions. One exception is the initiation
process which involves more factors and appears to be more complicated in eukaryotes than in prokaryotes. The steps of elongation and
termination appear to be quite similar and this chapter will summarize
some aspects of peptide chain elongation and termination in euka- o
ryotes.

The process of elongation involves the stepwise repetition of
three rections:

1) the codon specific binding of an aminoacyl-tRNA (AA-tRNA) to the
   ribosome;
2) the formation of the peptide bond; and
3) the movement or translocation of the growing peptide chain from
   one site on the ribosome (A site) to another (P site).

In this manner the growing peptide chain is elongated one amino
acid at a time until a terminator codon(s) is reached at which time
the completed protein is released from the ribosome. These three
reactions of elongation are catalyzed by elongation factor 1 (EF-1),
peptidyl transferase (peptide bond formation) and elongation factor
2 (EF-2), respectively, while the termination reaction requires the
presence of a protein which is called release factor (RF). Except
for the peptidyl transferase reaction, the factors involved in these
reactions are soluble, have been highly purified and require GTP.
However, the protein(s) involved in peptide bond formation are ribosomal proteins and no exogenous high energy compound is required for

Table 1. Proteins Involved in Elongation and Termination of Protein
Synthesis in Prokaryotes and Eukaryotes.

Prokaryotes	Eukaryotes	Reaction
EF-Tu	EF-1$_L$ [1]	Binding of aminoacyl-tRNA to the ribosome
EF-Ts	EF-1$_\beta$ [2]	Catalyzes GDP-GTP exchange
Peptidyl transferase	Peptidyl transferase	Peptide bond formation, termination
EF-G	EF-2	Translocation
RF-1, RF-2	RF	Termination

[1]. Also referred to as EF-1$_\alpha$ or eEF-Tu
[2]. Also referred to as eEF-Ts.

this reaction. Table 1 compares the factors involved in the elongation and termination reactions in both prokaryotes and eukaryotes.

BINDING OF AMINOACYL-tRNA TO THE RIBOSOME

Subsequent to the formation of the 80S initiation complex, the
information in the messenger RNA is now available for chain elongation. The first step in this process is the binding of the appropriate AA-tRNA to the A site of the ribosome. This reaction requires
GTP and is catalyzed by a protein which was originally called transferase 1 and is now referred to as elongation factor 1 (EF-1). This
protein functions in eukaryotes very similarly to the prokaryotic
factor EF-Tu.

Characteristics of EF-1

EF-1, which was first reported by Arlinghaus *et al.* (1) to be
present in reticulocyte lysates, has been purified from reticulocytes
(2-4), silkworm (5), calf brain (6), rat liver (7, 8), calf and pig
liver (9-11), ascites cells (12), wheat germ (13, 14) and *Artemia
salina* (15). It was originally observed during the purification of
EF-1 from rat liver, that activity was present in fractions which
ranged in molecular weight from $1 \times 10^5 - 4 \times 10^5$ (7). The protein,
as isolated from a variety of sources, has the peculiar characteristic
of being present as multiple species with molecular weights as high
as 2×10^6. These heavier forms of the factor have been called
EF-1$_H$ and appear to be aggregates. In spite of the large differences
in molecular weights of these various forms of EF-1, it first appeared

that the heavy form of EF-1 represented aggregates of a light form with a molecular weight of about 50, 000. This light species has been called EF-1$_L$ or EF-1$_\alpha$. As will be discussed below, the heavy form of EF-1 also appear to contain at least one other polypeptide (EF-1$_\beta$) which is thought to be responsible for its aggregation properties. In addition, purified EF-1$_H$ preparations from some sources have been found to contain varying amounts of phospholipids and cholesterol (6). The relationship between EF-1$_H$ and EF-1$_L$ (EF-1$_\alpha$) came from studies in which it was shown that a major protein band of about 50,000 daltons was observed after sodium dodecyl sulfate (NaDodSO$_4$) electrophoresis of the purified heavy form of the enzyme (6, 9, 10, 12, 15). The same molecular weight was observed regardless of the original size of the aggregate. These results suggested that EF-1$_H$ was composed of multiple copies of the 50,000 dalton protein. Indeed it was reported that the amino composition of EF-1$_H$ and EF-1$_L$ from calf liver was very similar (9). However, EF-1$_H$ now appears to contain not only the 50,000 dalton component but another protein (EF-1$_\beta$) with a molecular weight of about 30,000 (16-19). As will be discussed below, EF-1$_\beta$ appears to have a function similar to that of EF-Ts in prokaryotes and the heavy form of EF-1 may be analogous to the prokaryotic complex of EF-Tu and EF-Ts (20). The early failure to detect EF-1$_\beta$ in the EF-1$_H$ preparations from various sources (12, 21) is most likely due to the presence of smaller amounts of EF-1$_\beta$ in these tissues, and/or losses during the purification procedure. Although the major form of EF-1 from most sources has been observed to exist as large molecular weight aggregates, it has been shown that the low molecular weight form is very unstable but can be stabilized by the presence of glycerol (22). This suggests that in the absence of glycerol, EF-1$_L$ may be selectively inactivated. Nagata *et al.* (23) have shown that when crude extracts of pig and rat liver, rabbit reticulocytes and *Artemia salina* cysts were prepared in the presence of glycerol and chromatographed on Sephadex G-200, 70-90% of the EF-1 activity was present as a low molecular weight species.

The physiological significance of the heavy forms of EF-1 is, however, still not clear. Although the above results suggest that EF-1$_H$ may be an artifact of the preparation of the extracts, it has been reported that the relative amounts of EF-1$_H$ and EF-1$_L$ change during the development of *Artemia salina* (24) and during aging (25). However, it should be noted that both elastase (26) and carboxypeptidase (27) can cause the disaggregation of EF-1$_H$ to EF-1$_L$. Thus any change in the activity of these enzymes could be responsible for the observed disaggregation of EF-1$_H$ during development and aging. Another study however failed to find any changes in the pattern of EF-1 from liver under a variety of physiological conditions (9). Since EF-1$_L$ appears to be the active form of the enzyme (see below), it has been suggested that the function of EF-1$_H$ may be to serve as a storage form of the enzyme (20). It is not clear whether *in vivo* proteolysis may play a regulatory role in the conversion of EF-1$_H$ to EF-1$_L$.

Assay of EF-1

The assay for EF-1 activity is either based on the ability of the factor to bind Phe-tRNA to 80S ribosomes or to catalyze the synthesis of polyphenylalanine. A typical incubation for the binding assay contains in a final volume of 50 µl: 50 mM Tris-Cl pH 7.9, 50 mM NH_4Cl, 16 mM $MgCl_2$, 0.6 A_{260} unit of 80S ribosomes, 5 µg poly(U), 12 pmoles (^{14}C)Phe-tRNA, 0.1 mM GTP, and EF-1. After 5 min at 37° the incubations are diluted with cold buffer and passed through a nitrocellulose filter. The filter is then assayed for radioactivity. One unit of activity is defined as the amount of protein required to bind 1 pmole of Phe-tRNA under the standard conditions.

Purification of EF-1

a) $\underline{EF-1_H}$. The heavy forms of EF-1 have been purified from a number of sources using ammonium sulphate fractionation and chromatography on hydroxyapatite, phosphocellulose and Sepharose 6B (2-15). Table 2 summarizes the approximate molecular weights of $EF-1_H$ purified from various sources. It can be seen that although there is a wide range in the size of $EF-1_H$, the molecular weight of the major protein under denaturing conditions is about 50,000 from all sources. It now appears that $EF-1_H$ is composed of two additional proteins of about 50,000 and 26,000 daltons which are present in varying amounts (15, 17, 19, 20, 28).

b) $\underline{EF-1_L}$. $EF-1_L$ has been purified to homogeneity using a variety of different methods. A relatively rapid and convenient procedure utilized an aqueous two phase separation containing polyethylene glycol and dextran T-500. An ammonium sulfate extraction separated $EF-1_H$ and $EF-1_L$ and the latter protein was then further purified by CM and DEAE-Sephadex chromatography (11, 15). The purified protein from pig liver consists of a single polypeptide chain of 53,000 daltons. The amino acid composition shows that there is about

Table 2. Molecular Weight of EF-1 from Various Sources.

Source	Aggregate	Monomer
	Daltons	
Silk Worm	$2-6 \times 10^5$	60,000
Reticulocyte	$2-3 \times 10^5$	53,000
Wheat germ	$2-3 \times 10^5$	50,000
Krebs ascites	$2-3 \times 10^5$	47,000
Artemia salina	$2-3 \times 10^5$	50,000
Liver	$4-7 \times 10^5$	53,000
Brain	$1-2 \times 10^6$	50,000

Table 3. Amino Acid Composition of EF-1$_L$ from Pig Liver. [a]

Residue	Mol%
Aspartic acid	9.79
Threonine	5.61
Serine	4.73
Glutamic acid	9.22
Proline	5.66
Glycine	9.54
Alanine	8.46
Half-cystine	1.05
Valine	9.62
Methionine	2.12
Isoleucine	6.68
Leucine	6.15
Tyrosine	2.44
Phenylalanine	3.44
Histidine	2.48
Lysine	8.50
Arginine	3.69
Tryptophan	0.81

(a) From Nagata *et al.* (11)

1.1 mol% of half cystine residues which indicate that there are about 6 half cystine residues per mole of EF-1$_L$ (Table 3). In addition, it has been also shown that four of these residues contain sulfhydryl groups. One or more of these sulfhydryl groups are required for activity since when the enzyme was treated with 6 mM pCMB, 50% of its activity was lost (11).

c)   EF-1$_\beta$. EF-1$_\beta$ which stimulates the exchange of GTP or GDP with EF-1$_L$.GDP$^\beta$ has been purified from pig liver (29), silkworm (17), *Artemia salina* (18) and Krebs ascites cells (19) by standard techniques of protein purification. Although it was originally reported that this factor, as isolated from pig liver (29), had a molecular weight of 90,000 and consisted of two subunits of about 55,000 and 30,000 daltons, it has clearly been demonstrated that only the smaller protein is required for activity (30). The function of the larger protein is not known. A similar situation exists in *Artemia* (18) and silk gland (17) where, although the larger protein is also present in purified preparations of EF-1$_\beta$, activity only resides in proteins of 26,000 and 30,000 daltons, respectively. It is of interest to note the similarity in molecular weights of EF-1$_\beta$, isolated from a number of sources, and EF-Ts from *E. coli* (31).$^\beta$ These proteins appear to catalyze analogous reactions. The amino acid

composition of both *Artemia* and pig liver EF-1$_\beta$ reveals a rather high
content of acidic amino acids (18, 30).

Further studies on the characteristics of EF-1$_\beta$ has shown that
this protein has a tendency to form aggregates. Thus, when the
protein was chromatographed on Ultrogel ACA 44, EF-1$_\beta$ activity eluted
at about molecular weights corresponding to 200,000, 100,000-150,000
and 30,000 (18). This had led to the suggestion (18) that the
multiple aggregates that have been observed with EF-1$_H$ preparations
contain varying amounts of EF-1$_L$.EF-1$_\beta$ complexes and that it is the
EF-1$_\beta$ in this complex which is responsible for the aggregation.
Purified EF-1$_L$ does not appear to form high-molecular aggregates (10,
15).

Interaction of EF-1 with Guanosine Nucleotide and AA-tRNA

The ability of guanosine nucleotides to bind to both the heavy
and light forms of EF-1 has been studied in a number of systems using
a Millipore filter technique. The reaction takes place at both 0°
and 37° and is complete within two minutes. The binary complex has
also been demonstrated by Sephadex G-150 chromatography. EF-1$_H$ and
EF-1$_L$ appear to bind both GTP and GDP although GTP is bound slightly
better (22, 32, 33). This is in contrast to the prokaryotic situation
where the dissociation constant for GDP is 100 times lower than that
for GTP (34). The K_m values for GTP and GDP for pig liver EF-1$_L$ were
found to be 2.0 x 10^{-7} and 5.0 x 10^{-7} M, respectively (22). It has
been observed that EF-1$_L$ from *Artemia* (15) and calf brain (33) has a
much greater ability to bind GTP than does EF-1$_H$. *Artemia* EF-1$_L$ bound
about ten-fold more GTP than EF-1$_H$ while about five times more GTP
was bound to calf brain EF-1$_L$ than EF-1$_H$. The EF-1$_H$.GTP (or GDP)
complex in *Artemia*, however, appears to be somewhat more labile than
the analogous light form since and EF-1$_H$.GTP complex could not be
isolated by Sephadex chromatography (35). Krebs ascites EF-1$_H$,
however, did not form stable GTP or GDP complexes that could be
detected by either Millipore filtration or Sephadex chromatography,
but a binary complex was shown to occur by equilibrium dialysis. An
association constant for a binary complex between Krebs ascites EF-1
and either GTP or GDP was found to be 92 mM^{-1} (36). Although in
earlier studies (35) only about 10% of the EF-1 incubated formed a
binary complex with GTP, other studies by Nagata *et al.* (22) have
shown the isolation of a complex containing stoichiometric amounts
of EF-1$_L$ and GTP. Since the above experiments were carried out in
the presence of glycerol, the failure of other investigators to
obtain this stoichiometry may be due to the lability of the enzyme in
the absence of glycerol.

The EF-1$_L$.GTP binary complex reacts with AA-tRNA to form a
ternary complex:

$$EF\text{-}1_L.GTP + AA\text{-}tRNA \rightarrow AA\text{-}tRNA.EF\text{-}1_L.GTP$$

A convenient assay to monitor the above reaction is based on the observation that the ternary complex does not bind to a nitrocellulose filter while EF-1$_L$.GTP is retained by the filter. In addition, Sephadex chromatography has also been used to detect the formation of the ternary complex. Using these assays, it has been shown that EF-1$_H$.GTP reacts very poorly (21) or not at all (35) with AA-tRNA. However, the binary complex containing the light form of the enzyme from calf brain is about five-six times more efficient than the heavy form (21). In addition, although EF-1$_H$ appeared to form a ternary complex when assayed by the Millipore technique only EF-1$_L$ was found to be present in the ternary complex when it was isolated by Sephadex chromatography. These results suggested that either EF-1$_H$ was converted to EF-1$_L$ during the reaction or that a ternary complex containing EF-1$_H$ was very unstable. That the former explanation is probably the correct one is indicated by studies in wheat germ (14) and reticulocytes (3) which showed that EF-1$_H$ is converted to EF-1$_L$ after incubation with GTP and AA-tRNA. Nagata *et al.* (22) have observed that when pig liver EF-1$_L$ is incubated with GTP and Phe-tRNA and chromatographed on Sephadex G-75 a ternary complex can be isolated which contains stoichiometric amounts of all three components. The formation of the ternary complex does not occur with EF-1$_L$.GDP and deacylated tRNA or N-blocked AA-tRNA cannot substitute for AA-tRNA.

Interaction of the Ternary Complex with Ribosomes

In experiments designed to investigate the fate of the various components of the ternary complex when it was incubated with ribosomes, Weissbach *et al.* (37) prepared the ternary complex labeled with stoichiometric amounts of (^{3}H), (γ-^{32}P)GTP and (^{14}C)Phe-tRNA. When this complex was incubated with ribosomes and poly(U), Phe-tRNA was bound to the ribosome, GTP was hydrolyzed and stoichiometric amounts of an EF-1.GDP binary complex were formed. It was also found that when poly(U) was omitted from the incubation the intact ternary complex still bound to the ribosome but there was no GTP hydrolysis. However, when poly(U) was subsequently added to these incubations GTP was rapidly hydrolyzed. The data suggest that the ternary complex has a high affinity for the ribosome but only in the presence of the poly(U) is it properly positioned so that GTP hydrolysis occurs. Other studies with EF-1 from reticulocytes have also shown that the hydrolysis of GTP was dependent upon the presence of ribosomes and AA-tRNA (38). Slobin and Möller (35) have demonstrated that when *Artemia* EF-1 was incubated with ribosomes, Phe-tRNA, poly(U) and GTP, the amount of GTP hydrolyzed was about stoichiometric to the amount of EF-1 and Phe-tRNA present in the incubation. The hydrolysis was dependent upon the presence of both ribosomal subunits as well as Phe-tRNA. Nolan *et al.* (39) investigated the ribosome binding of radiolabeled EF-1 from Krebs ascites cells. They found, using a centrifugation technique, that the interaction of EF-1 with the ribosome required poly(U), AA-tRNA, and a guanosine nucleotide containing three phosphate groups. When GTP was the nucleotide, it was hydro-

lyzed and EF-1 was released from the ribosome. In contrast when the
nonhydrolyzable GTP analogue 5'-guanylylmethylenediphosphonate (GDPCP)
was used, EF-1 remained tightly bound to the ribosome. These series
of reactions are very similar to those observed with prokaryotic
factors.

Recycling of EF-1

     As noted above, an EF-1.GDP complex is formed as a result of the
interaction of the AA-tRNA.EF-1.GTP ternary complex with the ribosome.
Since the binary complex containing GDP cannot react with an AA-tRNA,
it must be converted to EF-1.GTP so that EF-1 can function catalytically. In prokaryotes, elongation factor Ts is responsible for the
regeneration of EF-Tu.GTP according to the following reaction:

$$a) \quad \text{EF-Tu.GDP} + \text{EF-Ts} \rightleftarrows \text{EF-Tu.EF-Ts} + \text{GDP}$$

$$b) \quad \text{EF-Tu.EF-Ts} + \text{GTP} \rightleftarrows \text{EF-Tu.GTP} + \text{EF-Ts}$$

     Although early studies failed to find a protein with EF-Ts like
activity in eukaryotes, such a factor has now been observed and
purified from pig liver (29, 30), *Artemia* (18), silk gland (17), and
Krebs ascites cells (19). The protein (EF-1$_\beta$ or eEF-Ts) has been
shown to have a marked capacity to stimulate an exchange between
EF-1.GDP and GTP according to the following reaction:

$$a) \quad \text{EF-1}_L\text{.GDP} + \text{EF-1}_\beta \rightleftarrows \text{EF-1}_1\text{.EF-1}_\beta + \text{GDP}$$

$$b) \quad \text{EF-1}_L\text{.EF-1}_\beta + \text{GTP} \rightleftarrows \text{EF-1}_L\text{.GTP} + \text{EF-1}_\beta$$

     The formation of an EF-1$_L$.EF-1$_\beta$ complex has been observed when
the two purified proteins were incubated together (18), and this
complex is dissociated when GTP is added to the incubations. Although
it is now thought that the multiple high molecular weight aggregate
of EF-1$_H$ which have been observed in most tissues are composed of
EF-1$_1$.EF-1$_\beta$ complexes no evidence of such large aggregates was
observed with the *in vitro* reconstituted complexes (18, 19).

     It is of interest to note that Slobin and Möller have recently
reported the detection of an EF-1$_1$.EF-1$_\beta$.GTP complex (18). This
complex which was found to be stable only at 0° is a hypothetical
intermediate in the EF-1$_\beta$ stimulated exchange reaction. In addition
to catalyzing a guanosine nucleotide exchange, EF-1$_\beta$ also stimulates
the EF-1$_L$ dependent binding of AA-tRNA to the ribosome and the EF-1
and EF-2 dependent synthesis of polyphenylalanine. This stimulation
of the latter two reactions is probably related to the ability of
the factor to accelerate the conversion of EF-1.GDP to EF-1.GTP.

     In summary, the reactions involved in the EF-1 dependent binding
of AA-tRNA to the ribosome in a variety of eukaryotes appear to be

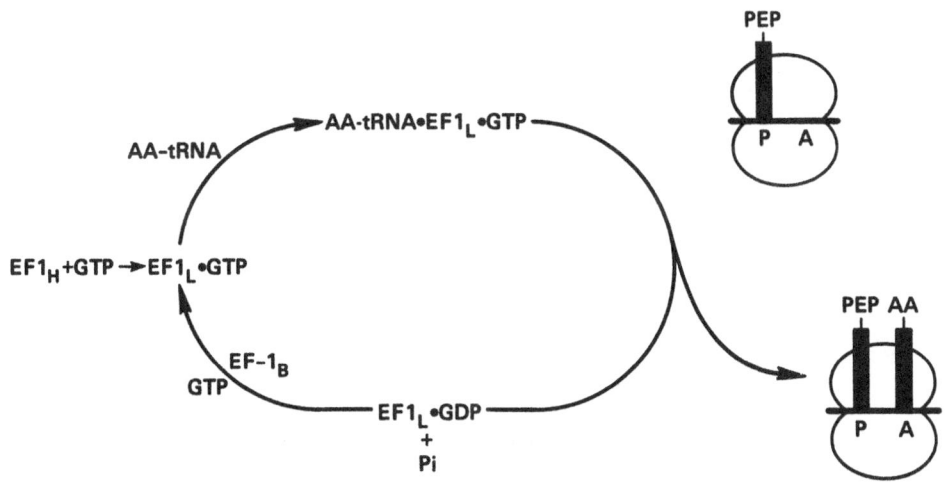

Fig. 1.   The EF-1 dependent binding of aminoacyl-tRNA to the ribosome.
          PEP - Peptidyl-tRNA; AA - Aminoacyl-tRNA.

very similar to those observed in the prokaryote system. The steps
are shown in Fig. 1. The heavy form of EF-1, which has been found
to be present in most tissues, is thought to react with GTP resulting
in the formation of EF-1$_L$.GTP. A ternary complex is then formed as a
result of the interaction of EF-1$_L$.GTP with an AA-tRNA. The AA-tRNA
from this complex is then bound to the A site of the ribosome.
During this process GTP is hydrolyzed and EF-1$_L$.GDP is released from
the ribosome. The regeneration of EF-1$_L$.GTP is accomplieshed by a
GTP.GDP exchange reaction catalyzed by EF-1$_\beta$.

PEPTIDE BOND FORMATION

    In contrast to all of the other reactions of protein synthesis
which require a soluble factor, the formation of the peptide bond is
catalyzed by a protein(s) that is a structural component of the large
ribosomal subunit. This enzyme (peptidyl transferase) is involved in
the formation of a peptide bond between the peptidyl-tRNA bound to
the P site of the ribosome and an aminoacyl-tRNA in the A site (Fig.
2). During this process the ester linkage of the peptidyl-tRNA is
broken and a new peptide bond is formed with the amino group of the
aminoacyl-tRNA. This reaction serves to elongate the growing peptide
chain by one amino acid. No additional protein factors or exogenous
external energy source are required for the reaction.

    Although in the bacterial system a number of ribosomal proteins
have been reported to be involved in peptide bond formation (40),
almost nothing is known about the proteins involved in eukaryotic

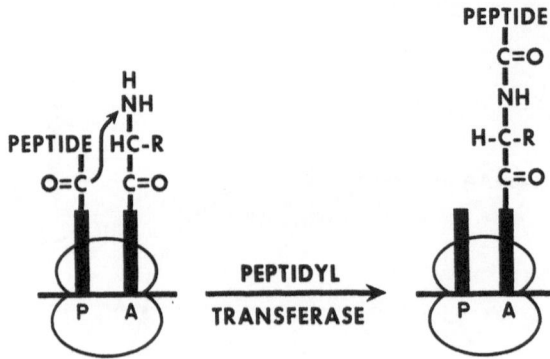

Fig. 2. Peptide bond formation. Solid bars represent tRNA.

peptide bond formation. It is known that the catalytic activity
resides in the 60S ribosomal subunit of the eukaryotic ribosome and
is inhibited by various antibiotics such as lincomycin, sparsomycin
and anisomycin as well as a number of other compounds. A more detailed discussion can be found in Chapter 2 and 13 of this volume. For
a comprehensive summary the reader is referred to an earlier review
by Pestka (41). As will be discussed below, peptidyl transferase
also participates in the reactions involved in the termination of
protein synthesis. This has been concluded from studies in which it
has been shown that the same antibiotics that inhibit the peptidyl
transferase reaction also inhibit the termination reaction.

TRANSLOCATION

    The process by which the ribosome moves along the messenger RNA
is probably one of the least understood reactions in protein synthesis. With the addition of each amino acid the ribosome moves along
the messenger RNA, precisely one codon (three bases) at a time, from
the initiation site near the 5'-end of the messenger to the termination codon at the 3'-end. As a result of translocation the nascent
peptide chain on the A site of the ribosome moves to the P site
(Fig. 3), the deacylated tRNA bound to the P site is ejected, and
the ribosome moves 3 bases closer to the 3'-end of the messenger RNA.

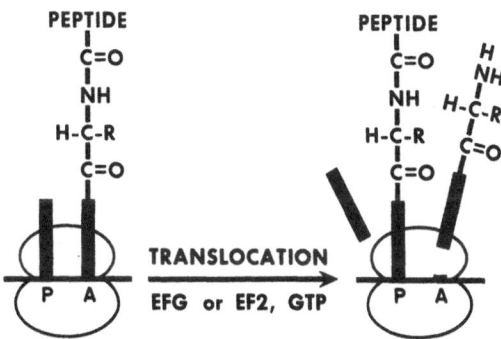

Fig. 3.   The process of translocation.   Solid bars represent tRNA.

This results in a new code work appearing in the A site which allows
a new AA-tRNA to bind to the now empty A site.   During the translocation reaction, GTP is hydrolyzed.   In both prokaryotes and eukaryotes, supernatant proteins referred to as elongation factor G and
elongation factor 2 (EF-2), respectively, are required.

Elongation Factor 2

   Purification and Properties.   EF-2 was one of the two factors
from rabbit reticulocytes which was initially shown to be required
for protein synthesis (1).   Highly purified preparations of EF-2 have
been obtained from rat liver (42, 43), pig liver (44), and silk gland
(45).   Raeburn et al. (43) using a modification of the procedure of
Galasinski and Moldave (44) purified EF-2 from rat liver about
hundred-fold.   The purification procedure which utilized DEAE-cellulose, Sephadex G-150 and phosphocellulose chromatography, followed
by isoelectric focusing yielded a protein which was about 85% pure.
A second isoelectric focusing step was later used (46) to remove a
contaminating transphosphorylase.

   The molecular weight of EF-2 from rat liver and pig liver has
been estimated to be about 110,000 (42) and 100,000 (47), respectively.   Wheat germ EF-2 appears to have a somewhat lower molecular
weight of 70,000 (48).   Rat liver EF-2 contains 18 sulfhydryl groups
and 2 disulfide bonds per mole of enzyme (49).   One of the unique
characteristics of EF-2, which was first observed by Honjo et al.
(50) is its ability to be ADP-ribosylated in the presence of diphtheria toxin and NAD.   The very high specificity of the reaction
allows its use as an assay for detection of EF-2 (discussed below).

   Assay.   The presence of EF-2 activity can be detected in a
number of ways:

1) formation of a peptidyl puromycin product using ribosomes or polysomes containing nascent protein chains (51-53);
2) EF-2 dependent polypeptide synthesis;
3) GTP hydrolysis dependent on EF-2 and ribosomes; and
4) ADP ribosylation of EF-2 in the presence of diphtheria toxin.

For a detailed description of the assays the reader is referred to Raeburn *et al.* (42).

One of the more specific assays for detecting the presence of EF-2 is the ADP ribosylation reaction which is carried out as follows:

Diphtheria toxin (2.9 µg) is incubated with varying amounts of EF-2 in a reaction mixture (150 µl) containing 5mM Tris-Cl, pH 7.2, 1 mM DTT, 0.1mM EDTA, 405 pmoles of (^{3}H)NAD (labeled in the adenosine moiety) and 0.25 to 0.5 mg of albumin to stabilize the enzyme. After an incubation at 30° for 1 hr, the protein is precipitated with 5% trichloroacetic acid and heated at 90°. The precipitate is washed onto glass fiber filters and the amount of radioactivity bound to the filter is a measure of the amount of EF-2 present.

Interaction of EF-2 with Guanosine Nucleotides and Ribosomes. The similarity between prokaryotic and eukaryotic elongation factors once again is seen with EF-G and EF-2. Both factors interact with GTP, GDP and fusidic acid. Skogerson and Moldave (54) first observed a GTP dependent binding of rat liver EF-2 to ribosomes. The binding required a sulfhydryl component and the ribosomes isolated from an incubation containing EF-2 and GTP were active in protein synthesis in the absence of added EF-2. These results indicated that EF-2 remained bound to the ribosomes. It was subsequently shown (52, 53) that both GDP and GDPCP were also effective in the binding of EF-2 to the ribosomes; however, only GTP yielded a ribosome EF-2 complex that was active in stimulating AA-tRNA binding to the ribosome. This suggested that in the presence of EF-2 and GTP, the A site on the ribosome became available for an AA-tRNA to bind, presumably due to the EF-2 dependent translocation of a peptidyl-tRNA on the ribosome from the A site to the P site. EF-2 has also been demonstrated to be associated with the ribosomes *in vivo*. Smulson and Rideau (55) observed that EF-2 was bound to HeLa cell monosomes but not to heavy or light polyribosomes. In addition, under certain growth conditions (amino acid, starvation or puromycin treatment), the amount of EF-2 on the monosomes increased.

Purified preparations of EF-2 have been shown to form stable complexes with various guanosine nucleotides. Although no stable EF-G.guanosine nucleotide complex has been isolated from prokaryotes, its formation has been shown by equilibrium dialysis studies (56). An EF-2.GTP complex has been observed with EF-2 preparations from

liver (46, 57, 58), calf brain (59), reticulocytes (60), and human
tonsil (61). A convenient and rapid assay for the formation of an
EF-2.GTP complex involves the selective retention of the complex on
a nitrocellulose filter. The complex has also been isolated by gel
filtration. In a similar manner both GDP and GDPCP have also been
shown to form stable complexes with EF-2. A single EF-2 guanosine
nucleotide binding site has been reported (46, 58, 62) and dissociation constants for EF-2.GTP and EF-2.GDP from pig liver were found to
be 1.4×10^{-5} M and 4.1×10^{-7} M, respectively (63). Henrikson et $al.$
(46), however, using equilibrium dialysis have shown the Kd values
for GTP, GDP, or GDPCP for rat liver EF-2 binary complexes to be
2×10^{-6} M, 4×10^{-7} M and 2×10^{-5} M, respectively.

When EF-2 and GTP are incubated in the presence of ribosomes, a
ribosome EF-2.GDP ternary complex is formed (57, 59-64) containing
stoichiometric amounts of each of the components (64). Thus, since
a stable EF-2.GTP complex can be isolated, the interaction of the
ribosomes with this complex results in the hydrolysis of GTP. The
results suggest the following sequence of reactions:

a)   EF-2 + GTP $\rightleftharpoons$ EF-2.GTP

b)   EF-2.GTP + Rib $\rightarrow$ Rib.EF-2.GDP + Pi

It has been shown, however, that the formation of the ternary
complex occurs when GDP or GDPCP is substituted for GTP, indicating
that GTP hydrolysis is not required for this reaction.

The interaction of a ribosome containing a peptidyl-tRNA on the
A site with EF-2.GTP, results in the translocation of the peptidyl-
tRNA to the P site of the ribosome with the concomittant hydrolysis
of the GTP. It had originally been thought that the energy released
from the hydrolysis of the GTP was required for translocation. However, studies first with EF-G (65-69) and subsequently with EF-2 (70)
have shown that translocation also occurred in the presence of non-
hydrolyzable analogues of GTP but not with GDP. These and other
results suggest that the binding of EF-2.GTP (or GDPCP) to the ribosome is sufficient for translocation to occur and that the hydrolysis
of GTP is required for the $recycling$ of EF-2.

Fusidic acid is a steroidal antibiotic which is known to form a
very stable fusidic acid.ribosome.EF-G.GDP complex in prokaryotes.
This stable complex explains how fusidic acid exerts its antibiotics
action since it prevents EF-G from being released from the ribosome
and acting catalytically. In eukaryotes the mechanism of fusidic acid
inhibition is identical to that observed in prokaryotes. Fusidic acid
forms a complex with EF-2 bound to the ribosome according to the
following sequence of reactions:

a)   EF-2 + GTP $\rightleftarrows$ EF-2.GTP

b)   EF-2.GTP + ribosome → Rib.EF-2.GDP + Pi

c)   Rib.EF-2.GDP $\xrightarrow[\text{Acid}]{\text{Fusidic}}$ Fusidic.Rib.EF-2.GDP

The sequence of events that occurs during protein synthesis suggests that EF-2.GDP is released from the ribosome subsequent to the translocation reaction. The released EF-2.GDP is then capable of reacting with another mole of GTP and, in this fashion, function catalytically. It is to be noted that unlike EF-1.GDP, EF-2.GDP can exchange with GTP to form an EF-2.GTP in the absence of additional protein factors (63). In the presence of fusidic acid EF-2.GDP is trapped on the ribosome and thereby inhibits the subsequent binding of AA-tRNA since EF-1 cannot bind to ribosomes that contain EF-2 (71, 72). The interaction of an EF-2.GTP complex with the ribosome, and GTP hydrolysis, appears to require only the 60S subunit although GTP hydrolysis was somewhat stimulated when the 40S subunit was added to an incubation containing the large subunit (73). This is again very similar to what is observed in the prokaryotic system.

The Inhibition of EF-2 Activity by Diphtheria Toxin

Diphtharia toxin which is produced by *Corynebacterium diphtheriae* is a protein consisting of a single polypeptide chain of about 62,000 daltons and is extremely lethal, in doses of 50-100 ng/Kg or less. When the protein is cleaved at both a specific peptide bond and a disulfide linkage, two fragments are released. Fragment A is derived from the amino terminal end of the toxin and has a molecular weight of 21,100 while the carboxyl terminal fragment B, is about 40,000 daltons (74, 75). It is now clear that the toxic effects of diph-theria toxin are related to its ability to inactivate EF-2. Strauss and Hendee (76), first demonstrated that protein synthesis was spe-cifically inhibited within two hours after exposure of HeLa cells to the toxin. It is now known that the cytotoxicity of the toxin is a two-step process. The initial reaction involves the binding of the toxin to specific receptors on the membrane and is a function of fragment B of the toxin. Fragment A is then released into the cyto-plasm of the cell where it catalyzes the inactivation of EF-2. It appears that the *in vivo* resistance to the effect of diphtheria toxin is related to the inability of the toxin to bind to the plasma membrane. Collier and Pappenheimer (77) first showed that the toxin inhibited protein synthesis in cell free extracts of HeLa cells and reticulocytes and that NAD was required for the inhibition. It was later found that the effect of the toxin and NAD was to catalyze the inactivation of EF-2 (78-81). Honjo *et al.* (50) observed that the mechanism of inhibition of the toxin involves the covalent transfer of the adenosine diphosphate ribose moiety of NAD^+ to EF-2 as shown by the following reaction:

$$\text{NAD}^+ + \text{EF-2} \xrightarrow{\text{toxin}} \text{ADP-ribosyl-EF-2} + \text{nicotinamide} + \text{H}^+$$

The reaction is essentially *irreversible* at pH7 and physiological concentrations of the reactants, however, if NAD$^+$ is removed, the pH lowered and excess nicotinamide added, EF-2 activity can be restored in the presence of toxin and the inactive ADP-ribosylated EF-2.

The ADP-ribosylation of EF-2 by the toxin is accompanied by an inhibition in the ability of the enzyme to catalyze polyphenylalanine synthesis and ribosome dependent GTP hydrolysis (59, 81). The inactive enzyme can still form a binary complex with GTP (59, 61, 81), however contradictory results have been reported concerning the ability of the ribosylated factor to interact with the ribosome to form a ternary complex (59, 61, 81).

The ADP ribosylation of EF-2 catalyzed by diphtheria toxin is very specific for EF-2 isolated from a wide variety of sources including plants and yeast (82). EF-G from bacteria or mitochondria is not affected by the toxin and no substrate, other than EF-2, has been found in mammalian extracts. The inactivation of eukaryotic EF-2 from different sources suggest that this protein contains a unique feature which is essential for diphtheria toxin recognition. The basis of the specificity was investigated by Robinson *et al.* (83) who purified and sequenced a peptide obtained after tryptic digestion of ADP-ribosylated EF-2 from rat liver. The peptide contained 15 amino acids with the following sequence:

Phe-Asp-Val-His-Asp-Val-Thr-Leu-His-Ala-Asp-Ala-Ile-X-Arg

They found that the ADP-ribose was linked to an unknown and unusual weakly basic amino X. Recently, Brown and Bodley (84) have found an analogous tryptic peptide from EF-2 isolated from beef liver, yeast and wheat germ which contained considerable sequence homology including the presence of amino acid X. The nature of this amino acid has recently been characterized by NMR spectroscopy as 2-(3-carboxyamido-3-(trimethylammonio)-propyl)histidine (85) (Fig. 4) and has been given the trivial name *dipthamide*. These studies suggest that dipthamide is ribosylated via one of the nitrogens of the imidiazole ring of the histidine. The steps leading to the synthesis of dipthamide are unknown but would appear to involve a post-translational modification of histidine. It would be interesting to speculate that since this amino acid derivative is found in EF-2 of all eukaryotes studied that it is important for the function of this protein. It is to be noted, however, that mutants have been isolated which lack this residue and do not seem to be defective in protein synthesis (86). However, other studies have shown that the toxin cannot ADP-ribosylate the 15 amino acid peptide from EF-2 that contains dipthamide (87). Thus the presence of peptide bond-dipthamide is insufficient for ADP-ribosylation and some additional characteristic of

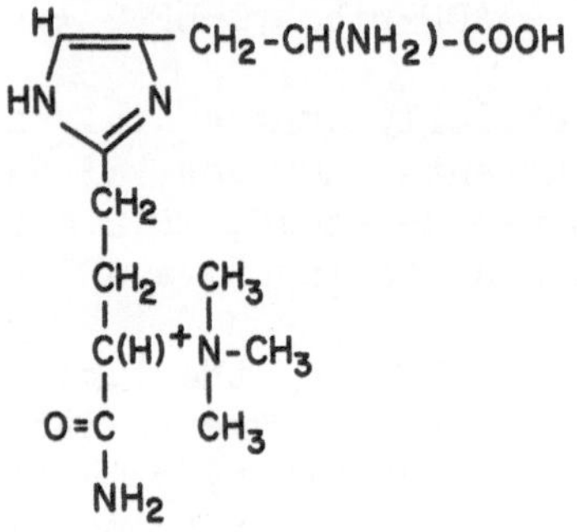

Fig. 4.   Structure of Dipthamide.   2-(3-carboxyamido-3-(trimethyl-
           ammonio)-propyl)histidine.

EF-2 may be required for the reaction to occur. The role and synthesis of this unique residue in EF-2 will be of great interest.

TERMINATION

     The final step in the synthesis of a protein is the release of
the completed polypeptide chain from the ribosome-mRNA complex. This
process, which is called termination, requires a termination codon,
a specific protein called release factor (RF) and GTP. The process
in prokaryotes is very similar except that there are two separate
release factors (RF-1, RF-2) and a third protein (RF-3) which stimulates this reaction. The requirement for GTP for termination in the
bacterial system is still not clear.

     Goldstein et $al.$ (88) first presented evidence for a release
factor from rabbit reticulocytes using an assay which involved the
release of formylmethionine (fMet) from a fMet-tRNA.ribosome complex.
The release of fMet was dependent on RF, GTP and tetranucleotides
UAAA, UAGA, or UGAA (89). GDP could not substitute for GTP and GDPCP
inhibited the reaction. These results suggest that GTP hydrolysis is
required for the reaction. Release factor activity has also been
observed and purified from insects (90) and rat liver (91). The
reticulocyte factor has a molecular weight of about 115,000 as
determined by Sephadex G-200 chromatography and sedimentation equilibrium, but NaDodSO$_4$ polyacrylamide gel analysis shows that the
protein consists of two subunits of about 56,000 daltons.

The binding of the purified release factor to the ribosome has been studied utilizing (^{3}H)UAAA. The binding of the protein to the ribosome can be assayed by the retention of the labeled terminator codon by a nitrocellulose filter. The binding requires GTP (or GDPCP) and the factor recognizes each of the tetranucleotides equally well since the binding of (^{3}H)UAAA is effectively competed with UAGA and UGAA. In addition to its role in binding RF, the ribosome also participates in the termination reaction through the peptidyl transferase site. Thus it has been shown that antibiotics that inhibit the peptidyl transferase reaction also inhibit the RF dependent release of fMet from the ribosome (88). These antibiotics, however, do not effect the binding of RF to the ribosome. It has been suggested that peptidyl transferase catalyzes the hydrolysis of peptidyl-tRNA in the presence of RF, and thereby releases the completed polypeptide chain from the ribosome.

The present knowledge of the process of termination suggests the following sequence of events. Subsequent to the formation of the last peptide bond the completed protein is translocated to the P site of the ribosome and a termination codon now appears in the A site of the ribosome. In the presence of GTP the release factor binds to the ribosome with the resulting hydrolysis of the ester bond between the completed protein chain and the tRNA catalyzed by the peptidyl transferase system. The hydrolysis of GTP permits the dissociation of the release factor from the ribosome.

REFERENCES

1. Arlinghaus, R., Favelukes, G. and Schweet, R. (1963), Biochem. Biophys. Res. Commun. 11, 92-96.
2. McKeehan, W.L. and Hardesty, B. (1969), J. Biol. Chem. 244, 4330-4339.
3. Prather, N., Ravel, J.M., Hardesty, B. and Shive, W. (1974), Biochem. Biophys. Res. Commun. 57, 578-583.
4. Kemper, W.M., Merrick, W.C., Redfield, B., Liu, C.K. and Weissbach, H. (1976), Arch. Biochem. Biophys. 174, 603-612.
5. Ejiri, S., Taira, H. and Shimura, K. (1973), J. Biochem. 74, 195-197.
6. Moon, H.M., Redfield, B., Millard, S., Vane, F. and Weissbach, H. (1973), Proc. Natl. Acad. Sci. U.S.A. 70, 3282-3286.
7. Schneir, M. and Moldave, K. (1968), Biochim. Biophys. Acta 166, 58-67.
8. Collins, J.F., Moon, H.M. and Maxwell, E.S. (1972), Biochemistry 11, 4187-4194.

9. Liu, C.K., Legocki, A.B. and Weissbach, H. (1974), in: Lipmann
    Symposium: *Energy Biosynthesis and Regulation in Molecular
    Biology*, (D. Richter, ed.), pp. 384-398, Walter de Gruyter
    Verlag, Berlin and New York.

10. Iwasaki, K., Nagata, S., Mizumoto, K. and Kaziro, Y. (1974), J.
    Biol. Chem. 249, 5008-5010.

11. Nagata, S., Iwasaki, K. and Kaziro, Y. (1977), J. Biochem. 82,
    1633-1646.

12. Drews, J., Bednarik, K. and Grasmuk, H. (1974), Eur. J. Biochem.
    41, 217-227.

13. Golinska, B. and Legocki, A.B. (1973), Biochim. Biophys. Acta
    324, 156-170.

14. Bollini, R., Soffientini, A.N., Bertani, A. and Lanzani, G.A.
    (1974), Biochemistry 13, 5431-5445.

15. Slobin, L.I. and Möller, W. (1976), Eur. J. Biochem. 69, 351-
    366.

16. Nagata, S., Motoyoshi, K. and Iwasaki, K. (1976), Biochem.
    Biophys. Res. Commun. 71, 933-938.

17. Ejiri, S., Murakani, K. and Katsumata, T. (1977), FEBS Lett. 82,
    111-114.

18. Slobin, L.I. and Möller, W. (1978), Eur. J. Biochem. 84, 69-77.

19. Grasmuk, H., Nolan, R.D. and Drews, J. (1978), Eur. J. Biochem.
    92, 479-490.

20. Slobin, L.I. and Möller, W. (1977), Biochem. Biophys. Res.
    Commun. 74, 356-365.

21. Moon, H.M., Redfield, B. and Weissbach, H. (1972), Proc. Natl.
    Acad. Sci. U.S.A. 69, 1249-1252.

22. Nagata, S., Iwasaki, K. and Kaziro, Y. (1976), Arch. Biochem.
    Biophys. 172, 168-177.

23. Nagata, S., Izasaki, K. and Kaziro, Y. (1976), J. Biochem. 80,
    73-77.

24. Slobin, L.I. and Möller, W. (1975) Nature, London, 258, 252-254.

25. Bolla, R. and Brot, N. (1975), Arch. Biochem. Biophys. 169,
    227-236.

26. Twardowski, T., Redfield, B., Kemper, W.M., Merrick, W.C. and
    Weissbach, H. (1976), Biochem. Biophys. Res. Commun. 71, 272-
    279.

27. Twardowski, T., Hill, J. and Weissbach, H. (1977), Arch. Biochem.
    Biophys. 180, 444-451.

28. Iwasaki, K., Mizumoto, K., Tanaka, M. and Kaziro, Y. (1973), J.
    Biochem. 74, 849-852.

29. Motoyoshi, K., Iwasaki, K. and Kaziro, Y. (1977), J. Biochem.
    82, 145-155.

30. Motoyoshi, K. and Iwasaki, K. (1977), J. Biochem. 82, 703-708.

31. Hachmann, J., Miller, D.L. and Weissbach, H. (1971), Arch.
    Biochem. Biophys. 147, 457-466.

32. Moon, H.M. and Weissbach, H. (1972), Biochem. Biophys. Res.
    Commun. 46, 254-262.

33. Legocki, A.B., Redfield, B. and Weissbach, H. (1974), Arch. Biochem. Biophys. 161, 709-712.

34. Miller, D.L. and Weissbach, H. (1970), Arch. Biochem. Biophys. 141, 26-37.

35. Slobin, L.I. and Möller, W. (1976), Eur. J. Biochem. 69, 367-375.

36. Nolan, R.K., Grasmuk, H., Hogenauer, G. and Drews, J. (1974), Eur. J. Biochem. 45, 601-609.

37. Weissbach, H., Redfield, B. and Moon, H.M. (1973), Arch. Biochem. Biophys. 156, 267-275.

38. Lin, S.Y., Mc Keehan, W.L., Culp, W. and Hardesty, B. (1969), J. Biol. Chem. 244, 4340-4350.

39. Nolan, R.D., Grasmuk, H. and Drews, J. (1975), Eur. J. Biochem. 50, 391-402.

40. Harris, R.J. and Pestka, S. (1977), in: *Molecular Mechanisms of Protein Biosynthesis* (H. Weissbach and S. Pestka, eds.) pp. 413-465, Academic Press, New York.

41. Pestka, S. (1977), in: *Molecular Mechanisms of Protein Biosynthesis* (H. Weissbach and S. Pestka eds.) pp. 467-553, Academic Press, New York.

42. Raeburn, S., Collins, J.F., Moon, H.M. and Maxwell, E.S. (1971), J. Biol. Chem. 246, 1041-1048.

43. Galasinski, W. and Moldave, K. (1969), J. Biol. Chem. 244, 6527-6532.

44. Mizumoto, K., Iwasaki, K., Tanaka, M. and Kaziro, Y. (1974), J. Biochem. 75, 1047-1056.

45. Taira, H., Ejiri, S. and Shimura, K. (1972), J. Biochem. 72, 1527-1535.

46. Henriksen, O., Robinson, E.A. and Maxwell, E.S. (1975), J. Biol. Chem. 250, 720-724.

47. Muzumoto, K., Iwasaki, K., Kaziro, Y., Nojiri, C. and Yamada, Y. (1974), J. Biochem. 75, 1057-1062.

48. Twardowski, T. and Legocki, A.B. (1973), Biochim. Biophys. Acta 324, 171-183.

49. Robinson, E.A. and Maxwell, E.S. (1972), J. Biol. Chem. 247, 7023-7028.

50. Honjo, T., Nishizuka, Y., Hayaishi, O. and Kato, I. (1968), J. Biol. Chem. 243, 3553-3555.

51. Schneider, J.A., Raeburn, S. and Maxwell, E.S. (1968), Biochem. Biophys. Res. Commun. 33, 177-181.

52. Skogerson, L. and Moldave, K. (1968), J. Biol. Chem. 243, 5354-5360.

53. Skogerson, L. and Moldave, K. (1968), J. Biol. Chem. 243, 5361-5367.

54. Skogerson, L. and Moldave, K. (1967), Biochem. Biophys. Res. Commun. 27, 568-572.

55. Smulson, M.E. and Rideau, C. (1970), J. Biol. Chem. 245, 5350-5353.

56. Arai, N., Arai, K. and Kaziro, Y. (1977), J. Biochem. 82, 687-694.

57. Baligo, B.S. and Munro, H.N. (1972), Biochim. Biophys. Acta 277,
     368-383.
58. Mizumoto, K., Iwasaki, K. and Kaziro, Y. (1974), J. Biochem. 76,
     1269-1280.
59. Chaung, D.M. and Weissbach, H. (1972), Arch. Biochem. Biophys.
     152, 114-124.
60. Bodley, J.W. and Lin, L. (1970), Nature, London, 227, 60-62.
61. Bermek, E. and Matthaei, H. (1971), Biochemistry 10, 4906-4912.
62. Montanaro, L., Sperti, S. and Mattioli, A. (1971), Biochim.
     Biophys. Acta 238, 493-497.
63. Mizumoto, K., Iwasaki, K. and Kaziro, Y. (1974), J. Biochem. 76,
     1269-1280.
64. Henriksen, O., Robinson, E.A. and Maxwell, E.S. (1975), J. Biol.
     Chem. 250, 725-730.
65. Inoue-Yokosawa, N., Ishikawa, C. and Kaziro, Y. (1974), J. Biol.
     Chem. 249, 4321-4323.
66. Belitsina, N.V., Glukhove, M.A. and Spirin, A.S. (1975), FEBS
     Lett. 54, 35-38.
67. Modolell, J., Girbes, T. and Bazquez, D. (1975), FEBS Lett. 60,
     109-113.
68. Belitsina, N.V., Glukhova, M.A. and Spirin, A.S. (1976), J. Mol.
     Biol. 108, 609-613.
69. Girbes, T., Vasquez, D. and Modolell, J. (1976), Eur. J. Biochim.
     67, 257-265.
70. Tanaka, M., Iwasaki, K. and Kaziro, Y. (1977), J. Biochem. 82,
     1035-1043.
71. Baliga, B.S., Schechtman, M.G. and Munro, H.N. (1973), Biochem.
     Biophys. Res. Commun. 51, 406-413.
72. Nombela, C. and Ochoa, S. (1973), Proc. Natl. Acad. Sci. U.S.A.
     70, 3556-3560.
73. Mc Keehan, W. (1972), Biochem. Biophys. Res. Commun. 48, 1117-
     1122.
74. Collier, R.J. (1975), Bacteriol. Rev. 39, 54-85.
75. Pappenheimer, A.M., Jr. (1977), Ann. Rev. Biochem. 46, 69-94.
76. Strauss, N. and Hendee, D.D. (19597, J. Exp. Med. 109, 145-163.
77. Collier, R.J. and Pappenheimer, A.M., Jr. (1964), J. Exp. Med.
     120, 1019-1039.
78. Collier, R.J. (1967), J. Mol. Biol. 25, 83-98.
79. Goor, R.S. and Pappenheimer, A.M., Jr. (1967), J. Exp. Med. 126,
     899-912.
80. Goor, R.S. and Pappenheimer, A.M., Jr. (1967), J. Exp. Med. 126,
     913-921.
81. Raeburn, S., Goor, R.S., Schneider, J.A. and Maxwell, E.S. (1968)
     Proc. Natl. Acad. Sci. U.S.A. 61, 1428-1434.
82. Van Ness, B.G., Howard, J.B. and Bodley, J.W. (1978), J. Biol.
     Chem. 253, 8687-8690.
83. Robinson, E.A., Henriksen, D. and Maxwell, E.S. (1974), J. Biol.
     Chem. 249, 5088-5093.
84. Brown, B.A. and Bodley, J.W. (1979), FEBS Lett. 103, 253-255.

85. Van Ness, B.G., Howard, J.B. and Bodley, J.W. (1980), J. Biol.
     Chem., in press.
86. Moehring, J.M., Moehring, J.J. and Danley, D.E. (1980), Proc.
     Natl. Acad. Sci. U.S.A. $\underline{77}$, 1010-1014.
87. Van Ness, B.G. and Bodley, J.W. (1980), FEBS Lett., in press.
88. Goldstein, J.L., Beaudet, A.L. and Caskey, C.T. (1970), Proc.
     Natl. Acad. Sci. U.S.A. $\underline{67}$, 99-105.
89. Beaudet, A.L. and Caskey, C.T. (1971) Proc. Natl. Acad. Sci.
     U.S.A. $\underline{68}$, 619-624.
90. Ilan, J. (1973), J. Mol. Biol. $\underline{77}$, 437-448.
91. Innanen, V.T. and Nicholls, D.M. (1973), Biochim. Biophys. Acta
     $\underline{324}$, 533-544.

BIOSYNTHESIS, MODIFICATION AND

PROCESSING OF VIRAL POLYPROTEINS

Gebhard Koch, Friedrich Koch, John A. Bilello
Eckhard Hiller, Claudia Schârli, Gûnther Warnecke
and Carsten Weber

Abt. Molekularbiologie, Physiol. Chem. Inst.
der Universität Hamburg
Grindelallee 117, D-2000 Hamburg 13, FRG

INTRODUCTION

The biosynthesis of several cellular and many viral proteins is accompanied by modification (e.g. phosphorylation, hydroxylation, glycosylation, ADP ribosylation, poly ADP ribosylation, acetylation) and proteolytic processing of the primary translation product. Both modification and processing can take place during polypeptide chain elongation, as well as long after the completion of protein synthesis (for reviews see 1-12). By way of example, glycosylation (13), especially of membrane proteins and hydroxylation (collagen, ref. 14) occurs on nascent polypeptide chains. Similarly, forms of specific proteolytic processing, e.g. removal of the leader sequence from membrane and secretory proteins (15-20), and cleavage of the picorna-virus polyproteins to three distinct precursor proteins occur during elongation (11b, c, d).

Examples of modifications that take place long after the completion of polypeptide synthesis are:

a)   phosphorylation and the resulting activation and (reversible) inactivation of many enzymes within the major biochemical pathways (21) or of ribosomal proteins and initiation factors which function in controlling protein synthesis (22-28);

b)   the conversion of zymogens into active enzymes by limited proteolysis (e.g. digestive proteases, 29; blood clotting, 30; complement, 31);

c)   morphogenetic cleavages of viral coat proteins during assembly (11a, 32-52);

d)   cleavage of neurohormone precursors to a number of smaller
functionally distinct proteins (1b, 12).

Modification and processing of a polypeptide can occur concomitantly and in some instances may be interdependent (53-60). Both processes often have important regulatory functions (activation, inactivation, transport, compartmentalization, secretion, virus assembly).
In order to illustrate just a few of the many problems, questions
regulatory principles and present day experimental approaches to
this wide, expanding and important subfield of protein synthesis, we
have chosen two experimental systems, the picornaviruses and RNA
tumor-viruses, which we will discuss in detail

Functionally distinct types of proteolytic clevage are involved
in the processing of viral proteins (1-5, 7-12). The *formative*
cleavages serve to convert primary precursors to functional proteins.
The second type of proteolytic processing are *morphogenetic* cleavages,
which are intimately associated with the intermediary and final steps
of virus assembly. These might serve to convert the assembly reactions in which they participate to irreversible processes (36-48). A
possible third type may involve the initiation of the viral replicative function for RNA synthesis.

Studies of the picornaviruses led to the discovery of the
important principle of the nascent and post translational processing
of larger precursor polyproteins to a number of functionally distinct
secondary polypeptides (1, 2, 4, 5, 7-10, 11b, c, d, 61-85). Subsequently, animal virus model systems have been developed (2, 12,
47-50, 53-57, 86-116) which have led to the observations that processing and modification play an important role in many biological
processes. For instance, processing of precursor proteins plays an
important role in the synthesis and maturation of the RNA tumor
viruses (47-50) and thus this sytem will serve to illustrate the
severe consequences of a disturbance during synthesis of viral
proteins on subsequent processing (12, 116) in the various cleavage
steps involved in virus maturation. It should be kept in mind that
different proteases of host cell and viral origin are utilized. Many
viruses, notably reoviruses, the myxo and paramyxoviruses, rely on
host cell proteases only for cleavage of virus specific proteins
(2, 117-120). These viruses have apparently evolved such that they
can utilize proteolytic mechanisms already present in host cells,
rather than investing their own limited genetic capacity for this
purpose. Other viruses may utilize their own proteases in addition
to host protease systems. The picornaviruses, for example, appear to
utilize two distinct virus specified proteases (79, 80). Some viruses
incorporate proteases in their particles during maturation and these
proteases can be purified with the virus particles. Among the animal
viruses the RNA tumor viruses are the best studied example of virion
associated proteases (2, 100, 121).

Accordingly, we thought germane to the purpose of this review
to consider in some detail the mechanism of processing of picorna-
virus-directed proteins as a first example, and the postranscriptional
modifications of oncornavirus proteins to illustrate a second way of
virus maturation.

THE PROTEOLYTIC PROCESSING OF PICORNAVIRUS PROTEINS

Picornavirus-Directed Protein Synthesis

In a pioneering study, which introduced the now widely used SDS-
PAGE technique, Summers et al. (61) identified 14 different virus
specific polypeptides in poliovirus infected HeLa cells. The combined
molecular.weight of the 14 virus specific proteins exceeded 500 kd,
while the coding capacity of poliovirus RNA is only sufficient to
code for approximately 250 kd of protein. In subsequent experiments
it was shown that certain virus specific peptides are preferentially
labelled during a short pulse of radioactive amino acids. During
subsequent incubation of pulse labelled cells with excess unlabelled
amino acids (*chase*) radioactivity previously associated with certain
larger polypeptides was found to be present in several lower molecular

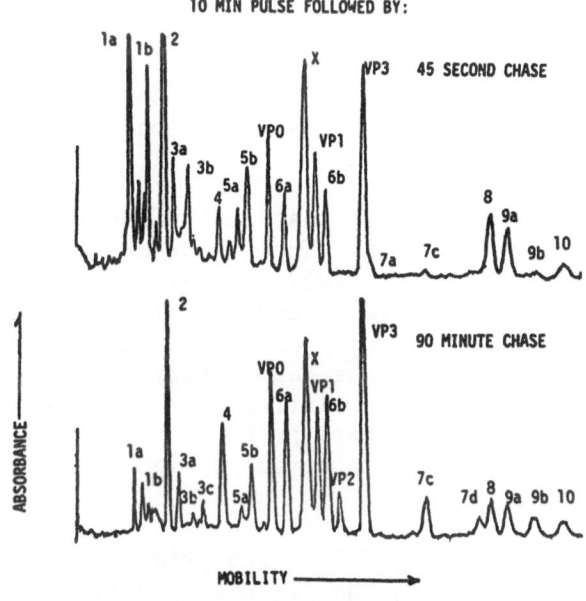

Fig. 1.   Densitometer tracing from autoradiograph of an electrophero-
gram of poliovirus-infected HeLa cells. At 210 minutes postinfection the cells were pulsed 10 minutes with ^{35}S-methio-
nine. The label was chased by adding excess unlabelled methionine and cells were lysed at the indicated chase period
and electrophoresed on an SDS-polyacrylamide slab gel.
(Taken with permission of the authors from reference 116)

weight proteins. These observations were taken as an indication that
the smaller peptides arose by cleavage of larger primary products
(62-64). Fig. 1 depicts an example of the electrophoretic analysis
of the viral peptides of poliovirus infected cells labelled by a
brief pulse with ^{35}S-methionine followed by a "chase" with a 10,000-
fold excess of cold pethionine. The protein pattern illustrates the
formation of smaller polypeptides by secondary cleavage of a large
primary product. A large polypeptide with a molecular weight over
200 Kd representing approximately the entire coding capacity of the
viral RNA was subsequently identified after incubation of infected
HeLa cells in the presence of several amino acid analogs (65) or by
incubation at high temperature (66). This protein was classified
as a *polyprotein* and specified as Non Capsid Virus Protein (NCVP00)
(36, 65). The processing of poliovirus precursor proteins is also
inhibited by incubation of cells infected with temperature sensitive
mutants at non-permissive temperature (66, 75, 77).

The time required for a ribosome to traverse a molecule of poliovirus RNA has been estimated to be in the order of 10 minutes (122).
In short pulse labelling experiments (up to 10 minutes), three major
primary virus specific peptides were typically observed: NCVPla,
NCVP3b and NCVPlb (see below and Fig. 1 and 2). These peptides must
arise therefore by cleavage of the ultimate precursor, the polyprotein, during its synthesis on polyribosomes.

It should be mentioned here that these studies led also to the
concept that poliovirus mRNA (and in turn all eukaryotic mRNAs) contain only one functional site for polypeptide chain initiation, close
to the 5'end of the RNA.

The question then arose as to what is the gene order of the
primary products on the viral RNA. The construction of a genetic map
was achieved by genetic analysis (123), as well as by analysis of the
effects of the drug pactamycin (124-126), or elevated medium
osmolarity (specific inhibitors of polypeptide chain initiation) on
the pattern of labelling of virus specific peptides (122). Further
details of the precursor product relationships of the three primary
gene products and the many secondary products (Fig. 2) were obtained
by determining and comparing the tryptic peptide maps of the different
viral proteins (127-129).

a) <u>Processing of NCVPla and NCVP3b</u>. The primary translation product
closest to the 5'end is NCVPla, the precursor to the capsid proteins
VP1-4. NCVPlb is the precursor to the viral proteins of the replicase
complex (see Fig. 2) and probably contains also the viral protease
(80).

The transit time of the ribosome through the capsid protein gene
region takes about 5 minutes. As the ribosomes leave the region of
the virion structural protein gene, the first cleavage of the nascent

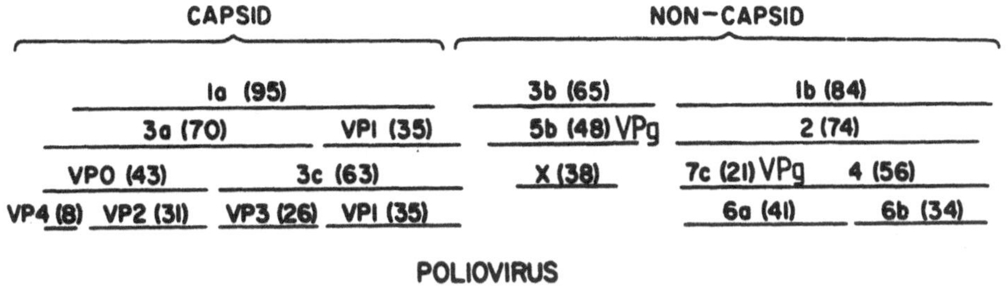

CAPSID			NON–CAPSID	
Ia (95)			3b (65)	Ib (84)
3a (70)		VPI (35)	5b (48) VPg	2 (74)
VPO (43)		3c (63)	X (38)	7c (21) VPg 4 (56)
VP4 (8) VP2 (31)	VP3 (26)	VPI (35)		6a (41) 6b (34)

POLIOVIRUS

Fig. 2.   Proposed processing map of the major poliovirus coded
          polypeptides.

chain occurs liberating NCVPla, a protein, which is subsequently
cleaved to the viral capsid proteins. Nevertheless, the ribosomes
proceed further along the RNA - normally without termination and
without re-initiation - and translate the next region on the mRNA.

The second primary translation product, NCVP3b, is rapidly
processed to NCVP5b and then to NCVPX, which is probably a protease
(79). A second *nascent* cleavage occurs, and again the ribosomes
proceed to translate the last gene, coding for the replicase proteins
yielding the primary translation product NCVPlb. This protein is
again rapidly converted to NCVP2 and then more slowly processed
further (see below).

The *primary* cleavages are normally so fast that special inhibitory treatments are required clearly to demonostrate the larger
primary translation products (36, 65). Only one paper reports the
presence of a giant picornavirus protein under physiological conditions (64).

Synthesis of prohormones, secretory proteins (1b, 12, 130-132)
and of picornavirus proteins takes place on *membrane bound* polysomes (7, 11, 133-135). Picornaviruses, therefore, may well utilize
the same processing system as cellular proteins. Comparative studies
on the effect of various protease inhibitors on the processing of
cellular and viral proteins should prove or disprove this suggestion.

The processing of the viral coat protein precursor is associated
with the assembly steps, i.e. these are *morphogenetic* cleavages.
The capsid precursor protein NCVPla is assembled into pentamers,
called protomers (36, 41, 43). This is accompanied (or rapidly
followed) by cleavage of NCVPla into VPO, VPl, VP3 (40), which
appears to commit the protomers to assemble. The cleavage
of VPO into VP4 and VP2, which accompanies or follows the encapsidation of virion RNA, results in a dramatic conformational shift
of the capsid (as revealed by a change in antigenicity) (39) and
completes poliovirus assembly and maturation. Comparable morho-
genetic cleavages of viral proteins were first discovered in phage
infected bacteria (3, 32-35) and later found in many viral systems

(2, 4, 5, 7-10, 36-50).

b)   NCVP1b, VPg and Replication of Viral RNA. The processing of
the replicase protein NCVP1b to NCVP2 and VPg, and NCVP2 to NCVP4
might be ultimately connected to the replication of viral RNA (11e,
i, j).

The genome RNA of picornaviruses is unique with regard to its
5'end: Eukaryotic mRNAs and other animal virus mRNAs are capped
at the 5'end (see chapters 5 and 8 of this volume and ref. 136).
Picornavirus mRNA is *not* capped (8, 137, 138), and the genome RNA
carries on its 5'end a covalently attached protein VPg (12, 139-142,
190).

Wimmer *et al.* (11e, 143) have mapped the viral protein VPg:
It is coded for by the region close to the 5'end of the replicase
gene. It has been suggested that NCVP1b (143, 144) might contain VPg.
The cleavage of VPg from NCVP1b could serve two functions: VPg may
provide a primer for the synthesis of new RNA and/or might activate
the replicase enzyme. According to this model (141), the synthesis of
every viral RNA strand (the minus as well as the plus strand), would
require a new molecule of VPg. RNA synthesis may be, therefore,
stringently dependent on the continued synthesis of viral proteins
coded for by the recplicase gene and the continued activity of a
protease.

Since viral RNA synthesis occurs on smooth endoplasmic membranes
(133-135) and viral proteins are synthesized on polyribosomes
attached to rough endoplasmic membranes (133-135), viral proteins
must be transported from their site of synthesis to their site of
function. It is conceivable that transport and processing are closely
connected or interdependent processes. The smooth membranes contain
besides the viral replicase another predominant viral protein, the
NCVPX which possesses protease activity (79). Hypothetically X might
be responsible for the liberation of VPg from its precursor. Recently, Palmenberger *et al.* (80) have observed that part of the primary
translation product of the replicase gene itself possesses proteolytic activity, which is responsible also for some of the processing
of coat proteins. Rueckert (12, 80) has proposed that the protease
of the replicase precursor might in addition cleave off autocata-
lytically both the protease itself and VPg, i.e. it may bite off its
own tail. Whether both proteases (X and one apparently derived from
the replicase protein precursor by cleavage) or only one of the two
are required for RNA synthesis is not yet clear.

Addition of 2 mM guanidine (11i, j; 63, 145, 146) rapidly blocks
viral RNA synthesis in picornavirus infected cells. Yet the mechanism
of this inhibition is not clear. *In vitro* guanidine interferes with
the initiation of viral RNA synthesis (147). Since guanidine affects
initiation of RNA synthesis and there is a stringent need for the

individual protease activities involved in making the VPg available for initiation to occur, the effect of guanidine on viral protein synthesis and processing will be re-examined here.

The Effect of Guanidine on the Processing of Viral Proteins

Interference with protein processing is possible in several independent ways:

1) by inhibition of proteases,
2) by alteration in primary, secondary or tertiary structure of the primary *translation* product or of primary *cleavage* products,
3) by the non-coordinate transport of viral proteases and their substrates, i.e. interference with compartmentalization,
4) furthermore, membrane alterations might cause changes in the secondary or tertiary structure of proteins.

Guanidine could exert an effect on either of these four sites or even on more than one site.

Experiments like the one described in Fig. 3 revealed that guanidine had no detectable effect on cellular protein synthesis, whereas poliovirus functions were altered to a considerable extent:

a)   In the presence of the drug, the virus-induced shut off of cellular protein synthesis was less efficient, and cell proteins can still be produced, at a rate of about 35% that of uninfected cells.

b)   In the presence of guanidine, NCVP1a was more rapidly cleaved into NCVP3a and 3c and then somewhat slowlier into the further products VP0, 1 and 3. However, the overall processing of the viral coat protein in the presence of guanidine is altered to only a small extent. For the time being, the early appearance of a protein which migrates like VP2 in the guanidine treated culture has no explanation.

c)   In contrast, the further processing of the other viral proteins, NCVP1b and the precursor to NCVPX are drastically affected by guanidine. The viral replicase precursor protein NCVP1b is cleaved in at least two alternate pathways one leading to NCVP6a and 6b, the other to the cleavage products NCVP4 and 7. Both these pathways must be inhibited by guanidine since only small amounts of NCVP4 and 6b are detectable. In addition, NCVPX appears in reduced amounts if at all.

d)   The guanidine induced alterations in the processing of viral proteins is better demonstrated by the appearance of several protein bands which are neither present in *uninfected* cells, nor in infected cells incubated in the *absence* of guanidine. These proteins which are identified by the superscript G in Fig. 3A and b migrate with

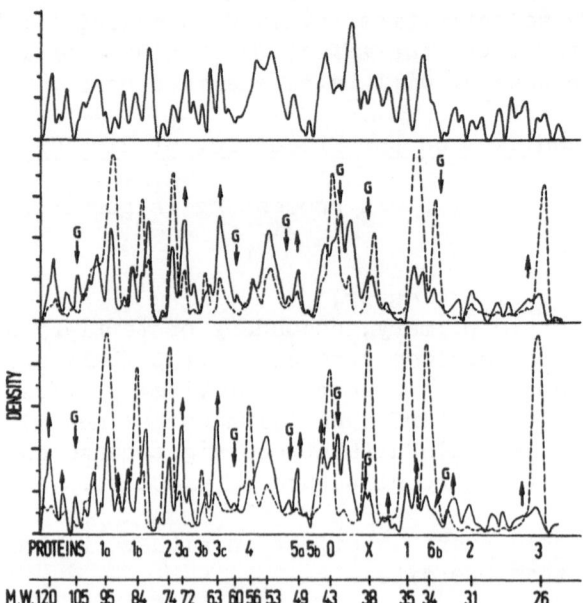

Fig. 3.   Guanidine mediated alterations in the processing of polio-
          virus proteins. HeLa cells suspended at a density of 1x10[7]
          in serum-free minimal essential medium supplemented with
          25 mM HEPES and 5% of the normal methionine concentration
          were incubated for 30 min at room temperature with 500 plaque
          forming units of poliovirus/cell. The temperature was then
          raised to 37°C and the cell suspension was diluted at the
          same time to a density of 2.5x10[6] cells/ml with minimal
          medium containing 7.5% fetal calf serum, 25 mM HEPES and
          5% of the normal methionine concentration. Mock infected
          and infected cell cultures were then incubated for 75 and
          95 min in the absence or presence of 2 mM guanidine.
          The cell suspensions were exposed to hypertonic medium for
          25 min. Isotonicity was then restored, and 6 minutes later a
          4-min pulse with 250 μCi ^{35}S-methionine/ml was carried out,
          followed by a 4-min chase in minimal medium containing
          10[4]-fold excess of unlabelled methionine and 25 mM HEPES.
          Proteins were separated as described in Table 1. The scan
          in the *upper* part (a) shows the protein pattern of *uninfected*
          HeLa cells.

          The two superimposed scans in the *lower* part of the figure (b)
          show the migration of ^{35}S labelled proteins from (---) polio-
          virus infected HeLa cells and (——) poliovirus infected HeLa
          cells incubated with 2 mM guanidine. The arrow ↑ indicates
          protein peaks which are higher in infected cells incubated
          with guanidine, compared to uninfected (upper part) or
          infected cells without guanidine (lower part). Peaks marked
          with ↓ G appear only in poliovirus infected cells incubated
          in the presence of guanidine.

apparent molecular weights of 105, 60, 50, 42, 39 and 33 Kd.

e)   Some proteins (M.W. 120, 110, 90, 49, 36.5, 32.5 and 27 Kd)
appear to label to a higher extent in the presence of guanidine:
Whether they represent authentic viral peptides will remain a matter
of speculation until peptide mapping experiments clarify this issue.

f)   It seems possible that the missing NCVPX may be present in the
1 Kd larger new protein: In fact, the normal precursor to NCVPX is
a 48 Kd protein, and there is an increase in the production of a
49 Kd protein.

    The guanidine-induced changes can be tentatively quantitated by
comparing the areas under individual peaks in the densitometric
tracings as described in Fig. 4 (Due account should be taken in this
case of the fact that in the presence of guanidine the virus-induced
shut-off is less efficient, and cellular proteins continue to be
produced to a non negligeable extent.)

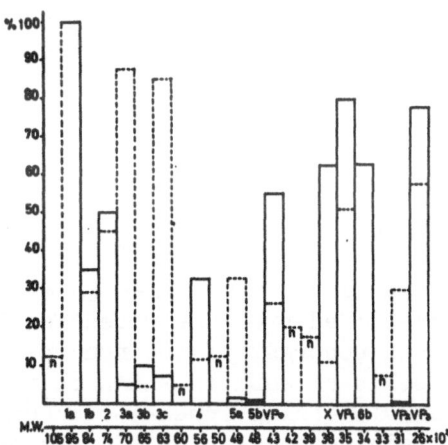

Fig. 4.   Quantitation of guanidine induced alteration in the proces-
          sing of viral proteins. The ratio of polioproteins synthe-
sized in the presence (---) or absence (——) of 2 mM guanidine was
established by comparison of the areas under each individual peak in
Fig. 3B. The letter n indicates newly synthesized proteins which are
neither found in proteins isolated from uninfected cells or infected
cells incubated in the presence of guanidine (proteins G of Fig. 3).
The areas of NCVP1a with or without guanidine was arbitrarily set as
100. In poliovirus infected cells the shut-off of host cell protein
synthesis is almost complete, whereas in cells infected *in the presence* of guanidine, HeLa protein synthesis continues at about 35% of
the rate of uninfected cells. An attempt was made to correct the cellular continuation of viral proteins: Accordingly, 35% of the corresponding host cell protein peaks are substracted from the areas of
viral proteins, or those newly synthesized under the influence of guanidine.

The general conclusion to be drawn so far is that viral proteins synthesized in the presence and absence of guanidine are processed differently. Guanidine might alter the structure of the primary translation product or its interaction with cellular membranes. Alternatively guanidine might interfere with viral protease(s). However, since the sensitivity or resistance of picornaviruses to guanidine is genetically determined, it is unlikely – though not yet excluded – that guanidine acts on components *of the host cell* required for viral growth.

Non-Uniform Synthesis and/or Accumulation of Poliovirus Proteins Under Conditions of Restricted Polypeptide Chain Initiation and at Early Times after Infection

Several years ago, we analyzed the effect of medium hypertonic-ity on the synthesis of poliovirus specific proteins in infected HeLa cells (149). No significant differences were observed in the ratios of the individual viral proteins when their synthesis was analyzed by pulse labelling with ^{32}S-methionine under isotonic and hypertonic conditions. We concluded that the hypertonic initiation block had no effect on the processing of viral proteins (122).

In a later study we performed similar experiments with poliovirus infected BSC1 cells (150). To our surprise the labeling pattern of viral proteins changed considerably when the osmolarity in the medium was increased.

Hypertonic initiation block (HIB) leads to a preferential accumulation of virus specified proteins coded for by the 3'end of the genome. This is in contrast to the non-uniform synthesis in Mengovirus infected cells at earlier times in the replication cycle, when the viral *coat* proteins appear to accumulate at a higher rate than the virus proteins coded for by RNA regions near the 3'end (81). We, therefore, re-examined viral protein synthesis in HeLa cells incubated in media with different osmolarities (see details in Table 1). Under restrictive conditions of polypeptide chain initiation, the ratio of NCVP1a to other proteins changes considerably in favor of 1b, 2 and X by 87, 50 and 136%, respectively. This result mimics our previous observation obtained with BSC1 cells (150) and suggests either the existance of two independent initiation sites on viral mRNA(s) or a preferential degradation of NCVP1a when initiation of protein synthesis is inhibited by HIB.

These experiments were performed at a time in the viral replication cycle when relatively large amounts of viral proteins are synthesized, the hypothetical existance of two independent initiation sties on viral mRNA(s) called for an analysis of viral protein synthesis at early times after infection (see Fig. 5). Only two viral proteins were clearly detectable by this procedure in the infected cells at early times of infection: NCVP2 and NCVPX. In

Table 1. Differential Accumulation of Viral Proteins under Normal
          and Restricted Conditions of Polypeptide Chain Initiation

Labelling conditions	Viral proteins			
	NCVP1a	NCVP1b	NCVP2	NCVPX
Isotonic (a)	100	100	100	100
Hypertonic (b)	100	187	150	236

HeLa cell monolayers were infected with 300 PFU/cell of poliovirus.
2.5 h post infection the medium was replaced by MEM containing 7.5%
FCS but no methionine (a). The medium for (b) contained in addition
150 mM extra NaCl. 25 min later the hypertonic medium in (b) was
replaced by medium containing 40 mM extra NaCl. 6 min later both
cultures were labelled with ^{35}S-methionine for 6 min and chased with
a 1,000-fold excess of cold methionine for 5 min. The media were
removed, the cells were immediately lysed by addition of sample
buffer (32) and boiled for 2 min. Proteins were separated by SDS-
PAGE on slab gels as described (148). The autoradiographs were
scanned and the peak areas of viral proteins labelled under isotonic
and hypertonic conditions were compared (the areas of all viral
proteins labelled under isotonic conditions (and only NCVP1a labelled
under hypertonic conditions) were arbitrarily set as 100).

addition two other viral proteins, NCVP1b and NCVP3, might be present.
Again these results suggested that NCVP1a was either not synthesized
or preferentially degraded at early times after infection. A com-
parision of the two densitometer tracings also reveal the differential effect of the virus induced shut-off on the synthesis of
individual cellular proteins.

     To sum up, a non-uniform synthesis and/or accumulation of various
viral proteins seems to take place early in infection, and under
conditions of restricted polypeptide chain initiation. These findings can be hardly reconciled with the classical view of translation
of picornavirus RNA. In principle, it is possible to visualize a
mechanism of initiation of viral protein synthesis on more than one
viral mRNA, or on more than one initiation site on one viral mRNA,
or to preferential degradation of proteins coded for by the 5'
terminal region of the viral mRNA. The question is still unsettled,
and calls for further experiments to decide between these alternatives.

Further Characterization of Proteases Using Viral Proteins as
Substrate: Studies in Cell-Free Systems

     The most direct evidence for virus-coded proteases is obtained
in studies with cell-free systems after allowing viral mRNA to direct

the synthesis of polyproteins (151-159). Those studies have been
performed with picornavirus RNA and with the RNA from several RNA
tumor viruses. Pelham (155) showed that encephalomyocarditis (EMC)
RNA can be translated completely in rabbit reticulocyte extracts.
The translation of EMC RNA yielded an active proteolytic processing
enzyme. Although the processing of the translation products could
be followed over time, only a small proportion of the precursor was
cleaved. This observation was not surprising, since there is a
general agreement that in infected cells the primary cleavages were
performed by *host* proteases. Another reason for the lack of extensive
processing is the observation that in infected cells viral proteins
remain membrane associated, while protein synthesis in reticulocyte
lysates proceeds on free ribosomes.

Shih *et al.* (157) reported that poliovirus RNA could be translated completely in rabbit reticulocyte lysates and that the products
were extensively processed. In a later study, they extended these
studies to EMC RNA. Figure 6 shows the genetic map of EMC virus. In
the cell-free system, the viral RNA is translated into three proteins:
A, F and C. A_1, which is analogous to poliovirus NCVP1a (the precursor to coat protein) is converted by a protease of the reticulocyte
to A. C carries the replicase function and is processed via D to E,
and protein F corresponds to the poliovirus protein NCVPX. The viral
coat protein precursor A is further processed to a slightly smaller

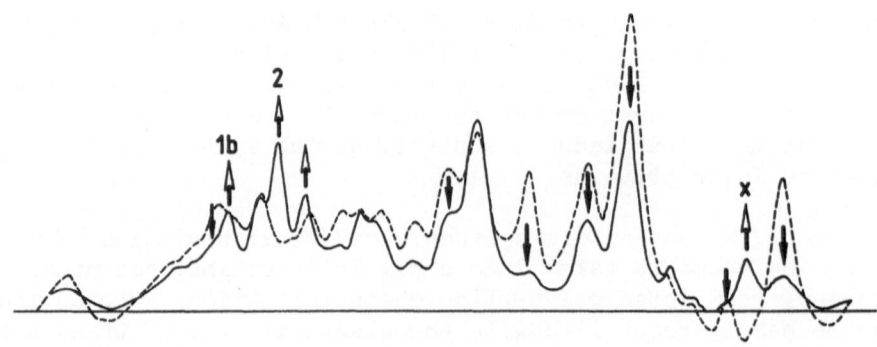

Fig. 5. Differential inhibition of individual host proteins by polio-
         virus. HeLa cells were infected as described in Table 1.
         At 105 min post infection cells were incubated for 15 min in
         medium containing 7.5% FCS but no methionine. At 120 min
         post infection the infected cells and the uninfected control
         cells were labelled with 250 μCi ^{35}S-methionine for 5 min
         and chased for 4 min with a thousand-fold excess of cold
         methionine. The incorporation of ^{35}S-methionine into
         individual protein bands was determined as described in
         (148).

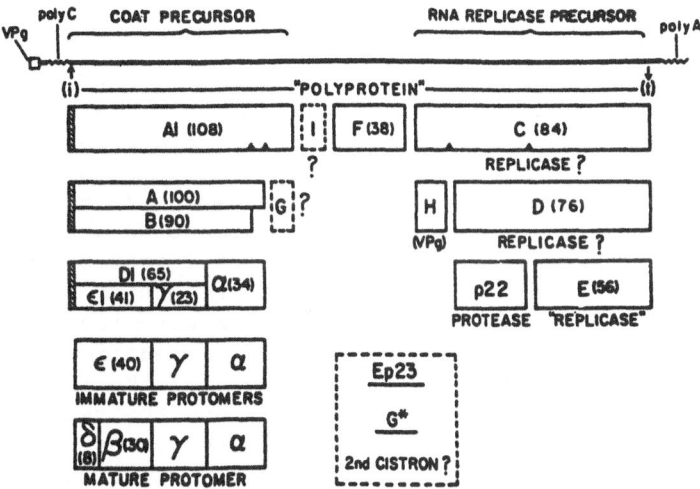

Fig. 6. Processing map illustrating functional relationship of EMC
         viral proteins. Translation of the RNA, about 7,500 bases
         long, is from left (5'-end) to right. Arrows (top line)
         indicate approximate location of the initiation (i) and
         termination (t) sites for the polyprotein cistron (6,200-
         6,500 bases). Products of the "second cistron" have so far
         been detected only in cell-free translation systems (see
         text). By occupying a reading frame different from that of
         the polyprotein cistron, a second cistron ($\sim$ 620 bases)
         might be located anywhere on the RNA.
         (taken with permission of the authors from Rückert,
         Palmenberg and Pallansch, in press (in ref. 12).

product designed B, which was also present in infected cells. Only
when the translation was permitted to proceed past the coat protein
region, a further processing of coat precursor proteins was observed.
This kinetic analysis revealed that the gene for the putative
protease must be located near the middle of the genome. Another,
less striking explanation for the observed processing could be
activation of a pre-existing reticulocyte protease by a viral
protein.

In continuation of this study, Palmenberg *et al.*(80) purified
the protease synthesized in reticulocyte lysates in response to the
addition of EMC RNA. The major proportion of the purified virus
specific protease migrated on slab gels with an apparent molecular
weight of 22,000 daltons. This protein, p22, shares tryptic
peptides with the protein D. The tryptic peptides derived from the
E protein complement the ones from the p22, and together they
represent all tryptic peptides from protein D. The authors have
tentatively assigned p22 to the amino terminal site of the D precursor. This assignment does not agree with the observation reported

in the previous paper (156) that E is completed before D, but is
supported by the statement that a new proteolytic activity appeared
after about 20 minutes of translation. In fact, the appearance of
proteolytic activity at 20 minutes, which was the time required to
complete the translation of the F region, suggested that protein F
might possess proteolytic activity also. This suggestion gains
support from the findings of Korant (79) which indicate that the
poliovirus specified protein X (which corresponds to the EMC virus
protein F) was a protease. Reports from other laboratories have
assigned protease activity also to the coat protein region of picornaviruses (76). Perhaps picornaviruses code for more than one protease,
each of which may be responsible for a certain specific cleavage(s).
The X (poliovirus) or F (EMC) proteins may cleave the p22 from the
replicase precursor and thereby activate p22 and the replicase. p22
in turn may be responsible for coat protein processing. Sequential
activation of protease was previously found in activation of complement and in conversion of fibrogen.

POSTRANSLATIONAL MODIFICATIONS OF ONCORNAVIRUS-DIRECTED PROTEINS

Proteases Specific of RNA Tumor Virus

The widespread interest in the biological functions of RNA tumor
viruses has accelerated the elucidation of the molecular biology of
their growth cycle. The synthesis of their structural and non-
structural proteins has been intensively studied in several laboratories and with many different RNA tumor viruses. The initial reports
of proteolytic processing of viral precursor proteins in avian
oncornavirus infected cells (2, 86, 87) have been confirmed and
extended to murine, feline and primate RNA tumor virus systems (2, 87,
94, 103, 104).

The genome RNA of oncornaviruses is approximately the same size
as the picornavirus RNA, but the virus particle has a far more
complex structure, consisting of a ribonucleoprotein core and a lipid
membrane envelope, with additional virus-specific membrane glycoproteins.

The core peptides or gag proteins are named according to their
molecular weight as p15, p12, p30 and p10. gp70 and p15E constitute
the envelope proteins of the *murine* retroviruses.

All oncornavirus particles contain in addition to the *structural*
proteins listed above an RNA-dependent DNA polymerase (reverse
transcriptase).

Results from many laboratories indicate that the order of the
viral genes from the 5' to the 3'end of the RNA is as follows: core
proteins, reverse transcriptase, envelope proteins and, in case of
Rous sarcoma virus, the transformation gene (src" or "onco"-gene).

A number of laboratories have established that the viral
structural proteins of murine leukemia viruses are first synthesized
as high molecular weight precursors, which are subsequently cleaved
and processed to give rise to the mature viral proteins (56, 57, 97).
These studies have also indicated that the major viral glycoprotein
and major internal core proteins are initially synthesized as part
of two distinct separate precursors, referred to as the env and gag
gene products, respectively. Additionally, it appears that the viral
polymerase is initially synthesized as part of a common gag-pol
precursor of about 200 Kd (98).

The analysis of oncornavirus protein synthesis is hindered by
the relatively small amounts of virus specific proteins which are
synthesized even in productive infection, and usually amount to only
1% of the total protein synthesis of the infected cells. Experimental
conditions which lead to an increase in the level of virus specific
protein synthesis in these systems have not yet been reported.

The use of the Hypertonic Initiation Block (HIB) technique,
which leads to the amplification of viral protein synthesis in nearly
all tissue culture cells infected by a cytopathogenetic virus (149,
150), fails to increase viral specific protein synthesis in RNA tumor
virus infected cells. This is most likely due to the fact that the
oncornavirus mRNAs (in contrast to cytopathogenic viruses) posess
only low affinities for ribosomes and/or initiation factors and are
therefore *inhibited* rather than *amplified* by all conditions which
interfere with polypeptide chain initiation (150). Therefore, RNA
tumor virus specific proteins have to be isolated from infected cell
lysates with the help of specific antibodies and adsorption of the
ensuing protein antibody complexes to *Staphylococcus aureus* (161).

The genomic RNA of oncornaviruses can direct protein synthesis
in cell free extracts. Although the viral RNA contain the genetic
information for all viral proteins, the *in vitro* translation products
ordinarily are only those coded for by the 5'end of the RNA representing all core proteins in one polyprotein, the gag translation product
with a molecular weight of 67 Kd for the *murine* virus, 76 Kd for the
avian viruses. Sometimes, however, a larger product was observed,
with a molecular weight of 180-195 Kd, which contains the reverse
transcriptase in addition to the gag proteins. Termination codons
at the end of the gag gene prevent the translational machinery to
read through and translate the polymerase gene (163).

A special mRNA from which the internal termination codons have
been removed by splicing, codes for a 180 Kd or 195 Kd precursor
protein which contains both the core and the reverse transcriptase
gene products. Other studies have shown that the envelope glygo-
protein is synthesized from a subgenomic spliced 23S mRNA which has
the same 5'end as does the gag and gag pol mRNA but the translation-
able part of the mRNA contains sequences related to the 3'end. This

mRNA gives rise to a 85 Kd glycoprotein precursor which codes for the envelope glycoprotein gp70 and p15E. Thus the core proteins, the envelope proteins, and probably also the polymerase are translated from individual mRNA species. Recently it was suggested that the gag precursor protein is translated from four different mRNAs. Processing of viral precursor proteins takes place in infected cells as well as in cell-free protein synthesizing systems and has been studied in many systems.

Synthesis and Processing of Viral Proteins in Friend Erythroleukemia Cell Lines

In many RNA tumor virus transforming systems, two viruses inter- act simultaneously with host cells, a replication-defective transform- ing virus, and a replicating-helper virus, which provides functions needed for the replication of both viruses. Defective viruses usually contain deletions in viral replicative genes, and in some instances insertions of cellular genes or gene regions.

In the mouse, infection of hematopoietic cells with Friend virus infected proerythroblast (a hematopoietic stem cell committed to erythropoietic differentiation) is no longer responsive to erythropoietin and proliferates instead of differentiating *in vivo*.

Friend virus transformed mouse erythroleukemic cell lines have attained widespread use as a model for the *in vitro* study of dif- ferentiation (164, 165). The Friend cells are morphologically similar to proerythroblasts (166) and are presumably arrested at this stage of differentiation by a virus-induced transforming event. Further- more, many of these cell lines release virus particles into the growth medium (47, 166).

Terminal differentiation can be induced in erythroleukemia cells by a great variety of compounds (for review see 167) including planar polar compounds such as hexamethylene bisacetamide (HMBA), purines and purine derivates and pyrimidine analogs short chain fatty acids, hemin, ouabain, DMSO and hypertonic salt (168). Induction of dif- ferentiation in these cells is assayed most conveniently by synthesis of hemoglobin (164). The terminal differentiation pathway also involves synthesis of heme pathway enzymes, erythrocyte membrane proteins and a number of changes in cell architecture (169, 170). The appearance of all these differentiation markers is only detect- able after a certain latent period, which varies with the genetic history of the cells, culture conditions, and with the particular inducer utilized. The cell line FSD1/F4 (F4) was established from the spleens of DBA/2 mice infected with the NB tropic BALB/c adapted Friend virus complex by Ostertag *et al.* (165). While growing in culture, these cells release biochemically active Friend virus complex: the lymphatic leukemia helper virus and the replication- defective erythroid cell transforming spleen focus forming virus (171).

   A 5-bromodeoxyuridine resistant and thymidine-kinase negative
(TK⁻) subclone of line FSD1/F4, clone B8, was found by Ostertag
et al. (165) to release 1,000- to 100,000-fold reduced amounts of
Friend virus, as compared with the TK⁺ parental cell clone F4.
Interestingly, morphological studies by electron microscopy revealed
that the non-producer B8 cells contained an abundance of intracis-
ternal A type particles, many more so than the virus producing F4
cells (173). In spite of these differences, however, both cell
clones do share that characteristic property of most Friend cell

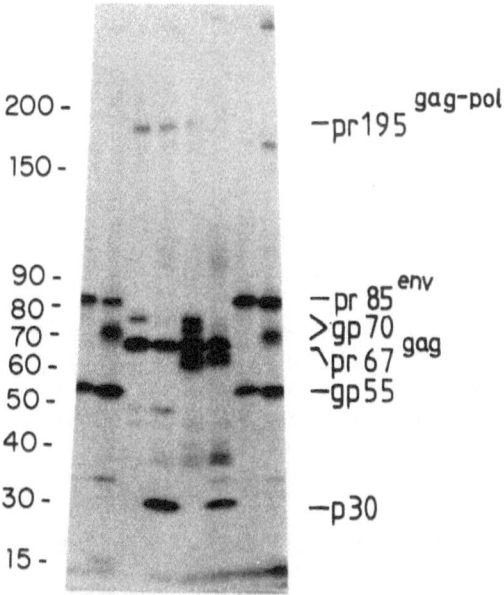

Fig. 7. Autoradiograph of SDS-gelelectrophoresis of immuno-
         precipitates derived from ³⁵S-labelled viral proteins from
         F4 and 745 Friend cells. Cells were pulse labelled with
         ³⁵S-methionine for 5 minutes and cell extracts prepared
         either immediately or after a 40-minute chase in the present
         of excess unlabelled methionine. Viral proteins were
         isolated by immunoprecipitation with anti-gp70 serum (a, b,
         g, h) or anti-p30 serum (e, f) and the antigen-antibody
         complexes adsorbed to Staphylococcus aureus strain Cowan I.
         The proteins were separated on 20 cm long 5-20% exponential
         gradient polyacrylamide gel. For experimental details see
         102.

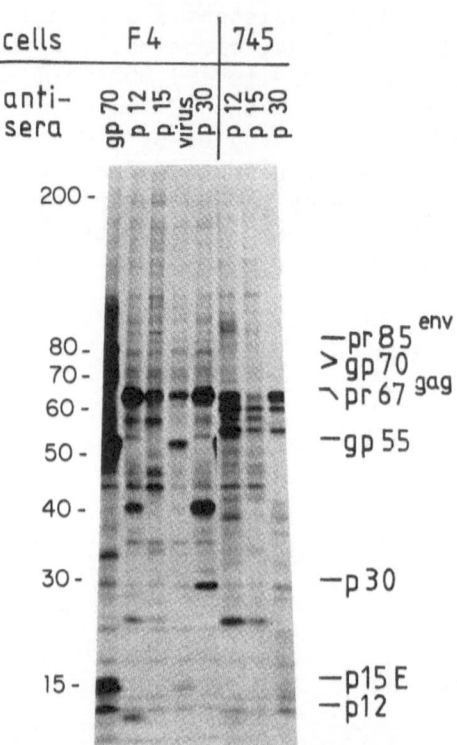

Fig. 8.   Detection of possible processing intermediates in F4 and 745
cells.   Cells were labeled with ^{32}S-methionine for 90
minutes and the viral proteins isolated by immunoprecipitation with the indicated sera processed as described in
Fig. 7.

lines: the capacity to be induced to differentiate upon exposure
to various agents (168, 174-177).

A unique aspect of the intracellular viral protein synthesis
observed in the Friend cells in the presence of large quantities of
a glycosylated protein of 55 Kd precipitable with anti-gp70 serum,
as shown in Fig. 7. For a more detailed discussion of the precursor→
product relationship among the protein synthesized by Friend cells
upon induction of differentiation, the reader is referred to (102).

Detection of Processing Intermediates

We observed that upon immunoprecipitation of cell extracts with
monospecific antisera to p30, p12, and p15, the 67 Kd precursor was
precipitated by all (Fig. 8) indicating that it contained antigenic
reactivities to all three viral polypeptides as has been reported

by Stephenson *et al.* (92) for the Rauscher virus system. In addition lower molecular weight species intermediate in size between the gag precursor polyprotein and the mature viral proteins could be specifically precipitated by two or all three of the antisera. Immunoprecipitates of F4 cell extracts revealed a species of 41 Kd that was precipitated by both anti-p12 serum and anti-p-30 serum. In addition, a polypeptide of about 25 Kd was recognized by both anti-p12 serum and anti-p-15 serum. Analysis of 745 cell extracts showed that both a 55 Kd protein and pr67^{gag} are recognized by all three antisera (anti-p12, p15, and p30). Similar to what was observed in F4 cells, a 25 Kd protein was precipitated by anti-p12 serum and anti-p15 serum. These polypeptides which share antigenic determinants with two or more viral structural proteins probably represent processing intermediates. The difference in the size of these intermediates in the two cell lines probably reflects the differences in host cells specified processing pathways.

Modification and Processing of Viral Precursor Polypeptides During the Induced Differentiation of Friend Cells

Since Friend eythroleukemia cell lines are capable of differentiation upon exposure to a number of agents, the question arose as to what proteins are synthesized during this process. Analysis of the intracellular proteins (performed as described in Fig. 9 and ref. 102), revealed that:

a)   processing of the gag precursor protein, as detected by immunoprecipitation with anti-p30 serum, showed marked alterations at 24 hours post induction as compared to the processing in a duplicate control culture.

b)   The changes observed in the culture induced for 24 hours were an increase in pr67^{gag}, the disappearance of an 58 Kd intermediate, and an increase in the intensity of a 40 Kd intermediate.

c)   By 72 hours post induction, the pattern once more resembled that of the control culture.

d)   Analysis of the extracts with anti-gp70 serum again revealed major changes at 24 hours post induction, as compared to the control culture. A new band migrating faster than gp70 appeared in the 24-hour-culture, and there was no noticeable quantitative reduction in the amount of p15E.

No increase in virus production (as measured by reverse transcriptase activity in the extra-cellular medium) was noted in the induced cell cultures. Virus production was found to be directly related to cell growth, as previously reported (101, 178), and in fact the induced cell cultures produced *less* virus, since cell proliferation was slowed down by the inducing agent.

Synthesis and Processing of Viral Precursor Proteins under
Conditions Inducing Terminal Differentiation

    As discussed in the preceding section, the pattern of processing
of viral proteins is markedly altered after 24 hours induction with
hypoxanthine.

    Little is known about the early events following the addition
of an inducer to Friend cells. Since the block in differentiation
in these cells is related to infection with the Friend spleen focus
forming virus (SFFV), and might depend on the continued expression

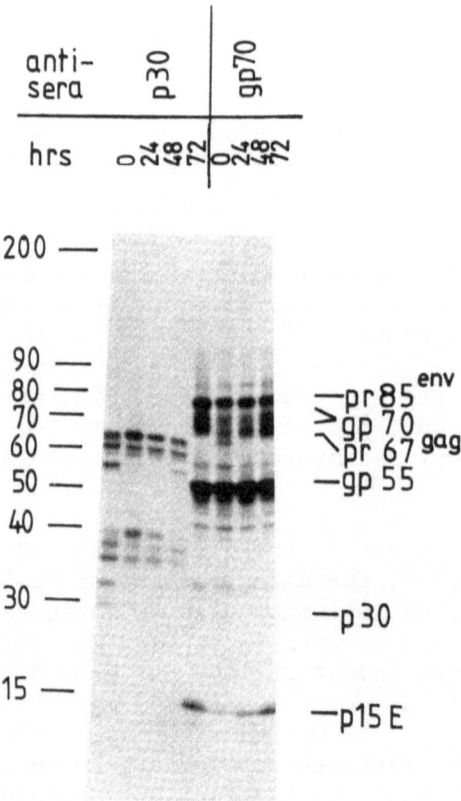

Fig. 9. Analysis of viral precursor processing during induced
        differentiation. Terminal differentiation was induced in
        745 cells by addition of 4 mM hypoxanthine. Cells were
        labeled with ^{35}S-methionine for 90 minutes at the indicated
        times after addition of the inducer. Cell extracts were
        prepared and immunoprecipitation with anti-p30 and anti-
        gp70 sera were performed as described (102).

of its genetic information, we asked whether inducers of differentiation might interfere with viral gene expression. SFFV specific RNA
synthesis increases after induction of differentiation; therefore
we turned our attention to the control of protein synthesis in these
cells. These studies seemed particularly relevant since DMSO (one
of the best inducers of differentiation) is known to interfere with
initiation of protein synthesis in cultured cells (179) and also in
Friend erythroleukemia cells (180, 181).

AMPLIFICACTION OF TRANSLATIONAL CONTROL OF GENE EXPRESSION DURING
THE DIFFERENTIATION OF FRIEND ERYTHROLEUKEMIC CELLS

a)   Inducers of differentiation inhibit protein synthesis. The
effect of a number of different inducers on Friend cell protein
synthesis are summarized in Table 2. In Friend cell lines protein
synthesis, as measured by incorporation of ^{35}S-methionine into hot
TCA insoluble material, is inhibited throughout the process of the
induction. Incorporation of labelled amino acids into proteins is
influenced by the size of the pool in the cells. Within 30 minutes
after the induction of differentiation there is a detectable shift
in the distribution of ribosomes in polysomes and monosomes. The
number of ribosomes in polysomes decreases, and there is a concomitant
increase in the number of monosomes. This pattern of ribosome
distribution (which is indicative of reduced initiation) can be
observed throughout the course of induction of terminal differentiation in Friend cells.

Table 2. Effect of Inducers of Friend Erythroleukemic Cell Protein
          Synthesis in vivo

Addition	mM	Hemoglobin+ concentration µg/10^8 cells	Protein synthesis (% control) 24 hr	72 hr
None	–	14	100	100
DMSO	192	155	34	33
HMBA	5	239	51	34
Butyric acid	1	178	42	37
Hemin	0.1	189	69	nd
Hypoxanthine	2.5	230	65	nd
Ouabain	0.1	121	56	nd
Excess NaCl	75	81	53	33

+ Hemoglobin production was measured in cell extracts 3 days after
  addition of the compound:

  Protein synthesis was determined by incorporation of ^{35}S-methionine
  into hot TCA insoluble material 24 or 72 hr after induction.

b)   <u>Differential effect of inhibitors of polypeptide chain initiation</u>
     <u>on the synthesis of viral and cellular proteins</u>. Studies with
temperature sensitive mutants of avian erythroblastosis virus suggest
that the block in erythroid differentiation depends upon the *continued*
expression of a viral gene product (182). As mentioned earlier, the
block in differentiation in these cells is related to infection with
the Friend spleen focus forming virus (SFFV) and might depend on
continued expression of its genetic information. Experiments were
performed to determine whether inducers of differentiation interfere
with viral gene expression. When the relative translational efficiency (RTE) of viral mRNAs was determined to see whether any of the viral
gene products was preferentially inhibited by the hypertonic initiation block, it was regularly observed that, in contrast to viral
RNAs of cytopatogenic viruses, the RNA of tumor viruses do not possess
a high RTE (Table 3). The RTE for viral mRNA coding for the gag or
virion core protein (p30 related) is considerably *lower* than that for
the average cellular mRNA. Also, the mRNA coding for the gp70 and
related proteins have a lower RTE than mRNAs coding for actin, which
was heretofore considered as the mRNA with the lowest RTE.

Table 3.   RTE of mRNAs in Friend Erythroleukemia Cells

Protein	Molecular weight (Kd)	RTE
Cellular	44	0.4
	22	2.0
	20	1.0
	19	1.0
Retroviral		
gp70 related	85	0.9
	70	0.3
	55	0.3
p30 related	67	0.5
	61	0.5
	52	1.4
	30	1.9

NaCl concentration dependent changes in the relative synthesis of
viral and cellular proteins. After a 1-hr-pulse with ^{35}S-methionine
in the presence and absence of 100 mM excess NaCl (100 mM NaCl
affects 80% inhibition of overall protein synthesis) immunoprecipitates were prepared and analyzed as described (102).

RTEs were calculated from planometry of densiometer tracings from
the resulting autoradiographs as described earlier (148).

Of special interest is the low RTE of the mRNA coding for the gp55 proteins. All the available evidence suggests that gp55 is the gene product of the transforming Friend spleen focus forming virus (102, 183-186). Other studies with transforming avian and mammalian defective viruses suggest that viral coded transformation proteins are required for the maintenance of the transformed state (187). In the context of a model for induced differentiation of Friend ertythro-leukemic cells we have suggested that the synthesis of a viral protein responsible for maintaining the differentiation block, is decreased during differentiation (188). The primary candidate for such a protein is the gp70 related gp55 protein which possesses a "low" RTE (188).

EFFECT OF INHIBITORS OF POLYPEPTIDE CHAIN INITIATION ON THE MODIFICATION AND PROCESSING OF VIRAL POLYPROTEINS

Processing of precursors can be altered by changes in glyco-sylation and/or intracellular transport. Both the envelope and some gag primary gene products are glycosylated. Evidence has also been presented which indicates that tunicamycin inhibits both glycosylation and processing of precursor proteins and interferes with virus maturation (60). One point is obvious from the data presented in table III. The RTEs for the primary gene products and their cleavage products often do not agree and are markedly altered upon induction. It is readily apparent that the precursor $pr67^{gag}$ and its product p30 have different RTEs; 0.5 and 1.9 respectively, while the same effect was seen for the $pr85^{env}$ (RTE 0.9) and gp70 (RTE 0.3; Table 3). We could only rationalize these results by postulating that the presence of high salt during the pulse labelling period had an effect on viral protein *processing*. Experiments like the one depicted in Fig. 10 demonstrated that this is indeed the case. The conditions of the chase, i.e. incubation under isotonic or hypertonic conditions, leads to major alterations in the processing pathways. Moreover, the fate (among other examples) of $pr85^{env}$ (which is processed to gp70 at an 8-fold higher rate), and that of $pr67^{gag}$ (whose processing to p30 shows marked alterations), make a strong case to support the contemption that not only do the condi-tions of processing have an influence on the kintetics, but the conditions under which *synthesis* occurred do also have a pronounced effect on subsequent processing events (see for instance, Table 4).

In further studies, in which Friend cells were labelled with ^{35}S-methionine and the *chase* was carried out under conditions which interfere with polypeptide chain *initiation* (higher concentrations of NaCl or KCl, DMSO, growth at elevated temperatures) it was observed that:

a)   DMSO and elevated temperatures (which inhibits polypeptide chain initiation to a comparable extent as excess salt in the medium), have no effect on the processing of the $pr67^{gag}$.

b)   An *increase* in the osmolarity of the medium (with either NaCl or KCl) accelerates the processing of $pr67^{gag}$ to $pr50^{gag}$ (Fig. 11A).

c)   At high salt concentrations, an additional intermediate of 42-44 KCl appears, both species being detected in very limited amounts in cells incubated under isotonic conditions.

As far as the processing of $pr85^{env}$ is concerned, the same studies indicated that:

a)   Excess of NaCl or KCl inhibits the processing to gp70 (Fig. 11B): With 100 mM excess of salt, 6-8-fold more label is retained in the $pr85^{env}$

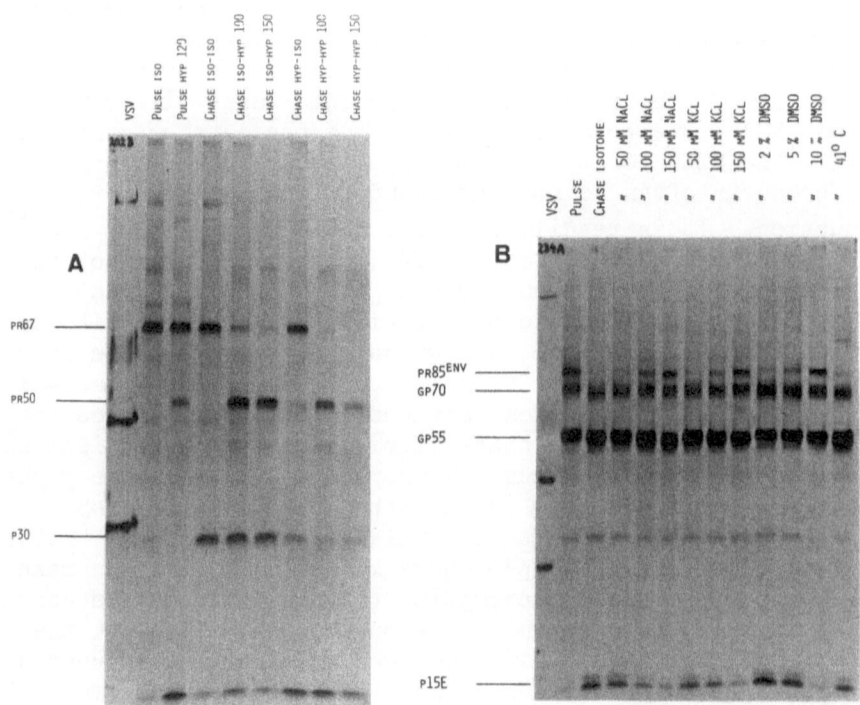

Fig. 10. Effect of hyperosmolarity on the synthesis and processing of viral polyproteins. Cells were labelled for 30 minutes with ^{35}S-methionine under isotonic and hypertonic conditions. Cell extracts were prepared at that time or after a one-hour chase under isotonic or hypertonic conditions. Immuno-precipitation with anti-p30 (A) or anti-gp70 (B) sera was as described (102).

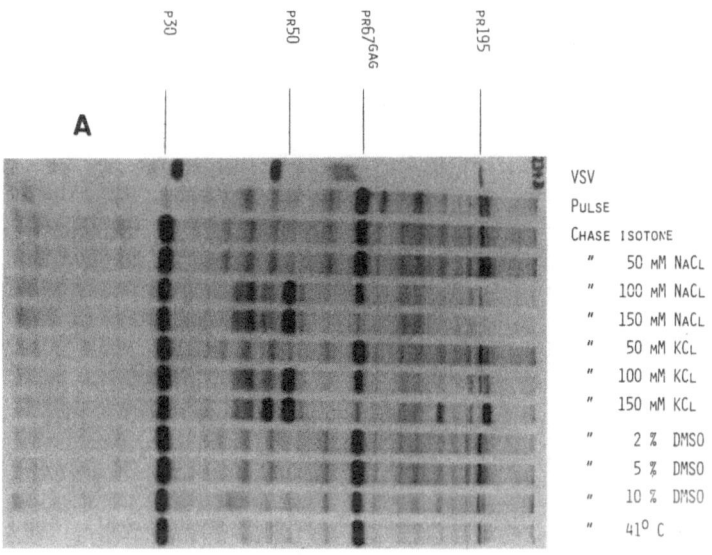

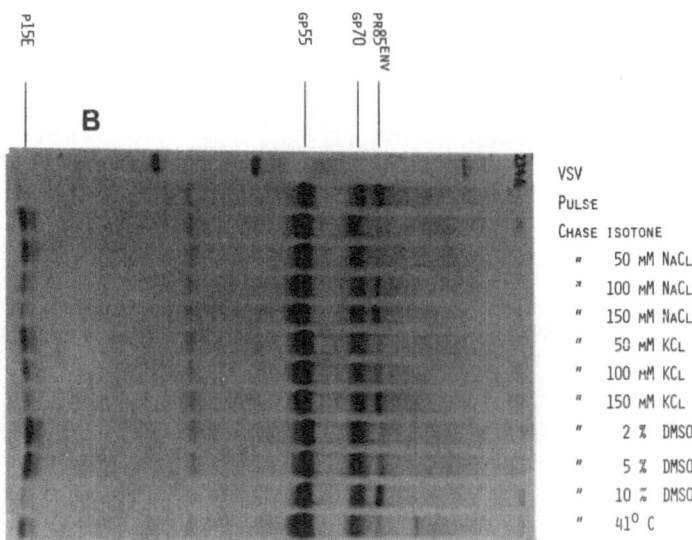

Fig. 11. Effect of hyperosmolarity, DMSO and incubation at 41°C
on the processing of viral polyproteins. Friend cells were
labelled for 30 minutes with ^{35}S-methionine, aliquoted,
centrifuged and suspended in medium containing a hundred-
fold excess of unlabelled methionine and the indicated
additions. Cell extracts were prepared and immunoprecipitation with anti-p30 (A) or anti gp70 (B) sera was performed
as described (102).

Table 4. Distribution of Label in Virus Specific Proteins

	gp70 specific proteins			p30 specific proteins		
	pr85^{env}	gp70	p15E	pr67^{gag}	pr50	p30
Pulse 30' isotone	14.5	11.3	0	52.0	1	1
Chase 60' isotone	1.6	10.6	3.8	44.9	0	(16.8)
Chase 60' hypertone	11.9	9.7	1	5.4	17.7	20.0
Pulse 30' hypertone	12.7	28.8	0	55.3	9.6	0
Chase 60' isotone	5.2	23.0	2.8	37.8	6.7	6.2
Chase 60' hypertone	9.8	21.0	0	1	15.0	10.8

in arbitrary units as determined by densitometry (see reference 148).

b)   Addition of 2-10% DMSO clearly *inhibits* the processing of the
precursor.

It is worth mention here that concentration of DMSO capable of
inducing differentiation (1-2%) concomitantly alter both, the processing of gp55 and membrane transport functions (89, 150, 186).

CONCLUDING REMARKS

Modification and/or processing of viral and cellular polypeptides
have a profound effect on, and may be are required for, their
biological activity. Incorporation of amino acid analogs into nascent
polypeptides can prevent concomitantly and subsequently occuring
cleavages. This is probably not solely due to alterations in primary
structure, but to induced changes in secondary and tertiary structure
as well. In this review, this point was illustrated with two examples: the processing of poliovirus polyprotein during exposure of
infected cells to guanidine, and the changes induced by hyperosmolarity on the fate of proteins of RNA tumour viruses.

Cellular protease can perform the primary cleavage of the
polioviral polyprotein (68, 72, 75). The further specific processing
of viral proteins seem to depend on viral protease(s). The experimental evidence reviewed here suggests that the primary cleavage pattern
exerted by cellular proteases might be altered by guanidine. The
secondary and tertiary structure of the polyprotein has a profound
influence on both the primary and secondary cleavages. Guanidine may
induce structural alterations in the viral proteins, interfere with
the interaction of viral proteins with cellular components (for

instance membranes), or inhibit virus specific proteases. None or
only small amounts of authentic NCVPX, a viral protease candidate
(79), are detectable in guanidine treated cells. Guanidine interferes with the processing of virus specific protease(s) and/or
the viral protein(s) engaged in the replication of viral RNA. Every
newly initiated viral RNA may not only depend on VPg as a primer
molecule but also require one protease activated protein as a
component of the replicase (144).

There is presently some evidence indicating that the viral
replicase function is not stable, and depends on the continued
synthesis (and correct processing) of new virus specific proteins.
Our observations also offer, therefore, a plausible explanation for
the effect of guanidine when added at later times.

Secondary and tertiary structure of nascent polypeptides may
be altered not only by the presence of guanidine but also by an
increase in the medium osmolarity. The latter may induce changes in
protein processing by interference with glycosylation. Dependence
of processing on prior glycosylation has been observed to occur in
several systems (57-60).

Analysis of the processing of preformed RNA tumor virus-coded
polyproteins which were chased under different experimental conditions
revealed the preferential processing or inhibition of various processing steps in the pathway of some polypeptides, and the accentuation
in the accumulation of certain intermediates. The conditions under
which synthesis has occurred also have a profound effect on subsequent
processing events: In fact poliovirus polyproteins synthesized at
41°C were shown to remain stable and are not further processed upon
return to 37°C. Since after shift down only newly synthesized
proteins are processed properly, it can be excluded an effect on
protease(s). Such observation is more compatible with a persisting
conformational alteration (or compartmentalization) such that protease
and substrate cannot interact with each other. With regard to
conformation and compartmentalization, it should be recalled that both
the synthesis of proteins and the assembly during maturation of
poliovirus and RNA tumor viruses, are membrane associated processes.

Until the cellular (?) protease activities responsible for
cleavage of viral precursor proteins are isolated it is not possible
clearly to determine whether hypertonicity affects the activity of
the protease or modifies the conformation of the precursor so that
its fate is altered. Our experience in the analysis of several
systems leads us to conclude that one can not generalize about the
effects on processing or secretion induced by hyperosmolarity.
Indeed, while the effect of HIB on the secretion of IgG polypeptides
in MCP-11 cells was negligeable, glycosylation and secretion of
chicken liver cell proteins were markedly decreased under HIB
conditions (Grieninger, McFarland and Koch, unpublished).

ACKNOWLEDGEMENT

The work performed in the authors' laboratory and reviewed here,
was partly supported by Deutsche Forschungsgemeinschaft and Stiftung
Volkswagenwerk. We thank Dr. Rueckert for permission to reprint
illustrations from reference 116 and 12.

REFERENCES

1.   *Proteases and Biological Control* (E. Reich, D. Rifkin, E. Shaw,
     eds.), Cold Spring Harbor Lab., Cold Spring Harbor 1974.
     a) Korant, B.D.: Regulation of animal virus replication by
     protein cleavage, p. 621-644.
     b) Steiner, D.F., Kemmler, W., Tager, H.S., Rubenstein, A.H.,
     Lernmark, A. and Zühlke, H.: Proteolytic mechanism in the
     biosynthesis of polypeptide hormones, p. 531-549.
2.   Korant, B.D. (1978), in *Molecular Basis of Biological Degradating
     Processes*, (R. Berlin, H. Herrmann, I. Lepow, G. Tanzer, eds.),
     pp. 171-220; Academic Press, N. York.
3.   Showe, M. and Kellenberger, E. (1975), in *Control Processes in
     Viral Multiplication* (D.C. Burke, W. Russel, eds.), pp. 407-422,
     Cambridge University Press, London.
4.   Shatkin, A.J. (1974), Ann. Rev. Biochem. 43, 643-665.
5.   Hershko, A. and Fry, M. (1975), Ann. Rev. Biochem. 44, 775-797.
6.   Neurath, H. and Walsh, K.A. (1976), Proc. Natl. Acad. Sci. USA
     73, 3825-3832.
7.   Levintow, L. (1974), in *The Comprehensive Virology* (H. Fraenkel-
     Conrat, and R.R. Wagner, eds.), Vol. 2, pp. 109-169, Plenum
     Press, New York.
8.   Rueckert, R.R. (1976), in *The Comprehensive Virology* (H. Fraenkel-
     Conrat and R.R. Wagner, eds.), Vol. 6, pp. 131-213, Plenum
     Press, New York.
9.   Rekosh, D. (1977), in *The Molecular Biology of Animal Viruses*
     (D.P. Nayak, ed.), pp. 64-110, Marcel Dekker Inc., New York.
10. Butterworth, B.E. (1977), in *Current Topics in Microbiology and
     Immunology*, Vol. 77, pp. 1-41, Springer Verlag, New York.
11. *The Molecular Biology of Picornaviruses* (R. Perez-Bercoff, ed.),
     Plenum Press, New York and London (1979).
     a) Scraba, D.G.: The picornavirion: structure and assembly,
     pp. 1-23.
     b) Rueckert, R.R., Matthews, T.J., Kew, O.M., Pallansch, M.,
     McLean, C. and Omilianowski, D.: Synthesis and Processing of
     picornaviral polyproteins, pp. 113-125.
     c) Lucas-Lenard, J.M.: Virus-directed protein synthesis,
     pp. 127-147.
     d) Korant, B.D.: Role of cellular and viral proteases in the
     processing of picornavirus proteins, pp. 149-173.
     e) Wimmer, E.: The genome-linked protein of picornaviruses:
     discovery, properties and possible function, pp. 175-190.
     f) Jackson, R.J.: The mechanism and cytoplasmic control of

mammalian protein synthesis, pp. 191-222.

g) Ehrenfeld, E.: In vitro translation of picornavirus RNA, pp. 223-238.

h) Revel, M., Kimchi, A., Schmidt, A., Shulman, L. and Zilber-stein, A.: The interferon system: studies on the molecular mechanism of interferon action, pp. 239-253.

i) Baglioni, C., **Maroney**, P.A., Chatterjee, G.E. and Minks, M.A.: Interferon-induced activation of an endonuclease by 2'5' oligo (A), pp. 279-292.

j) Perez-Bercoff, R.: The mechanism of replication of picorna-virus RNA, pp. 293-318.

12. *Biosynthesis, Modification and Processing of Cellular and Viral Polyproteins* (G. Koch, D. Richter, eds.), 1980 Acad.Press,N.York.

13. Behrens, N.H. (1974), in *Biology and Chemistry of Eukaryotic Cell Surfaces* (E.Y.C; Less, E.D. Smith, eds.), pp. 159-180.

14. Prockop, D.J., Kiwirikko, K.I., Tuderman, L. and Gucman, N.A. (1979), New Engl. J. Med. 301, 13-23.

15. Milstein, C., Brownlee, G.G., Harrison, T.M. and Mathews, M.B. (1972), Nature New Biol. 239, 117-120.

16. Blobel, G. and Dobberstein, B. (1975), J. Cell Biol. 67, 835-851.

17. Katz, N.F., Rothman, J.E., Lingappa, V.R., Blobel, G. and Lodish, H.F. (1977), Proc. Nat. Acad. Sci. USA 74, 3278-3282.

18. Blobel, G. and Dobberstein, B. (1975), J. Cell Biol. 67, 852-862.

19. Lingappa, V.A;, Katz, F.N., Lodish, H.F. and Blobel, G. (1978), J. Biol. Chem. 253, 8667-8670.

20. Wickner, W. (1979), Ann. Rev. Biochem. 48, 23-45.

21. Krebs, E.G. and Beavo, J.A. (1979), Ann. Rev. Biochem. 48, 923-959.

22. Kaerlein, M. and Horak; I. (1976), Nature, London, 259, 150-151.

23. Kruppa, J. and Martini, O.H.W. (1978), Biochem. Biophys. Res. Commun. 85, 428-435.

24. Wool, I.G. (1979), Ann. Rev. Biochem. 48, 719-754.

25. Kramer, G., Cimadevilla, J.M. and Hardesty, B. (1976), Proc. Natl. Acad. Sci. USA 73, 3079-3082.

26. Levin, D.H., Ranu, R.S., Ernst, V. and London, I.M. (1976), Proc. Natl. Acad. Sci. USA 73, 3112-3116.

27. Benne, R., Edman, J., Traut, R.R. and Hershey, J.W.B. (1978), Proc. Natl. Acad. Sci. USA 75, 108-112.

28. James, L.A. and Tershak, D.R., (1981), Can. J. Microbiol. 27, 28-35.

29. Kassell, B. and Kay, J. (1973), Science 180, 1022-1027.

30. Doolittle, R.F. (1973), Advan. Protein Chem. 27, 1-109.

31a. Schultz, D.R. (1976), in *Proteolysis and Physiological Regulation* (D.W. Ribbons and K. Brew eds.), Miami Winter Symposia, Vol. 11, pp. 143-167.

31b. Cooper, N.R. and Ziccardi, R.J. (1976), in *Proteolysis and Physiological Regulation* (D.W. Ribbons and K. Brew eds.), Miami Winter Symposia, Vol. 11, pp. 167-187.

32. Laemmli, U.K. (1970), Nature, London, 227, 680-682.
33. Goldstein, J. and Champe, S.P. (1974), J. Virol. 13, 419-427.
34. Goldstein Giri, J., McCullough, J.E. and Champe, S.P. (1976), J. Virol. 18, 894-903.
35. Bachrach, U. and Benchetrit, L. (1974), Virology 59, 51-58.
36. Baltimore, D. (1969), in *The Biochemistry of Viruses* (H.B. Levy ed.), pp. 103-176, Marcel Dekker, New York.
37. Jacobson, M.F. and Baltimore, D. (1968), J. Mol. Biol. 33, 369-378.
38. Fernandez-Tomas, C.B. and Baltimore, D. (1973), J. Virol. 12, 1122-1130.
39. Mandel, B. (1971), Virology 44, 554-568.
40. Ghendron, Y., Yacobsen, E. and Mikhenjeva, A. (1972), J. Virol. 10, 261-266.
41. McGregor, S., Hall, L. and Rueckert, R.R. (1975), J. Virol. 15, 1107-1120.
42. Phillips, B.A. and Wiemert, S. (1978), Virology 88, 92-104.
43. Phillips, B.A. (1972), Current Topics in Microbiology and Immunology, 58, 156-174.
44. Phillips, B.A. (1969), Virology, 39, 811-821.
45. McGregor, S. and Ruckert, R.R. (1977), J. Virol. 21, 548-553.
46. La Colla, P., Marongiu, M.E. and Sau, M., in press.
47. Ussery, M.A., Ramirez-Mitchell, R. and Hardesty, B.A. (1976), J. Virol. 17, 453-461.
48. Robertson, D.L., Yau, P., Dobberton, D.C., Sweeney, T.K., Thach, S.S., Brendler, T. and Thach, R.E. (1976), J. Virol. 18, 344-355.
49. Luftig, R.B., Conscience, J.-F., Skoultchi, A., McMillan, P., Revel, M. and Ruddle, F.H. (1977), J. Virol. 23, 799.
50. Yoshinaka, Y., Ishigame, K., Ohno, T., Kageyama, S., Shibata, K. and Luftig, R.B. (1980), Virology 100, 130-140.
51. Tattaersall, P., Cawte, P.J., Shatkin, A.J. and Ward, D.C.,(1976) J. Virol. 20, 273-289.
52. Everitt, E., Meadoer, S.A. and Levine, A.S. (1977), J. Virol. 21, 199-214.
53. Dickson, C., Puma, J.P. and Nandi, S. (1976), J. Virol. 17, 275-282.
54. Famulari, N.G., Buchhagen, D.L., Klenk, H.-D. and Fleissner, E. (1976), J. Virol. 20, 501-508.
55. Rochhammer, J., Nexo, B.A. and Vaughan, H., jr., (1976), Virology, 71, 134-142.
56. Naso, R., Arcement, L.J., Karshin, W.L., Jamjoom, G.A. and Arlinghaus, R.B. (1976), Proc. Natl. Acad. Sci. USA 73, 2326-2330.
57. Shapiro, S.Z., Strand, M. and August, J.T. (1976), J. Mol. Biol. 107, 459-477.
58. Garoff, H. and Schwarz, R.T. (1978), Nature, London, 274, 487-490.
59. Katz, F.N., Rothman, J.E., Knipe, D.M. and Lodish, H.F. (1977), J. Supramolecular Structure 7, 353-370.

60. Leavitt, R., Schlesinger, S. and Kornfeld, S. (1977), J. Virol. 21, 375-385.

61. Summers, D.F., Maizel, J.V. and Darnell, D.E. (1965), Proc. Natl. Acad. Sci. USA, 54, 505.

62. Summers, D.C. and Maizel, J.V. (1968), Proc. Natl. Acad. Sci. USA, 59, 966-971.

63. Jacobson, M.F. and Baltimore, D. (1968), Proc. Natl. Acad. Sci. USA, 61, 77-84.

64. Holland, J.J. and Kiehn, E.D. (1968), Proc. Natl. Acad. Sci. USA, 60, 1015-1022.

65. Jacobson, M.F., Asso, J. and Baltimore, D. (1970), J. Mol. Biol. 49, 657-669.

66. Garfinkle, B.D. and Tershak, D.R. (1971), J. Mol. Biol. 59, 537-541.

67. Butterworth, B.E., Hall, L., Stoltzfus, C.M. and Rueckert, R.R., (1971), Proc. Natl. Acad. Sci. USA, 68, 3083-3087.

68. Korant, B.D. (1972), J. Virol. 10, 751-759.

69. Dobos, P. and Martin, E.M. (1972), J. Gen. Virol. 17, 197-212.

70. Ginevskaya, V.A., Scarlat, I.V., Kalinina, N.O. and Agol, V.I. (1972), Arch. Ges. Virusforsch. 39, 98-107.

71. Korant, B.D. (1973), J. Virol. 12, 556-563.

72. Butterworth, B.E. and Korant, B.D. (1974), J. Virol. 14, 282-291.

73. Lucas-Lenard, J. (1974), J. Virol. 14, 261-269.

74. Korant, B.D., Kauer, J.C. and Butterworth, B.E. (1974), Nature, London, 248, 588-590.

75. Korant, B.D. (1975), in *In vitro Transcription and Translation of Viral Genomes* (A.L. Haenni, and G. Beaud, eds.), pp. 273-279, INSERM, Paris.

76. Lawrance, C. and Thach, R.E. (1975), J. Virol. 15, 918-928.

77. Cooper, P.D., Geissler, D., Scotti, P.D. and Tannock, G.A., (1971), in *Ciba Symposium: The Strategy of the Viral Genome* (Wolstenholm and O'Connor eds.), pp. 75-100, Churchill Livingston, London.

78. Nakai, K. and Lucas-Lenard, J. (1976), J. Virol. 18, 918-925.

79. Korant, B., Chow, N., Lively, M. and Powers, J. (1979), Proc. Natl. Acad. Sci. USA, 76, 2992-2995.

80. Palmenberg, A.C., Pallansch, M.A. and Rueckert, R.R. (1979), J. Virol. 32, 770-778.

81. Paucha, E., Seehafer, J. and Colter, J.S. (1974), Virology, 61, 315-326.

82. McLean, C., Matthews, T.J. and Rueckert, R.R. (1976), J. Virol. 19, 903-914.

83. Cooper, P.D. and Bennett, D.J. (1973), J. Gen. Virol. 20, 151-160.

84. Korant, B.D. (1977), Acta Biol. Med. Germ. 36, 1565-1573.

85. Summers, D.F., Shaw, E.N., Stewart, M.L. and Maizel, J.V. (1972), J. Virol. 10, 880-884.

86. Vogt, V.M. and Eisenman, R. (1973), Proc. Natl. Acad. Sci. USA, 70, 1734-1738.

87. Moennig, V., Frank, H., Hunsmann, G., Schneider, I. and Schäfer, W. (1974), Virology, 61, 100-111.
88. Halpern, M.S., Bolognesi, D.P. and Lewandowski; L.J. (1974), Proc. Natl. Acad. Sci. USA, 71, 2342-2346.
89. Ihle, J.N., Hanna, M.G., jr., (1975), Virology, 63, 60-67.
90. Jamjoom, G., Karshin, W.L., Naso, R.B., Argcment, L.J. and Arlinghaus, R.B. (1975), Virology, 68, 135-145.
91. Ikeda, H., Hardy, W., jr., Tress, E. and Fleissner, E. (1975), J. Virol. 16, 53-61.
92. Stephenson, J.R., Tronick, S.R. and Aaronson, S.A. (1975), Cell, 6, 543-548.
93. Van Zaane, D., Gielkins, A.L.J., Dekker-Michielson, M.J.A. and Bloemers, H.P.J. (1975), Virology, 67, 544-552.
94. Eisenman, R., Vogt, V.M. and Diggelman, H. (1975), Cold Spring Harbor Symp. Quant. Biol. 39, 1067-1075.
95. Barbacid, M., Stephenson, J.R. and Aaronson, S.A. (1976), Nature, London, 262, 554-559.
96. Van Zaane, D., Dekker-Michielsen, M.J.A. and Bloemers, H.P.J. (1976), Virology, 75, 113-129.
97. Krantz, M.J., Strand, M. and August, J.T. (1977), J. Virol. 22, 804-81ç.
98. Jamjoom, G.A., Naso, R.B. and Arlinghaus, R.B. (1977), Virology 78, 11-34.
99. Evans, L.H., Dresler, S. and Kabat, D. (1977), J. Virol. 24, 865-874.
100. v.d. Helm, K. (1977), Proc. Natl. Acad. Sci. USA, 74, 911-815.
101. Racevskis, J. and Koch, G. (1977), J. Virol. 21, 328-337.
102. Racevskis, J. and Koch, G. (1978), Virology, 87, 354.
103. Sherr, C.J., Sen, A., Todaro, G.J., Sliski, A. and Essex, M. (1978), Proc. Natl. Acad. Sci. USA, 75, 1505-1509.
104. Kawakami, T.G., Huff, S.D., Buckley, P.M., Dungworth, D.C. and Snyder, J.P. (1972), Nature, London, New Biol. 235, 170-171
105. Steeves, R.A., Strand, M. and August, J.T. (1974), J. Virol. 14, 187-189.
106. Strand, M. and August, J.T. (1974), J. Virol. 13, 171-180.
107. Bolognesi, D.P., Collins, J.J., Leis, J.P., Moennig, V., Schaefer, W. and Atkinson, P.H. (1975), J. Virol. 16, 1453-1463.
108. Pinter, A. and Fleissner, E. (1977), Virology, 83, 417-422.
109. Racevskis, J. and Sarkar, N.H. (1978), J. Virol. 25, 374-383.
110. Beemon, K. and Hunter, T. (1978), J. Virol. 28, 551-566.
111. Beemon, K. and Hunter, T. (1977), Proc. Natl. Acad. Sci. USA, 74, 3302-3306.
112. v.d. Helm, K. and Duesberg, P.H. (1975), Proc. Natl. Acad. Sci. USA, 72, 614-618.
113. Papkoff, J., Hunter, T. and Beemon, K. (1980), Virology, 101, 91-103.
114. Yoshida, M. and Toyoshima, K. (1980), Virology, 100, 484-487.
115. Lai, M.M.C., Neil, J.C. and Vogt, P.K. (1980), Virology, 100, 475-483.
116. Warnecke, G., Bilello, J.A., Koch, G., Weber, C. (1981), in press.

117. Homma, M. (1971), J. Virol. 8, 619-629.

118. Lazarowitz, S.G., Goldberg, A.R. and Chopping, P.W. (1973), Virology, 56, 172-180.

119. Scheid, A. and Choppin, P.W. (1974), Virology, 57, 475-490.

120. Ohuchi, M. and Homma, M. (1976), Virol. 18, 1147-1150.

121. Yoshinaka, Y. and Luftig, R.B. (1977b), Biochem. Biophys. Res. Commun. 76, 54-63.

122. Saborio, J.L., Pong, S.S. and Koch, G. (1974), J. Mol. Biol. 85, 194-211.

123. Cooper, P.D. and Bennett, D.J. (1973), J. Gen. Virol. 20, 151-160.

124. Taber, R., Rekosh, D. and Baltimore, D. (1971), J. Virol. 8, 395-401.

125. Summers, D.F. and Maizel, J.V. (1971), Proc. Natl. Acad. Sci. USA, 68, 2852-2856.

126. Rekosh, D. (1972), J. Virol. 9, 479-487.

127. Dubos, P. and Plourde, J.Y. (1973), Eur. J. Biochem. 39, 463-469.

128. Abraham, G. and Copper, P.D. (1975), J. Gen. Virol. 29, 199-213.

129. Beckman, L.D., Caliguiri, L.A. and Lilly, L.S. (1976), Virology 73, 216-227.

130. Blobel, G. and Sabatini, D.D. (1971), Biomembranes, 2, 193-195.

131. Jackson, R.C. and Blobel, G. (1977), Proc. Natl. Acad. Sci. USA, 74, 5598-5602.

132. Spielman, L.L. and Bancroft, F.C. (1977), Endocrinology, 101, 651-658.

133. Caliguiri, L.A. and Tamm, I. (1969), Science, 166, 885-886.

134. Caliguiri, L.A. and Tamm, I. (1970), Virology, 42, 100-122.

135. Girard, M., Baltimore, D. and Darnell, J.E. (1967), J. Mol. Biol. 24, 59-74.

136. Shatkin, A.J. (1976), Cell, 9, 645-653.

137. Nomoto, A., Lee, Y.J. and Wimmer, E. (1976), Proc. Nat. Acad. Sci. USA, 73, 375-380.

138. Hewlett, M.J., Rose, J.K. and Baltimore, D. (1976), Proc. Natl. Acad. Sci., USA, 73, 327-330.

139. Lee, Y.F., Nomoto, A., Detjen, B.M. and Wimmer, E. (1977), Proc. Natl. Acad. Sci. USA, 74, 59-63.

140. Flanegan, J.B., Pettersson, R.F., Ambros, V., Heweltt, M.J. and Baltimore, D. (1977), Proc. Natl. Acad. Sci. USA, 74, 961-965.

141. Nomoto, A., Detjen, B., Pozzatti, R. and Wimmer, E. (1977a), Nature, London, 268, 208-213.

142. Fernandez-Munoz, R. and Lavi, U. (1977), J. Virol. 21, 820-824.

143. Kitamura, N., Adler, C;, Martinko, J., Nathenson, S.G. and Wimmer E. (1980), Cell, 21, 295-302.

144. Pallansch, M.A., Kew, O.M., Palmenberg, A.C., Golini, F., Wimmer, E. and Rueckert, R.R., (1980), J. Virol. 35, 414-419.

145. Cords, C.E. and Holland, J.J. (1964), Proc. Natl. Acad. Sci. USA, 51, 1080-1082.

146. Ikegami, N., Eggers, H.J. and Tamm, I. (1964), Proc. Natl. Acad. Sci. USA, 52, 1419-1426.

147. Tershak, D.R., Yin, F.H. and Korant, B.D. (1981), in *Handbook of Experimental Pharmacology* (L. Caliguiri, P. Came, eds.), Springer Verlag, in press.

148. Oppermann, H. and Koch, G. (1976), Arch. Virol. 52, 123.

149. Nuss, D.L., Oppermann, H. and Koch, G. (1975), Proc. Natl. Acad. Sci. USA, 72, 1258-1262.

150. Koch, G., Bilello, J.A., Kruppa, J., Koch, F. and Oppermann, H. (1980), Ann. N.Y. Acad. Sci., 339, 280-306.

151. Eggen, K.L. and Shatkin, A.J. (1972), J. Virol. 9, 636-645.

152. Svitkin, Y.V., Ugarrova, T.Y., Ginevskaya, V.A., Kalinina, N.O., Scarlat, I.V. and Agol, V.I. (1974), Intervirology, 4, 214-220.

153. Esteban, M. and Kerr, I.M. (1974), Eur. J. Biochem. 45, 567-576.

154. Villa-Komaroff, L., Guttman, N., Baltimore, D. and Lodish, H.F. (1975), Proc. Natl. Acad. Sci. USA, 72, 4157-4164.

155. Pelham, H.R.B. (1978), Eur. J. Biochem. 74, 457-462.

156. Hackett, P.B., Egberts, E. and Traub, P. (1978), Eur. J. Biochem. 83, 353-361.

157. Shih, D.S., Shih, C.T., Kew, O., Pallansch, M., Rueckert, R.R. and Kaesberg, P. (1978), Proc. Natl. Acad. Sci. USA, 75, 5807-5811.

158. Shih, D.S., Shih, C.T., Zimmern, D., Rueckert, R.R. and Kaesberg P. (1979), J. Virol. 30, 472-480.

159. Svitken, Y.V. and Agol, V.I. (1978), FEBS Lett. 87, 7-11.

160. Papkoff, J., Hunter, T. and Beemon, K. (1980), Virology, 101, 91-103.

161. Kessler, S.W. (1977), J. Immunol. 15, 1617-1624.

162. Philipson, L., Andersson, P., Olshevsky, U., Weinberg, R., Baltimore, D. and Gesteland, R. (1978), Cell, 13, 189-199.

163. Oppermann, H., Bishop, J.M., Varmus, H.E., Levintow, L. (1977), Cell, 10, 993-1005.

164. Friend, C., Scher, W., Holland, J.G. and Sato, T., Proc. Natl. Acad. Sci. USA, 63, 378-382.

165. Ostertag, W., Melderis, H., Steinheider, G., Kluge, N. and Dube S. (1972), Nature, London, New Biol. 239, 231-234.

166. Ostertag, W., Cole, T., Crozier, T., Gaedicke, G., Kind, J., Kluge, N., Krieg, J.C., Roesler, G., Steinheider, B., Weimann, J. and Dube, S.K. (1974), in *Differentiation and Control of Malignancy in Tumor Cells*, pp. 493-520, University of Tokyo Press, Tokyo.

167. Marks, P.A. and Rifkind, R.A. (1978), Ann. Rev. Biochem. 47, 419.

168. Koch, G., Warnecke, G., Kühne, J. and Bilello, J.A. (1979), Adv. Ophthal. 38, 222-233.

169. Sassa, S. (1976), J. Exptl. Med. 143, 305.

170. Eisen, H., Bach, R. and Emry, R. (1977), Proc. Natl. Acad. Sci. USA, 74, 3898-3902.

171. Dube, S.K., Pragnell, I.B., Kluge, N., Gaedicke, G., Steinheider G. and Ostertag, W. (1975), Proc. Natl. Acad. Sci. USA, 72, 1863-1867.

172. Ostertag, W., Crozier, T., Kluge, N., Melderis., H. and Dube, S. (1973), Nature, London, New Biol. <u>243</u>, 203-205.
173. Ostertag, W., Roesler, G., Krieg, C.J., King, J., Cole, T., Crozier, T., Gaedicke, G., Steinheider, G., Kluge, N. and Dube, S. (1974), Proc. Natl. Acad. Sci. USA, <u>71</u>, 4980-4985.
174. Leder, A. and Leder, P. (1975), Cell, <u>5</u>, 319-322.
175. Reuben, R.C., Wife, R.L., Breslow, R., Rifkind, R.A. and Marks P.A. (1976), Proc. Natl. Acad. Sci. USA, <u>73</u>, 862-866.
176. Grusella, J.F. and Housman, D. (1976), Cell, <u>8</u>, 263-269.
177. Takahashi, E., Yamada, M., Saito, M., Kuboyama,; M. and Ogasa, K. (1975), Gann, <u>66</u>, 577-580.
178. Sherton, C.C., Evans, L.H., Polonoff, E. and Karat, D. (1976), J. Virol. <u>19</u>, 118-125.
179. Saborio, J.L. and Koch, G. (1973), J. Biol. Chem. <u>248</u>, 8343-8347.
180. Bilello, J.A., Warnecke, G. and Koch, G. (1978), in *Modern Trends in Human Leukemia III* (R. Neth, R.C. Gallo, P.-H. Hofschneider and K. Mannweiler, eds.), pp. 303-306, Springer Verlag, Berlin, Heidelberg, New York.
181. Bilello, J.A., Kühne, J., Warnecke, G. and Koch, G. (1980), in *In Vivo and in Vitro Erythropoiesis: the Friend system* (G.B. Rossi, ed.), pp. 229-238, Elsevier/North Holland Biomedical Press, Amsterdam, N. York.
182. Graf, T., Ade, N. and Beug, H. (1978), Nature, London, <u>275</u>, 496-501.
183. Dresler, S., Ruta, M., Murray, M.J. and Kabat, D. (1979), J. Virol, <u>30</u>, 564-575.
184. Ruscetti, S., Linemeyer, D., Feild, J., Troxler, D. and Scolnick, E. (1978), J. Exptl. Med. <u>148</u>, 654-663.
185. Ruscetti, S.K., Linemeyer, D., Feild, J., Troxler, D. and Scolnick, E. (1979), J. Virol. <u>30</u>, 787-798.
186. Bilello, J.A., Colletta, G., Warnecke, G., Koch, G., Frisby, D., Pragnell, I.B. and Ostertag, W. (1980), Virology, <u>107</u>, 331-344.
187. Bishop, J.M. (1978), Ann. Rev. Biochem. <u>47</u>, 35-88.
188. Koch, G., Weber, C. and Bilello, J.A. (1979), in *Modern Trends in Human Leukemia III* (R. Neth, R.C. Gallo, P.-H. Hofschneider, K. Mannweiler, eds.), pp. 295-301, Springer Verlag, Berlin, Heidelberg, Nez York.
189. Bernstein, A., Boyd, A.S., Crichley, V. and Lamb, V. (1976), in *Biogenesis and Turnover of Membrane Macromolecules* (J.S. Cook, ed.), pp. 145-159 Raven Press, New York.
190. Pérez-Bercoff, R. and Gander, M. (1978), FEBS Lett, <u>96</u>, 306-312.

SECTION V:

INHIBIITON OF PROTEIN BIOSYNTHESIS AT SELECTED LEVELS

ACTION OF INHIBITORS OF PROTEIN BIOSYNTHESIS

David Vazquez, Eulalio Zaera, Humberto Dölz and
Antonio Jiménez

Centro de Biologia Molecular, C S.I.C. and U.A.M.
Canto Blanco, Madrid-34, Spain

INTRODUCTION

For didactic purposes, the process of protein biosynthesis can be arbitrarily divided into: a) events taking place *prior* to translation and b) those that take place at the ribosome level.

Inhibitors of steps taking place *prior* to translation are of little relevance in eukaryotes due mainly to their lack of selective action. This is the case of borrelidin (an inhibitor of treonyl-tRNA synthesis) and furanomycin (an inhibitor of leucyl-tRNA synthesis) which are active on bacterial and mammalian cells. Therefore, in this review we will refer only to inhibitors of the translation process.

Translation inhibitors have been very useful in the last thirty years to elucidate the overall mechanism of translation and to study different individual steps in the process. Although the protein synthesizing machinery of prokaryotic and eukaryotic organisms differ to a considerable extent, there are a number of features common to both systems. Therefore translation inhibitors acting on eukaryotic systems can be broadly classified according to their specificity into those affecting selectively a) both systems i.e.: prokaryotic and eukaryotic-type inhibitors (Table 1) and, b) those active only in eukaryotic systems (Table 2). In general, antibiotics affecting prokaryotic-type systems are active in bacteria, blue-green algae, mitochondria and chloroplasts, whereas those acting selectively on eukaryotic-type systems are active in higher cells, known to have 80S-type cytoplasmic ribosomes.

Table 1.   Inhibitors of Translation Acting on Prokaryotic and
           Eukaryotic Systems

Actinobolin	Guanylyl-imido-diphosphate
Adrenochrome	Nucleocidin
AHR-1811	Pactamycin
Amicetin group:	Polydextran sulphate
Amicetin	Polyvinyl sulphate
Bamicetin	Puromycin
Plicacetin	Pyrochatechol violet
Anthelmycin	Showdomycin
Aurintricarboxylic acid	Sparsomycin
Blasticidin S	Tetracycline group:
Chartreusin	Chlortetracycline
Deoxystreptamine-containing	
antibiotics	Doxycycline
Edeine A$_1$	Oxytetracycline
Fusidic Acid	Tetracycline
Gougerotin	Tosylphenylalanylchloromethane
Guanylyl-methylene-diphosphate	

This study will be concerned with the present state of
the problem of translation inhibitors active on *eukaryotic* systems
and our own recent unpublished observations in this field.   An
extensive survey of the literature will not be possible in this
contribution and other reviews (1-7) can be useful to readers interested in some specific aspects of the problem in a more comprehensive
manner.

TRANSLATION OF mRNA

The overall reactions taking place in the process of translation
by eukaryotic ribosomes can be represented schematically according
to the translocation model with one or two entry sites on the small
subunit.   Following this model, the process of translation can be
arbitrarily divided into the phases of *initiation, elongation* and
termination (Fig. 1).   The most recent developments concerning the
mechanism of translation in eukaryotes and the action of inhibitors
have been extensively reviewed recently (refs. 6-9 and different
contributions in this volume).

INHIBITORS OF INITIATION

The process of initiation of the polypeptide chain involves
four sequential steps required for the attachment of the initiator
substrate Met-tRNA$_f$ to the correct position on the 80S ribosome.
The steps that we will consider are: (A) the recognition of eIF-3
by the 40S subunit and binding of the eIF-2.Met-tRNA$_f$.GTP ternary
complex;   (B) recognition of eIF-1, eIF-4A, eIF-4B, eIF-4C and the

Table 2. Inhibitors of Translation Acting on Eukaryotic Systems

Abrin	Narciclasine group:
Alpha sarcin	Haemanthamine
Anisomycin	Narciclasine
5-azacytidine	Pretazettine
Bruceantin	PAP
Crotins	PA toxin
Curcins	Pederine
Diphteria toxin	Phenomycin
Emetine group:	PR toxin
Emetine	Ricin
Tubolosine	Sodium fluoride
Enomycin	Tenuazonic acid
Glutarimide group:	Trichotecene antibiotics:
Actiphenol	Trichodermin group:
Cycloheximide	Fusarenon-X
Streptimidone	Trichodermin
Streptovitacin A	Trichodermol
Harringtonine group:	Trichothecin
Harringtonine	Verrucarin A group:
Homoharringtonine	Deacetoxyscirpenol
Isoharringtonine	Nivalenol
Holacanthone	Toxin T-2
Lycorine group:	Verrucarin A
Lycorine	Tylophora alkaloids:
Pseudolycorine	Cryptopleurine
MDMP	Tylocrebrine
Modeccin	Tylophorine

mRNA initiation triplet; (C_1) attachment of the 60S ribosomal sub-unit to the initiation complex formed in step (B) and, (C_2) inter-action of the 3' end of Met-tRNA$_f$ with the donor site of the peptidyl transferase center (Fig. 1).

Inhibitors of Recognition of the Initiator Substrate (step A)

Step A in eukaryotes involves the following reactions: inter-action of the 40S subunit with eIF-3, formation of the eIF-2.Met-tRNA$_f$.GTP ternary complex and interaction of this complex with the 40S subunit with eIF-3 bound.

There is ample experimental evidence showing that kasugamycin blocks the interaction of the initiator substrate fMet-tRNA$_f$ with the 30S ribosomal subunit in *bacterial* systems. Similar studies have not been performed in eukaryotic systems, although the anti-biotic might be expected to act similarly in bacterial and euka-ryotic systems (refs. 5, 6 and references therein).

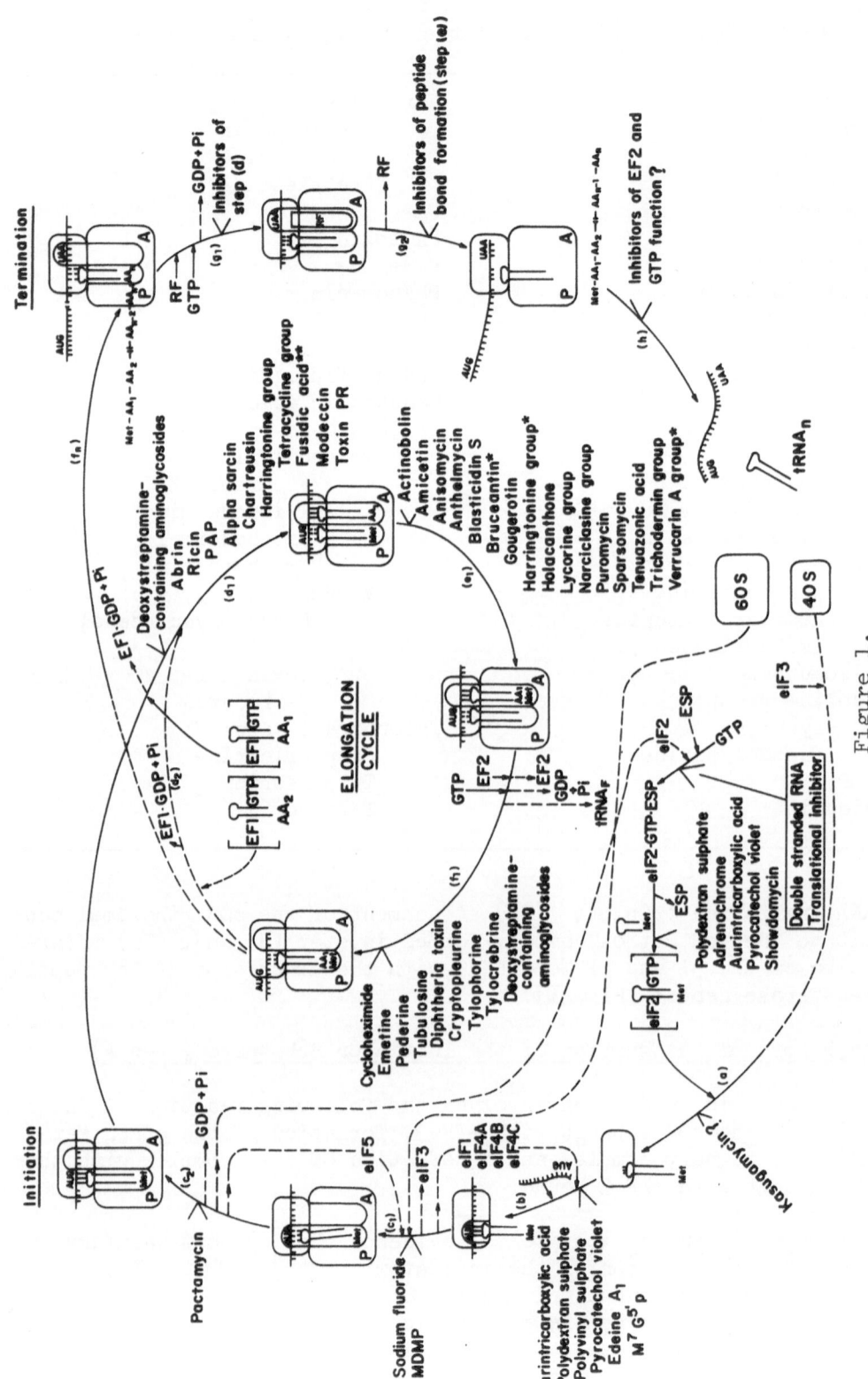

Figure 1.

Showdomycin (an analogue of N-ethyl-maleimide), adrenochrome
and triphenylmethane dyes (such as aurintricarboxylic acid, pyro-
chatechol violet, azure blue B, fuchsin acidic and fuchsin basic)
inhibit the formation of the eIF-2-Met.tRNA$_f$.GTP ternary complex in
cell-free systems (refs. 5, 6 and references therein). However these
drugs are not very useful in studies with intact cells. Indeed
there is a permeability barrier for the triphenylmethane dyes and
showdomycin interferes with cellular process others than protein
synthesis (reviewed in 10).

Step A is also the target of the haem-controlled repressor and
that of the double-stranded RNA-activated inhibitor as well. The
effects of interferon on translation seem to be quite more complex,
though there is evidence pointing to step A being affected by interferon treatment. The mechanism of action of these translational inhibitors is reviewed in depth elsewhere in this volume.

Inhibitors of mRNA Recognition (step B)

Within this group of inhibitors we can include a number of
triphenylmethane derivatives (the best known of which is aurintricarboxylic acid), some aminochromes (the best known of which is
adrenochrome), poly dextran sulphate and polyvinyl sulphate. All
these compounds have a rather complex mode of action involving more
than one interaction and inhibiting more than one step. Indeed all
of them appear to interact with the 40S ribosomal subunit and therefore prevent mRNA recognition in cell-free systems, but they are
hardly used in intact cells due to the permeability barrier. As
indicated above these compounds also inhibit recognition of the
initiator (step A) by preventing the formation of the ternary complex
eIF-2.Met-tRNA$_f$.GTP in a reaction in which the ribosome is not
involved (5, 6 reviews).

Edeine A$_1$ is the best known member of the edeine complex of
related antibiotics. In intact bacteria edeine A$_1$ at low concentrations inhibits reversibly DNA synthesis but in cell-free systems
the antibiotic is also an effective inhibitor of initiation of
translation, by interacting with the smaller ribosomal subunit (6
and references therein). The initial finding that edeine A$_1$ spe-

---

Fig. 1. Translation process in eukaryotic cells. Site of action of
translation inhibitors[*]. Do not interact with polysomes.
Therefore bind to free ribosomes and ribosome subunits and
prevent only the first few rounds of peptide elongation[**].
Is an inhibitor of aminoacyl-tRNA binding in intact cells
or in integrated systems in which elongation is proceeding
in the presence of EF-1 and EF-2. Does not inhibit aminoacyl-tRNA binding in resolved systems in the absence of
EF-2. Can inhibit translocation in cell-free systems.

cifically inhibits the formation of the functional eIF-2.Met-tRNA$_f$.
GTP.40S complex (11) is supported by experimental evidence with other
systems. Indeed the antibiotic interferes with the shift of labelled
mRNA from the 40S to the 80S in a wheat germ system (12) and in a
rabbit reticulocyte system (13) and interferes with the AUG recognition of reovirus mRNA (14). Furthermore a direct inhibition of
step B (Fig. 1) by edeine A$_1$ has recently been confirmed also in
other systems (15, 16). Other experimental evidence suggests that
edeine A$_1$ freezes the 40S initiation complex by preventing release
of factors and therefore blocking step C$_1$ (68).

The compound m^{7}G$^{5'}$p(7-methyl-guanosine-5'-monophosphate) and
some other analogs have been found to block translation of mRNA from
different systems where mRNA is provided with the sequence m^{7}G^{5}ppp...
("cap") at the 5' end (6, review). It has been shown, however, that
the translation of some *uncapped* mRNAs is also inhibited *in vitro*
by cap analogues (17, 18). Indeed it appears that, although the
translation of encephalomyocardites (EMC) virus RNA (an *uncapped*
mRNA) is inhibited by m^{7}Gp and m^{7}Gppp, the inhibitory effects of
these cap analogues are stronger in the translation of a capped
messenger such as globin mRNA (19). Furthermore the same preferential
inhibitory effects were observed with inhibitors of mRNA binding to
ribosomes such as aurintricarboxylic acid but not with inhibitors of
the later steps in initiation of protein synthesis such as pactamycin
(19). Therefore, it was suggested that the differences in the inhibition by cap analogues between capped or uncapped mRNAs are due
only to the different affinities of mRNA for the components of the
initiation machinery independent of the presence of cap in the 5'
terminus; it was further observed that 2'-O-mGppp cannot be considered
only as a cap analogue because it affects some step prior to mRNA
binding to the ribosome (19).

Inhibitors of Subunit Joining (step C$_1$)

Fluoride Na$^+$ and K$^+$ salts and the D-stereoisomer of 2-(4-methyl-
2,6-dinitroanilino)-N-methyl-propionamide (D-MDMP) have been proposed
as inhibitors of the joining of the 60S ribosomal subunit to the
initiation complex formed on the 40S ribosomal subunit (6, 69 and
references therein).

Inhibitors of the Positioning of the Initiator in the Donor Site (step C$_2$)

The antibiotic pactamycin at low concentrations acts on the
small ribosomal subunit of both prokaryotic and eukaryotic ribosomes
and therefore prevents *initiation* of the polypeptide chain in both
systems by blocking the step C$_2$ (Fig. 1). At higher concentrations
the antibiotic also interacts with the larger ribosomal subunit and
therefore has a secondary effect on *elongation* (5, 6 and references

therein). The major inhibitory effect of pactamycin on step C_2 has been confirmed recently in rabbit reticulocyte cell-free systems using native 40S ribosomal subunits (34).

Unclassified Inhibitors of Initiation

The base analog 5-azacytidine primarily blocks translation in several eukaryotic systems at the level of initiation of the polypeptide chain. Since the drug is incorporated into RNA and does not inhibit initiation of translation in cell-free systems, it has been postulated that 5-azacytidine might be incorporated in a new RNA species, not yet described, somehow involved in the process of initiation (6 and 20 references therein).

The infection of an animal cell by a cytolytic virus results in the specific inhibition of host cell protein synthesis (the shut-off phenomenon reviewed by G. Koch in this volume). The inhibition of translation after viral infection is exerted at some step on the initiation level. It has been suggested that viral infection interferes with some membrane function, thus altering the ionic concentration inside the cell and that this ionic concentration might differentially affect viral and cellular protein synthesis (21). In support of this suggestion it has been shown that viral and mRNA have different optima of monovalent ions for translation (22-24). Moreover the ionic concentration in the cell does drastically change after viral infection (22, 25, 26, 27) and these changes are correlated with the shut-off of host protein synthesis.

Some ionophore antibiotics such as valinomycin, nigericin and A23187 are potent inhibitors of protein synthesis in intact cells (6 and references therein). Similarly as occurs after viral infection the inhibition of translation by ionophores is a membrane mediate process in which the ionic unbalance in the cell cytoplasm causes the inhibition. This is reinforced by the finding that increased concentrations of potassium ions in the culture medium reverse the inhibition exerted by nigericin on yeast cells (28).

Hypertonicity resulting from an increased NaCl concentration in the growth medium causes an inhibition of protein synthesis in cell cultures. It was postulated that hypertonicity lead to an inhibition of peptide chain initiation since it causes polysome breakdown. It was further observed in different virus-cell systems that cellular protein synthesis is more sensitive to hypertonic media than viral protein synthesis. This preferential inhibitory effect of hypertonicity might be due simply to a change in ionic concentration, since it has been shown in cell-free systems that sodium ions inhibit cellular protein synthesis and enhance viral protein synthesis (6 and references therein). Inhibition of protein synthesis due to hypertonicity is considered more extensively in G. Koch's contribution in this volume).

When exponentially growing cultures of bacteria and eukaryotic
cells are cooled to below 8°C, (temperature shift-down) initiation
of protein synthesis is blocked. The compounds 3-methyleneoxindole
and 2'(3'), 5'-ADP have also been proposed as inhibitors of polypeptide chain initiation (6 and references therein).

INHIBITORS OF ELONGATION

Polypeptide elongation takes place in eukaryotic systems by
repeated cycles composed of three sequential steps: EF-1-dependent
aminoacyl-tRNA binding (step D), ribosome catalyzed peptide bond
formation (step E) and EF-2-dependent translocation (step F).
Besides the initiation complex formed in step C_2, two molecules of
GTP, the specific aminoacyl-tRNA and the elongation factor EF-2 are
required in the first elongation cycle.

There is ample experimental evidence supporting a single entry
site in intact cells for GTP-and EF-1-dependent binding of aminoacyl-tRNA on the A-site of the ribosome. Non-enzymic binding of
aminoacyl-tRNA to the A-site can also be studied in cell-free
systems in the absence of EF-1.

Peptide bond formation (step 2) is catalyzed by the peptidyl
transferase center, which is an integral part of the 60S ribosomal
subunit.

Translocation (step F) requires elongation factor EF-2 and
involves movement of the peptidyl-tRNA from the A- to the P-site,
coupled to the movement of one triplet of the mRNA on the ribosome.
Hydrolysis of one molecule of GTP is required for the release of
EF-2. Spontaneous non-enzymic translocation takes place slowly in
cell-free systems in the absence of elongation factors and GTP. This
non-enzymic translocation is considerably accelerated in the presence
of high concentrations of K^+ or NH_4^+.

The two elongation factors (EF-1 and EF-2) complexed with GTP
appear to interact with a common or overlapping area of the larger
ribosomal subunit and therefore do not interact simultaneously but
only alternately with the ribosome. Thus the interaction of one of
the elongation factors takes place after the release of the other.

Compounds Interfering with Aminoacyl-tRNA Recognition (step D)

Within this important category of inhibitors we can distinguish
at least two groups of compounds. One of these groups includes the
inhibitors of EF-1-dependent binding of aminoacyl-tRNA and the other
one clusters the misreading compounds that may prevent protein
synthesis by favoring the erroneous interaction of aminoacyl-tRNA
in the acceptor site of the ribosome.

a)   Inhibitors of EF-1-dependent binding of aminoacyl-tRNA. The
antibiotics of the tetracyline group are strong inhibitors in cell-
free systems of EF-1-dependent binding of aminoacyl-tRNA. These
antibiotics appear to have a number of binding sites with both ribosomal subunits, at least in bacteria, but their interaction with the
smaller ribosomal subunit appear to be more relevant for their mode
of action (6 and references therein). Activities of tetracylines
in eukaryotic cells is very small compared with their activities in
bacteria. This is so because only bacteria has an active transport
for these antibiotics and they are concentrated in their cytoplasm.
A number of bacterial mutants resistant to tetracyclines have been
found to have a modification in one of the proteins of the outer
membrane (29-31).

       Fusidic acid can inhibit the step of translocation in cell-free
systems when added in a large excess or alternatively a very small
amount of EF-2 is added to the system. This is because a GTP.ribo-
some.fusidic acid.EF-2 complex is formed and therefore the antibiotic
can sequester all EF-2 available, but the effect of fusidic acid in
such systems is abolished by saturating concentrations of EF-2.
However in intact cells and integrated systems for translation the
antibiotic is really an inhibitor of EF-1-dependent aminoacyl-tRNA
binding since this reaction cannot take place with the ribosome.
fusidic acid.EF-2.GTP complex. Indeed this complex is rather stable
and the GTP.EF-1.aminoacyl-tRNA complex cannot interact with its
ribosomal site which is overlapping with the site occupied by EF-2.
GTP (6 and references therein).

       There are an increasing number of protein and glycoprotein
toxins that are known to block the elongation cycle in translation
by catalytic inactivation of the 60S subunit of the eukaryotic ribosome (Table 3). At least two of these compounds ricin and abrin,
and probably some others are of glycoprotein nature and consist of
two subunits A and B.

       Subunit B is involved on the interaction with the cell membrane
and facilitates the entrance of subunit A that inactivates the ribosome. Three of the compounds (alpha sarcin, mitogillin and restrictocin) have been shown to cleave 320 nucleotides from the 3' terminal
end of the larger ribosomal RNA from the 60S subunit. In all cases
the catalytic inactivation of the 60S ribosomal subunit results in
a decrease in the affinity of elongation factors EF-1 and EF-2 for
the ribosome. Therefore an inhibition of aminoacyl-tRNA binding and
(or) translocation by these toxins has been reported (Table 3). Inhibition of translocation is not observed in all cases since in the
case of some toxins or under certain experimental conditions the
pretranslocated peptidyl-tRNA bound to the A-site prevents the
interaction of the EF-2.GTP complex (6 and references therein).
Alpha sarcin cross-reacts with mitogillin and restrictocin and the
three toxins cause the cleavage of the larger RNA in polysomes (32).

Table 3. The Effects of Glycoprotein and Protein Toxins Acting on the Eukaryotic Ribosome on Model Reactions (data are taken from references 6, 32, 35-39)

	EF-1-dependent		EF-2-dependent			Non-enzymic translocation
	Binding of aminoacyl-tRNA	GTP hydrolysis	Formation of the EF-2-GTP-ribosome complex	Uncoupled GTP Hydrolysis	Translocation	
Ricin A-chain	Inhibition	Inhibition	Inhibition	No effect or inhibition	No effect or inhibition or stimulation	No effect
Abrin A-chain	Inhibition	Inhibition	Inhibition	No effect or inhibition	No effect or stimulation	No effect
Alpha sarcin	Inhibition	Inhibition	Inhibition	Inhibition	No effect	---
Mitogillin	---	---	---	---	---	---
Restrictocin	---	---	---	---	---	---
PAP	Inhibition	Inhibition	Inhibition	No effect or stimulation	No effect or stimulation	---
Crotin II	Inhibition	Inhibition	inhibition	No effect	No effect or inhibition	---
Modeccin	Inhibition or no effect	---	Inhibition	---	No effect or inhibition	---
Enomycin	Inhibition	---	No effect	Inhibition	No effect	---
PR toxin	?	---	?	---	---	---

Permeability mutants of abrin or ricin show cross resistance to both
drugs (33). Identity of ricin and *Ricinus communis* agglutinin has
been reported (34). However there are significant differences between
the different toxins presented in Table 3. Not all of them are of
glycoprotein nature or have two subunits. Furthermore some of them
do not inhibit translocation since they cannot interact with ribosomes
when peptidyl-tRNA is bound to the A-site (6, 38; reviews). On the
other hand some others like modeccin have been reported to inhibit
translocation (37) although it has also been observed that the
presence of EF-2 strongly increase the rate of ribosome inactivation
by modeccin (35). Moreover ribosomes might not be the *only* target
of modeccin, and some other toxins of this group, in intact cells
(39). This type of toxins appear to be widely distributed in nature
(6, 40).

The antibiotic chartreusin, the mucopolysaccharide heparin, and
a number of catechols and aminochromes, including adrenochrome, have
been also reported as inhibitors of EF-2-dependent binding of aminoacyl-tRNA to 80S ribosomes (5, 6 and references therein).

b)   Misreading compounds. The high fidelity of translation is a
well established feature of the eukaryotic ribosomes: In fact, no
significant misreading was detected when mammalian cell-free systems
were incubated under a wide variety of conditions (including the
presence of polyamines, streptamine-containing aminoglycosides, some
organic solvents, increased concentrations of Mg^{++} ions, and suboptimal temperatures of incubation) known to induce misreading in
bacterial cell-free systems (6 and references therein). However an
increasing number of aminoglycoside antibiotics that contain the
deosystreptamine or some deoxystreptamine-derived moiety cause
misreading in bacterial and eukaryotic cells and cell-free systems.
Since these antibiotics also cause an inhibition of *translocation*
we will discuss their effects when dealing with inhibitors of this
step.

Inhibitors of Peptide Bond Formation (step E)

Puromycin is the most universally used and best known inhibitor
of peptide bond formation. The antibiotic is well known to act as
an analogue of the 3'-end of aminoacyl-tRNA (aminoacyl-adenosine)
at the acceptor site of the peptidyl transferase center on the larger
ribosomal subunit of both prokaryotic and eukaryotic ribosomes.
Hence, the use of puromycin provides a simplified method for the
study of peptide bond formation in a reaction in which the $-NH_2$
group of puromycin becomes linked to the C-terminal end of the
relevant amino acid when either f-Met-tRNA$_f$ or peptidyl-tRNA ("puromycin reaction") or f-Met- or Ac-Phe- or Ac-Leu-oligonucleotide
("fragment reaction") are used as donor substrates. Studies concerning structure-activity of puromycin as an acceptor substrate
are well documented but the site of interaction of the antibiotic

on the 60S ribosomal subunit is not yet well resolved. Thus the
treatment of rat liver ribosomes with iodoacetyl-puromycin leads to
labelling of proteins L28 and L29 whereas other studies suggest that
proteins L21, L26, L31 (but neither L28 and L29) are involved in the
peptidyl transferase center that from the functional point of view
is supposed to be the site of action of puromycin (for a comprehensive
discussion see chapter 2, and ref. 6).

All the other inhibitors of peptide bond formation act like
puromycin on the 60S ribosomal subunit. They can be widely classified
in two groups including in one of them those compounds that are active
in both prokaryotic and eukaryotic ribosomes and in the second group
those inhibitors that are active only in eukaryotic ribosomes. Inhibitors of the first group (amicetin, gougerotin, blasticidin S,
actinobolin, anthelmycin and sparsomycin) act on the acceptor site
of the peptidyl transferase center. They are mutually exclusive in
their binding site(s) to the ribosome, and also prevent puromycin
interaction but, unlike this antibiotic, do not have acceptor activity
in peptide bond formation (6, 41). Concerning the very wide group of
peptide bond formation inhibitors that selectively act on eukaryotic
ribosomes they differ in their binding site, mode of action, inhibitory spectra and cross-resistance. Thus anisomycin, tenuazonic acid,
bruceantin and inhibitors of the narciclasine, harringtonine, trichodermin, verrucarin A and lycorine groups (Table 2) are mutually
exclusive in their binding sites. However bruceantin, harringtonine
and verrucarin A compounds only block the first few rounds of peptide
bond formation prior to polysome formation, compounds of the lycorine
group do not show cross-resistance with trichodermin, and tenuazonic
acid is *inactive* on yeast ribosomes (6 and references therein).
Holacanthone is chemically related to bruceantin (42) and possess high
antitumor activity. Similarly to bruceantin, holacanthone inhibits
protein synthesis by blocking specifically peptide bond formation.
Polyphenylalanine synthesis (7×10^{-5}), polypetide synthesis (10^{-4})
and the fragment reaction (3×10^{-6}) are inhibited by the compound
(figures in brackets indicate final M concentrations of holacanthone
giving 50% inhibition). 80S ribosomes from a yeast mutant isolated
for resistance to narciclasine (Nar 2b) are also resistant to anthelmycin and inhibitors of the trichodermin and narciclasine groups but
are sensitive to gougerotin and blasticidin S (members of the 4-
aminohexoxyl-cytosine group of antibiotics like anthelmycin), (43).
This pleitropic phenotype is expressed by a single nuclear mutation.
Therefore anthelmycin, which differs chemically from gougerotin and
blasticidin S in lacking an aminoacyl moiety, must have a common or
overlapping site of action to the one for narciclasine and trichodermin antibiotics, but not closely related to the site(s) of
action of gougerotin and blasticidin S. Neither gougerotin nor
blasticidin S nor anthelmycin affect the binding of either (^{14}C) trichodermin, (^{3}H) narciclasine or (^{3}H)anisomycin to eukaryotic ribosome
(6, 43). Indeed there are drastic alterations in the ribosomes from
the yeast mutant Nar 2b resistant to narciclasine since the patterns

of ribosomal proteins labelled with the aminoacyl-tRNA analogue p-nitrophenylcarbamyl-phenylalanyl-tRNA are very different in the yeast wild strain and the Nar 2b mutant resistant to narciclasine and anthelmycin (43a).

Inhibitors of Translocation (step F)

We distinguish three classes of inhibitors within this group including inhibitors acting of elongation factor EF-2, those acting on the 40S ribosomal subunit, and finally the inhibitors blocking translocation by interacting on the 60S subunit.

Diphteria toxin and PA toxin, produced by *Corynebacterium diphteriae* $(\beta)^{tox+}$ and *Pseudomonas aeruginosa*, respectively, inhibit translocation due to the enzymatic inactivation of EF-2 in the presence of NAD^+. The reaction involves formation of ADP-ribosyl-EF-2 and nicotinamide (see M. Brot's contribution in this volume). ADP-ribosyl-EF-2 forms a complex with GTP that interacts with the ribosome, but its affinity and the rate of hydrolysis of GTP is very low and therefore the efficiency of ADP-ribosyl-EF-2 in catalyzing translocation is very low indeed. Diphtheria toxin (MW 60,000) is active in intact cells and has two subunits, one of them fragment B (MW 38,000-39,000) being responsible for the interaction with the cell membrane and the entrance of the A fragment (MW 21,000-24,000) that is required in a free form to catalyze ADP-ribosylation of EF-2(5, 6 and references therein). A subunit of MW 26,000 has been prepared from PA toxin (MW 60,000) and cannot enter in intact cells but also causes ADP-ribosylation of EF-2; however this subunit is not identical to fragment A of diphtheria toxin since it is not neutralized by fragment A-specific antiserum (44, 45).

Antibiotics containing deoxystreptamine, aminoglycoside and cryptopleurine, tylocrebrine, tylophorine, emetine and tubulosine are included between the inhibitors of translocation, their major action being exerted at the level of the 40S ribosomal subunit. However deoxystreptamine-containing antibiotics not only inhibit translocation but also induce misreading and therefore we will review both effects here. Streptomycin-induced phenotypic suppression has been reported in a *his 1* missense mutant of *Saccharomyces cerevisiae* (47) and should be confirmed by other workers. Aminoglycoside antibiotics other than streptomycin can induce high levels of misreading in *in vitro* systems from wheat germ (48), cultured human cells (49) and *Tetrahymena* (50). Moreover a number of aminoglycoside antibiotics induce phenotypic suppression of nonsense mutations in yeast cell(51, 52). The effects of these antibiotics on both phenotypic suppression and misreading in eukaryotic organisms and cell-free systems are summaryzed in Fig. 2 and Table 4. We have tested the effects of a number of aminoglycoside antibiotics on polypeptide and polyphenylalanine synthesis and misreading of poly(U) by yeast cell-free systems (Fig. 2 and 3). From our data and

Table 4. Effects of Minoglycoside Antibiotics in Eukaryotes: Induction of Misreading *in vitro* and Phenotypic Suppression *in vivo*

Antibiotic group	Misreading Active	Misreading Inactive	Phenotypic suppression Active	Phenotypic suppression Inactive
Monosubstituted deoxystreptamine antibiotics	Hygromycin B Destomycin A Apramycin		Hygromycin B Destomycin A Apramycin	
4,6-disubstituted deoxystreptamine antibiotics	Kanamycins A,B and C, Tobramycin, Gentamicins C_1,C_{1a} and C_2 Sisomicin	Gentamicin B	Kanamycins A and B, Tobramycin, Gentamicins A_1,C_{1a},C_2,X_2 Sisomicin	Kanamycin C Gentamicins A,B,C_1,C_2
4,5-disubstituted deoxystreptamine antibiotics	Neomycin, Paramomycin, Lividomycins A and B Butirosin A and B		Neomycins B and C Paromomycin Lividomycins A and B	Butirosins A and B
Others		Streptomycin Bluensomycin		Streptomycin Bluensomycin Kanamycin Fortimycin A

This table is essentially taken from ref. 51 collecting also data presented in references 48-53 and our unpublished observations.

It did not show any misreading effect on wheat germ, although it induced misreading in yeast cell-free systems.

Only active in missense *his 1* mutants (47).

Not tested in eukaryotes. On the basis of its chemical structure and inhibitory effects on *E. coli* ribosomes (53a) we assume that it might be an inhibitor of translocation and an inducer of misreading in eukaryotes

previous reports (48, 51, 52) it is possible to conclude that the antibiotics are poor inhibitors of polypeptide and polyphenylalanine synthesis in a number of eukaryotic cell-free systems. However hygromycin B blocks drastically polypeptide synthesis promoted by

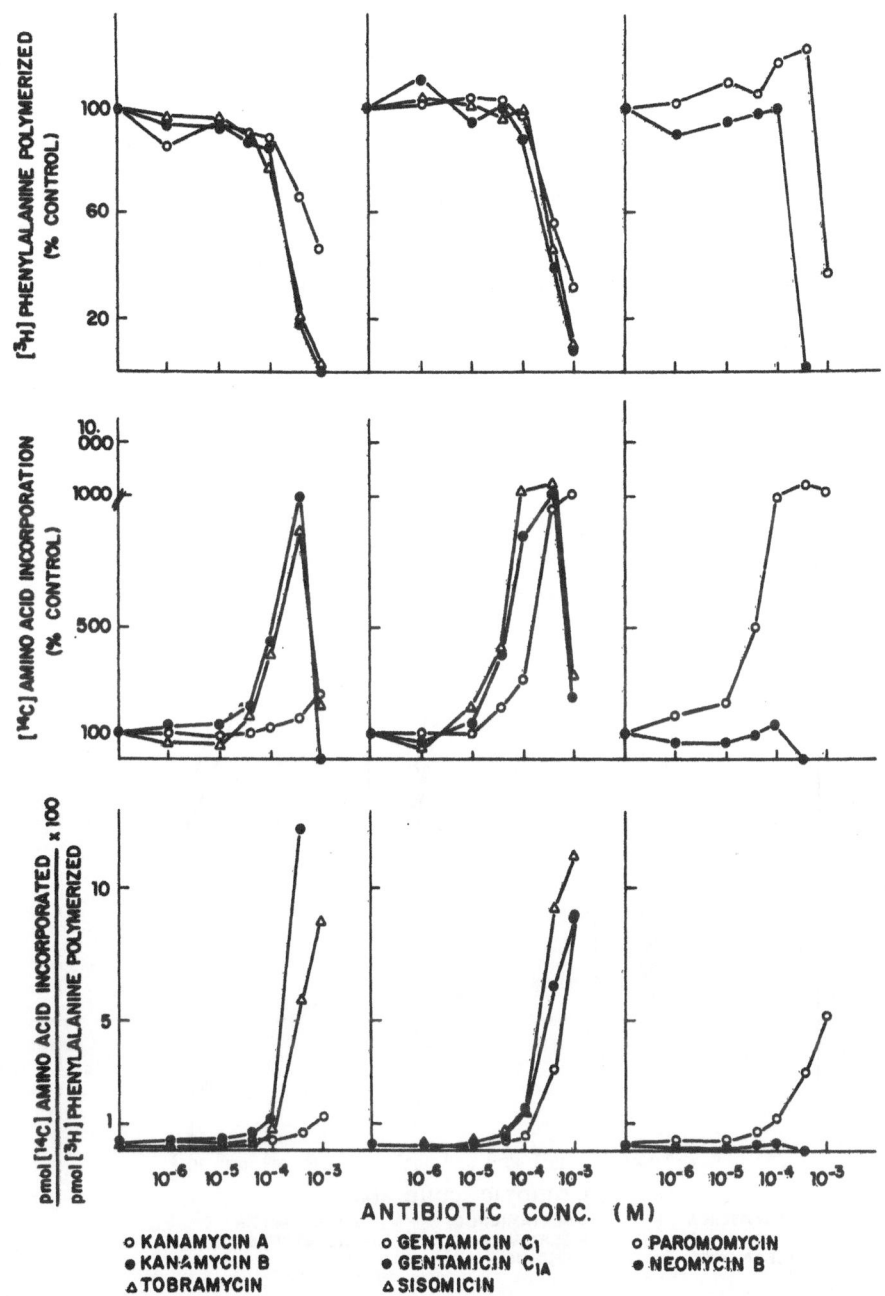

Fig. 2. Effects of aminoglycoside antibiotics on misreading and amino acid incorporation. Poly(U) directed incorporation of (^{3}H)phenylalanine, (^{14}C)serine, (^{14}C)tyrosine and (^{14}C)isoleucine by yeast ribosomes. Misreading is expressed by the ratio (^{14}C)amino acid (^{3}H)phenylalanine incorporation.

endogenous mRNA on yeast polysomes and BMV RNA in a wheat germ
system (53).   Also gentamicin and paromomycin cause a strong inhibition
of polypeptide synthesis in cell-free extracts, from *Tetrahymena* (50).
The strongest promoter of misreading on 80S ribosomes from yeasts is
hygromycin B (53).   In fact, on the basis of total amount of amino
acids incorporated no inhibition of total polypeptide synthesis
occurs, even at the highest concentration of the antibiotic.   Omission
of non cognate amino acids from the incubations did not modify the
incorporation of phenylalanine, suggesting that the incorporation of
non cognate amino acids did not replace that of phenylalanine but
took place *in addition* to it.   Similar results have been obtained in
E. coli cell-free systems (54).

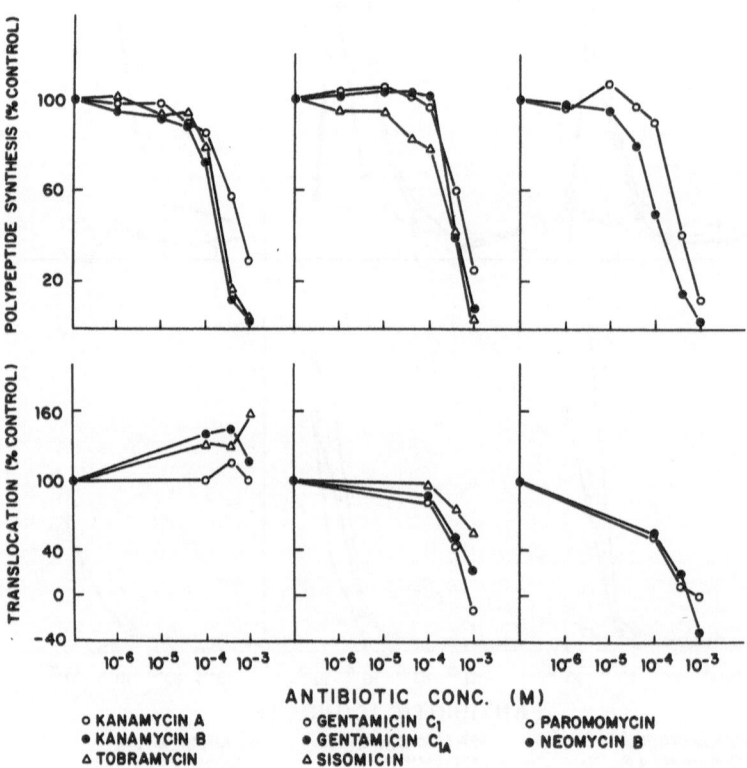

Fig. 3.   Effects of aminoglycoside antibiotics on polypeptide
          synthesis and translocation by yeast polysomes. Polypeptide
          synthesis was followed by studying (^{14}C)phenylalanine
          incorporation. Translocation was calculated by studying
          peptidyl-(^{3}H)puromycin formation prior and after transloca-
          tion.

Although there have been suggestion (45) that only those amino-glycoside antibiotics that bear a 6'-hydroxyl group on the aminosugar linked to the 4-position of 2-deoxystreptamine are highly active in promoting misreading in 80S ribosomes, there is no clear evidence to sustain this statement. Thus, sisomicin and kanamycin B bear a 6'-NH_2 group and both induce a strong misreading, while kanamycin A that has a 6'-hydroxyl group induces misreading very poorly (Fig. 2). Gentamicin Cl has a 6'-CH_3-NH- group and also promotes misreading very efficiently (48 and Fig. 2). Moreover hygromycin B lacks the structural requirement initially proposed as essential for misreading and is the most potent inducer of mistranslation *in vitro* (our un-published observations). The structural requirements proposed neither hold true for the reported effects of the aminoglycoside antibiotics on phenotypic suppression, if one considers that this phenomenon is promoted by mistranslation caused in yeasts by the antibiotics. Thus kanamycin B is clearly more active than kanamycin C and sisomicin and hygromycin B are also very active (51 and Fig. 2). Activities of different amino-glycoside antibiotics in intact cells and cell-free systems cannot be compared since there are very important differences in permeability for closely related antibiotics (54a).

The situation found in bacterial systems rose the expectation that amino-glycoside antibiotics active on eukaryotes would interact with 80S ribosomes. Though, there is so far direct evidence in this sense only in the case of paromomycin (48). Taking into account that misreading increases with drug concentrations (51) as in the case of *E. coli* ribosomes (56), it is likely that multiple binding sites for all these drugs are available on 80S ribosomes. Indeed antibiotics that induce misreading such as (^{14}C) tuberactionomycin (56) (^{3}H) kanamycin A (57) neomycin (58), viomicin (59) and acetyl tobramycin (59a) have been shown to bind to both 30S and 50S subunits from *E. coli*. There is also genetic evidence showing an action of different aminoglycoside antibiotics on both ribosomal subunits (59b, 59c).

Recent studies have shown that hygromycin B blocks the EF-2 and GTP-dependent translocation of peptidyl-tRNA from the acceptor site to the donor ribosomal site (53 and Fig. 2). In addition other aminoglycoside antiobiotics, namely paromomycin, neomycin, gentamicins C_1 and C_{1A} and sisomicin also appear to inhibit translocation (Figs. 2 and 3). Although these inhibitory effects are much lower than that promoted by hygromycin B, they parallel the observed patterns of inhibition of polyphenylalanine synthesis (Fig. 2). At first glance, the effects of the antibiotics of the kanamycin group on translo-cation, do not support the contemption that they block this reaction, but the opposite appears to be true. However these results are misleading, and most probably due to the stimulation of the puromycin reaction that is induced by these antibiotics (Table 5). Otherwise peptide bond formation does not seem to be drastically affected by other aminoglycoside antibiotics, and only neomycin and, to a lower extent paromomycin and gentamicin C_1 show a discrete inhibition of

Table 5.   Effects of Aminoglycoside Antibiotics on Enzymic Trans-
          location and Peptide Bond Formation by Yeast Polysomes.
          Data concerning enzymic translocation are taken
          from Fig. 3. Data on peptide bond formation with polysomes
          are obtained by I. Correas (personal communication).

| | % Control | | | |
| | Enzymic translocation | | Peptide bond formation | |
	4×10^{-4} M	10^{-3} M	4×10^{-4} M	10^{-3} M
Hygromycin B	15	15	100	100
Gentamicin C_1	45	-12	103	48
Gentamicin C_{1A}	50	22	118	107
Sisomicin	76	57	135	111
Paromomycin	9	0	89	64
Neomycin	29	-32	53	34
Kanamycin A	118	101	194	229
Kanamycin B	114	116	236	171
Tobramycin	131	158	260	282

the puromycin reaction (Table 5). However when the synthesis of
peptidyl-(^{3}H) puromycin depends upon translocation of peptidyl-tRNA,
the inhibition is considerably higher, a situation which indicates
that it is the translocation step the main target of the antibiotics.

Hygromycin B differs from the other aminoglycoside antibiotics
not only in chemical structure and a higher activity as an inducer
of mistranslation, but also in its effects on protein synthesis in
eukaryotic cell-free systems (6). Thus the inhibitory effects of
hygromycin B on either polypeptide synthesis or translocation can be
reversed by increasing concentrations of elongation factor EF-2 (53)
while this effect is not observed for the other antibiotics (our
unpublished observations). This would suggest that hygromycin B
might impair either the functioning or the binding of elongation
factor EF-2 to the 80S ribosome.

In bacterial systems, it has been shown that neomycins interact
with both ribosomal subunits. The interaction of these antibiotics
with the 30S ribosomal subunits causes misreading, whereas their
inhibitory effects on translocation are due to interaction with both
30S and 50S subunits. Moreover, their stabilising effects on the
guanosine nucleotide.EFG.ribosome (leading to inhibition of elongation) are due to interaction with the 50S subunit (58). It is
difficult in eukaryotic systems to correlate the effects of the

aminoglycoside antibiotics on misreading, translocation and protein
synthesis. Thus paromomycin, gentamicin C_{1A} and sisomicin promote
mistranslation at concentrations which do not inhibit polyphenylalanine synthesis (Fig. 2). On the other hand in the presence of
neomycin, kanamycins A and B, tobramycin and gentamicin C_1 misreading
of poly(U) and inhibition of polyphenylalanine synthesis in yeast
cell-free systems have parallel dependences on antibiotic concentrations (ref. 52, and Fig. 2). Nevertheless, an excess of elongation
factor EF-2 may eliminate the inhibitory effect of hygromycin B on
both polypeptide synthesis and translocation (53) but has no effect
on the drug-promoted misreading in a reticulocyte cell-free system
(A. Gonzalez, personal communication) suggesting that it might not
be a strict relationship between misreading and inhibition of translocation by the antibiotic. It might be possible, however, that some
ribosomal components involved in translocation or aminoacyl-tRNA
binding are differentially affected by the action of the antibiotics,
since elongation factor EF-2 was shown not to revert the inhibitory
effect of kanamycins A and B, tobramycin, gentamycins C_1 and C_{1A} and
sisomicin on polyphenylalanine synthesis (our unpublished observa-

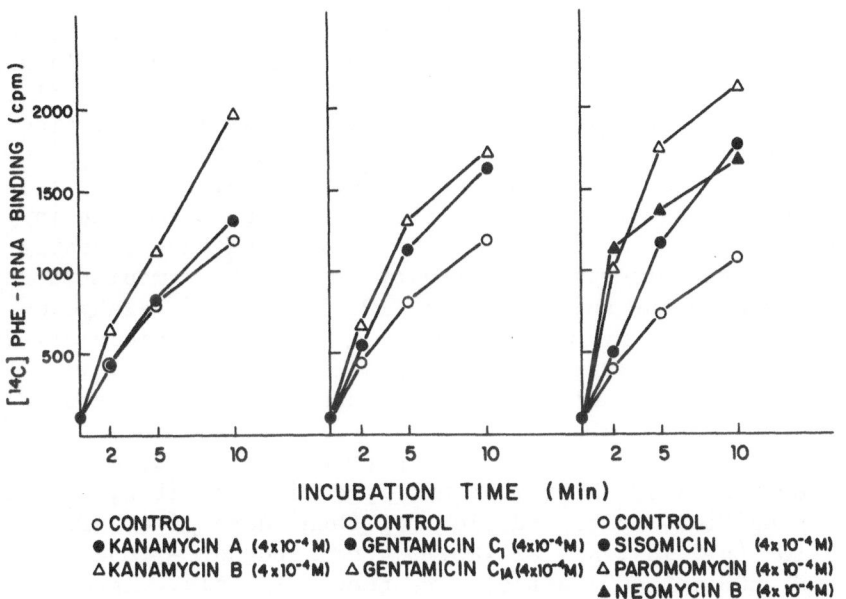

Fig. 4. Time course effects of aminoglycoside antibiotics on EF1-
         dependent binding of (^{14}C)phenylalanyl-tRNA (958 cpm/pmol)
         directed by poly(U) to rabbit reticulocyte ribosomes.

tions). It has been reported that hygromycin B stabilizes both the enzymic and the non-enzymic binding of Phe-tRNA to the 80S ribosomal acceptor site and therefore blocks translocation (53). A similar effect on the binding of N-Ac-Phe-tRNA to 80S ribosomes is also observed for kanamycin B, tobramycin, gentamicin C_{1A}, sisomicin, neomycin and paromomycin, all of which (except neomycin) are highly active in promoting misreading, kanamycin A is a rather poor inducer of misreading, and a poor stabilizer of the N-Ac-Phe-tRNA ribosome complex. However, gentamicin C_1 which induces misreading (Fig. 2) does not stabilizes the binding of N-Ac-Phe-tRNA to 80S ribosomes. Moreover hygromycin B (53), kanamycin B, tobramycin, gentamicin C_1, gentamicin C_{1A}, sisomicin, paromomycin and neomycin strongly enhance the rate of EF-1-dependent binding of Phe-tRNA to rabbit reticulocyte ribosomes (Fig. 4). Therefore stabilization of peptidyl-tRNA in the *acceptor* site by the antibiotics, that could result in a stronger interaction of both cognate and non-cognate aminoacyl-tRNA, might have some relationship with their effects on misreading and trans- location but would neither explain by itself the misreading effect nor the inhibition of translocation by all the aminoglycoside anti- biotics.

There is genetic and biochemical evidence that cryptopleurine, tylophorine, tylocrebrine, emetine and tubulosine act on the 40S ribosomal subunit at identical or overlapping sites. All of these compounds inhibit EF-2- and GTP-dependent translocation but this effect can be overcome in cell-free systems by the addition of a large excess of EF-2. Cryptopleurine also inhibits peptide bond formation in cell-free systems at concentrations more than hundred-fold higher than that required to block translocation (6 and references therein). On the basis of spectrofluorometric measurements, multiple binding sites for cryptopleurine on yeast ribosomes and their 60S and 40S subunit have been reported (60). Their existence, however, is hard to reconcile with the finding of one-step nuclear mutations in yeast confering resistance to the drug (6, review). We have recently purified (^{3}H)cryptopleurine prepared by the $NaB(^3H)_4$-mediated reduct- ion of the perchlorate salt of 9, 11, 12, 13, 14, 15 hexahydro-$^{10-14a}$ dehydro-2,3,6 trimethoxy-phenanthro (9, 10-b) quinolizidinium. We have observed that there is a leveling off of the binding when 2 molecules of drug interact per 80S ribosome, while the onset of binding to the 60S and 40S ribosomal subunits takes place at about 1 molecule of (^{3}H)cryptopleurine per particle (61). Drug concentra- tions higher than $5x10^{-7}$M promote multiple binding sites, throught to be *non* specific. Scatchard plots of these data give values of $Kd^{80}=8.46x10^{-8}$ (n=2.15), $Kd^{60}=16.20x10^{-8}$ (n=1) and $Kd^{40}=3.65x10^{-8}$ (n=0.84) for the 80S, 60S and 40S particles respectively (61). In addition there are multiple binding sites of very low affinities on 80S, 60S and 40S ribosomal particles and on *E. coli* 70S ribosomes as well. The two molecules of (^{3}H)cryptopleurine bound per 80S ribosome could account for the dual effect of the inhibitor on translocation at low

concentrations and peptide bond formation at higher concentrations
(61). In agreement with this proposal we have observed that one
single molecule of (^{3}H) cryptopleurine interacts with high affinity
with one 40S ribosomal subunit. This interaction might promote the
inhibition of translocation; this is in agreement with the existence
of one step cell mutants resistant to cryptopleurine that express the
mutation in the 40S subunit, at the level of enzymic translocation
(6, review) and the finding of an altered 40S ribosomal protein (S20)
in a CHO cell mutant resistant to emetine (62, 63). The single
molecule of (^{3}H)cryptopleurine bound to each 60S ribosomal subunit,
with an affinity four times lower than that for the 40S one, might
account for the inhibition of peptide bond formation at higher concentrations of the drug (6, review). This might explain the observation that resistance to cryptopleurine both *in vivo* and *in vitro* is
never complete, and a small increase in drug concentration has
always a slowing down effect either on growth of cry^r cells, or on
protein synthesis in cell-free systems (64). In further studies we
have shown that the extent of (^{3}H) cryptopleurine binding to ribosomes decreased in preparations from yeast mutants resistant to the
drug. This was due to the smaller extent of interaction of the drug
with the higher affinity to the ribosome (our unpublished observations). Furthermore, 40S subunits from this resistant mutant have
a lower affinity for the drug (our unpublished observations).

As indicated above, emetine and tubulosine are thought to share
a common mechanism of action on translocation with cryptopleurine,
although they do not affect peptide bond formation even at high
concentrations. Accordingly, emetine and tubulosine inhibit (^{3}H)-
cryptopleurine interaction with its target site(s) to either 80S
ribosomes or polysomes from yeasts (our unpublished observations).
The results would suggest that the presence of peptidyl-tRNA on the
ribosome does not alter the binding parameters of emetine and
tubulosine to their target site(s). Fifty percent inhibition of the
binding of (^{3}H)cryptopleurine to 80S ribosomes requires hundred times
higher concentrations of tubulosine and emetine than those of
unlabelled cryptopleurine (our unpublished observations). These
results might suggest lower binding affinities for emetine and
tubulosine than for cryptopleurine, and an identical or overlapping
binding site for the three drugs on the 40S ribosomal subunit.

Pederine and antibiotics of the glutarimide group are included
within the inhibitors of translocation that interact with the 60S
ribosomal subunit. The best known antibiotic of the glutarimide group
is cycloheximide. Both pederine and cycloheximide block not only
the EF-2-dependent translocation but also the non-enzymic translocation that takes place in the presence of rather high concentration
of either K$^+$ or NH$_4^+$. However pederine is active on cycloheximide
resistant ribosomes, suggesting that both do not interact precisely
with the same binding site on the ribosome (6, review).

INHIBITORS OF TERMINATION

The termination phase of translation begins when a chain-terminating codon of the mRNA (nonsense codon, either UAA, UAG or UGA) is recognized, thus determining the interaction of the release factors with the ribosome. A single release factor (RF) has been isolated from eukaryotic systems, GTP being required in the termination step (G_1). After recognition of the release factor(s) the peptidyl transferase of the larger ribosomal subunit exercises its peptidyl hydrolase activity and cleaves the bond between the peptidyl and the tRNA moieties of peptidyl-tRNA so that the peptide chain is released (step G_2). A further reaction is required for termination to be completed in which the unchanged tRNA is released (step H). This step is not yet resolved in eukaryotic systems (6, review).

Protein synthesis inhibitors blocking translation specifically in the termination phase are not known. A number of the inhibitors described above, however inhibit termination as consequence of their interaction with the ribosome, besides inhibiting also other steps of translation. Thus it is most likely that tetracyclines and some other inhibitors of step D are also inhibitors of termination, with the selectivity indicated in Tables 1 and 2. (6, review).

Peptidyl transferase is involved not only in peptide bond formation (step E) but also in the peptidyl-tRNA hydrolysis reaction (step G_2). In fact, all the peptidyl transferase inhibitors which have been tested block termination with the selectivity indicated in Tables 1 and 2. The only exception is anisomycin that does not inhibit this step in certain model systems (6, review).

As indicated in Figure 1 EF-2 and GTP appear to be required in step H. It is therefore likely that inhibitors that interfere with EF-2 activity and certain analogues of GTP also inhibit step H (6, review).

GTP ANALOGS

A number of GTP analogs have been developed to inhibit translation by competing with GTP on the ribosome (6, review) since this compound is required in the initiation, elongation and termination phases of the translation process (Fig. 1). The most interesting coumpounds of these series are guanylyl-methylene-diphosphophate and guanylyl-imido-diphosphate that have a methylene and imido group respectively replacing the oxygen between the β and γ-phosphorus atoms, thus preventing enzymic cleavage at this position. Guanylyl-methylene-diphosphonate and guanlylyl-imido-diphosphate are very effective inhibitors of initiation, elongation of termination since GTP hydrolysis is required in the three cases for fast release of the protein factors from the ribosome. Inhibition of this release results in slowing down the subsequent step in translation (6,review).

SELECTIVITY OF THE INHIBITORS

Eukaryotic ribosomes have been very conservative in the evolution and their sensitivity to translation inhibitors usually follows the broad scope mentioned above and presented in Tables 1 and 2. However, there are some cases in which certain antibiotics appear to have a narrower spectra of selectivity. Thus tenuazonic acid inhibits protein synthesis by mammalian but not by yeast ribosomes; cyclo-heximide is inactive on ribosomes of some wild type strains of *Kluyveromyces fragilis* and *Kluyveromyces lactis*, while fusidic acid is *inactive* on ribosomes from *Neurospora crassa* mitochondria (6, review). Differences are more important when intact cells are con-sidered, due to the permeability barrier. Thus triphenylmethane dyes are very poor inhibitors of intact cells. Similarly, glycoprotein and protein toxins that inactivate the ribosome are very inactive when deprived of the subunit β required for permeability. Yeast are very well known for their resistance to many of the inhibitors presented in Tables 1 and 2 due to their permeability barrier although this difficulty can be overcome in some cases by changing the pH of the media. Cell permeability can be also changed in some cases by membrane alterations caused by cell infection or by the use of drugs. Thus a number of inhibitors that are inactive on tissue culture cells such as guanylyl-methylene-diphosphonate, blasticidin S, gougerotin and edeine A_1 (65) are active in picornavirus-infected cells (66, 67).

On the other hand, in a few cases a wider spectrum of activity that than indicated in Tables 1 and 2 has been observed. Thus, sensitivity of mitochondrial ribosomes to emetine, cryptopleurine, tylophorine and tylocrebrine has been reported, and a yeast mutant with cytoplasmic 80S-type ribosomes sensitive to streptomycin has also been described (6, review).

SPECIFICITY

Although most of the translation inhibitors indicated in Tables 1 and 2 are considered to act preventing specifically protein synthesis, there are a number of cases in which their action is less specific. Thus, in intact cells, edeine A_1 is mainly an inhibitor of DNA synthesis, and cycloheximide can inhibit not only translation but also DNA and RNA synthesis. Furthermore, triphenylmethane dyes and aminoglycoside antibiotics are known to have many interactions in cell-free systems which might result in a number of effects on intact cells other than inhibition of protein synthesis. Moreover most glycoprotein toxins that inactivate catalytically the ribosome have one subunit (subunit B) that interacts with the membrane, somehow affecting some functions of this cell structure (6, review).

The above considerations should be taken into account when using translation inhibitors as biochemical tools in intact cells. Obvious-ly the risks of erroneous interpretations are smaller when the in-

hibitors are used in cell-free systems to investigate defined
reactions.

REFERENCES

1. Gale, E.F., Cundliffe, E., Reynolds, P.E., Richmond, M.H. and
    Waring, M.J. (1972), In *The Molecular Basis of Antibiotic Action*,
    pp. 278-379, John Wiley and Sons, London.
2. Kaji, A. (1973), In *Progress in Molecular and Subcellular
    Biology* (F.E. Hahn, ed.), Vol. 3, pp. 85-158, Springer-Verlag
    Berlin-Heidelberg-New York.
3. Vazquez, D. (1974), *Inhibitors of Protein Synthesis*, FEBS
    Letters 40 (Suppl.), S63-S84.
4. Pestka, S. (1977), In *Molecular Mechanisms of Protein Bio-
    synthesis* (H. Weisbach and S. Pestka, eds.) pp. 447-553, Academic
    Press, New York-San Francisco-London.
5. Vazquez, D. (1978), In *International Review of Biochemistry*,
    Vol. 18, Amino Acid and Protein Biosynthesis II (H.R.V.
    Arnstein, ed.), pp. 169-232 , University Park Press, Baltimore.
6. Vazquez, D. (1979), *Inhibitors of Protein Biosynthesis*, Springer-
    Verlag, Berlin-Heidelberg-New York.
7. Vazquez, D. and Jiménez, A. (1980), In *Ribosomes: Structure,
    Function and Genetics* (G. Chambliss, G.R. Craven, J. Davies, L.
    Kahan and M. Nomura, eds.) pp. 847-869, University Park Press,
    Baltimore.
8. Weisbach, H. and Pestka, S. (1977), *Molecular Mechanisms of
    Protein Biosynthesis*, Academic Press, New York-San Francisco-
    London.
9. Safer, B. and Anderson, W.F. (1978), CRC Critical Rev. Biochem.
    pp. 261-290.
10. Visser, D.W. and Roy-Burman, S. (1979), In *Antibiotics V-2
    Mechanism of Action of Antieukaryotic and Antiviral Compounds*
    (F.E. Hahn, ed.), pp. 363-371, Springer-Verlag, Berlin-Heidel-
    berg-New York.
11. Fresno, M., Carrasco, L. and Vazquez, D. (1976), Eur. J. Bio-
    chem. $\underline{68}$, 355-364.
12. Hunter, A., Jackson, R. and Hunt, T. (1977), Eur. J. Biochem.
    $\underline{75}$, 159-170.
13. Santon, J.B. and Stanley, W.M. (1978), Biochem. Biophys. Res.
    Commun. $\underline{84}$, 985-992.
14. Kozak, M. (1979), J. Biol. Chem. $\underline{254}$, 4731-4738.
15. Odom, O.W., Kramer, G., Henderson, A.B., Pinphanichakarn, P.
    and Hardesty, B. (1978), J. Biol. Chem. $\underline{253}$, 1807-1816.
16. Fresno, M. and Vazquez, D. (1978), Eur. J. Biochem. $\underline{83}$, 169-178.
17. Seal, S.N., Schmidt, A., Tomaszewski, M. and Marcus, A. (1978),
    Biochem. Biophys. Res. Commun. $\underline{82}$, 553-559.
18. Bergmann, J.E. and Lodish, H.F. (1979), J. Biol. Chem. $\underline{254}$, 459-
    468.
19. Fresno, M. and Vazquez, D. (1980), Eur. J. Biochem. 103, 125-132.

20. Grunberger, D. and Grunberger, G. (1979), In *Antibiotics V-2 Mechanism of Action of Antieukaryotic and Antiviral Compounds* (F.E. Hahhn, ed.) pp. 110-123, Springer-Verlag, Berlin-Heidelberg-New York.

21. Carrasco, L. (1977), FEBS Letters, 76, 11-15.

22. Carrasco, L. and Smith, A.E. (1976), Nature, London, 264, 807-807.

23. Carrasco, L., Harvey, R., Blanchard, C. and Smith, A.E. (1979), In *Modern Trends in Human Leukemia III* (R. Neth, R.C. Gallo, P.H. Hofchneider and K. Mannweiler, eds.), pp. 277-281, Springer-Verlag, Berlin.

24. Cherney, C.S. and Wilheln, J.M. (1979), J. Virol. 30, 533-542.

25. Egberts, E., Hackett, P.H. and Traub, P. (1977), J. Virol. 22, 591-597.

26. Garry, R.F., Bishop, J.M., Parker, S., Westbrook, K., Lewis, G. and Waite, M.R.F. (1979), Virology 96, 108-120.

27. Nair, C.N., Stowers, J.W. and Singfield, B. (1979), J. Virol. 31, 184-189.

28. Alonso, M.A., Vazquez, D. and Carrasco, L. (1979), Antim. Agents. Chemoth. 16, 750-756.

29. Tait, R.C. and Boyer, H.W. (1978), Cell 13, 73-81.

30. Chopra, I. and Ecles, S.J. (1978), Biochem. Biophys. Res. Commun. 33, 550-557.

31. Sompolinsky, D., Hammerman, I., Assaf, O. and Wojdani, A. (1978), FEMS Microbiology Letters 4, 23-26.

32. Conde, F.P., Fernandez-Puentes, C., Montero, M.T.V. and Vazquez, D. (1978), FEMS Microbiology Letters 4, 349-355.

33. Olsnes, S. and Refsnes, K. (1978), Eur. J. Biochem. 88, 7-15.

34. Cawley, D.B., Hedblom, M.L. and Houston, L.L. (1978), Arch. Biochem. Biophys. 190, 744-755.

35. Olsnes, S. and Abraham, A.K. (1979), Eur. J. Biochem. 93, 447-452.

36. Montanaro, L., Sperti, S., Zamboni, M.C., Denaro, M., Testoni, G., Gasperi-Campani, A. and Stirpe, F. (1978), Biochem. J. 176, 371-379.

37. Sperti, S. and Montanaro, L. (1979), Biochem. J. 178, 233-236.

38. Vazquez, D. (1979), In *Antibiotics V-2 Mechanism of Action of Antieukaryotic and Antiviral Compounds* (F.E. Hahn, ed.) pp. 341-352, Springer-Verlag, Berlin-Heidelberg-New York.

39. Sperti, S., Montanaro, L., Derenzini, M., Gasperi-Campani, A. and Stirpe, F. (1979), Biochim. Biophys. Acta. 562, 495-503.

40. Barbieri, L., Lorenzoni, E. and Stirpe, F. (1979), Biochem. J. 182, 633-635.

41. Gonzalez, A., Vazquez, D and Jiménez, A. (1979), Biochim. Biophys. Acta 561, 403-409.

42. Wall, M.E., Wani, M.C. and Taylor, H. (1976), Cancer Treatment Rep. 60, 1011-1030.

43. Gonzalez, A., Santamaria, F., Vazquez, D. and Jiménez, A. (1980) Molec. Gen. Genetics, submitted.

43a. Leon, G., Perez-Gosalbez, M. and Ballesta, J.P.G. (1979), FEBS
     Letters, (in press).

44. Chung, D.W. and Collier, R.J. (1977), Infection and Immunity 16,
     832-841.

45. Chung, D.W. and Collier, R.J. (1977), Biochim. Biophys. Acta
     483, 248-257.

46. Collier, R.J. (1979), In *Antibiotics V-2 Mechanism of Action of
     Antieukaryotic and Antiviral Compounds* (F.E. Hahn, ed.), pp.
     155-172, Springer-Verlag, Berlin-Heidelberg-New York.

47. Bayliss, F.T. and Ingraham, J.L. (1974), J. Bacteriol. 118, 319.

48. Wilhelm, J.M., Pettit, S.E. and Jessop, J.J. (1978), Biochemistry
     17, 1143-1149.

49. Wilhelm, J.M., Jessop, J.J. and Pettit, S.E. (1978), Biochemistry
     17, 1149-1153.

50. Palmer, E. and Wilhelm, J.M. (1978), Cell 13, 329-334.

51. Singh, A., Ursic, D. and Davies, J. (1979), Nature, London, 277,
     146-148.

52. Palmer, E., Wilhelm, J.M. and Sherman, F. (1979), Nature, London,
     277, 148-149.

53. Gonzalez, A., Jiménez, A., Vazquez, D., Davies, J.E. and
     Schindler, D. (1978), Biochim. Biophys. Acta 521, 459-469.

53a. Perzynski, S., Cannon, M., Cundliffe, E., Chabwala, S.B. and
     Davies, J. (1979), Eur. J. Biochem. 99, 623-628.

54. Cabanas, M.J., Vazquez, D. and Modolell, J. (1978), Eur. J.
     Biochem. 87, 21-27.

54a. Lee, B.K., Condon, R.G., Munayyer, M. and Weinstein, M.J. (1978)
     J. Antibiotics 31, 141-146.

55. Davies, J. and Davis, B.D. (1968), J. Biol. Chem. 243, 3312-3316.

56. Misumi, M., Tanaka, N. and Shiba, T. (1978), Biochem. Biophys.
     Res. Comm. 82, 971-976.

57. Misumi, M., Nishimura, T., Komai, T. and Tanaka, N. (1978),
     Biochem. Biophys. Res. Comm. 84, 358-365.

58. Campuzano, S., Vazquez, D. and Modolell, J. (1979), Biochem.
     Biophys. Res. Comm. 87, 960-966.

59. Choi, E.C., Misumi, M., Nishimura, T., Tanaka, N., Nomoto, S.
     Teshima, T. and Shiba, T. (1979), Biochem. Biophys. Res. Comm.
     79, 904-910.

59a. Le Goffic, F., Capmau, M.L., Tangy, F. and Baillarge, M. (1979),
     Eur. J. Biochem. 102, 73-81.

59b. Buckel, A., Buchberger, A., Bock, A. and Wittmann, H.G. (1979),
     Mol. Gen. Genetics 158, 47-54.

59c. Piepersberg, W., Noseda, V. and Bock, A. (1979), Mol. Gen.
     Genetics, 171, 23-34.

60. Bucher, K. and Skogerson, L. (1976), Biochemistry, 15, 4755-4759.

61. Dolz, H., Söllhuber, M., Trigo, G.G., Vazquez, D. and Jiménez,
     A. (1980), Anal. Biochem., submitted.

62. Boersma, D., McGill, S.M., Mollenkamp, J.W. and Roufa, D.J.
     (1979), Proc. Nat. Acad. Sci. U.S.A. 76, 415-419.

63. Boersma, D., McGill, S.M., Mollenkamp, J.W. and Roufa, D.J.
     (1979), J. Biol. Chem. 254, 559-567.

64. Grant, P., Sanchez, L. and Jiménez, A. (1974), J. Bacteriol.
    120, 1308-1314.
65. Contreras, A., Vazquez, D. and Carrasco, L. (1978), J. Anti-
    biotics 6, 151-155.
66. Contreras, A. and Carrasco, L. (1979), J. Virology, 29, 114-122.
67. Carrasco, L. (1978), Nature, London, 264, 807-809.
68. Thomas, A., Goumans, H., Voorma, H.O. and Benne, R. (1980), Eur.
    J. Biochem., (in press).
69. Holland, R.I. (1979), Cell Biol. Intern. Rep. 3, 701-705.

VIRUS-INDUCED SHUT-OFF

OF HOST SPECIFIC PROTEIN SYNTHESIS

Friedrich Koch, Gebhard Koch, and Joachim Kruppa

Abt. Molekularbiologie
Physiol. Chem. Inst. Univ. Hamburg
Grindelallee 117
D-2000 Hamburg 13, FRG

INTRODUCTION

Infection of cells by viruses is often followed by a rapid decline in the synthesis of host cell specific macromolecules, predominantly of protein and RNA (1-3). This phenomenon, called *shut-off*, was first intensively analyzed in phage infected bacteria. It was shown that adsorption of the phage protein shell (protein ghost) to *E. coli* leads to a rapid change in membrane permeability followed by cessation of macromolecular synthesis and death of the host cell (4, 5). Even soluble fractions of the phage coat are able to kill *E. coli* cells (6). Membrane leakiness is rapidly repaired in productively infected *E. coli* due to phage coded proteins.

The shut-off phenomenon was subsequently found to be induced by many other cytopathogenic animal viruses early during infection (for review see 1-3). The virus induced inhibition of protein synthesis affects the translation of host mRNA to a different extent. The severity of this inhibition and the rate with which it occurs depends on the multiplicity of infection and on the host cell and its nutritional state. Translation of the viral RNA is generally required to induce the shut-off (see below). Addition of inhibitors of protein synthesis interrupts the process of shut-off. Upon removal of reversible inhibitors, protein synthesis resumes at the level the cells showed at the time of addition of the drug. In contrast, inhibition of viral RNA synthesis by guanidine does not stop the shut-off (7). Since viral RNA synthesis is not required for the shut-off, it is likely that the synthesis of double-stranded RNA plays no essential role (8).

Infection by vaccinia virus results in inhibition of cellular protein synthesis even though UV inactivated viruses are used or synthesis of all vaccinia mRNA is blocked (9). However, infection of cells by UV inactivated poliovirus does *not* trigger the shut-off phenomenon. The UV inactivation of the shut-off ability follows one hit kinetics indicating that one hit anywhere in the genome is sufficient to abolish the induction of the shut-off (10). As long as we do not know to what extent UV inactivated RNA is still able to function as mRNA, this result is difficult to interpret, especially in the light of experiments with temperature sensitive mutants of poliovirus defective in the structural proteins (see below). These mutants were unable to inhibit cellular protein synthesis at the non-permissive temperature (11). Temperature sensitive mutants with defects in the non-structural proteins (including those involved in RNA replication) were able to induce the shut-off. It appears that only part of the structural protein gene is required for shut-off. Cole and Baltimore (12) have isolated defective interfering (DI) particles of poliovirus with deletions in the 5' region of the viral RNA. These DI particles lack one third of the gene coding for the coat precursor protein. The corresponding coat precursor protein which is synthesized after infection by DI particles is unstable and rapidly degraded. Further experiments are called for to prove whether the unstable two thirds of the capsid precursors or part of them persist long enough in the cell to induce the shut-off.

Detailed information about shut-off by animal viruses has come from studies of picornavirus infections. Already within one hour after picornavirus infection there is a dramatic reduction in the overall rate of amino acid incorporation into proteins. Maximal inhibition of protein synthesis takes place at approximately two to three hours post infection. At this point incorporation of amino acids into *host* proteins is reduced by more than 90 %. Thereafter, the rate of total protein synthesis increases. Results from poly-acrylamide gel electrophoresis (PAGE) of pulse labelled proteins have revealed that this increase is due solely to the synthesis of viral proteins (see Fig. 1). Inhibition of protein synthesis is accompanied by disaggregation of polysomes indicating an effect on polypeptide chain initiation (13, 14, 15). Host cell mRNAs are not destroyed or irreparably damaged after infection: both the 3' (poly A) and 5' (cap) ends are normal, the cellular mRNA is not detectably modified, and it retains the ability to prime protein synthesis in cell-free extracts (16, 17, 18). The fate of histone mRNA was studied after infection of Ehrlich ascites tumor cells with mengovirus (19). At 4 hrs. after infection the histone mRNA content had declined to half of the control, the translation of histone mRNA, however, had declined to 10% by that time. Histone mRNA was found in polysomes even at 8 hrs. post infection.

The shut-off phenomenon does not only effect cellular proteins, it can also trigger the inhibition of the translation of mRNA of

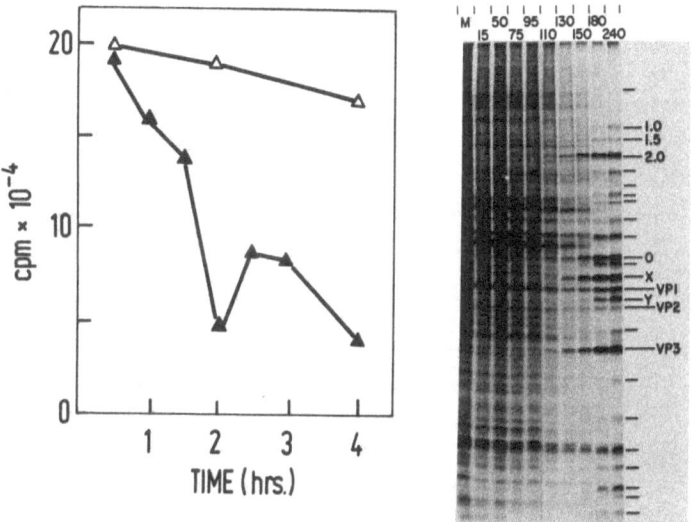

Fig. 1.   <u>Protein synthesis in mock-infected and poliovirus-infected</u>
<u>HeLa cells</u>.   HeLa S-3 cells grown in suspension, were
collected by centrifugation and resuspended at a density of
10^7 cells/ml in serum-free, Joklik-modified minimum essential
medium containing only 1/20 the normal concentration of
methionine and leucine.   Adsorption of poliovirus type I,
Mahoney strain was allowed to proceed at room temperature
for 20 min.   Mock infected and infected cultures were further
incubated at 37°C and at a density of 4×10^6 cells/ml in
the presence of 25 mM Hepes.   2.5 µCi of (^{35}S)-methionine
and 7.5 µCi of (^{3}H)-leucine were added to 5-ml aliquots of
the culture at the indicated times.   After 15-min incubation
at 37°C the incorporation of labelled aminoacids into the
acid-precipitable fraction was determined.   (   ) mock-
infected; (   ) poliovirus-infected.   Labelled proteins
taken from cells at the indicated times were analyzed by
polyacrylamide gel electrophoresis (right).

another virus.   This phenomenon of viral interference has been
observed in experiments with various viruses but was studied in
detail by superinfection of Vesicular Stomatitis Virus (VSV) infected
cells by poliovirus (20, 21).   By 2 hrs. after poliovirus superinfection, VSV specific polysomes disaggregate.   PAGE analysis of
pulse labelled proteins revealed a change in the pattern of protein
synthesis from VSV specific to poliovirus specific proteins.   Nevertheless, the synthesis of VSV mRNA continued unabated.   Physically
and biochemically intact VSV mRNAs accumulate in the cell.   Their
translation *in vivo* (but not *in vitro*) was abolished.   It appeared
that VSV mRNA translation is inhibited like cellular mRNA

Table 1. Possible Mechanisms Involved in the Shut-Off

	Activation of negativ pleiotropic response	Competition of mRNA	Membrane leakiness	Inactivation of initiation components	Interaction of viral proteins with cellular components
Polio (ss RNA+)	+	+	+	+	+
Sendai	+	+	+	?	?
VSV (ss RNA-)	+	+	?	?	?
Reo (ds RNA)	-	+	?	?	?
SV40 (DNA)	-	+	?	?	+
Vaccine (DNA)	-	+	?	?	+

translation by the poliovirus induced shut-off. Although the shut-off phenomenon has been under investigation for over 20 years and important details have been observed and described, the exact mechanism (1) by which the inhibition takes place is still unclear.

A priori, there is no reason to expect only one unique mechanism for these virus induced shut-off effects. Indeed, it is known that phages kill their host cells by several independent mechanisms (22). Similarly, a number of theories have been developed to explain these shut-off effects. For the sake of clarity, we shall discuss some of the proposed mechanisms for the shut-off separately. However, it is important to note that these mechanisms are not mutually exclusive (as has occasionally been assumed) moreover, that several or all of them may act in a concerted manner. Different mechanisms may be operating with different viruses (see Table 1).

In the course of this review we shall discuss:

a)   One model put forward by our group (15, 23, 24) which proposes that the virus somehow induces a normal host cell regulatory mechanism for the inhibition of the synthesis of cellular macromolecules.

b)   Competition between viral and host cell mRNAs and between different viral mRNAs. This may play an important role, especially late in infection.

c)   Another model postulates that an increase in the intracellular Na$^+$ concentration is the cause for the inhibition of host cell protein synthesis (25, 26) (membrane leakiness).

d)   Suppression of host mRNA translation might be exerted by alterations of the protein synthesizing machinery which result in preferential interaction with viral mRNA, i.e. initiation factors and ribosomal proteins (phosphorylation, dephosphorylation).

e)   Viral proteins might interact with cellular constituents and thereby directly or indireclty interfere with the translation of host mRNAs (27, 28).

Finally, we shall suggest that the cell membrane may play a central role in mediating some of these mechanisms.

PHYSIOLOGAL REGULATION OF PROTEIN SYNTHESIS AT THE LEVEL OF TRANSLATION

Animal cells respond rapidly to external signals with increased or decreased synthesis of macromolecules. Many external signals first interact with the cell membrane, which functions as both a receiver and a transmitter of regulatory signals. A great number of studies

have been performed to elucidate the role of the cell membrane in
the induction of the proliferative response of cells to mitogens (29).
The binding of mitogens triggers: a) changes in membrane permeability
for ions, b) alterations in the level of cyclic nucleotides, c)
increased transport of ions, nucleosides, glucose, and amino acids
(24). These very early events are soon followed by elevation in the
rate of protein synthesis. This elevation takes place by employing
pre-existing mRNA and pre-existing ribosomes. The corresponding
negative response is activated by a number of unfavorable growth
conditions, such as amino acid (30) or glucose deprivation (31), high
cell density (32), ATP depletion (33), exposure to elevated temperature (34), and incubation of cells in medium with increased medium
osmolarity (35, Table 2).

During the course of our work on this subject, it became
increasingly clear that the modulating effect upon protein synthesis
was not unilateral, in that the synthesis of all proteins was not
equally affected. Under *optimal* growth conditions, protein synthesis
in tissue culture cells appears to be primarily regulated at the
level of *transcription*: or by the amount of an individual mRNA
available for translation. In contrast, when cells are transferred
to *unfavorable* growth conditions, control at the *translational* level
is amplified, and the translation of each mRNA species proceeds with
its own characteristic efficiency (23). This amplification of
translational control is most readily achieved in tissue culture cells
by increasing the osmolarity of the medium (35).

As will be shown below, viral infection and medium hyperosmolarity cause comparable alterations in the synthesis of individual
cellular proteins (23, 24). We propose that infection with certain
viruses triggers such a normal cellular regulatory mechanism on the
cell membrane. This induced negative response may be a mechanism
leading to the shut-off. Studies of the shut-off phenomenon might
provide valuable insight in the mechanism whereby protein synthesis
is regulated in uninfected cells.

Table 2. Physiological Regulation of Protein Synthesis

Enhancement (proliferative response)	Inhibition (negative pleiotropic response)
Mitogens	Nutrient Starvation
Serum	Serum Depletion
Protease	Medium Hyperosmolarity
Low Cell Densities	High Cell Densities
	Virus infection

DIFFERENTIAL INHIBITION OF mRNA TRANSLATION BY HYPERTONIC INITIATION BLOCK (HIB)

The growth of animal cells in culture is optimal within a narrow range of osmolarity in the medium. The mebrane of animal cells is more permeable to water than to ions. Variations in the osmolarity of the growth medium will, therefore, result in a rapid flow of water into or out of the cell, causing alterations in the intracellular osmolarity. An increase in the tonicity of the growth medium results in a specific inhibition of polypeptide chain initiation (HIB). Inhibition is independent of the solute used to increase the osmolarity of the growth medium. NaCl, KCl, NH_4Cl or even sucrose are equally suitable for this purpose (36). In the experiments discussed below, the solute used to increase tonicity was NaCl.

The mouse plasmacytoma cell line MPC-11 was used as a model system to investigate the effect of HIB in an *uninfected* cell line. MPC-11 cells synthesize an immunoglobulin gamma heavy (H) chain of approximately 55 Kd and a 23 Kd kappa light (L) chain (Table 3). The L and H chains account for as much as 20% of the newly synthesized polypeptides (37). A comparison of the distribution of ^{35}S-methionine incorporation into MPC-11 polypeptides labelled under isotonic and hypertonic conditions revealed an increase in the relative incorporation into the L chain under hypertonic conditions when compared to that observed under isotonic conditions. Likewise, there was an increase in the relative incorporation of ^{35}S-methionine into the H chain polypeptide. The percentage of total ^{35}S-methionine incorporation, which was associated with the L chain, increased from a value of 6.9% under isotonic conditions to a value of 27.2% when cells were pulse labelled under hypertonic conditions. Under the same conditions this value increased from 8.8% to 12.8% for the H peptide. Amino acid starvation caused similar differential inhibitory effects on the synthesis of L and H chains and other cellular proteins (38). These results suggested that the mRNAs coding for the specialized IgG polypeptides were more efficient messengers when compared to those mRNA species coding for other cellular proteins.

The ratio of the synthesis of light and heavy chain is 1.6 under isotonic conditions. This ratio was changed to a value above 6 when total protein synthesis was inhibited by more than 98% by inhibitors of polypeptide chain initiation. In contrast the L/H ratio remained essentially unaltered when protein synthesis was reduced in response to inhibitors of polypeptide chain elongation (39). For the more general implication of these findings, the reader is referred to chapter 17 of this volume.

Table 3. Resistance of Light- and Heavy-Chain Synthesis to Hypertonic Initiation Block

Excess NaCl (mM)	^{35}S-methionine incorporation				% of total ^{35}S-incorporation			
	Total (% of control)	Light (% of control)	Heavy (% of control)	Non-IgG (% of control)	L	H	L+H	L/H
0	100	100	100	100	6.9	9.5	16.3	1.60
20	92	102	95	90	7.7	9.9	17.5	1.71
40	64	81	73	58	8.6	10.8	19.4	1.76
60	45	68	60	41	10.4	12.5	22.9	1.82
100	20	41	27	17	14.0	12.9	27.0	2.38
120	11	29	15	9	17.8	12.7	30.6	3.08
140	6	19	8	5	21.6	13.2	34.7	3.60

Radioactivity in the total gel lanes and the individual IgG bands were determined directly from the corresponding gel. The relative sensitivity of the L, H, and non-IgG polypeptide synthesis to HIB is presented as a percentage of the control (isotonic) value. The counts recovered from the control gel lane were: 167,000 cts/min, total; 11,400 cts/min, L; 15,800 cts/min, H; and 139,000 cts/min, non-IgG. Also shown is the percentage of total ^{35}S-methionine incorporation which was found in the L and H bands separately and combined. In addition, the value for the ratio of the synthesis of L to H (L/H) with increasing hypertonicity is included. This value was obtained by correcting for the relative number of methionine residues in the L and H chains determined by measuring the ratio of radioactivity in L and H chains derived from H_2L_2 molecules recovered from polyacrylamide gels.

COMPARISON OF THE EFFECTS OF HIB AND VIRAL INFECTION ON THE RELATIVE
SYNTHESIS OF INDIVIDUAL CELLULAR PROTEINS IN HOST CELLS

Virus induced suppression of host mRNA translation, like HIB,
induces an indiscriminate reduction in the rate of polypeptide chain
initiation (40). This reduction could provide a translational
advantage for viral mRNAs with a relative high translational
efficiency. Were this the case, virus infection and HIB should cause
comparable alterations in the pattern of protein synthesis. A
suitably system for these studies was provided by Ig-G producing
myeloma cells, infected with VSV. Therefore, we compared the effects
of HIB and virus infection on the relative synthesis of IgG and non-
IgG proteins in myeloma cells. Infection of myeloma cells with VSV
resulted in rapid inhibition of total protein synthesis (23) and
alterations in the distribution of labelled amino acids in L, H, and
non-IgG polypeptides that were similar to those observed following
exposure of uninfected cells to HIB (Table 4). Thus in VSV infected
myeloma cells, host mRNAs with high relative translation efficiency
(RTE) - i.e. those for eavy and ight IgG - were more resistant to
virus-directed suppression than other cellular mRNAs.

A virus may favor its own replication by lowering the overall
rate of polypeptide chain initiation for the translation of all
mRNA, in a fashion similar to HIB. Experiments with SV40-infected
BSC-1 cells (Oppermann and Koch, unpublished) demonstrated that the
synthesis of actin (mRNA with low RTE) was preferentially suppressed
by virus infection. This conclusion was further substantiated by
studies on poliovirus-infected HeLa-cells (Fig. 2).

When the incorporation of ^{35}S-methionine into individual proteins
was evaluated by autoradiography and densitometry, it was evident
that the synthesis of each cellular peptide was affected to a
different degree. Whereas the synthesis of histones proceeded
unabated, the synthesis of actin was preferentially suppressed. The
other cellular proteins were also differentially inhibited but were
only characterized by size and we do not know their functions in the
cell. Bientz and coworkers (42), however, did not observe a
differential inhibition of cellular protein synthesis after infection
of HEp2 cells by poliovirus. It is conceivable that shut-off is
brought about by different mechanisms in different cells, or else
that the degree of variation of translational efficiencies of mRNAs
is considerably lower in certain cells.

Yet, translational control can be more. complex, as was first
indicated by our studies of the translation of vaccinia virus mRNA
in L- or HeLa-cells (43). Late in the infection cycle, vaccinia
virus mRNAs with close to average RTEs are preferentially translated
under standard growth conditions whereas "early" virus mRNAs with
high RTEs which are still present are not translated. Similar
observations were obtained with cells infected with frog virus 3 (44).

Table 4. Relative Synthesis of L, H, and Non-IgG Proteins after Infection of MPC-11 Cells by VSV

Time post-infection (min)	^{35}S-methionine incorporation (cpm x 10^{-3})						L/H
	L	% of control	H	% of control	Non-IgG	% of control	
Mock	9.3	(100)	11.7	(100)	144	(100)	1.7
60	8.1	(88)	7.7	(66)	71	(49)	2.3
120	6.1	(66)	5.6	(48)	52	(36)	2.4
180	5.2	(56)	4.4	(38)	44	(31)	2.6
240	3.7	(41)	3.4	(29)	31	(22)	2.4
300	2.3	(25)	2.0	(17)	21	(15)	2.5

The bands corresponding to the IgG L chain, IgG H chain, and VSV proteins, were excised, dissolved in 15% hydrogen peroxide, and counted in Aquasol scintillation fluid. The remainder of each gel lane was then excised in 0.5-cm fractions and subjected to the same process and is reported as ^{35}S cpm non-IgG proteins. The numbers in the columns designated by L, H, and non-IgG represent ^{35}S cpm recovered in the corresponding gel regions. The numbers in parentheses represent the remaining level of synthesis presented as the percentage of the level observed in the mock-infected cells pulse labelled at 60 min. postinfection. Also included is the value for the ratio of L to H synthesis after infection. This value was obtained by correcting for the relative number of methionine residues in the L and H chains, which was determined by measuring the ratio of radio-methionine activity in L and H derived from H_2L_2 molecules recovered from polyacrylamide gels (41).

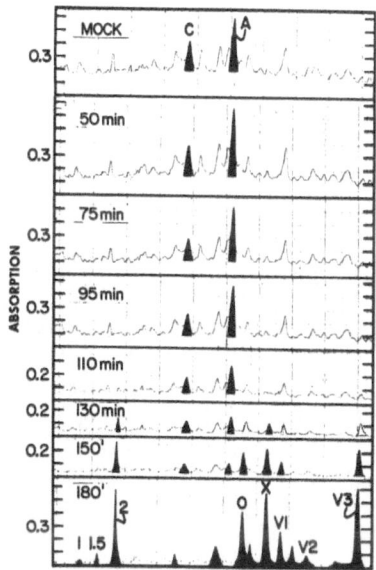

Fig. 2. <u>Alterations in protein synthesis following poliovirus
         infection of HeLa cells.</u> HeLa cells were pulse-labelled
         with (^{35}S)-methionine for 15 min at different times after
         infection, and the proteins were analyzed by polyacrylamide
         gel electrophoresis as in Fig. 1. Densitometric tracings
         of the autoradiographs. Shaded areas indicate the relative
         amounts of actin (A) and a chosen protein (C).

These results indicated that during the viral replication cycle
other alterations in the translational control mechanisms may occur.

COMPETITION BETWEEN VIRAL AND HOST mRNAs

        Since the subject of mRNA competition will be covered in full
length in another section of this volume, we shall limit our discussion here to the case of competition between viral and host mRNAs.
Competition between viral and host mRNAs for ribosomes and initiation
factors is likely to have an impact on host protein synthesis and may
be an adequate explanation for the inhibition of host protein synthesis in all infected cells in which total protein synthesis continues
at the rate observed prior to infection. This is the case in reovirus
infected BHK-cells (40). Also infection of HeLa-cells by VSV results
in only a slight overall reduction in the rate of protein synthesis.
Nevertheless, in both cell systems, synthesis of host proteins con-
tinuously declines, whereas synthesis of viral proteins increases
with time after infection.   Viruses which contain RNA polymerases as
integral components of the virus particle are able to synthesize large
amounts of virus specific mRNA without prior viral protein synthesis.
Such viral mRNA is, therefore, able to outcompete cellular mRNA just
by number. However,both VSV and reovirus mRNAs are high translational

efficiency messengers and, therefore, preferentially translated under
conditions of restricted polypeptide chain initiation (24, 40).

It is not surprising that viral mRNA outcompetes cellular mRNA in
cell-free systems when both mRNAs are added in identical amounts (45).
Golini *et al.* (46) showed that the initiation step of translation
is the site of competition. The effectiveness of binding of the viral
and cellular mRNAs to the 40S ribosomal subunits appears to be unequal
for both mRNA classes suggesting that the components responsible for
competition may be the initiation factors themselves. This suggestion
was supported by the observation that addition of excess crude initiation factors from reticulocytes or plasmacytoma cells could relieve
the competition, thus allowing the translation of cellular mRNAs (10S
RNA or globin mRNA) (46). Since the concentration of initiation
factors in cells is not yet known, it is a matter of speculation to
what extent these observations are relevant for the *in vivo* situation.

EFFECT OF NUTRITIONAL CONDITION ON VIRUS INDUCED SHUT-OFF

The severity of virus induced shut-off is dependent on the
multiplicity of infection and on the nutritional state of the cell.
In the absence of serum, shut-off occurs faster and is more severe.
Interestingly replication of poliovirus, i.e. synthesis of viral
protein and RNA are delayed by approximately 30 minutes after addition
of 5% serum to poliovirus infected cells. The delay in both shut-off
and virus replication exerted by addition of serum suggests that
efficient virus replication depends on interference with macromolecular synthesis in the host cell. This view is supported by studies on
the sensitization of cells for infection by isolated viral RNA.

ROLE OF INITIATION FACTORS IN SHUT-OFF OF HOST PROTEIN SYNTHESIS

Crude initiation factor preparations from poliovirus infected
cells are inactive in stimulating polypeptide chain initiation with
cellular mRNAs although they do support the initiation of translation
of poliovirus RNA (47, 48).

Poliovirus superinfection of VSV-infected cells results in inhibition of VSV as well as host protein synthesis (20, 21). It was,
therefore, predicted that initiation factor preparations from
poliovirus infected cells should be unable to support translation of
VSV mRNA and cellular mRNA but should promote initiation of poliovirus
mRNA translation. Initiation factors from uninfected cells should
allow the translation of all mRNAs. This prediction was experimentally verified by Brown and Ehrenfeld (49). Later on Trachsel *et al.*
(50) showed that initiation factor preparations from poliovirus
infected cells lack a functional cap binding protein which was
previously characterized by Sonenberg *et al.* (see chapter 8 and ref.
51) as a protein with a molecular weight of 24,000. James and
Tershak (52) have recently shown that poliovirus infected cells

contain a 24 kd protein which is higher phosphorylated than a
comparable protein from uninfected cells. It is still to be shown
whether this 24 kd protein is identical with the cap binding protein
mentioned above. Lack of cap binding protein activity, however, is
not the *only* difference between the initiation factors present in
uninfected and poliovirus-infected cells.

Inhibition of host mRNA and VSV mRNA binding to 40S ^{met}tRNA
complexes by ribosomal salt washes or crude initiation factor preparations from poliovirus infected cells has recently been reported
by Brown and Ehrenfeld (49): Poliovirus-infected cells appear to
contain a factor which actively intereferes with the binding of host
mRNA to 40S ^{met}tRNA complexes. Conceivably, such a factor inactivates
or inhibits the function of the cap binding protein. Since host cell
or VSV mRNAs are capped, while the polio mRNA is not, inactivation
of the cap binding protein would provide polio RNA with a translational advantage, and could explain the observed poliovirus induced
shut-off of host cell and VSV mRNAs. However, other viruses (such
as VSV) which do have capped messenger RNAs, also induce the shut-
off phenomenon, though to a different extent in various cell systems
(24). It is possible that such viruses rely on different strategies
for shut-off.

ALTERATION IN THE PHOSPHORYLATION STATE OF RIBOSOMAL AND CYTOPLASMIC
PROTEINS AFTER HIB TREATMENT AND VIRUS INFECTION

Chemical modification of ribosomal proteins and of initiation
factors, notably by phosphorylation, is another mechanism for affecting the control of translation. Globin synthesis in rabbit reticulocytes may be regulated by phosphorylation of the initiation factor
eIF2 (53, 54, 55). SV40-transformed African Green Monkey kidney
cells contain a ribosome-bound, highly phosphorylated protein that
is absent in untransformed cells (56). Correlated with the binding
of the phospho-protein to ribosomes is the ability of transformed
cells to translate some late adenovirus mRNAs while untransformed
cells fail this ability: Polysome and ribosome preparations from
poliovirus-infected HeLa-cells contain 10-15 fold more protein kinase
activity *in vitro* than similar preparations from uninfected cells.
(Tershak, personal communication). Therefore, the phosphorylation
state of ribosomal proteins was analysed after incubation of cells
in isotonic and hypertonic media before and after glucose starvation
and before and after viral infection (57, 58).

In MPC-11 cells a positive correlation between the phosphorylation
of the small subunit protein S6 and the amount of ribosomes engaged
in protein synthesis was observed (57). Raising the tonicity of the
growth medium by addition of 100 mM excess NaCl resulted in a large
increase of single ribosomes, with a concomitant reduction in the
phsophorylation of protein S6 in single ribosomes and in residual
polysomes (Table 5) (58). Phosphoprotein S6 was almost completely

Table 5.   Phosphorylation of S6 of Messenger-Free Ribosomes and
Polysomes from Control and Salt-Treated Cells

| Extra NaCl (mM) | Ribosomes (%) | Polysomes (%) | Specific activity of protein S3 | |
			Ribosomes (Cpm/mg)	Polysomes (Cpm/mg)
0	21	79	900	1840
100	70	30	100	260

The proportion of messenger-free ribosomes and polysomes were
determined from sucrose gradients in high salt buffer. Single ribosomes bound to mRNA have been included in the polysome fraction.
700-900 μg of ribosomal protein were used for two-dimensional gels.
After autoradiography the protein spot S6 was cut out of the gels,
and the radioactivity was measured by Cerenkov counting in a liquid
spectrometer. The cpm-values were normalized for reasons of
comparison.

dephosphorylated after glucose starvation of Ehrlich ascites cells
(59).

It is tempting to speculate that the massive dephosphorylation
of a specific small subunit protein accompanying starvation and a
tonicity shift affects specifically the initiation process of
translation. Several ribosomal proteins within the large subunit
also incorporated ^{32}P, but the changes in the degree of phosphorylation were minimal upon raising the medium tonicity.

Vaccinia virus infected HeLa-cells (60) and VSV infected mouse
L-cells (61) contain ribosomal phosphoproteins not present in
uninfected cells. Changes in the phosphorylation state of 60S
ribosomal proteins are observed in cells 1 1/2 hours after poliovirus
infection. Whether the observed quantitative changes in the ribosomal protein phosphorylation are cause or consequence of virus
induced protein synthesis inhibition remains to be determined.

CHANGES IN PERMEABILITY OF CELL MEMBRANES AFTER VIRUS ADSORPTION

Early interactions of viruses with cellular membranes were reviewed by Kohn (1). During the first few minutes following phage
infection of *E. coli*, the bacterial membrane becomes permeable for
potassium, magnesium, and also to proteins up to the size of β-
galactosidase (62, 4, 5). At the same time infected cells become
permeable for acridine (63). Phage infected bacteria reseal their

leaky membranes and reestablish their norma K^+-concentration gradient within two minutes in a process that requires phage-coded protein synthesis. Similar changes in membrane permeability are exerted by colicins, which are lethal to bacteria. Colicins affect the micro-viscosity of the cell membrane (64) and cause a rapid membrane de-polarization (65). The death of bacteria after adsorption of phage ghosts or colicins is due to loss of K^+ and a decrease in the intra-cellular K^+-concentration below the physiological level needed to sustain protein synthesis and glycolysis (66, 67). Bacteria survive lethal effect of colicin in media containing 100 mM K^+ (68).

Changes in membrane permeability in animal cells after virus ad-sorption were first observed by Klemperer (69) in studies with New-castle disease virus (NDV) and HeLa,Krebs 2 ascites tumor cells and fowl erythrocytes. Hemolysis induced by NDV adsorption was preceeded by loss of potassium and increased cell volume. More intensive investi-gations were performed with Sendai virus. Virus adsorption causes in a number of different tissue culture cells loss of K^+, a transient depolarization of the cell membrane (69, 70, 71), an increased permeability for Na^+ (72), and swelling of cells.

In a recent study on the nature of the Sendai virus mediated changes in membrane permeability Impraim et al. (73) arrived at the following conclusions: a) Small molecules that are accumulated inside the cell by phosphorylation are not retained because of leakage of the phosphorylated metabolites out of the cells. b) Substances that are transported by linkage to the Na^+ gradient are no longer accumulated because of a collapse of the gradient due to an increased permeability to Na^+. c) Increased permeability to Na^+ and K^+ results in membrane depolarization and cell swelling. d) Ca^{++} inhibits the leakage of small molecules. Leakage of cellular constituents as a result of adsorption of different viruses (also UV inactivated viruses) to various host cells has also been demonstrated by gas chromatography studies (74).

Carrasco and Smith (26) and Carrasco (25) have proposed that selective permeability changes to monovalent ions occur after infection of cells by animal viruses and lead to an increase in the intracellular concentration of Na^+ and a decrease in the concentration of K^+. They further speculated that this change in the intracellular environment results in a preferential synthesis of viral proteins while the synthesis of host proteins is suppressed. A study by Egberts et al. (75), however, revealed a *drop* in the intracellular content of Na^+ and a concomitant rise in the content of K^+ early after infection of Ehrlich ascites cells with Mengovirus. Support for the hypothesis of Carrasco was recently reported by Garry et al. (76): They studied early events in cells after infection by Sindbis virus. Whereas protein synthesis in *uninfected* cells was rapidly inhibited when the monovalent cation concentration in the medium was either reduced or elevated, protein synthesis in *infected* cells

proceeded unabated by these changes in the tonicity of the growth
medium. Furthermore, protein synthesis in virus infected cells - in
contrast to uninfected cells - was found to be resistant to exposure
to ouabain (77). The analysis of Na^+ and K^+ content revealed indeed
a rise in Na^+ after infection. However, the rise in Na^+ might not
be caused by alterations in the permeability of the membrane for ions,
but due to a virus-induced inhibition of the Na^+K^+-pump (76). Virus-
induced inhibition of the pump should result in a decrease in membrane
potential. However, the membrane potential is not altered after
infection of HeLa-cells by poliovirus (R. Hiller and Koch, unpublished).

     Changes in membrane fluidity after adsorption of picorna-myxo-
and arboviruses were observed by fluorescence depolarization of DPH
(1,6 diphenyl 1,3,5 hexatriene) (78, 79, 80, 81). These changes
in membrane fluidity were acccompanied by a redistribution of some
membrane proteins (82) and by capping of the adsorbed viruses (83).
Viruses may interfere with interactions between the cytoskeleton and
the membrane (80), promoting alterations in several membrane functions
notably the transport of metabolites and the permeability barrier for
ions (26, 84). The induction of membrane leakiness by many cytopathogenic viruses does not only result in loss of small molecules
from the infected cells but also allows the penetration of a number
of specific inhibitors of protein synthesis ($GppCH_2p$, edeine,
blasticidin S) into the cells which do not enter uninfected cells
(85, 86). Similarly, herpes virus infected cells, but not their
uninfected counterparts, are permeable for nucleotide triphosphates.
Analogs of nucleotide triphosphates can, therefore, be used as
specific antiherpes drugs (87, 88).

INHIBITION OF AMINO ACID TRANSPORT IN CELLS UPON VIRUS INFECTION

     An increase in amino acid transport has been observed as an early
event following the stimulation of cell proliferation in several
systems (89-93). Sander and Pardee (94) found cell cycle-related
changes in amino acid transport. While neither the mechanism nor
the significance of these changes is well understood, it has been
proposed that alterations in membrane transport of small molecules
may play a role in the control of cell proliferation (95, 96).
Treatment of cells with hypertonic medium (HIB), or exposure to
ethanol, DMSO, or DEAE-dextran (97-99) which sensitize cells for
infection by isolated viral RNA, inhibit transport of amino acids.
In order to determine whether similar effects also occur subsequent
to viral infection, we analysed the effect of poliovirus infection
of HeLa-cells on the uptake of ^{35}S-methionine. The experiments and
results are depicted in Fig. 3: After infection with poliovirus
there is a rapid decline in the uptake of amino acids by He La cells,
which is partially reversed from about 90 min. postinfection. The
inhibition of amino acid uptake is maximal at a time when protein
synthesis in infected cells has reached a minimum, that is, the rate

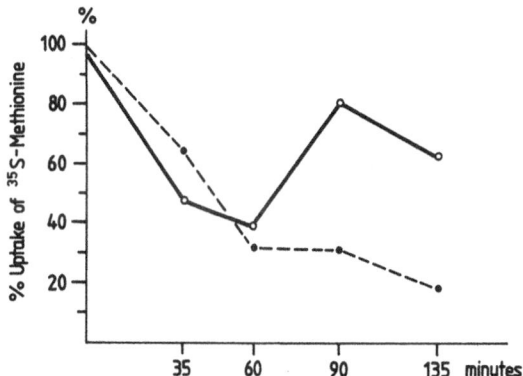

Fig. 3. <u>Effect of poliovirus infection on aminoacid uptake</u>. HeLa S-3
         cells were concentrated to 5 x 10^6 cells/ml and divided
         into replicate cultures: one half was infected with 200
         PFU/cell of poliovirus type I Mahoney strain, while the
         remaining one was mock infected. After 30 min adsorption
         at 37°C the cells were diluted to 5 x 10^5 cells/ml and
         monitored for amino acid uptake at the indicated time points.
         Aminoacid uptake was followed in duplicate 2-ml aliquots
         of controls and infected cells. Replicates were labelled
         for 10 min with (^{35}S)-methionine and the cells in duplicate
         200-µl samples were collected onto glass-fibers filters
         (Whatman GF/c). After washing, the cell-associated radio-
         label was determined ( ). At the same time, 100-µl
         aliquots were used to determine the incorporation of the
         label into the TCA-precipitable material (0).

of amino acid uptake and the rate of protein synthesis decline
concomitantly.

    Presently, we do not know whether this inhibition of aminoacid
transport is the *cause* (as is the case with amino acid starvation)
or *consequence* of reduced protein synthesis, via a feedback control
of aminoacid uptake (112).

    As already mentioned, in Ehrlich ascites tumor cells the intra-
cellular level of K$^+$ rises concomitantly with a drop in the level
of Na$^+$ following infection by Mengovirus (75). Transport of several
amino acids in tissue culture cells is coupled to transport of Na$^+$;
wether a correlation exists between reduced aminoacid uptake and a
decreased content of Na$^+$ in infected cells is a goal of future investigations.

ROLE OF CELL MEMBRANE IN MEDIATING THE PLEIOTROPIC RESPONSE

There are a number of arguments that implicate that the cell
membrane plays an important role in mediating, both the negative
pleiotropic response and the virus-induced shut-off phenomenon (23,
24, 111). A wide variety of substances inhibit protein synthesis
in intact cells but not in cell-free extracts from these cells,
implicating a mediatory role of the cell membrane in the inhibitory
process (23). The membrane functions as a barrier between the cell
and its environment, thus environmental perturbations are necessarily
communicated to the cell interior via the membrane. For example,
virus adsorption to cell surface receptors causes conformational
alterations in the cell membrane, as manifest by changes in membrane
fluidity and permeability to cations and other low molecular weight
substances (1, 25, 26).   Conditions mentioned above which inhibit
protein synthesis at the level of initiation all lead to a
reduced uptake of amino acids (24).   As already discussed,
infection with poliovirus leads to a rapid decline in the uptake of
amino acids in HeLa-cells, concomitantly with a reduction in the
rate of protein synthesis. The mechanism whereby such apparent
membrane mediated inhibition of protein synthesis is affected, is
still uncertain. This process may involve the activation of a
normal host cell mechanism used to regulated protein synthesis at
the translational level. No detectable endogenous protein synthesis
in vitro is present in crude extracts prepared from cells 2 to 3
hours after infection with poliovirus, 45 minutes after transfer to
amino acid deficient medium, or from cells exposed to HIB for 15
minutes. Protein synthesis in these extracts can be *partially*
restored upon Sephadex G-25 gel filtration of the extracts, indicating
that neither ribosomes nor mRNA are irreversibly inactivated by
exposure of cells to these three conditions. Late eluting fractions
from the Sephadex G-25 column contain low molecular weight substances
(including amino acids) which inhibit protein synthesis in cell-free
extracts from untreated cells.

The inhibitor of protein synthesis isolated by gel filtration
appears to have the following properties: Upon dialysis of the cell
extract the inhibitor is found in the dialysate. Gel filtration over
G-15 and G-10 Sephadex reveals a molecular weight *below* 1,000. The
inhibitor does not give the ninhydrin reaction, indicating absence
of free primary or secondary amino groups. During high voltage paper
electrophoresis at pH 3.5, the inhibitor remains at the origin,
indicating the absence of anionic groups e.g. carboxyl or phosphate
residues. The compound is quite hydrophobic because it migrates on
cellulose thin layer plates using butanol as solvent and is bound to
XAD2 columns at pH 2.0 in the presence of 2M $CaCl_2$. It also binds
strongly to glass surfaces, thus hampering the isolation and
purification. By adding the compound to an *in vitro* protein
synthesizing system its strong inhibitory activity is apparent. The
inhibitor binds to ribosomes during sucrose gradient centrifugation.

We propose that the inhibitor which acts at the level of initiation is released or activated by a membrane mediated event and is reversibly bound to ribosomes. Ribosomes with bound inhibitor are less active or inactive in the formation of initiation complexes.

INTERACTION BETWEEN VIRAL PROTEINS AND CELLULAR CONSTITUENTS

One function of the viral coat is to protect the viral RNA from attack by RNA-degrading enzymes, in addition the protein coat is required for the efficient adsorption of viruses to their host cells. When polioviruses are exposed to diethylpyrocarbonate (DEP, Baycovine, Naftone Inc., New York) both these functions are affected and viruses are no longer adsorbed to their host cells. Loss of adsorbability can be restored by addition of DEAE-Dextran, but the infectivity of DEP-exposed poliovirus remains the same as that of isolated viral RNA (100, 101). Provided that neither loss of RNAse resistance nor adsorbability are responsible for the comparatively low infectivity of naked RNA and of virus particles inactivated by heat or DEP, the viral coat proteins must have yet another function, somehow in supporting the efficiency of infection. Coat proteins of infecting parental virus particles are found in association with virus specific polysomes (102). Matthews *et al.* (106) and Wright and Cooper (27) proposed that association of viral proteins with host ribosomes are responsible for suppression of host mRNA translation. Racevskis *et al.* (28) observed an inhibition of initiation of protein synthesis in reticulocyte lysates after addition of intact polioviruses. Interference with host protein synthesis might, therefore, be an important third function of the viral coat proteins.

ACKNOWLEDGEMENT

Work of the authors was supported in part by Deutsche Forschungsgemeinschaft.

REFERENCES

1. Kohn, A. (1979), in *Advances in Virus Research* (M.A. Laufer, F.B. Bang, K. Maramorosch and K.M. Smith, eds.), pp. 233-276, Academic Press, New York.
2. Bablanian, R. (1975), in *Progress in Medical Virology* (J.L. Melnick, ed.), pp. 40-83, Karger Verlag, Basel.
3. Lucas-Lenard, J.M. (1979), in *The Molecular Biology of Picornaviruses* (R. Perez-Bercoff, ed.), pp. 73-93, Plenum Press, New York.
4. Puck, T.T. and Lee, H.H. (1954), J. Exptl. Med. 99, 481-494.
5. Puck, T.T. and Lee, H.H. (1955), J. Exptl. Med. 101, 151-175.

6. Koch, G. and Jordan, E.M. (1957), Biochim. Biophys. Acta 25, 437.

7. Tershak, D.R., Yin, F.H. and Korant, B.D. (1981), in *Handbook of Experimental Pharmacology* (L. Calaguiri and P. Came eds.), Springer Verlag, Berlin-N. York.

8. Ehrenfeld, E. and Hunt, T. (1971), Proc. Natl. Acad. Sci. USA 68, 1075-1078.

9. Moss, B. (1968), J. Virol. 2, 1028-1037.

10. Helentjaris, T. and Ehrenfeld, E. (1977), J. Virol. 21, 259-267.

11. Steiner-Pryor, A. and Cooper, P. (1973), J. Gen. Virol. 21, 215-225.

12. Cole, C.N. and Baltimore, D. (1973), J. Mol. Biol. 76, 325-343.

13. Franklin, R.M. and Baltimore, D. (1962), Cold Spring Harbor Symp. Quant. Biol. 27, 175-198.

14. Penman, S., Scherrer, K., Becker, Y. and Darnell, J.E. (1963), Proc. Natl. Acad. Sci. USA 49, 654-661.

15. Lawrence, C. and Thach, R. (1974), J. Virol. 14, 598-610.

16. Colby, D.S., Finnerty, V. and Lucas-Lenard, J. (1974), J. Virol. 13, 858-869.

17. Fernandez-Munoz, R. and Darnell, J.E. (1976), J. Virol. 18, 719-726.

18. Koschel, K. (1974), J. Virol. 13, 1061-1066.

19. Gallwitz, D., Traub, U. and Traub, P. (1977), Eur. J. Biochem. 81, 387-393.

20. Doyle, M. and Holland, J.J. (1972), J. Virol. 9, 22-28.

21. Ehrenfeld, E. and Lund, H. (1977), Virology 80, 297-308.

22. Schweïger, M., Wagner, E.F., Hirsch-Kaufmann, E.F., Ponta, H. and Herrlich, P. (1978), in *Gene Expression* (B.F.C. Clark, H. Klenow and J. Zeuthen, eds.), pp. 171   , Pergamon Press, New York.

23. Koch, G., Oppermann, H., Bilello, P., Koch, F. and Nuss, D. (1976) in *Modern Trends in Human Leukemia II* (R. Neth, R.C. Gallo and K. Mannweiler, eds.), pp. 541-555, J.F. Lehmanns Verlag, München.

24. Koch, G., Bilello, J.A., Kruppa, J., Koch, F. and Oppermann, H. (1980), Ann. N.Y. Acad. Sci., 344, 280-306.

25. Carrasco, L. (1977), FEBS Lett. 76, 11-15.

26. Carrasco, L. and Smith, A.E. (1976), Nature, London, 264, 807-809.

27. Wright, P.J. and Cooper, P.D. (1974), Virology, 59, 1-20.

28. Racevskis, J., Kerwar, S.S. and Koch, G. (1976), J. Gen. Virol. 31, 135-138.

29. *Control of Proliferation in Animal Cells, Cold Spring Harbor Conferences on Cell Proliferation*, Vol. 1 (B. Clarkson and R. Baserga, eds.), Cold Spring Harbor Laboratory, 1974.

30. Vaughan, N.P., Pawlowski, P. and Forchhammer, J. (1971), Proc. Natl. Acad. Sci. USA 68, 2057-2061.

31. Van Venrooij, W.J.W., Henshaw, E.C. and Hirsch, C.A. (1972), Biochem. Biophys. Acta 259, 127-137.

32. Levine, E.M., Becker, Y., Boone, C.W. and Eagle, H. (1965), Proc. Natl. Acad. Sci. USA 53, 350-355.

33. Giloh, H. and Mager, J. (1975), Biochem. Biophys. Acta 414, 293-308.

34. McCormick, W. and Penman, S. (1969), J. Mol. Biol. 39, 315-333.
35. Saborio, J.L., Pong, S.S. and Koch, G. (1974), J. Mol. Biol. 85, 195-211.
36. Oppermann, H., Saborio, J., Zarucki, T. and Koch, G. (1973), Fed. Proc. 32, 531 A.
37. Laskov, R. and Scharff, M.D. (1974), J. Exptl. Med. 140, 1112-1116.
38. Sonenshein, G.E. and Brawerman, G. (1976), Biochemistry 15, 5497-5501.
39. Nuss, D.L. and Koch, G. (1976), J. Mol. Biol. 102, 601-612.
40. Nuss, D.L., Oppermann, H. and Koch, G. (1975), Proc. Natl. Acad. Sci. USA 72, 1258-1262.
41. Nuss, D.L. and Koch, G. (1976), J. Virol. 17, 282-286.
42. Bienz, K., Egger, D., Rasser, Y. and Bossart, W. (1980), Virology 100, 390-399.
43. Oppermann, H. and Koch, G. (1976), J. Gen. Virol. 32.
44. Willis, D.B., Goorha, R., Miles, M. and Granoff, A. (1977), J. Virol. 24, 326-342.
45. Abreu, S. and Lucas-Lenard, J. (1976), J. Virol. 18, 184-192.
46. Golini, F., Thach, S.S., Birge, C.H., Safer, B., Merrich, W.C. and Thach, R.E. (1976), Proc. Natl. Acad. Sci. USA 73, 3040-3044.
47. Helentjaris, T.G. and Ehrenfeld, E. (1978), J. Virol. 26, 510-521.
48. Helentjaris, T.G., Ehrenfeld, E., Brown-Lueidi, M.L. and Hershey, J.W.B. (1979), J. Biol. Chem. 254, 10973-10978.

49. Brown, B.A. and Ehrenfeld, E. (1980), Virology 103, 327-339.
50. Trachsel, H., Sonenberg, N., Shatkin, A.J., Rose, J.K., Leong, K., Bergmann, J.E., Gordon, J. and Baltimore (1980), Proc. Natl. Acad. Sci. USA 77, 770-774.
51. Sonnenberg, N., Rupprecht, K.M., Hecht, S.M. and Shatkin, A.J. (1979), Proc. Natl. Acad. Sci. USA 76, 4345-4349.
52. James, L.A. and Tershak, D.R. (1981), Can. J. Microbiol., in press
53. Benne, R., Edman, J., Traut, R.R. and Hershey, J.W.B. (1978), Proc. Natl. Acad. Sci., USA 75, 108-112.
54. Kramer, G., Cimadevilla, J.M. and Hardesty, B. (1976), Proc. Natl. Acad. Sci. USA 73, 3078-3082.
55. Levin, D.H., Ranu, R.S., Ernst, V. and London, I.M. (1976), Proc. Natl. Acad. Sci. USA 73, 3112-2116.
56. Segawa, K., Yamaguchi, N. and Oda, K. (1977), J. Virol. 22, 679-693.
57. Martini, O.H.W. and Kruppa, J. (1979), Eur. J. Biochem. 95, 349-358.
58. Kruppa, J. and Martini, O.H.W. (1978), Biochem. Biophys. Res. Commun. 85, 428-435.
59. Kruppa, J. (1979), Membrangebundene Biosynthese zellulärer und viraler Proteine in Säugetierzellen. Habilitationsschrift, Hamburg University, Hamburg.
60. Kaerlein, M. and Horak, I. (1976), Nature, London, 259, 150-151.
61. Marvaldi, J. and Lucas-Lenard, J. (1977), Biochemistry 16, 4320-4327.

62. Doerman, A.H. (1948), J. Bacteriol. 55, 257-276.
63. Hessler, A.Y., Baylor, M.B. and Baird, J.P. (1967), J. Virol. 1, 543-549.
64. Helgerson, S.L., Cramer, W.A., Harris, J.M. and Lytle, F.E. (1974), Biochemistry, 13, 3057-3061.
65. Tokuda, H. and Konisky, J. (1978), Proc. Natl. Acad. Sci. USA 75, 2579-2583.
66. Shapira, A., Giberman, E. and Kohn, A. (1974), J. Gen. Virol. 23, 159-171.
67. Gould, J.M. and Cramer, W.A. (1977), J. Biol. Chem. 252, 5491-5497.
68. Kopecky, A.L., Copeland, D.L. and Lusk, J.E. (1975), Proc. Natl. Acad. Sci. USA 72, 4631-4634.
69. Klemperer, H.G. (1960), Virology 12, 540-552.
70. Pasternak, C.A. and Micklem, K.J. (1973), J. Membr. Biol. 14, 293-303.
71. Spegelstein, P.F., Haimsohn, M., Gitelman, J. and Kohn, A. (1978), J. Cell. Physiol. 95, 223-233.
72. Poste, G. and Pasternak, C.A; (1978), Cell Surf. Rev. 5, 305-367.
73. Impraim, C.C., Foster, K.A., Micklem, K.J. and Pasternak, C.A. (1980), Biochem. J. 186, 847-860.
74. Levanon, A., Klibansky, Y. and Kohn, A. (1977), J. Med. Virol. 1, 227-237.
75. Egberts, E., Hackett, P.B. and Traub, P. (1977), J. Virol. 22, 591-597.
76. Garry, R.F., Bishop, J.M., Parker, S., Westbrook, K., Lewis, G. and Waite, M.R. (1979), Virology 96, 108-120.
77. Garry, R.F., Westbrook, K. and Waite, M.R.F. (1979), Virology 99, 179-182.
78. Levanon, A., Kohn, A. and Inbar, M. (1977), J. Virol. 22, 353-360.
79. Levanon, A. and Kohn, A. (1978), FEBS Lett. 85, 245-248.
80. Lyles, D.S. and Landsberger, F.R. (1977), Proc. Natl. Acad. Sci. USA 74, 1918-1922.
81. Moore, N.F., Patzer, E.J., Shaw, J.M., Thompson, T.E. and Wagner, R.R. (1978), J. Virol. 27, 320-329.
82. Bächi, T., Aguet, M. and Howe, C. (1973), J. Virol. 11, 1004-1012.
83. Gschwender, H.H. and Traub, P. (1979), J. Gen. Virol. 42, 439-442.
84. Mickle,; K.J. and Pasternak, C.A. (1977), Biochem. J. 162, 405-410.
85. Contreras, A. and Carrasco, L. (1979), J. Virol. 29, 114-122.
86. Lacal, J.C., Vasquez, D., Fernandez-Sousa, J.M. and Carrasco, L. (1980), J. antibiotics, 33, in press.
87. Gauri, K.K. and Albrecht, G. (1978), in *Advances in Ophtalmology* (K.K. Gauri, ed.), pp. 64-71, Karger, Basel.
88. Koch, G. (1981), in preparation.
89. Costlow, M. and Baserga, R. (1973), J. Cell. Physiol. 82, 411-419.
90. Vaheri, A., Ruoslahti, E., Hovi, T. and Nordling, S. (1973), J. Cell. Physiol. 81, 355-364.

91. Foster, D.O. and Pardee, A.B. (1969), J. Viol. Chem. <u>244</u>, 2675-2681.
92. Isselbacher, K.J. (1972), Proc. Natl. Acad. Sci. USA, <u>69</u>, 585-593.
93. Villereal, M.L. and Cook, J.S. (1977), J. Supramol. Struct. <u>6</u>, 179-189.
94. Sander, G. and Pardee, A.B. (1972), J. Cell. Physiol. <u>80</u>, 267-271.
95. Holley, R.W. (1972), Proc. Natl. Acad. Sci. USA, <u>69</u>, 2840-2841.
96. Pardee, A.B. (1964), Nat. Cancer Inst. Monogr. <u>14</u>, 7-20.
97. Saborio, J.L., Wiegers, K.J. and Koch, G. (1975), Arch. Virol. <u>49</u>, 81-87.
98. England, J.M., Howett, M.K. and Tan, K.B. (1975), J. Virol. <u>16</u>, 1101-1107.
99. Koch, F., Pawlowski, P.J. and Lukens, L.N. (1977), Arch. Biochem. Biophys. <u>178</u>, 373-380.
100. Oberg, B. (1970), Biochem. Biophys. Acta <u>204</u>, 430-440.
101. Breindl, M. and Koch, G. (1972), Virology <u>48</u>, 136-144.
102. Franklin, R.M. and Wecker, E. (1959), Nature, London, <u>174</u>, 343-345.
103. Bishop, J.M. and Koch, G. (1969), in *Fundamental Techniques in Virology* (K. Habel and N.E. Salzman, eds.) mmP 131-145, Academic Press, New York.
104. Borgert, K., Koschel, K., Täuber, H. and Wecker, E. (1971), J. Virol. <u>8</u>, 1-6.
105. Habermehel, K.O., Diefenthal, W. and Buchholz, M. (1974), in *Advances in the Biosciences* (G: Raspé, ed.), pp. 41-63, Pergamon Press, Vieweg, Oxford.
106. Matthews, T.J., Butterworth, B.E., Chaffin, L. and Rueckert, R.R. (1973), Fed. Proc. <u>32</u>, 461.
107. Koch, G. (1973), Current Topics in Microbiol. Immunol. <u>62</u>, 89-138.
108. Saborio, J.L. and Koch, G. (1973), J. Biol. Chem. <u>248</u>, 8343-8347.
109. Koch, F. and Koch, G. (1974), Res. Commun. Chem. Path. Pharm. <u>9</u>, 219.
110. Koch, G. and Oppermann, H. (1975), Virology <u>63</u>, 395-403.
111. Koch, G., Bilello, P. and Kruppa, J. (1979), in *Regulation of Macromolecular Synthesis by Low Molecular Weight Mediators* (G. Koch and D. Richter, eds.), pp. 273-290, Academic Press, New York.
112. Ring, K. and Heinz, E. (1966), Biochem. Z. <u>344</u>, 446-461.

SECTION VI:

MECHANISMS OF REGULATION AND CONTROL

THE CYTOPLASMIC CONTROL OF PROTEIN SYNTHESIS

Richard J. Jackson

Department of Biochemistry, University of Cambridge

Cambridge, CB2 1QW, England

INITIAL REMARKS

After a long period during which to claim to be studying translational control was to invite a variety of negative responses from sympathetic looks to undisguised sneers, it has now become such a respectable topic that no fewer than five reviews published in the last two years (1-5) are entirely devoted to it or treat it as a major topic, and a sixth will shortly appear (6). In these circumstances, the last thing in the world that is needed is yet another review which does nothing more than cover the same ground once again. Through a stroke of lucky timing, I do have something new to say on the control of reticulocyte protein synthesis by sugar phosphates and reducing agents, which is discussed in Section VI. In order that those who are unfamiliar with this field should be able to see this control system in its context, I must summarise what is known about other reticulocyte control systems (Sections III-V); I hope that any reader who already feels saturated by this information will forgivingly skip these sections. Finally, I have felt compelled to make some comments on the control of translation in other systems (Sections II and VII), since it seems to me that the reticulocyte system may be attracting too much attention, with the result that regulatory phenomena in other cells are all too often interpreted solely in terms of the reticulocyte model instead of being examined with an open mind. This is not a comprehensive review in the normal sense; I have made no effort to digest the whole mass of literature and to record every contribution. It is my personal interpretation of the state of knowledge (or lack of it) and references are therefore given only to illustrate particular points and to provide the reader with an entry into the literature.

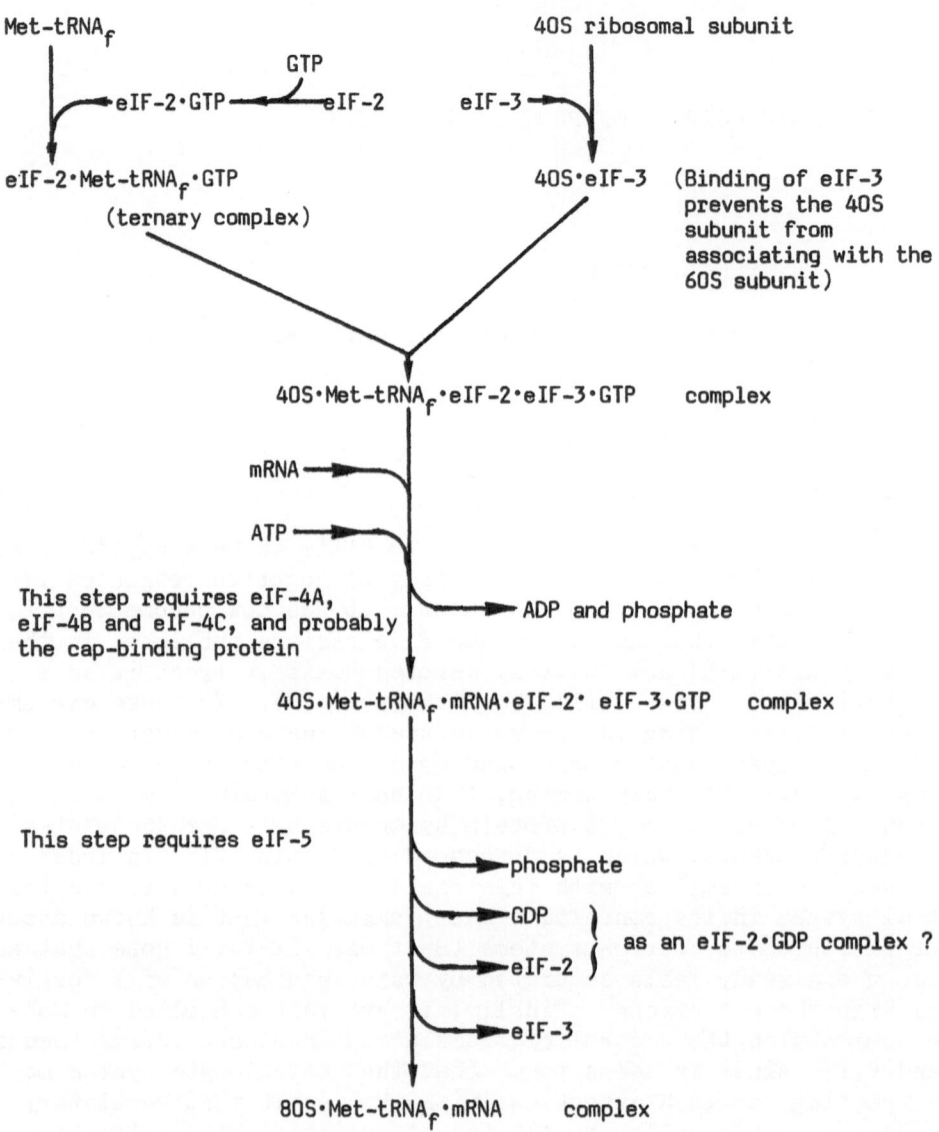

Fig. 1.   The probable sequence of events in the initiation of mam-
          malian protein synthesis. The initiation factor require-
          ments for each step are shown except that the function of
          eIF-1 is difficult to assign to any one step. The asso-
          ciation of initiation factors with the different interme-
          diate complexes is known with certainly only for eIF-2 and
          eIF-3.

I. THE MECHANISM OF INITIATION: OUTSTANDING PROBLEMS

Most of us interpret our experiments on the control of initiation in terms of a rather simple picture of the mechanism of initiation such as that shown in Fig. 1. This scheme is based on the work of the N.I.H. group (2) and of the groups of Staehelin (7, 8) and Hershey (9). Other chapters in this volume discuss this mechanism in more detail, but there remain several questions which those of us who are interested in the control of initiation would like to see answered by the experts.

Are Additional Initiation Factors Required ?

The first problem is really a two-part question. Why is the activity of the highly fractionated system used to study and assay the purified initiation factors so poor compared with the activity of the crude reticulocyte lysate from which the factors are derived (7) ? Are there yet more factors to be discovered, and if so, would these additional factors enhance the rate of initiation in the fractionated system, which is typically only 3-5% of the rate in the crude lysate ? As far as the existence of additional factors is concerned, it is already clear that the answer is affirmative. Gupta's group has described three preparations from rabbit reticulocytes which they call Co-eIF-2A, B and C, since these protein fractions stimulate or modulate the activity of eIF-2 in the formation of ternary complexes with Met-tRNA$_f$ and GTP (extensively reviewed elsewhere in this volume and summarised in ref. 4). Since these Co-eIF-2s could not readily by accommodated in the scheme shown in Fig. 1 there has been a tendency to ignore them, and to rationalise this stance by saying that the assay system which revealed the activity of these factors was rather remote from 'real' protein synthesis and therefore possibly subject to trivial artefacts. It is now evident that we can no longer adopt this ostrich-like attitude, at least in the case of Co-eIF-2A. An antibody raised against this protein has been shown to strongly inhibit protein synthesis in the highly active crude reticulocyte lysate (10). Co-eIF-2A, then, must play an important role in initiation. What this role is precisely, remains unclear. One would like to know what step in the initiation process (Fig. 1) is inhibited when lysates are supplemented with the antibody. The importance and incisiveness of this antibody test cannot be over-emphasised and one would like to see it extended to other factors such as Co-eIF-2B and C, which have not yet gained universal recognition as obligatory initiation factors.

Another new factor which needs to be accommodated in the scheme is the 24,000 dalton cap-binding protein, which appears to be necessary for initiation on capped mRNAs but not on uncapped (11, 12). This protein had been lurking as a contaminant in the eIF-3 preparations of some laboratories or in the eIF-4B preparation of the Basel group (11). Apart from the question of the exact role of this

protein in initiation, an important subsidiary point is whether this
'contaminant' is in fact the only active entity in eIF-4B preparations. Does the major protein of about 80,000 daltons (7, 9) have any
role at all ? If the cap-binding protein were the only active entity
in eIF-4B one would expect that eIF-4B would not be required for
initiation on uncapped mRNAs. Some years ago, the Basel group reported that eIF-4B was required for translation of the uncapped EMC RNA
in a fractionated system (13). On the other hand Thach and his colleagues found eIF-4B to be less important for EMC translation than
for the translation of globin mRNA or other cellular (capped) mRNAs
(14), but their system cannot reveal whether eIF-4B is totally redundant for the uncapped mRNAs. The question seems still open as to
whether initiation on uncapped mRNA has no requirement for eIF-4B, or
whether it merely requires lower amounts of this factor than is the
case with capped mRNAs.

Are Initiation Factors Specific with Respect to mRNA ?

Are there initiation factors specific for certain mRNA species or
for a restricted group of mRNAs (apart from the cap-binding protein) ?
The present evidence points *against* the idea of specific initiation
factors, and the same set of factors seems to be required for all
mRNAs. This question is of such crucial importance that one would
like to see it re-examined. If, as seems likely, all mRNAs (except
uncapped mRNAs) use the same factors, then it follows that any modification of the activity of an initiation factor must affect the rate
of initiation of all mRNA species, although the quantitative magnitude
of the change in the rate of initiation need not be the same for all
mRNAs (15).

The Recycling of Initiation Factors

Are there special proteins which aid the recycling of initiation
factors between successive initiation events and, in particular, how
does eIF-2 recycle ? It is common experience that GDP strongly inhibits the formation of ternary complexes between eIF-2, Met-tRNA$_f$
and GTP (2). Ternary complex formation is stimulated by the presence
of either pyruvate kinase plus phosphoenolpyruvate, or an ATP-
generating system plus nucleoside diphosphate kinase, both of which
convert any GDP contaminant or by-product to GTP (16). The affinity
of eIF-2 for GDP is approximately hundred-fold greater than for GTP
(16). In the normal conditions prevailing in the cell it seems likely
that the GTP/GDP ratio is sufficiently high that a significant
fraction of the eIF-2 would bind GTP (and hence also Met-tRNA$_f$) rather
than GDP. But a potential complication is introduced by the fact that
eIF-2 leaves the initiation complex at the step of subunit joining
(Fig. 1), the very step where GTP hydrolysis occurs. It is therefore
far from improbable that the eIF-2 leaves as an eIF-2.GDP complex, in
which case the dissociation of this complex might be the rate limiting

step in the recycling of eIF-2 for another round of initiation. Is
an additional protein needed to facilitate the dissociation of this
complex in the same way that EFT$_s$ catalyses the dissociation of
EFT$_u$.GDP complexes in bacteria ? There is little data bearing on
this question since eIF-2 does not recycle in most of the assays
currently in use, nor, to my knowledge, has anyone looked for proteins
which stimulate ternary complex formation starting from eIF-2.GDP
complexes. Clemens has discovered a GDP'ase which relieves the inhibitory effect of GDP and hence stimulates ternary complex formation
in conventional assays (17). However, the GDP'ase does not seem to
hydrolyse GDP bound to eIF-2, and so no stimulation of ternary complex
formation occurs when one starts from an eIF-2.GDP complex (M. Clemens
personal communication).

       This is an important question since eIF-2 seems to be the target
of many regulatory phenomena, and it is intriguing that the phosphorylation of eIF-2 occurs in the subunit which is thought to bind GTP
(18). Could phosphorylation of this subunit affect the recycling of
eIF-2 by interfering with the dissociation of eIF-2.GDP complexes ?
Clearly, we need to know more about the functional interactions of
eIF-2: is it released as an eIF-2.GDP complex, and do special
proteins catalyse the dissociation of this complex ? However, when
GDP was added to crude lysates it was *elongation* and not *initiation*
that was inhibited (19), so it may turn out that the significance
which we attach to the high affinity of eIF-2 for GDP is exaggerated,
but I don't think that this removes the necessity for further investigation into the mechanism of recycling of this factor.

The Selection of the Initiation Site on mRNA

       How is the correct initiation site on the mRNA recognised and
selected ? The *scanning ribosome* model proposed and reviewed by Kozak
in the appropriate section of this book (see also ref. 20), supported
as it is by very elegant experiments (21, 22), must clearly be a
very close approximation to the truth, but there are still some
puzzling details. Do the 40S ribosomal subunits recognise the 5'-end
of the mRNA and then slide or jump to the first AUG codon, or is it
a physical threading of the mRNA through the 40S subunit-initiation
factor complex ? In a *threading* model, it is hard to see why secondary structure in the mRNA between the 5'-end and the initiation site
should not be unwound, and this makes it difficult to invoke secondary structure to explain those cases (notably SV 40 VP1 mRNA) where
the first AUG codon is *not* the initiation site (20). Why do 5'
fragments of Brome Grass Mosaic Virus RNA -3 which comprise the 5'
cap but not the first AUG codon bind an 80S ribosome (23) when the
model would seem to predict the binding of only a 40S subunit ? A
similar problem is raised by the fact that Tobacco Mosaic Virus RNA
binds two 80S ribosomes, one at the initiation site and the other at
some point between the cap and the first AUG codon (24). Does
initiation on *uncapped* mRNAs proceed by a radically different

mechanism, or is it again a case of the ribosome binding to the 5'-end
and moving to the first AUG codon ? In the case of the cardiovirus
RNAs (EMC or Foot and Mouth Disease Virus), which contain a large
poly(C) tract near the 5'-end, the initiation site can be placed with
some certainty at a minimum of 500-1000 residues (depending on the
virus strain) from the 5'-end (25, 26). If ribosomes reach this site
by traversing the RNA from the 5'-end, one might expect to see queues
of ribosomes - or 40S subunits - building up on this long stretch, but
in our hands EMC RNA binds only one 80S ribosome in a conventional
ribosome binding assay (see Fig. 2).

     Perhaps the most important question for those interested in
translational control concerns the molecular basis underlying the
different initiation rates seen on different mRNAs. It is generally
accepted that initiation proceeds more efficiently on some mRNA
species (e.g. β-globin mRNA) than on others (α-globin mRNA), so that
when there is competition between mRNA species for a limited initiation capacity, initiation occurs more frequently on the more efficient
mRNA (15). In any model in which a nucleotide sequence is recognised
as such (e.g. the "Shine and Dalgarno" sequences of bacterial mRNA
initiation sites) it is conceptually easy to explain such differences
as residing in the actual nucleotide sequences concerned. What is
the mechanistic basis of these differences in the scanning ribosome
model ? Is the efficiency of cap recognition different for different
mRNAs; is it the rate of scanning that differs; or is it the efficiency with which the 40S subunit arrests at the first AUG codon and
becomes committed to initiation at that site ? (For an extensive
discussion on the competition of different mRNAs for eIF-2, the
reader is referred to the contribution of R. Kaempfer elsewhere in
this volume).

What is the Role of ATP in Initiation ?

     The requirement for ATP is a feature of eukaryotic initiation
not shared by the bacterial system. Unlike all the GTP-requiring
steps in protein synthesis where the use of non-hydrolysable analogues
allows at least some partial reaction to occur, non-hydrolysable
analogues of ATP seem to have no action whatsoever in substituting
for ATP in initiation (8, 13). Nor does the ATP seem to be required
for the phosphorylation of a protein (27). Two years ago at the
Maratea ASI on Picornaviruses I was rash enough to speculate that
ATP hydrolysis might be the driving force for the migration of the
ribosome from the 5' cap to the AUG initiation codon (47). This
prompted me to test whether I could show an ATP requirement for mRNA
binding to ribosomes in crude reticulocyte systems, and whether all
types of mRNA exhibited the same requirement. In an attempt to obtain
an ATP-free system I used gel-filtered lysates. These turn out to
retain a small ATP pool (about 40 µM) but if the preincubation with
sparsomycin (the customary first step in initiation complex formation
assays) is carried out in the absence of an energy generating system,

this ATP is converted to ADP and AMP. If GMPPCP (guanylyl, β, γ-methylene diphosphonate) and ^{35}S-Met-tRNA$_f$ are added after the pre-incubation step, sucrose gradient analysis shows that ^{35}S-Met-tRNA$_f$ binds almost exclusively to the native 40S subunits (Fig. 2). To test for the binding of mRNA to 40S subunits, various types of mRNA (particularly those large viral RNAs which form complexes with native 40S subunits that sediment considerably faster than 40S and so are readily distinguished from free 40S subunits) together with ATP or non-hydrolysable analogues were added immediately following the GMPPCP and ^{35}S-Met-tRNA$_f$, and 2 minutes later the samples were ana-lysed on sucrose gradients. Initiation complex formation on capped RNAs such as Tobacco Mosaic Virus (TMV) RNA or Papaya Mosaic Virus (PMV) RNA showed an absolute requirement for ATP which could not be fulfilled by AMP, AMPPCP, AMPPNP (Adenylyl imidodiphosphate) or AMPCPP (the α, β methylene analogue of ATP). (Although this type of experi-ment does not lend itself to detailed kinetic analysis, my results did not convince me that the non-hydrolysable analogues were acting competitively with respect to ATP.) Maximum initiation complex formation required approximately 0.4 mM ATP: a concentration of 0.1 mM gave about 60% of maximum yield, and 0.04 mM about 30%. The binding of ^{3}H-vaccinia mRNA (transcribed from corse *in vitro* in the presence of S-adenosyl methionine) was also absolutely dependent on ATP, and the binding of 40S subunits to the endogenous globin mRNA in polysomes was certainly highly dependent - if not absolutely dependent on ATP. On the other hand AMPPCP allowed binding of 40S subunits to the *uncapped* Cowpea Mosaic Virus (CPMV) RNA, and this binding was not greatly stimulated if ATP was used (Fig. 2). A similar ATP- *independence* of initiation complex formation seemed to hold true for EMC RNA and poly AUG although the interpretation of the data in these cases is less unambiguous.

The experiments were repeated using GTP in place of GMPPCP to allow the formation of 80S (and larger) initiation complexes. The use of GTP however, introduces a complication in that the nucleoside diphosphate kinase in the lysate transfers the γ-phosphate from GTP to the ADP present in the system after preincubation with sparso-mycin. Thus in these experiments the system is *not* ATP-free but contains about 0.04 mM ATP - a level which gave about 30% of maximum formation of 40S/Met-tRNA$_f$ complexes in the presence of GMPPCP. What is observed is that some 80S initiation complex formation takes place on TMV RNA, PMV RNA, globin mRNA or vaccinia mRNA even in the presence of AMPPCP or in the absence of any added adenine nucleotide, and one has a difficult decision of interpretation to make as to whether this is due to the endogenous 0.04 mM ATP or not. If additional ATP is added, 80S initiation complex formation is stimu-lated two-four fold on these mRNAs (Fig. 2); the stimulation seemed consistently higher with some RNAs (PMV RNA, for example) than others. On the other hand, with the uncapped mRNAs - EMC RNA, CPMV RNA and poly AUG - very efficient formation of 80S initiation complexes occurred in the absence of added nucleotides, and neither AMPPCP or

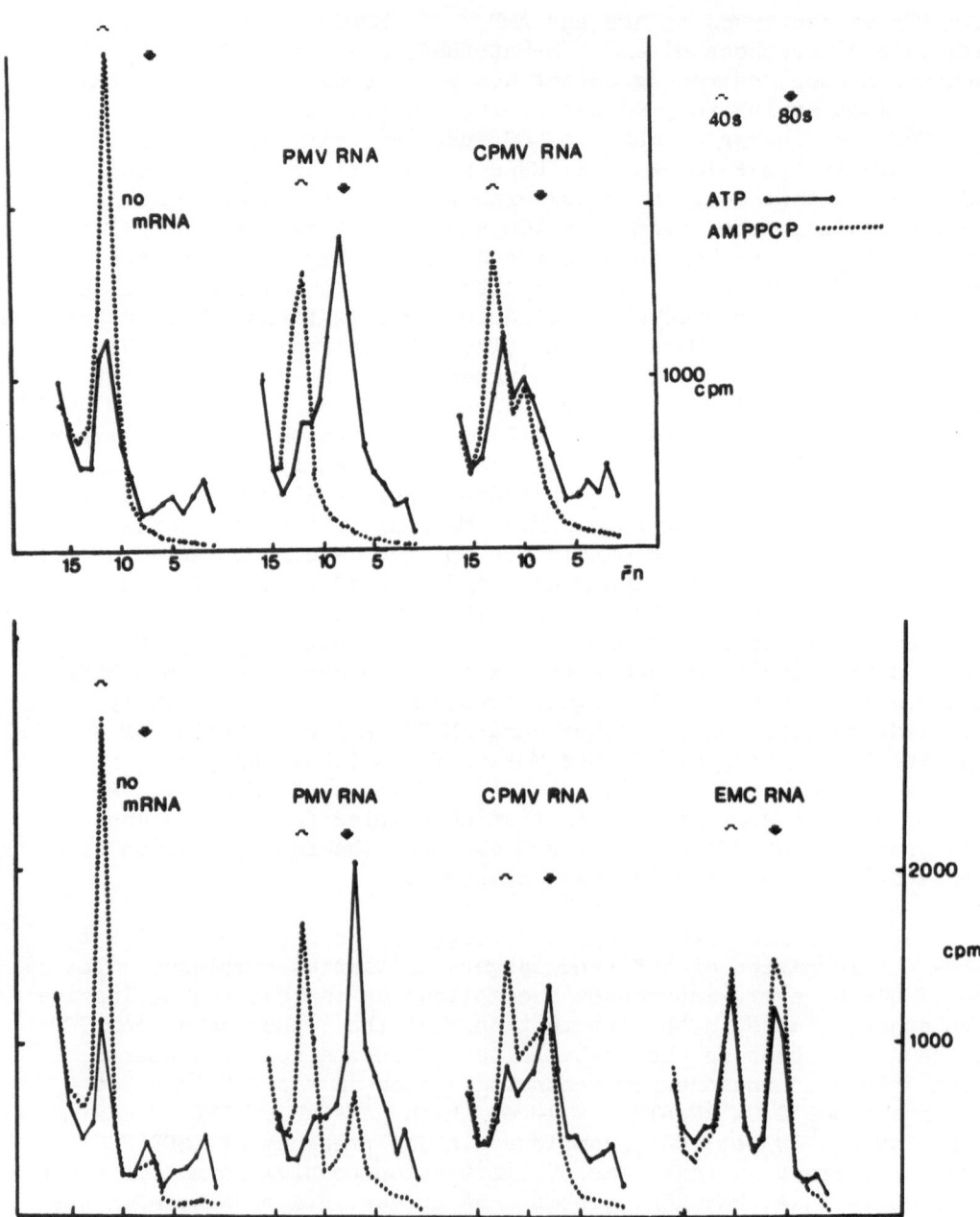

Fig. 2.   Samples (50 µl) of gel-filtered (Sephadex G-50) lysate were
          preincubated at 30° for 4 min under normal conditions for
          protein synthesis except that creatine phosphate, ATP and
          GTP were omitted, and 0.7 mM glucose 6-phosphate, 0.8 mM
          dithiothreitol and 0.2 mM sparsomycin were present. ^{35}S-Met-
          tRNA$_f$ and 0.6 mM GTP or GMPPCP were then added, followed
          immediately by 1.0 mM ATP or AMPPCP together with the follow-
          ing RNAs: PMV RNA 4 µg, mixed CPMV RNAs 4 µg total, and EMC

Fig. 2.   RNA 2.5 µg. After 2.5 min further incubation the samples
(cont.)   were diluted five-fold in ice-cold buffer and analysed on
          15-40% sucrose gradients centrifuged at 50,000 r.p.m. for
          2.5 hours in the SW 50.1 Beckmann rotor. Each fraction was
          counted but the counts in the top four fractions are not
          shown as they are off-scale. The top panel shows the results
          obtained with GMPPCP, the bottom with GTP.

ATP had any consistent effect on this (if anything ATP tended to inhibit slightly - see Fig. 2).

These results indicate that the ATP-dependence of initiation on
capped mRNAs is much higher than on uncapped. An extreme interpretation is that initiation on uncapped mRNAs is completely independent of ATP, whilst capped mRNAs are absolutely dependent, but
I don't think that this system can definitely resolve this question,
even if labelled mRNAs rather than ^{35}S-Met-tRNA$_f$ were used to make the
detection of initiation complexes containing mRNA clearer. I believe
that ultimately this whole question of the ATP-dependence of ribosome
binding to different mRNAs must be examined in fractionated systems.

II. CONTROL OF TRANSLATION: SPECIFIC EFFECTS

If, as seems likely, initiation on all species of capped mRNAs
requires the same initiation factors, and if the population of ribosomes in a cell is homogeneous, it follows that any modification of
ribosome activity or initiation factor activity cannot result in a
highly selective alteration in the rate of translation of a restricted group of mRNAs whilst the translation of the rest is totally unaffected. The translation of all mRNAs must be affected, although
the arguments advanced by Lodish (15) would allow for quantitative
differences in the changes in translation rate for different mRNAs
as a result of competitive effects. Are there highly specific and
selective changes of the type I have described, and, if so, how are
they brought about ? I will briefly examine a few cases which I
believe do show such specificity, although in almost all of them
there remains some slight room for doubt.

Poliovirus and Vaccinia Infection

In poliovirus infected cells the translation of host cell mRNAs
and of VSV mRNA is strongly inhibited whilst the translation of
poliovirus RNA proceeds reasonably efficiently (reviewed in ref. 28).
In cell-free extracts from uninfected cells, initiation on VSV mRNA
occurs in preference to poliovirus RNA, but cell-free extracts from
infected cells show the converse preference (11). The addition of
cap binding protein restores the competitive advantage of VSV mRNA
(11). It seems likely that poliovirus infection results in a shut-

off of the translation of host cell mRNAs whilst allowing selective
translation of poliovirus RNA because the cap binding protein (which
is probably redundant for initiation on poliovirus RNA) has been
destroyed or inactivated. The selective advantage of poliovirus RNA
therefore seems readily explicable in terms of the known properties
of initiation factors.

   The translation of host-cell mRNAs is shut-off in cells infected
with vaccinia virus, even under conditions where no viral mRNA
translation occurs: u.v. inactivated virus (29), pre-treatment of
the cells with interferon (30), or infection in the presence of
cordecypin or actinomycin (29, 31-33). The host-cell mRNA is not
destroyed (33), and shut-off is due to a decrease in the initiation
rate on host-cell mRNA with a secondary effect on elongation (32).
It is therefore generally supposed that under conditions of normal
infection when viral genes are expressed, the early vaccinia mRNA
is translated efficiently whilst the translation of host-cell mRNA
is largely shut-off as it is when vaccinia mRNA transcription or
translation is inhibited. This, in fact, strikes me as the weakest
point in the argument, and in my view needs further investigation.
How efficiently are early vaccinia mRNAs translated in comparison with
host-cell mRNAs under normal conditions of infection ? Is the efficiency of host-cell mRNA translation as low under conditions of
vaccinia gene expression as it is when viral gene expression is
blocked ? If the supposition does hold up, then we have a situation
in which one class of capped mRNAs is translated efficiently whilst
another set (the host-cell mRNAs) is translated inefficiently irrespective of whether the viral mRNAs are present or not. This
cannot easily be explained within the framework of our present understanding of the mechanism of initiation. The phosphorylation of
ribosomal protein S2 which has been observed in vaccinia infected
cells (34) appears to take place too late in the infectious cycle to
be able to account for the discrimination between the two classes of
mRNA. The recent development of cell-free systems from infected
cells which maintain at least some of the discrimination seen in the
intact cell (35) gives hope that this problem may be solved in the
near future.

Heat-Shock in Drosophila

   When Drosophila larvae or tissue culture cells are raised from
the normal temerature of 25° to 37°, the synthesis of some ten new
proteins (*heat-shock proteins*) is activated whilst the production of
those proteins which are synthesised at 25° (which I shall call
normal proteins) is diminished (reviewed in ref. 36). The first
description of this phenomenon in the Schneider cell line indicated
that the shut-down of normal protein synthesis occurred early after
heat-shock, before synthesis of heat-shock proteins had begun: polysomes disappeared immediately after heat-shock and then later poly_

somes making mainly heat-shock proteins re-formed (37). Moreover, inhibition of heat-shock protein synthesis by actinomycin did not prevent the polysome disappearance and the shut-off of normal protein synthesis (37). Since reports from another group indicated that the mRNAs coding for normal proteins survived in the heat-shocked cells (38), we appear to have a situation where a limited number of mRNA species (coding for heat-shock proteins) are translated efficiently whilst the translation of countless other mRNAs (the normal mRNAs) is suppressed by a mechanism quite independent of the presence of the heat shock mRNAs. This would then appear to be the equivalent situation to that believed to occur in vaccinia infected cells.

The problem seemed to me sufficiently interesting that in 1977 I took a short sabbatical leave in Alfred Tissières laboratory (Geneva) in order to study it in detail. Unfortunately I failed to throw any light on the problem - for the simple reason that the cell line used in Geneva (Echalier KCO line) seemed to behave quite differently. The shuf-off of normal protein synthesis was a gradual process occurring in parallel with the increasing rate of heat-shock protein synthesis, and actinomycin treatment before heat-shock failed to cause a rapid shut-off of normal protein synthesis. All the results pointed to a competitive process in which the increasing amounts of heat-shock mRNA outcompeted the normal mRNAs for the translation machinery. When the cells were lysed, the polysomes fractionated by sucrose gradient centrifugation, and the RNA from each fraction was translated *in vitro* I found that mRNAs coding for normal proteins (normal mRNAs) were associated with smaller polysomes at 37° than at 25°, and that exposure of the heat-shocked cells to low concentrations of cycloheximide for one hour before lysis resulted in a shift of these normal mRNAs into considerably larger polysomes, whilst there was only a small increase in the size of the polysomes translating heat-shock mRNAs. These are the classical diagnostic features of a competitive situation (15). Although the KCO line of cells exhibits a shut-off of normal protein synthesis which is readily explicable in terms of our understanding of the mechanism of initiation and competition between mRNAs at the initiation step, the fact remains that Schneider cells seem to consistently show a response (39) for which there is no obvious explanation, and further work on this topic is imperative.

Untranslated mRNAs in Eggs

It is generally accepted that in unfertilised eggs and ungerminated seeds the rate of protein synthesis is relatively quite low, and that only a small proportion of the total mRNA of the cell is being actively translated, whilst the majority is untranslated or *masked*. Fertilisation or germination results in an increase in the rate of protein synthesis arising not from a more efficient translation of those mRNAs which were being used before fertilisation, but rather from an *unmasking* or mobilisation of the previously untrans-

lated mRNAs, so that more mRNA is now being actively translated. How
specific is this masking-unmasking process ? Is post-fertilisation
protein synthesis merely producing the same proteins in larger amounts
as were synthesised before fertilisation ? Is the unmasking a random
event non-specific with respect to mRNA type, or is there a definite
programme in which different species of mRNA are unmasked at different
stages after fertilisation ? In the case of the sea urchin, which has
been the most frequently studied organism, these issues have been a
matter of controversy for some years, but the detailed analysis by
Brandhorst (40) suggests that there is little specificity in the
unmasking process. The activation of translation of stored maternal
mRNA following fertilisation seemed to be due to a quantitative rather
than a qualitative change in the population of mRNA available for
translation (40), and only very minor changes in the pattern of
protein products were noted after fertilisation. Thus the masking
process in sea urchins could be explained by a rather unsophisticated
and unselective mechanism of blocking mRNA from access to the translation machinery.

     On the other hand, recent analysis of the rates of synthesis of
different histones in *Xenopus* oocytes and egss has shown that the
changes in the rates of synthesis of different histones do not
parallel the changes in overall protein synthesis rate (41). The
common surf clam, *Spisula solidissima*, is another case which shows
selectivity in the masking and unmasking of mRNAs. The rate of
protein synthesis increases by two-four fold within a period of 10
minutes after fertilisation - a more modest increase than occurs in
some other organisms - but what is especially striking is the radical
change in the *types* of protein synthesised (42). The change is sufficiently marked that one-dimensional gel electrophoresis is sufficient to distinguish the pre-fertilization and post-fertilisation
patterns of protein synthesis. When *crude* extracts from unfertilised
eggs are mixed with reticulocyte mRNA-dependent lysate only the
proteins characteristic of the unfertilised state are synthesised,
yet when RNA is *extracted with phenol* from the very same crude extract
it can be translated *in vitro* into a mixture of the proteins characteristic of pre- and post-fertilisation states (42). Thus the mRNAs
which are translated in unfertilised eggs remain translatable when
crude homogenates of these eggs are prepared, whilst the mRNAs that
were masked *in vivo* remain undegraded but still masked in the crude
extract. This seems a particularly promising system for an investigation into the actual mechanism of masking-unmasking of mRNAs, and
into the basis of the specificity.

Untranslated mRNA in Somatic Cells

     The post-ribosomal supernatant from somatic cells also contains
untranslated mRNA in the form of mRNPs (messenger ribonucleoprotein
particles). The mRNA in these particles can be translated *in vitro*

if it is first isolated by phenol extraction, although the sub-cel-
lular location implies that it is not translated *in vivo*. These
mRNAs code for a small, unspecified, set of proteins in the case of
cytoplasmic mRNPs from Vero cells and from mouse sarcoma-180 ascites
cells (43, 44), and for duck globin and another unidentified protein
in duck reticulocytes (45, 46). As I have argued in detail elsewhere
(47) there are two explanations for the existence of cytoplasmic
mRNPs:

1)   they may contain mRNAs with extremely low efficiency initiation
sites which compete very poorly against other mRNAs and therefore
frequently have no ribosomes translating them at any one instant, or

2)   they may be a masked or repressed form of mRNA similar to the
masked mRNAs in eggs.

If the first of these explanations were correct we would expect the
mRNAs to be mobilised into polysomes and to be translated in the
presence of low concentrations of cycloheximide, since initiation is
no longer the rate-limiting step under these conditions and the
difference in initiation rate on mRNAs of low affinity and high affinity initiation sites is largely eliminated (15). This test does
not seem to have been made with the duck reticulocyte system, but
in Vero cells and in mouse ascites cells low cycloheximide *did*
promote mobilisation of mRNA from cytoplasmic mRNPs into polysomes
(43, 44). In the latter case, however, only some species of mRNA
were efficiently mobilised, whilst others remained as cytoplasmic
mRNPs even in the presence of low cycloheximide (44).

    If the specific 'masking' hypothesis were correct, mobilisation
of mRNA from cytoplasmic mRNPs into polysomes would not be expected
to occur in low cycloheximide, but should take place in response to
specific signals which unmask the mRNA. One might also expect that
when the cytoplasmic mRNPs were added as such (without deprotein-
isation) to *in vitro* translation systems, they would not be translated – unless the system happened to have the means of unmasking the
mRNA. The cytoplasmic globin mRNPs from duck reticulocytes, and
those mRNPs in mouse ascites cells that could not be mobilised by
cycloheximide treatment *in vivo*, were indeed not translated *in vitro*
even if the mRNPs had been isolated at quite high salt concentrations
(44, 46). Since many factors could conceivably affect translation
in vitro I am doubtful whether this test by itself (i.e. without
the corroborative evidence of the cycloheximide experiment) constitutes sufficient grounds for claiming a specific control mechanism
through 'masking' of mRNA. If cytoplasmic mRNPs were a specifically
masked form of certain types of mRNA, it should be possible to unmask
them and render them translated *in vivo* if the appropriate signal is
supplied. This would be manifest as a specific increase in the
synthesis of a certain protein, which is not inhibited by actinomycin and is not accompanied by an overall general increase in protein

synthesis. Two such cases are quite well documented: ferritin
synthesis in rat liver (48) and in HeLa cells (49) is stimulated by
iron administration, and α_{2u} globulin synthesis in the liver of male
hypophysectomised rats requires growth hormone administration (50) -
testosterone, glucocorticoids and thyroid hormone are required for
synthesis of α_{2u} globulin mRNA, but without growth hormone no synthesis of the protein takes place. In both cases the mRNAs occur as
cytoplasmic mRNPs in the absence of the appropriate stimulus (Fe^{++}
or growth hormone) and are mobilised into small polysomes in response
to the stimulus. Estela Ruiz in our group has shown that little or
no ferritin synthesis occurs when ferritin mRNPs from mouse liver are
added to cell-free translation systems, although RNA obtained from
the mRNPs by phenol extraction is efficiently translated. We have had
no success in attempts to unmask these mRNPs *in vitro*.

Are we really sure that these two cases represent a *specific*
control mechanism; that iron and growth hormone don't promote a
general increase in the rate of initiation on all mRNAs ? The fact
that the stimuli do not cause a general increase in the rate of
protein synthesis would seem to squash this objection, but one can
argue that a general increase in the initiation rate might not lead
to an increase in the rate of translation of the majority of mRNAs
if some other step - perhaps elongation - becomes rate-limiting. If
this premise is valid, a general unselective increase in the rate of
initiation would only increase the synthesis of those proteins coded
by mRNAs with low affinity initiation sites. As to the cytoplasmic
location of the ferritin and α_{2u} globulin mRNAs in unstimulated cells,
it may be significant that these are both rather small proteins, and
therefore a low initiation rate is more likely to result in the mRNA
having no ribosomes translating it at any one moment than would be
the case with longer mRNAs. It seems to me that the critical test
is to show that low concentrations of cycloheximide do *not* mobilise
these two mRNAs from cytoplasmic mRNPs into polysomes.

On balance I believe that the probabilities lie in favour of
specific control of at least some of these apparently masked mRNAs
which are found as cytoplasmic mRNPs. Since the mRNA isolated from
these particles by phenol extraction is efficiently translated *in
vitro*, it is presumably the proteins of the mRNPs that do the masking.
In this connection it is of interest that in duck reticulocytes two
different mRNPs containing two different mRNAs appear to have
distinct protein composition (45), which can be construed as supporting the idea of specificity of masking. One imagines that the
detachment of these proteins would result in the unmasking of the
mRNA: either the proteins dissociate when they bind a specific
effector molecule (as the lac repressor dissociates from DNA in the
presence of inducers), or some modification of the proteins causes
their dissociation. On the other hand, in some laboratories small
RNA molecules have been found in the cytoplasmic mRNPs, apparently
hydrogen-bonded to the mRNA, and it has been suggested that it is

this tcRNA which prevents translation of the mRNA (reviewed in ref. 1). The tcRNA isolated from cytoplasmic mRNPs containing myosin mRNA inhibited myosin mRNA translation *in vitro*, but also inhibited globin mRNA translation albeit to a somewhat lower extent (51). It is therefore hard to see how tcRNA could be the basis of a highly specific control mechanism. The issue is further complicated by the failure of other studies in other systems to reveal anything corresponding to tcRNA (46).

A number of different types of small RNA can be found in the cytoplasm of mammalian cells (52), not all of them associated with mRNA. When there is such association, it seems to be through RNA: RNA duplexes, and is mainly with the non-polysomal mRNA rather than with the mRNA located in polysomes (52). Fingerprinting of these RNAs suggests that there are a limited number of species, and the heterogeneity which would be expected of highly specific regulators is lacking (52). They seem closely related to similar-sized RNAs in the nucleus, although the fingerprints suggested some minor difference in modification of the RNA in the cytoplasm and nucleus (52). Since small RNAs in the nucleus have been suggested to play a role in splicing and processing of RNA (see chapters 4 and 5 of this volume, and ref. 53), is it possible that they remain associated with the mRNA during export to the cytoplasm and are only released when the mRNA is translated ? In this case the role of these RNAs would be not so much as regulators, but as part of the processing and transport of mRNA. In any case it is worth bearing in mind that even if small tcRNAs annealed to mRNA do block translation, it would surely need the action of a protein to melt off or destroy the tcRNA and to release the mRNA for translation. Thus the burden of control would once again fall upon proteins, as it does in the model in which the proteins do the actual masking of the mRNA.

III. CONTROL OF INITIATION IN RETICULOCYTE SYSTEMS

Control in Intact Cells

The first point that I would wish to draw to the attention of a newcomer to the field is that the rate of initiation is regulated in intact rabbit reticulocyte cells in a way very similar, if not identical, to cell-free extracts. Intact cells require either haemin, or Fe^{++} or Co^{++} (which they incorporate into metalloporphyrins) in order to maintain high rates of protein synthesis (reviewed in refs. 5, 55). In the absence of Fe^{++} or haemin, the rate of protein synthesis declines and polysomes disappear or are reduced in size, implying that the rate of initiation drops. There is a curious inverse temperature effect (56) in that the absence of haemin is less deleterious to protein synthesis at low temperatures (e.g. 25°) than at high temperatures ($37-40^{\circ}$), and it is reassuring to find the same response to temperature in cell-free systems (57). At still higher temperatures ($42-44^{\circ}$) the rate of initiation in intact cells

is reduced even in the presence of available haemin (56), and again
the same effect is seen in the cell-free extract (58, 59). Oxidation
of intracellular glutathione leads to inhibition of initiation in
intact cells (60), whilst addition of oxidised glutathione to the
lysate inhibits initiation, as I will discuss in Section VI. Thus
there are good grounds for believing that the control mechanisms
studied in cell-free extracts are indeed relevant to intact cells,
and are not some interesting artefact of the cell-free state. In
fact, recent experiments have demonstrated that incubation of intact
cells in the absence of Fe^{++} leads to the phosphorylation of eIF-2
(61), which is believed to be the basic cause of inhibition *in vitro*.
Double-stranded RNA is the only agent which regulates initiation in
lysates but has no effect on intact cells, and this disparity is
almost certainly due to the failure of dsRNA to enter the cells.

Control in Cell-Free Systems

Rabbit reticulocyte lysates are prepared by taking washed
reticulocyte cells, lysing them with $1-1\frac{1}{2}$ volumes of water per volume
of packed cells and then removing the mitochondria and stroma by
centrifugation at 10,000g. The resulting supernatant is the reticulocyte lysate and it is used as such, without dialysis or gel-
filtration - the importance of this will be discussed later (Section
VI). The lysate therefore contains a plethora of low molecular
weight compounds, many of which we have measured: typically 25 mM
K^+, 10 mM Na^+, 1.8 mM Mg^{++}, 0.5 mM ATP, 0.1 mM GTP, 0.4 mM spermidine,
0.05 mM spermine, 1.5-2.0 mM glutathione, 20 µM NADP + NADPH, and
unlabelled amino acid pools of various sizes. Some of these turn
out to be important for a high *rate* of protein synthesis (notably
spermidine and spermine) and others are important for the maintenance
of this high rate as discussed later (Section VI). It is only
necessary to add an ATP-generating system (creatine phosphate and
creatine kinase), 100 mM KCl and 0.5 mM $MgCl_2$ to compensate for the
dilution occurring on lysis of the cells, labelled and unlabelled
amino acids, and 20 µM haemin, in order to obtain a linear rate of
protein synthesis for about 60 minutes *at the same rate as in the
intact cell*. If haemin is omitted, protein synthesis starts at the
same rate but after about 5-10 minutes there is an abrupt transition
to a low rate, typically some 2-10% of the initial rate. Polysomes
are virtually absent during this phase of slow protein synthesis,
but if haemin is added after, say, 20 minutes incubation, protein
synthesis accelerates back to control rates and polysomes are reform-
ed (62) - the control is therefore *reversible*.

These observations became more interesting when it was found
that other conditions promote a similar inhibition of initiation even
in the presence of haemin:

1)   in the presence of low concentrations (0.1 - 100 ng/ml) of
      double-stranded RNA (Section V);

2)   when oxidised glutathione is added (63);
3)   if the lysate is preincubated at 42-44° (58, 59);
4)   if the lysate is subjected to high hydrostatic pressure (54);
5)   if the lysate is gel-filtered to remove endogenous low-
      molecular weight coumpounds - in such cases the addition of
      ATP, GTP and spermidine is necessary for maximum initial rates
      of protein synthesis, but these compounds do not prevent the
      early decline in initiation rate (55);
6)   if the lysate is isolated from reticulocytes that have been
      incubated anaerobically, or under other conditions which lead
      to depletion of the intracellular ATP pool (64, 65).

In spite of the diverse nature of these inhibitory agents and
conditions, the characteristics of inhibition are remarkably similar
in all cases:

1)   Protein synthesis proceeds at normal rates for a few minutes,
whereupon there is a fairly abrupt transition to a much lower rate,
about 2-10% of the initial rate, and the polysomes are converted to
80S ribosomes.

2)   Just before the transition to the inhibited state, 40S/Met-
tRNA$_f$/GTP complexes disappear, although the levels of 40S subunits,
GTP and Met-tRNA$_f$ are normal (66, 67). The conclusion generally
drawn from this is that the binding of Met-tRNA$_f$ to 40S subunits
is inhibited, but an alternative view has been advanced to the effect
that the loss of 40S/Met-tRNA$_f$ complexes is due to the activation
of a Met-tRNA$_f$ deacylase (68). As Hunt has described in detail, our
own observations are in complete disagreement with this view (5).

3)   The synthesis of all.proteins, including the synthesis of
proteins programmed by added mRNA is inhibited (reviewed in ref. 5).

4)   Inhibition can be prevented or reversed by the addition of
relatively large amounts (in relation to the probable concentration
in the lysate) of purified eIF-2; neither eIF-1; eIF-3, eIF-4A or
eIF-5 have the same effect (69). In our experience some preparations
of eIF-2 are distinctly better "rescuers" than others, even though
both may have the same activity in ternary complex formation. One
group has actually questioned whether the rescuing activity might
not be a contaminant of the eIF-2 preparations (70), but as far as
I know this dissenting voice is not echoed by other workers in the
field.

5)   High concentrations (5 mM) of 3':5' cAMP, 2-aminopurine, caffeine
and other purine derivatives prevent or reverse the inhibition (71).
At this concentration cAMP is the only phosphorylated derivative
which is active, but higher levels (10 mM) of cGMP give partial
protection against inhibition (64).

6)   No inhibition of the synthesis of chains initiated with formyl-
Met-tRNA$_f$ is observed (32, 68, 72). This observation has not found
a ready explanation except in terms of the Met-tRNA$_f$ deacylation
hypothesis (68).

The fact that these characteristics are shared by all the inhibitory conditions listed previously suggests that the proximal
mechanism of inhibition may be the same in all cases even if the
train of events which triggers this mechanism may differ in each
situation. Efforts at unravelling the triggering system and the
proximal cause of inhibition have been focused mainly on the effects
of haem-deficiency and of dsRNA, and have led to the hypothesis that
inhibition is due to the reversible phosphorylation of eIF-2 in the
smallest (35,000 daltons) of its three constituent subunits - often
termed eIF-2α. The evidence in support of this hypothesis is not
wholly satisfactory and although the general opinion seems to favour
the view that the model is basically correct but requires some as yet
uninvented modification to fit the facts, there has recently been some
more fundamental questioning of its validity (73), although no alternative model has been put forward. Lest the more recent dissent
should get disproportional attention I think it worthwhile summarising
the evidence in support of the phosphorylation model, and this is
done in the following section.

IV. HAEMIN CONTROLS eIF-2 PHOSPHORYLATION

Mechanism of Action of the Haem-Controlled Inhibitor

The model rests on two major points: 1) incubation of lysates
in the absence of haemin leads to the activation of an inhibitor of
initiation, and 2) this inhibitor is a protein kinase which is highly
specific for eIF-2α as substrate. The activation of the inhibitor in
reticulocyte post-ribosomal supernatant incubated in the absence of
haemin was first recognised by Rabinovitz and his colleagues (reviewed in refs. 5, 6, 54). It is identified and assayed by its ability
to inhibit protein synthesis in a test lysate containing haemin.
Two broad classes of inhibitor can be distinguished (74): a) *reversible* inhibitor, which when incubated with the test lysate causes only
a transient inhibition from which the system recovers back to control
rates of protein synthesis, and b) *irreversible* inhibitor, which
causes a more or less permanent shut-off. Reversible inhibitor is
formed by a brief incubation of post-ribosomal supernatant in the
absence of haemin, and if it is further incubated in the presence of
haemin before mixing with the test lysate, no inhibition occurs at
all (74, 75). Irreversible inhibitor - which has also been called
HCR (for Haem Controlled Repressor), forms on more prolonged incubation in the absence of haemin (74), but what appears to be the same
entity can also be obtained by brief treatment with N-ethylmaleimide
(NEM) irrespective of the presence of haemin (54, 76). A third type

intermediate inhibitor, has been distinguished (57): it resembles
reversible inhibitor in that it causes only transient inhibition in the
test lysate, but differs in that if it is preincubated with haemin
before assay it is not inactivated. These three different inhibitors
are believed to be related forms of the same protein (77), all
derived from the same non-inhibitory pro-inhibitor (Fig. 3). This is
a critical point in the argument since although it is the reversible
form which is probably responsible for the inhibition of initiation
in lysates incubated in the absence of haemin, the best studied form
is the irreversible, particularly the NEM-activated form of HCR, as
this is the most stable to purification in our experience. Are we
sure that NEM-activated HCR is the same protein, albeit in a slightly
different guise, as the reversible inhibitor, and that the mechanism
of action is the same ?

To answer this question I must first summarise the properties of
HCR which has been obtained at a purity in the range 50-100% in this
and other laboratories. It has three recognisable activities:

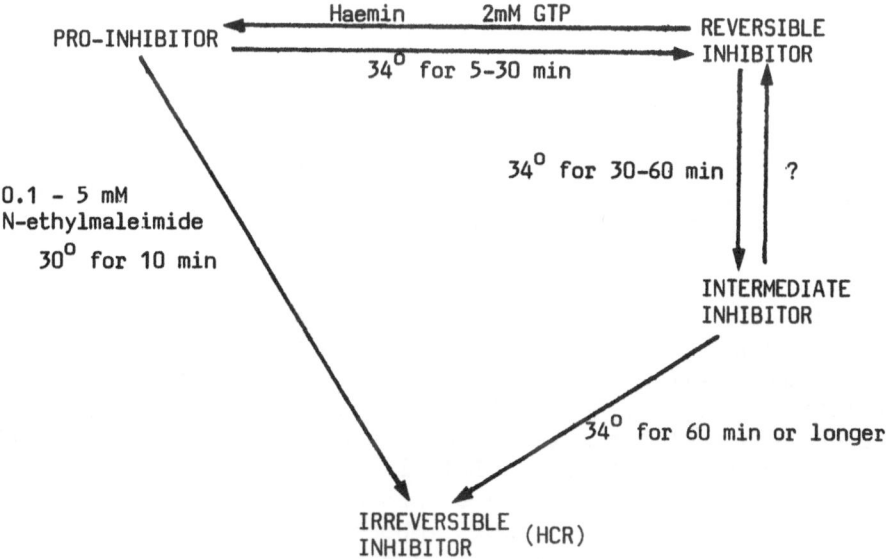

Fig. 3.   The interconversion of the different froms of the haem-
          controlled initiation inhibitor of rabbit reticulocytes.
          The incubation times shown are approximate and refer to the
          time required to generate each form of the inhibitor starting
          from pro-inhibitor. The irreversible form can first be
          detected after about 60 min incubation, but it requires 12-15
          hours at 34° to generate maximum levels of irreversible in-
          hibitor. For further details see refs 57, 76, 77, 81 and
          100.

1)   It inhibits protein synthesis as explained above.

2)   It phosphorylates the α-subunit of purified eIF-2 or the endogenous eIF-2 in crude ribosomes (78) in a cAMP-independent process
using ATP in strong (but not absolute) preference to GTP as phorphoryl
donor. The reaction is highly specific for eIF-2; other proteins
are phosphorylated at best very slowly in comparison with the preferred substrate (78). The kinase activity is strongly inhibited by
5 mM cAMP, 2-aminopurine and the other purine derivatives which
prevent inhibition of proteins synthesis. It seems likely that only
a single site is phosphorylated in eIF-2α, although there is still
room for doubt on this issue (see discussion in ref. 5).

3)   It phosphorylates itself in a process which appears to be a true
autokinase reaction since the rate is insensitive to dilution (54,
79). The phosphorylated polypeptide is about 85,000 daltons (precise
estimates cannot be given since the electrophoretic mobility relative
to standards is very much influenced by the gel composition). Hunt
has estimated a minimum of four sites phosphorylated (5), whilst
Gross proposes up to five phosphorylation sites (79). (This self-
phosphorylation greatly aids the purification of HCR: When HCR
obtained from a DEAE-cellulose column is incubated with ATP and then
rerun on DEAE-cellulose it elutes at a higher salt concentration and
is separated from most of the contaminants present after the first
column run). The native molecular weight is about 170,000 and cross-
linking experiments suggest that it is a homodimer of the 85,000
dalton polypeptide (5, 54). Its shape, however, must be very elongated since gel-filtration suggests a larger molecular weight, whilst
sucrose gradients give a lower figure (5, 54).

    Other forms of the inhibitor are identical to NEM-activated HCR
in their size (determined by gel filtration) and their eIF-2α kinase
activity (75), which is also sensitive to inhibition by cAMP and
other purine derivatives. Reversible inhibitor also shows self-
phosphorylation of the 85,000 dalton polypeptide (54, 75, 80). One
cannot be certain that the rate and extent of this self-phosphorylation is identical for both reversible inhibitor and HCR, since the
transient nature of the inhibition of protein synthesis caused by
reversible inhibitor makes it almost impossible to quantitate inhibitory activity in terms of a defined unit, as can be done with HCR
or irreversible inhibitor (5). It is also impossible to be absolutely certain that the self-phosphorylation observed in preparations of
reversible inhibitor is not due to traces of irreversible HCR; there
is no way of reliably estimating the amount of contaminating HCR in
a preparation of reversible inhibitor. All we can say at the present
moment is that reversible inhibitor preparations do show self-
phosphorylation activity.

    Antibodies raised against NEM-activated HCR cross-react with
heat-activated HCR and with the reversible form of the inhibitor,

whilst antibodies against the reversible form cross-react with HCR
(75, 81). (These experiments have so far been carried out using
crude anti-sera raised against inhibitor preparations which had been
extensively purified but probably not to homogeneity. In the near
future, monoclonal antibodies will probably be available to remove
the last shadow of doubt on this issue). The rate of appearance of
the different forms of the inhibitor is also consistent with the idea
of a sequential series of steps leading first to the reversible inhibitor, then to the intermediate form and finally to irreversible
HCR, as depicted in Fig. 3 (77). In short, the evidence favours the
view that all the different inhibitors are slighly different forms
of the same protein, with the same mechanism of action.

In discussing the properties of HCR and reversible inhibitor, I have
already presented most of the evidence in support of the eIF-2
phosphorylation hypothesis. A few additional points need to be made.

1)   Pro-inhibitor neither inhibits protein synthesis, nor does it
show eIF-2 kinase activity or self-phosphorylation. When proinhibitor
is converted to HCR, all three properties appear in parallel (54, 79),
and on purification of the HCR all three activities quantitatively
co-purify (78).

2)   One prediction of the hypothesis is that HCR should require ATP
in order to act as an inhibitor. Unfortunately this cannot be tested
in a protein synthesis assay since ATP is needed to sustain protein
synthesis, but we can study the effect of HCR on the binding of ^{35}S-
Met-tRNA$_f$ to native 40S subunits in crude ribosome preparations, as
this process requires GTP but not ATP. This binding is not inhibited
by HCR unless ATP is present, and is unaffected by ATP unless HCR is
also added (78). No inhibition is observed in the presence of HCR
and non-hydrolysable analogues of ATP, or if cAMP or 2-aminopurine
are added together with ATP and HCR (78). In this system inhibition
of the binding of Met-tRNA$_f$ to 40S subunits in the crude ribosome
preparation showed an absolute correlation with the phosphorylation
of endogenous eIF-2α assayed in parallel.

3)   By 2-dimensional gel electrophoresis, which can resolve the
phosphorylated and unphosphorylated forms of eIF-2α, it has been
shown that there is net phosphorylation of eIF-2α in crude lysates
incubated in the absence of haemin, or with HCR, dsRNA or oxidised
glutathione (82). The ratio of phosphorylated : unphosphorylated
protein never exceeded 2:1. We do not know whether this arises from
some artefact during sample preparation and analysis, or whether it
reflects the actual ratio in the lysate itself (see discussion in
ref. 82). More recently, haem-deficiency has been shown to result
in a five-fold increase in the labelling of eIF-2α in lysates incubated under conditions in which ^{32}P-ATP was maintained at constant
specific activity throughout the incubation (61).

This, then, is the evidence in favour of the hypothesis that inhibition of initiation in haem-deficient lysates is due to the activation of an inhibitory protein kinase which phosphorylates eIF-2α. Although much of the evidence is circumstantial, it is cumulatively convincing. Perhaps the weakest point in the argument so far is the supposition that it is the phosphorylation of eIF-2α rather than some other protein which is responsible for the inhibition of initiation. We have searched hard for other substrates without finding a plausible candidate. In the crude ribosome preparations which, as explained above, were sensitive to HCR inhibition provided ATP was present, eIF-2α seemed to be the only protein whose phosphorylation was dependent on HCR (78).

Is the Function of eIF-2 Impaired by Phosphorylation ?

The most serious setback to the credibility of the phosphorylation is the fact that phosphorylation (by HCR) of purified eIF-2 does not inhibit its function in either ternary complex formation, or the formation of 40S/Met-tRNA$_f$ complexes, or 80S/mRNA/Met-tRNA$_f$ complexes (83). Whilst some groups have reported a loss of activity (84) the general experience seems to be that increasing purification of the eIF-2 and other components leads to decreasing difference between phosphorylated and non-phosphorylated eIF-2. Before discussing possible explanations for this finding, I would like to point out an observation which suggests that inhibition by HCR is not so much a peculiarity of the crude lysate system as opposed to the highly fractionated system, but is a characteristic restricted to systems which exhibit a high rate of protein synthesis. If we carefully titrate the concentration of an elongation inhibitor (e.g. cycloheximide) to reduce the rate of protein synthesis in the crude lysate to about 5% of the normal rate (i.e. the lysate now functions at the same rate as the highly fractionated systems) neither haem-deficiency nor HCR inhibit protein synthesis (54). We have argued endlessly amongst ourselves as to the significance of this observation without coming to any useful conclusion, but I believe that it is a fact which needs to be taken into account when considering possible explanations for why phosphorylation of purified eIF-2 fails to inhibit its activity in the highly fractionated system.

Two types of explanation have been suggested. The first proposes that phosphorylated eIF-2 operates normally in a single-cycle process, but that it is the recycling which is impaired (85). If, for example, eIF-2 is released from the initiation complex as an eIF-2.GDP complex the dissociation of which requires the participation of another protein (see Section I), could it be that phosphorylation of the eIF-2 blocks the action of this protein which promotes release of GDP ? Whatever the actual details, *any* model in which it is the *recycling* of eIF-2 that is inhibited as a result of phosphorylation can be reconciled with the fact that purified phosphorylated eIF-2 shows full activity in highly fractionated systems, since little or no recycling occurs in these assay systems. The only situation in

which eIF-2 might recycle is when it is added to restore protein
synthesis in lysates that have been incubated without haemin. In
these experiments, it is usually observed that phsophorylated and unphosphorylated forms of eIF-2 restore protein synthesis activity to
the same extent (86), but since the lysate contains activated eIF-2
kinase (activated by the absence of haemin) as well as phosphatase
activity, it seems likely that the same steady state level of phosphorylation of the added eIF-2 is attained, regardless of whether it
was the phosphorylated or non-phosphorylated form that was added. The
critical question is how many moles of globin are synthesised per mole
of eIF-2 added. Until recently, the answer (86) was at most one mole
per mole (and was usually less), which would be compatible with the
idea that phosphorylation did not inhibit the first cycle of eIF-2
action, but did inhibit the recycling. There has now appeared a claim
that low amounts of eIF-2 promote the synthesis of up to 3 globin
chains per molecule of eIF-2 added to the haem-deficient lysate (73).

For an argument that depends so critically on numbers, it is
unfortunate that the basis of calculation of the globin chains
synthesised is not fully explained in this paper. The estimation of
the phosphorylation level of eIF-2 does seem to be based on a false
premise, however. In this experiment, where γ-^{32}P-ATP is added to
a lysate containing unlabelled ATP and creatine phosphate, the
specific activity of the ATP falls during the incubation and the
labelling of the phosphoproteins was therefore corrected for this
decrease (73). What has been overlooked here is that labelled
phosphate is not only lost from ATP by hydrolysis, but is also re-
distributed within the ATP when the AMP formed during protein synthesis is converted to ADP by myokinase (ATP + AMP = ADP), and the
ADP converted to ATP by creatine kinase. After a few minutes incubation, the surviving label in ATP is not confined to the γ-
position, as the authors seem to assume (73), but is more or less
equally distributed between the β- and γ- positions (87). The degree
of phosphorylation of eIF-2 has therefore been underestimated by a
factor of about 2. For myself, I think that this paper makes the
hypothesis that phosphorylation impairs recycling of eIF-2 somewhat
shaky, but I do not regard the case against it as proven until the
measurements have been carried out more rigorously.

In the other type of explanation, it is proposed that phosphorylation of eIF-2 inhibits its activity (even in a single cycle
assay) when, and only when, additional proteins are present. Ochoa's
group has discovered such a protein - called ESP (88), and Gupta's
group has a similar, if not identical, factor known as Co-eIF-2C
(for a detailed description of the properties of Co-eIF-2C and the
reactions involved, see reference 89 and the contribution by N.K.
Gupta in this volume). In both cases, the auxiliary protein stimulates eIF-2 activity in ternary complex formation (88, 89), the
stimulation being especially significant at low eIF-2 concentrations.
No stimulation is observed if phosphorylated eIF-2 is assayed under

these conditions, and so if all assays are carried out in the presence
of the auxiliary factor, phosphorylation of eIF-2 by HCR *does* impair
its function (88, 89). These additional proteins are unlikely to be
factors which catalyse the recycling of eIF-2, since ternary complex
formation is a single cycle process in which eIF-2 functions stoichiometrically. What we would like to know now is whether these auxiliary
proteins play a role in normal initiation - in crude lysates incubated
in the presence of haemin, for example. Experiments with antibodies
might resolve this issue. This is an important question since assays
of eIF-2, and especially phosphorylated eIF-2, are subject to artefacts caused by absorption of the eIF-2 to the walls of the vessel,
whether plastic or glass (90). (I myself have had bitter experience
of this when attempting to assay phosphatase activity using ^{32}P-eIF-2
as substrate and measuring the decrease in acid-precipitable radioactivity. What appeared to be a ubiquitous phosphatase activity,
turned out to be absorption of the labelled eIF-2 to the tube walls !)
Since ESP and Co-eIF-2C are defined proteins purified on the basis of
their stimulatory activity, it does seem very unlikely that they
serve merely as a non-specific competitor for the eIF-2 binding sites
on the plastic or glass surfaces, but even if this artefact is eliminated from consideration, we still urgently need to know whether
these proteins have a role in the process of initiation, and what
this role may be.

eIF-2 Phosphatase

    Since the inhibition of protein synthesis under a variety of
conditions ca not only be prevented by specific antagonists (e.g.
haemin or cAMP), but can also be reversed after onset of inhibition,
the hypothesis that phosphorylation of eIF-2α is the cause of inhibition demands that there is a phosphatase to dephosphorylate eIF-2.
If this phosphatase could be purified it might allow some critical
tests of the general hypothesis. The phosphatase is also of interest
since it, too, might be subject to regulation, and the overall phosphorylation state of eIF-2 might depend on controls exerted at both the
kinase and phsophatase steps.

    Some years ago, we tested for phosphatase activity in lysate by
adding purified ^{32}P-eIF-2 or crude ribosomes containing endogenous
labelled eIF-2 phosphorylated by incubation with HCR. Both substrates
were dephosphorylated with half-lives in the range of 5-10 minutes
(69). Fractionation of the lysate by gel-filtration on Sepharose 6B
or by chromatography on DEAE-cellulose showed a single peak of
phosphatase activity in both cases, but further purification was
frustrated by the instability of the activity. In two recent reports,
a much faster rate of dephosphorylation of purified ^{32}P-eIF-2 added
to crude lysates was observed than in our experiments: the half-
lives were 20 seconds at 30° (19) and 2.5 minutes at 25° (73). It may
be that the eIF-2 used in these experiments was in a better condition
("more native") than ours, but another reason for the faster rate

may be that whereas our eIF-2 substrate was labelled exclusively in the α-subunit, the 55,000 dalton subunit was also phorphorylated in both of these other studies - the HCR preparation used to phosphorylate the eIF-2 must have been contaminated with casein kinase which readily phosphorylates the 55,000 dalton subunit of eIF-2 (78). In one of these reports, various compounds were added to the lysate to test for an effect on phosphatase activity, but no physiologically significant inhibitors or activators were revealed (19).

Hardesty's group have resolved two reticulocyte phosphatases active against ^{32}P-eIF-2. The more potent of the two was partially purified and was found to be of rather broad specificity with respect to phosphoprotein substrate (91). The activity was significantly stimulated by Mn^{++} and it is an intriguing fact that the addition of 0.5-0.8 mM Mn^{++} to reticulocyte lysates partially prevents the inhibition of initiation caused by lack of haemin (69). On the other hand we found no convincing effect of Mn^{++} on the inhibition caused by dsRNA or HCR, and our tentative conclusion at the time was that Mn^{++} was more likely to be influencing the activation of the haem-controlled inhibitor than to be stimulating eIF-2 phosphatase activity. Whatever the significance of this effect of Mn^{++}, we clearly need to know a great deal more about the nature of the eIF-2 phosphatase(s) and the regulation of their activity.

Anti-Inhibitor Proteins

Throughout the last ten years there have surfaced occasional reports to be effect that reticulocyte post-ribosomal supernatant contains a protein which antagonises inhibition of protein synthesis in different conditions: oxidised glutathione (63, 92), haem-deficiency (65, 92), and lysates from cells subjected to anaerobiosis (65). Two groups have recently pursued this further and have isolated what is probably the same protein, called sRF by Gupta's group (93 and chapter of this volume) and "anti-inhibitor" by the Utrecth group (73, 94). This protein relieves the inhibition caused by haem-deficiency or by HCR, but both groups agree that it is not an eIF-2 phosphatase, nor does it influence the activity of purified eIF-2 in any of the standard assays. If a specific protein is needed to catalyse the dissociation of eIF-2.GDP complexes and to promote the recycling of eIF-2 (Section I), could this "anti-inhibitor" be the protein concerned ? It is interesting to note that in the purification of anti-inhibitor there were indications of a fairly tight association between anti-inhibitor and eIF-2 (94). Here again, the availability of antibodies might provide answers to the crucial questions of whether this protein plays an essential role in initiation in uninhibited lysates, whether it is needed for the recycling of eIF-2, and whether it is the same protein that is effective as an antagonist against inhibition in conditions other than haem-deficiency (63, 65, 92).

The Activation of the Haem-Controlled Inhibitor

The activation of HCR by N-ethylmaleimide occurs rapidly at very low concentrations (1 mM or less) and except in the presence of ATP, higher levels (10 mM) not only fail to activate HCR, but also seem to destroy the potential for subsequent activation by incubation in the absence of haemin or by treatment with 1 mM NEM (54). This surprised us since incubation of pre-activated HCR with 10 mM NEM causes only a slow inactivation of the inhibitory activity and the eIF-2 kinase activity. Pro-inhibitor is far more susceptible to inactivation by NEM than is fully activated HCR, and this points to an important conclusion: the conformations of the two states of the protein must be different. The fact that ATP protects pro-inhibitor against inactivation by 10 mM NEM and allows its conversion to HCR under these conditions, seems to imply that pro-inhibitor has a binding site for ATP even if it does not use ATP as a substrate. Activation by NEM of course suggests that certain sulphydryl groups are important in the activation mechanism. M.Gross finds that dithiothreitol inhibits activation, and will convert reversible inhibitor (but not HCR) back to harmless pro-inhibitor (95). We have not been able to reproduce these findings consistently, but Hunt has discovered two further methods of activation which are also consistent with a critical role of -SH groups: exposure to pH 9, and absorption to thiopropyl-Sepharose (54).

In respect of the quasi-physiological means of activation, the best known model is that proposed by Ochoa (96, 97) involving the conversion of pro-inhibitor to inhibitor as a consequence of the phosphorylation of the protein by cAMP-dependent protein kinase. The effect of haemin in preventing the activation is explained as being due to inhibition of the cAMP-dependent kinase by haemin. This hypothesis has now been tested in three different laboratories and all are agreed that it is invalid (54, 98, 99). In brief, the activity of endogenous cAMP-dependent protein kinase in crude lysates and the activity of beef heart cAMP-dependent protein kinase added to crude lysates are *not* inhibited by the addition of 20 μM haemin to the lysate. On the other hand, addition of the heat stable inhibitor protein to the lysate *does* inhibit the activity of endogenous cAMP-dependent protein kinase yet does not affect the inhibition of initiation caused by lack of haemin (54).

It is common experience that the rate of activation is not affected by dilution, so at least the rate-limiting step must be unimolecular, and since the conversion of pro-inhibitor into active inhibitor can be observed in crude gel-filtered post-ribosomal supernatant or in partially purified pro-inhibitor preparations, activation has no absolute requirement for low molecular weight metabolites. (The possibility that uncontrolled factors such as oxygen might be important is not eliminated by these experiments). Whilst no low molecular weight metabolites are required for activation (100), certain

compounds - haemin, GTP (1-2 mM) and in some laboratories, dithiothreitol - can prevent activation (79, 95, 101). Do these agents
block activation by a direct mechanism such as binding to the pro-
inhibitor itself (in the case of haemin and GTP) and acting as a
reducing agent for -SH groups in the pro-inhibitor in the case of
DTT, or is their effect indirect, acting on another protein which in
turn inhibits pro-inhibitor conversion to inhibitor ? Since no
homogeneously pure pro-inhibitor is available this question cannot
be answered definitively, but the fact that these agents still
influence the activation of partially purified pro-inhibitor favours
the view that they act directly on the pro-inhibitor itself (54, 79,
95). ATP antagonises the effect of GTP as an inhibitor of activation
(101), but on its own ATP does no more than marginally increase the
rate of activation in gel-filtered post-ribosomal supernatant or in
partially purified pro-inhibitor preparations (79, 101). Since pro-
inhibitor exhibits no self-phosphorylation activity (5, 54, 79), one
question which is frequently asked is whether self-phosphorylation is
an integral and essential part of the activation process. At first
sight the failure of ATP to accelerate activation seems to argue
against it, but a difficulty arises from the fact that ATP is needed
to assay the inhibitor (either as an inhibitor of protein synthesis
or as a protein kinase). Thus one cannot rule out the possibility,
for example, that two steps are needed for activation: a slow (rate-
limiting) conformational change followed by a rapid self-phosphorylation necessary to lock the protein in its active (inhibitory)
conformation.

     Gel-filtration experiments show that the conversion of pro-
inhibitor to active inhibitor is not accompanied by a change in
molecular weight (5, 100), so models which involve the dissociation
of a regulatory subunit can be ruled out unless the hypothetical
regulator is very small, e.g. less than 20,000 daltons. By a process
of elimination we come to the tentative conclusion that activation
probably involves a *conformational* change in the protein molecule.
The pro-inhibitor seems able to bind ATP, GTP (possibly at the same
site as ATP) and haemin. Binding of either of the last two ligands
inhibits activation, whilst the first has little effect except for
opposing the action of GTP. In addition the blocking, and possibly
the oxidation, of certain -SH groups can promote activation, although
it is far from certain whether oxidation of these -SH groups is
necessary for activation. If I have done my job properly, the reader
should be left with the impression that all these ideas are tentative
in the extreme !

     Before leaving this topic, I should mention that there is some
confusion in the literature over the effectiveness of 1-2 mM GTP as
an antagonist of inhibition of initiation in other conditions.
London's group reported that GTP counteracted the inhibition caused
by HCR and by dsRNA (102). In our experience GTP gives only partial
relief of inhibition in these conditions and it requires judicious

choice of borderline inhibitory doses of dsRNA or HCR to obtain a
reasonable effect (69), which may be the explanation for the results
presented by London's group (102). In our view GTP has two effects:
1) a rather weak antagonistic action against *all* inhibitory conditions, and 2) almost complete restoration of full activity in
certain specific conditions - haem-deficiency, high temperature, high
pressure, and probably gel-filtered lysates (Section VI). We would
interpret the latter as an effect of GTP in preventing activation of
the inhibitory kinase, or in promoting the reversion of the inhibitor
to harmless pro-inhibitor. The former seems more likely to be an
effect on the action of the kinases rather than on their activation.

High Pressure and High Temperature Effects

If lysates are preincubated at 42-44° in the presence of haemin,
or are subjected to high pressure in a pressure cell (also in the
presence of haemin), they show the characteristic inhibition of
initiation on incubation at 30° (54, 58, 59). (Since it may seem
a bizarre thing to subject lysates to hydrostatic pressure, I should
explain that we were lead to this by a consistent finding that post-
ribosomal supernatant prepared by high speed ultracentrifugation contained an inhibitor of initiation, which had apparently been activated
by the act of centrifugation). We believe that the inhibition in both
of these conditions arises from an alternative mode of activation of
the haem-controlled kinase, either the reversible or the intermediate
form of this inhibitory kinase,(Section IV). Our reasons for this
view are:

1)   Mixing a normal lysate with a 42° preincubated lysate or a
"pressed" lysate shows that both treated lysates *do* contain an inhibitor of initiation, but since the inhibition is frequently
transient and of short duration the inhibitor must be readily reversible (54, 58).

2)   GTP (2 mM) is very effective in preventing or reversing the inhibition caused by both treatments as happens with the "reversible"
form of the haem-controlled inhibitor (54, 58).

3)   When lysates subjected to either treatment are gel-filtered
using Sepharose 6B, an inhibitor of protein synthesis is found of
precisely the same size as HCR. Moreover, when the inhibitory
fractions are incubated with γ-^{32}P-ATP, one observes the phosphorylation of a protein identical to the self-phosphorylated component
of HCR (54).

Experiments carried out by Arnstein's group using antibodies
against HCR provided by Hardesty are in agreement with the view that
high temperature activates the haem-controlled kinase (S. Bonanou-
Tzedaki, personal communication).

Hardesty's group have used these antibodies to show that pressure does indeed lead to the activation of the haem-controlled inhibitor, but they also find a second inhibitor of lower molecular weight (103, 104). They have proposed that pressure initially activates a small heat-stable component known as HS. Activated HS then activates a heat-labile component, HL, a step which requires ATP and will not proceed in the presence of non-hydrolysable analogues. In the final step, activated HL is thought to activate HCR by a mechanism probably involving proteolysis, since HCR activation is inhibited by soybean trypsin inhibitor (104). Our own observations are at variance with these. We have failed to detect HS, nor do any of a wide range of inhibitors of proteolysis (soybean trypsin inhibitor, TPCK, leupeptin, chymostatin, trasylol, and PMSF) prevent the inhibition of protein synthesis caused by haem-deficiency, high pressure or high temperature. Indeed, the idea of a mechanism of inhibitor activation involving proteolysis seems hard to reconcile with the ease of reversal of protein synthesis inhibition in these conditions. It seems to me that whilst Hardesty's group may have discovered one mechanism of activation of HCR via HS and HL, this is not the way activation occurs in our hands. We have always envisaged that pressure-induced activation is either due to a change in the conformation and the volume of the pro-inhibitor, or might be due to the oxidation of -SH groups as has been reported for lactate dehydrogenase subjected to similar high pressures (105).

V. CONTROL BY DOUBLE-STRANDED RNA

The dsRNA-Activated eIF-2 Kinase

The addition of low concentration (0.1 - 100 ng/ml) of double-stranded RNA to lysates supplemented with haemin induces a delayed inhibition of initiation identical in its characteristics to that seen in the absence of haemin (67, 106). Consequently, it is of interest to know whether the addition of dsRNA to lysates leads to the phosphorylation of endogenous eIF-2, and this turns out to be the case (78, 82). There is a curious feature of the response to dsRNA in that whilst low concentrations inhibit initiation, high concentrations (10-20 µg/ml) do not (106). The explanation for this paradoxical concentration dependence is still obscure, but it is reassuring to find that high concentrations of dsRNA not only fail to inhibit initiation but also fail to cause eIF-2α phosphorylation in lysates (78, 82). Although the dsRNA-activated kinase has not yet been purified to the same extent as HCR, the information available so far shows that it is quite distinct from the haem-controlled kinase in respect of the following properties:

1) The ds-RNA-activated kinase is mainly associated with crude ribosomes (isolated in low salt buffers), whilst HCR is mainly in the post-ribosomal supernatant fraction (78).

2)   The activation of this kinase requires low concentration of dsRNA
and does not occur with high concentrations. Neither high nor low
levels of single-stranded RNA or DNA, double-stranded DNA, or DNA:RNA
hybrids activate the kinase, which correlates with the fact that none
of these polynucleotides inhibits Met-tRNA$_f$ binding to 40S ribosomal
subunits (78, 106). Activation is unaffected by the presence of
haemin; N-ethylmaleimide does not promote activation.

3)   Activation by dsRNA has an absolute requirement for ATP, which
cannot be fulfilled by non-hydrolysable analogues of ATP or by other
nucleoside triphosphates (69, 78).

4)   The activated dsRNA-controlled kinase shows no self-phosphorylation of an 85,000 dalton polypeptide. On the other hand the
phosphorylation of a 67,000 dalton protein is induced by dsRNA, and
this phosphorylation appears to be associated with the activation step
(78). In so far as the dsRNA-activated kinase has been purified,
this 67,000 dalton protein copurifies with it (107).

5)   Although the dsRNA-controlled kinase phosphorylates eIF-2α in
the same site(s) as does HCR (5, 78), it can be distinguished by its
rather broader substrate specificity which extends to certain
histones.

6)   Antibodies raised against purified haem-controlled inhibitor
fail to inhibit the kinase activity of the dsRNA-controlled kinase,
and fail to prevent the inhibition of protein synthesis by dsRNA
(108).

     These properties of the reticulocyte dsRNA-controlled kinase are
shared by a similar enzyme found in interferon-treated tissue culture
cells (reviewed by P. Lengyel in this book; see also ref. 109).
Protein synthesis activity in cell-free extracts from most nucleated
cells is rather insensitive to inhibition by dsRNA, and little, if
any, dsRNA-activated eIF-2 kinase activity can be found in these
extracts. Pre-treatment of the cells with homologous interferon
renders the cell-free extract sensitive to inhibition by dsRNA and
induces the dsRNA-dependent eIF-2α kinase (109). More work has been
done on the purification and characterisation of the interferon-
induced kinase than on the reticulocyte counterpart, although the
partial purification of the latter has recently been reported (107).
To my knowledge, all the information available so far indicates that
the two enzymes are identical, and since the interferon-induced
enzyme is discussed in detail elsewhere in this volume, I will refrain
from further discussion of the reticulocyte kinase.

The dsRNA-Activated Oligoisoadenylate Synthetase

     Reticulocytes and interferon-treated cells also share the common
feature of possessing the enzyme which synthesises ppp5'A2'p5'A2'p5'A

(and higher homologues) from ATP, provided dsRNA is present (reviewed in ref. 109). This enzyme, which is sometimes named oligoisoadenylate synthetase, appears to be identical in reticulocytes and interferon-treated cells, so I will again refer the reader to the relevant chapter in this volume for further details. Why reticulocytes should so closely resemble interferon-treated cells remains a total mystery, nor do we know at what stage in the maturation of these cells the two enzymes are acquired.

The oligoisoadenylate does not affect the initiation of protein synthesis but activates a latent endonuclease. This latent nuclease is present in reticulocytes and in interferon-treated cells, but is also found in non-treated cells albeit at slightly lower levels than after interferon treatment (109). The fact that the nuclease is present even if cells have not been exposed to interferon suggests that it might have a wider role than the anti-viral effect of interferon, and this in turn leads to speculation as to whether oligo-isoadenylate might be synthesised by another pathway (independent of the presence of dsRNA and the pretreatment of the cells with interferon) in response to particular signals (109). Unlike all the other enzymes discussed in this section, in respect of the nuclease there seems to be a minor difference between reticulocytes and other cells: the reticulocyte enzyme appears to need the tetranucleotide for efficient activation, whereas the trinucleotide is sufficient in other cases (110).

When dsRNA is added to reticulocyte lysates there are therefore two possible mechanisms of inhibition of protein synthesis: a) activation of the protein kinase leading to inhibition of initiation but no inhibition of the completion of nascent chains, and b) activation of the nuclease leading to scission of the mRNA which in turn results in the production of incomplete protein chains. There has been considerable argument and confusion over which system predominates. One simple distinguishing feature of predominating kinase action is whether high levels of dsRNA fail to inhibit protein synthesis: it is only the kinase system which shows a paradoxical concentration dependence, whilst the production of oligoisoadenylate seems to increase with increasing dsRNA levels (111). In our hands, *Penicillium chrysogenum* dsRNA is always much less inhibitory at 20 µg/ml than at 0.01 µg/ml, regardless of what type of mRNA is being translated. This result, together with an analysis of the size of the translation products points to the kinase system as the more important mode of inhibition in our experiments. On the other hand, the use of poly I:C and higher salt concentrations makes the oligo-isoadenylate-endonuclease system dominant (111). Given that there is also variability between different lysates (111) the whole argument seems somewhat sterile.

VI. CONTROL BY SUGAR PHOSPHATES AND REDUCING AGENTS

Introduction: The Nature of the Problem

The remaining three inhibitory conditions from the list in
Section III (gel-filtered lysates, lysates from cells subjected to
anaerobiosis, and the addition of oxidised glutathione to normal
lysates) have been grouped together because they share the common
feature that glucose 6-phosphate is a very effective antagonist of
inhibition. J. Mager was the first to examine the sugar phosphate
requirement in detail using lysates subjected to anaerobiosis or
to other treatments that caused depletion of the ATP pool (64).
These lysates could be reactivated by fructose 1:6 bisphosphate
(which I shall continue to abbreviate as FDP) together with dithiothreitol, or by G6P and NADP, although with severe ATP-depletion cAMP
was also necessary for significant reactivation (64). At about this
time we were studying gel-filtered lysates and we were struck by the
fact that they too required either FDP and DTT, or G6P in order to
prevent an early shut-off of protein synthesis, although some preparations were only fully activated if cAMP or 2-aminopurine was also
present. When tested on their own, the purine derivatives and cAMP
are at least partially effective in preventing inhibition of initiation in these conditions (64, 112), and 2 mM GTP has the same effect
in gel-filtered lysates (113). These observations could be taken
as indicative of a mechanism of inhibition involving phosphorylation
of eIF-2 possibly through the action of the haem-controlled kinase
(Section IV), yet there are other properties of these lysates which
argue against this view. In the case of all the other inhibitory
conditions described so far, mixing an inhibited lysate with normal
lysate leads to inhibition of protein synthesis in the mixture –
even if the inhibition is only transient, but with these three conditions one frequently finds no inhibition in this type of assay.
Although it has been reported that GSSG *does*. activate an inhibitor
similar, if not identical to the haem-controlled inhibitor (112),
in our experience it is not uncommon to obtain strong GSSG inhibition
in a lysate without finding any significant inhibition in the mixing
test (see also ref. 63). A dominant inhibitor could be detected in
lysates derived from cells subjected to anaerobiosis only if drastic
conditions for energy deprivation were used (65). So in all three
conditions it is possible to obtain a lysate in which initiation has
stopped but no inhibitor of protein synthesis can be found, as might
be expected if cessation of protein synthesis were due to activation
of an eIF-2α kinase. For myself, I don't think that these observations definitely eliminate the possibility that inhibition is caused
by activation of an inhibitory kinase, but they do show that any such
inhibitor must be readily reversible to a non-inhibitory form – even
more reversible than the classical "reversible" species generated by
brief incubation in the absence of haemin (Section IV).

The fact that inhibition of protein synthesis in these three conditions can be prevented by dithiothreitol or by NADPH-generating systems suggests that reducing power is an important element in maintaining high initiation rates. On the other hand, some authors have argued that the sugar phosphates play a direct role *per se*, and that NADPH plays at most a minor, subsidiary, role (113-115). Some support for this view comes from the finding that G6P, FDP and 2-deoxyglucose 6-phosphate, but not NADPH, enhance the binding of Met-tRNA$_f$ to native 40S ribosomal subunits (114, 115).

If *both* sugar phosphates *and* reducing power were needed for the maintenance of high initiation rates, it would be extremely difficult to reveal the dual requirement in crude lysates or gel-filtered lysates, since the addition of G6P to such lysates automatically results in the production of NADPH (through glucose 6-phosphate de-hydrogenase). Faced with the problem of how to dissociate the production of NADPH from the presence of G6P, we were on the point of trying to raise antibodies against purified glucose 6-phosphate de-hydrogenase, when our attention was drawn to reports that 2':5' ADP-Sepharose specifically absorbs most enzymes that utilise NADP and NADPH. By studying lysates which had been passed through 2':5' ADP-Sepharose we have been able to show that both sugar phosphates and reducing power are indeed required for the maintenance of high initia-tion rates. The sugar phosphate requirement can be met by glucose 6-phosphate, 2-deoxy glucose 6-phosphate and fructose 1:6 bis-phosphate, but not by 6-phosphogluconate. The reducing power requirement can be met by dithiothreitol or by NADPH generating systems. Dithiothreitol seems to affect the target directly, whilst NADPH works only through some intermediary which appears to be the thioredoxin reductase - thioredoxin system.

Before discussing the experiments which have lead us to these conclusions, I should mention that inhibition of initiation in other conditions can also be partially relieved by G6P (113). It is particularly effective in lysates supplemented with suboptimal con-centrations of haemin, and in some lysates maintained maximal rates of protein synthesis even in the complete absence of haemin (113). It was also reported to have some effect in lysates inhibited by HCR or dsRNA provided the dose of these inhibitors was chosen to give partial inhibition. It seems that G6P is a weak general antagonist of all inhibitory agents, but a highly efficient antagonist of the three conditions discussed in this section. In this respect, the effects of G6P show some parallel with those of GTP - see Section IV.

Properties of Gel-Filtered and 2':5' ADP Lysates

We gel-filter lysates using either Sephadex G-25 or G-50 in a buffer containing 25 mM KCl, 10 mM NaCl, 1.1 mM MgCl$_2$, 10 mM Hepes pH 7.2, 0.1 mM EDTA, which is thoroughly degassed. Since no DTT is added, we include EDTA in an attempt to avoid heavy-metal poisoning

of -SH groups, and it seems to have improved the consistency with
which active lysates are obtained. The lysates are supplemented with
haemin before gel-filtration, but as all the added haemin is tightly
bound to proteins, virtually none is removed by gel-filtration and
the addition of further haemin to the gel-filtered lysate does not
improve its activity. Another compound retained on gel filtration is
NADP + NADPH: the gel-filtered lysate has been estimated to contain
80% of the initial level of these cofactors (113), and although our
own figures are rather lower, gel-filtered lysates certainly contain
5-10 µM NADP + NADPH. On the other hand, ATP, GTP and spermidine are
removed by gel-filtration, and must be added back (at 0.5 mM, 0.1
mM and 0.4 mM respectively) to obtain efficient protein synthesis.

Although G-25 and G-50 lysates seem equally devoid of low
molecular weight metabolites, they differ because the lysate contains
about four small proteins (M.W. 10-15,000) which are marginally
included on Sephadex G-50 and are therefore largely absent from our
G-50 lysates. Tim Hunt has identified one of these as thioredoxin,
and another appears to be equivalent to glutaredoxin (116) since it
catalyses the glutathione-driven reduction of low molecular weight
disulphides:

Since these small proteins are only marginally included on Sephadex
G-50 it is doubtful whether we completely eliminate them from our
usual G-50 lysate preparations. In addition, Homgren (personal
communication) has noted that a proportion of thioredoxin in crude
extracts is always associated (probably through disulphide bonds)
with other proteins, and if this happens in reticulocyte lysates it
gives further cause to doubt that our G-50 lysates are completely
devoid of thioredoxin, though they certainly will have less than the
G-25 lysates.

We routinely pass up to five column volumes of lysate through columns of 2':5' ADP-Sepharose, using the same buffer as for gel-filtration. The gel-filtration effect is therefore negligible, and the resulting lysate (which I shall call 2':5' ADP lysate) contains all the low molecular weight metabolites found in the crude lysate. Glucose 6-phosphate dehydrogenase is quantitatively removed from the lysate; this constitutes a routine assay to monitor the performance of the column, which is necessary in our experience since the columns no longer work satisfactorily after a few runs. Glutathione reductase and 6-phosphogluconate dehydrogenase are also bound by the column, probably to the extent of complete removal of these enzymes for the lysate. On the other hand, NADP-utilising isocitrate dehydrogenase is not bound at all. Whilst we know that some thioredoxin reductase is absorbed to the column, it is extremely difficult to show if the total complement of this enzyme in the lysate is removed. Some of the properties of the 2':5' ADP lysate suggest that it is not. The enzymes bound to the column can be eluted with 1 mM NADP or NADPH, and SDS-polyacrylamide gel electrophoresis of this eluate confirms that the column binds a very select small number of proteins.

Whilst gel-filtered lysates and 2':5' ADP lysates exhibit a high initial rate of protein synthesis, we have had problems with lysates that have been subjected to both steps. Recently we have been able to obtain consistently high activity if the gel-filtration step is done first, before passage through 2':5' ADP-Sepharose, rather than vice-versa. Most of these preparations behave as though thioredoxin reductase was absent, or inactive, in contrast to the behaviour of lysates which have only been through 2'5' ADP-Sepharose.

Protein Synthesis in Gel-Filtered Lysates

The initial rate of protein synthesis in gel-filtered lysates is usually at least as high, if not higher, than that of the parent lysate (taking isotope dilution into account). After 15-20 minutes, protein synthesis stops (Fig. 4) with all the characteristics listed in Section III, except that the shut-off is rather later and less abrupt than, say, in a normal lysate incubated without haemin. The addition of more haemin does *not* prevent the shut-off. The compounds that *do* prevent the reduction in the rate of initiation are listed in Table I, which attempts to grade their relative efficiency. For convenience I shall refer to them as "reactivators", although strictly speaking what is normally tested is their ability to prevent the shut-off if they are added at the start of the incubation. In so far as has been investigated, these compounds will in fact restore protein synthesis activity to a gel-filtered lysate if they are added after protein synthesis has stopped. There is some variability between different preparations of gel-filtered lysates in terms of their response to different compounds, but this variability follows a semi-rational pattern, not unlike that seen with lysates from ATP-depleted

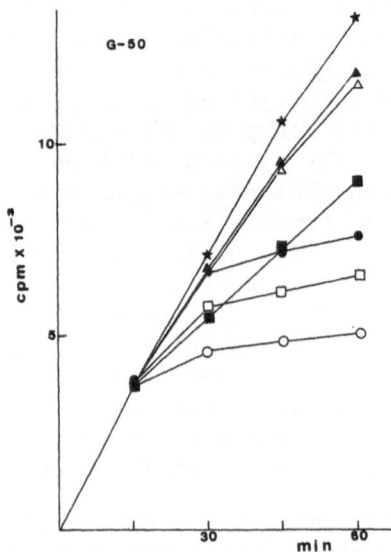

Fig. 4. Kinetics of protein synthesis in gel-filtered (Sephadex G-50)
         lysate incubated under standard conditions at 30° with ^{14}C-
         valine at 30 µCi/µmole (O). Incubations contained 0.2 mM
         G6P (★), 0.5 mM FDP (●), 0.5 mM DTT (■), FDP + DTT (△) 0.2
         mM isocitrate (▲), or 30 µM NADPH (□). Protein synthe-
         sis was determined in 5 µl samples.

cells (64). For example, although G6P normally gives complete
"reactivation" whilst DTT or FDP show only partial rescue, some
preparations show very little response to either DTT or FDP and are
only partially rescued by G6P, but G6P together with DTT gives complete reactivation. We have found no reproducible difference between
the response of G-25 lysates and G-50 lysates, except that G-50
lysates are more likely to need the synergistic action of two compounds for complete reactivation.

     Most of the reactivators are sugar phosphates (Table 1), but
there are three exceptions: DTT, isocitrate and NADPH. Isocitrate
could be acting directly as a cofactor for some process, or its
effect might be due to the production of NADPH and 2-oxoglutarate by
isocitrate dehydrogenase. The fact that 2-oxoglutarate has no effect
on protein synthesis eliminates one of these possibilities, and of
the other two we favour the view that it is NADPH production which

is important. On the other hand, NADPH itself gives poor stimulation over a wide concentration range. This observation would seem to mark the rapid death of any model in which NADPH was needed to maintain high initiation rates. If, however, the hypothetical NADPH-utilizing process were strongly inhibited by NADP, one might expect the addition of NADPH itself to be less effective than an NADPH-generating system such as isocitrate or G6P, because the addition of NADPH will eventually lead to higher NADP/NADPH ratios than when a system for continuous conversion of NADP to NADPH is present.

The best single reactivator is glucose 6-phosphate which at low concentrations gives linear protein synthesis for up to 60 minutes in nearly all preparations of gel-filtered lysate. Other compounds such as ribose 5-phosphate and ribulose 5-phosphate which can be converted to G6P in the lysate are equally effective. (To test for this conversion we add GSSG to gel-filtered lysates in the presence of potential generators of NADPH and then assay for the production of reduced glutathione). In most lysates, 6-phosphogluconate is almost as effective as G6P in the same concentration range, but we have had some G-50 preparations which show no response to 6PGA. As I will show later, 6PGA cannot supply the sugar phosphate requirement for the maintenance of high initiation rates, and its ability to reactivate these lysates is probably dependent on there being sufficient flow through the 6-phosphogluconate dehydrogenase step for G6P to be generated in adequate amounts. One observation consistent with this is that a G-50 lysate which failed to respond to 6PGA, did show partial reactivation when traces of NADP or oxidised glutathione were added to increase the flow through the 6-phosphogluconate dehydrogenase step.

Fructose 1:6 bisphosphate on its own gives partial reactivation, although much higher concentrations are needed than for any other sugar phosphate. The question therefore arises as to whether this effect might not be caused by G6P either present as a contaminant in commercial FDP or generated by metabolism of FDP in the lysate, but by assaying for GSSG reduction, as explained above, we have been able to rule out this possibility. If one uses gel-filtered lysates which show partial reactivation in response to DTT or isocitrate, we can usually obtain complete reactivation and linear rates of protein synthesis for 60 minutes if FDP is added together with DTT or isocitrate (i.e. there is synergism between the sugar phosphate and the source of reducing power). In these conditions the concentration of FDP required is much lower than in the absence of reducing agents, and is of the same order as that found with 2-deoxyglucose 6-phosphate.

Among the more significant inactive compounds not listed in Table I, are sorbitol 6-phosphate and all glycolytic intermediates beyond the glyceraldehyde 3-phosphate dehydrogenase step. Apart from the failure of NADPH to elicit a significant response, these results

Table 1.   The effectiveness of different 'reactivators' in gel-filtered lysates and 2':5' ADP lysates. The typical effect of
each compound has been scored on a scale ranging from 0 to
signify no stimulation of protein synthesis over the un-
supplemented control, to ++++ which signifies complete
reactivation and maximum rates of protein synthesis sustained for up to 60 min incubation. The concentrations given
in parentheses are the levels which typically give half-
maximal response; where two compounds are used together,
the concentration given refers to the level of sugar
phosphate needed for half-maximal response in the presence
of "saturating" concentrations of the other compound.

	Gel-filtered Lysates	2'5'-ADP Lysates	Gel-filtered 2'5'-ADP Lysates
DTT	+ + +	+ + + + (10μM)	+
NADPH	(+)	+ + + (5μM)	+
ISOCITRATE	+ + +(+)	+ + + + (4oμM)	+
G6P	+ + + + (10μM)	(+)	0
2d-G6P	+ + + + (70μM)	(+)	0
RIBULOSE-5P	+ + +(+)	n.t.	0
6PGA	+ + +(+)[†](20μM)	0	0
FDP	+ + ((500μM)	0	0
FDP + NADPH	+ +	+ + +	+ + +(+)
G6P + DTT	+ + + +	+ + + +	+ + + + (20μM)
2d-G6P + DTT	+ + + +	+ + + +	+ + + + (80μM)
RIBULOSE-SP DTT+	+ + +	n.t.	+ + + +
6PGA + DTT	n.t.	+ + + +	+
FDP + DTT	+ + + + (100μM)	+ + + +	+ + + +(100μM)
G6PDH	n.t.	+ + + +	
G6P + G6PDH	n.t.	+ + + +	+ + + +
G6P + ISOCITRATE	n.t.	+ + + +	+ + + +
G6P + NADPH	n.t.	+ + +(+)	+ + +(+)
GTP	+ + +	+ +	+ +
cAMP	+ + +	+ +	+ +
GTP + DTT	+ + +(+)	+ + + +	+ + + +
cAMP + DTT	+ + +(+)	+ + + +	+ + + +

Notes: † Whilst this is the typical response to 6PGA, some gel-
filtered lysates show almost zero reactivation even though
they respond well to other sugar phosphates - see text.

* With most preparations of gel-filtered 2':5' ADP lysates,
   this response is only seen when both thioredoxin and
   thioredoxin reductase from *E. coli* are added, otherwise the
   response is close to zero.

fit the following pattern: 1) sugar phosphates which *do not* supply
reducing power (e.g. FDP) give *partial* reactivation, 2) compounds
which supply reducing power (DTT or isocitrate) are rather better,
and 3) compounds which supply both functions (G6P or a combination
of FDP and DTT) give complete reactivation. Two reactivators that
do not conform to this pattern are GTP and cAMP, which are usually
fairly effective though not quite as consistenly good as G6P (Table
I).

Protein Synthesis in 2':5'ADP Lysates

The initial rate of protein synthesis in 2':5'ADP-lysates is
about the same as in the parent lysate, but after about 20 minutes
incubation there is a rapid decrease in the rate of initiation (Fig.
5) with the same characteristics as listed in Section III. These
lysates show little response to added sugar phosphates (Table I),
but they can be completely reactivated by several reagents which
share the common feature of supplying reducing power: DTT, NADPH
(for reasons which we don't understand NADPH is much more effective
in these lysates than in gel-filtered lysates), isocitrate and
glucose 6-phosphate dehydrogenase (G6PDH). Both commercial G6PDH

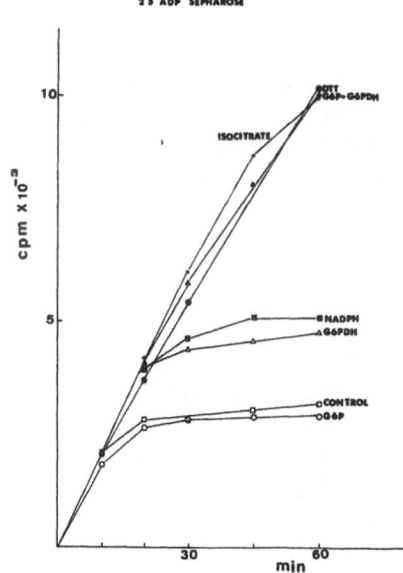

Fig. 5.
Kinetics of protein synthesis in 2':5'
ADP lysates determined as described in
Fig. 4, incubated with 0.2 mM G6P,
0.2 mM isocitrate, 0.5 mM DTT, 30 µM
NADPH, and reticolocyte glucose 6-
phosphate dehydrogenase as indicated
on the figure.

and reticulocyte G6PDH eluted from the 2':5'ADP-Sepharose column and subjected to further purification are effective.

These results provide clear evidence that either DTT or NADPH is needed for the maintenance of high rates of initiation. The fact that DTT is active suggests that the NADPH is not needed merely as a passive cofactor, but is required to donate reducing equivalents, probably to maintain an -SH group in the reduced state. What, then, is the enzyme system that utilises NADPH for this purpose ? Evidently it is present in 2':5' ADP-lysates and cannot have been absorbed to the 2':5' ADP-Sepharose column. A clue is provided by 2':5' ADP lysates which have been aged for 2-3 hours on ice: the ability to respond to isocitrate or G6PDH is diminished but the response to DTT is retained. The response to NADPH-generators (e.g. isocitrate) can be regained if the *E. coli* are needed, which is not altogether surprising since the bacterial thioredoxin reductase is specific for the homologous thioredoxin and will not reduce mammalian thioredoxin (117). (These purified *E. coli* components, which were kindly donated by Dr. A. Holmgren, are used in preference to the mammalian counterpart since they are much more stable.) This implies that the NADPH utilising system required to maintain high rates of protein synthesis

Fig. 6.   Reduction of insulin in (A) gel-filtered (Sephadex G-25) lysate, and (B) G-25 gel-filtered lysate passed through 2':5' ADP-Sepharose before assay. Samples of these lysates were incubated under conditions for protein synthesis assay (see Fig. 4) for 35 min, then made 0.5 mg/ml in insulin and incubated for a further 15 min at 30°. Samples (3 μl) of each assay were then incubated with 2 μl of 15 mM ^{14}C-N-ethylmaleimide for 20 min before electrophoresis on SDS-polyacrylamide gels (20%), which were fluorographed. For panel (A) the lysate samples were incubated with 0.2 mM G6P (tracks 2 and 9), 0.2 mM ribulose 5-phosphate (track 3), 0.2 mM 6-phosphogluconate (track 4), 0.5 mM FDP (tracks 5 and 10), 0.5 mM 2deoxyglucose 6-phosphate (track 6), 0.2 mM isocitrate (track 7), 30 μM NADPH (tracks 8-10), or 0.5 mM DTT (track 11). For panel (B) lysate samples were incubated with 0.2 mM G6P (tracks 5-11 inclusive), 0.5 mM DDT(tracks 2 and 6), 50 μM NADPH (tracks 4 and 78, reticulocyte glucose 6-phosphate dehydrogenase (tracks 3, 4, 7, 8 and 11), 9 μM thioredoxin and 0.5 μM thioredoxin reductase from *E. coli* (tracks 10 and 11). The assay in panel (B) track 9 contained material eluted by NADPH from a 2':5' ADP-Sepharose column through which rabbit liver post-ribosomal supernatant had been passed: this material contains NADPH (giving 50 μM in the lysate incubation), glucose 6-phosphate dehydrogenase and other enzymes including thioredoxin reductase. The solid arrow marks the position of the reduced (labelled) insulin, and the open arrow the position of the 24,000 dalton protein mentioned in the text.

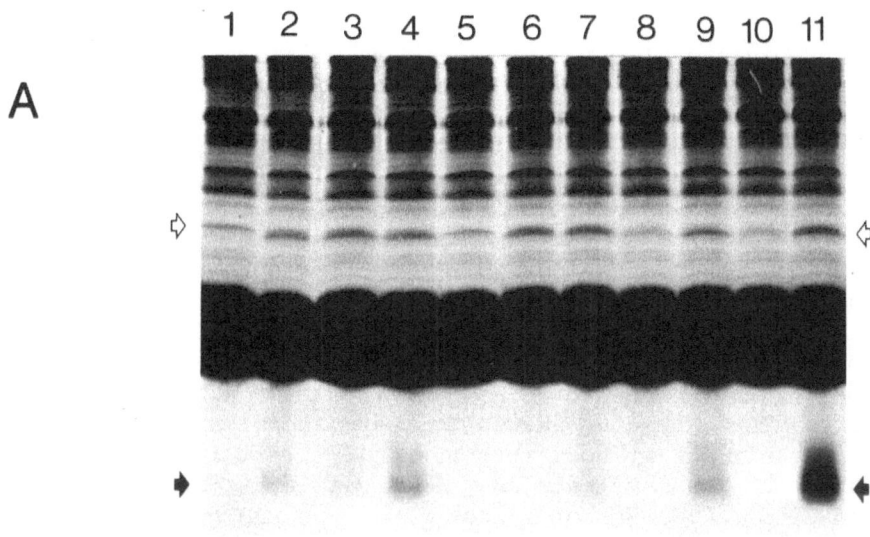

A

1 2 3 4 5 6 7 8 9 10 11

G-25

B

1 2 3 4 5 6 7 8 9 10 11

G-25 / 2 5 ADP

must be the thioredoxin system (or some system which has the same
function). Fresh 2':5' ADP lysates would seem to contain sufficient
thioredoxin reductase to maintain protein synthesis if NADPH is
supplied, but the thioredoxin reductase must be inactivated on "age-
ing" the lysate at 0°. (These findings led us to speculate whether
the reason why occasional rare batches of gel-filtered lysate failed
to show the normal response to the usual reactivators was because
the thioredoxin system was missing or inactivated. One such batch
obtained recently showed very little reactivation with G6P alone, but
did exhibit linear rates of protein synthesis if the *E. coli*
thioredoxin system was added in addition to G6P).

     Since 2':5' ADP lysates contain all the low molecular weight
metabolites of the parent lysate, it is possible that the maintenance of high initiation rates is dependent on other factors in
addition to the availability of NADPH. We therefore examined lysates
which had been gel-filtered (on G-25) and then passed through 2':5'
ADP-Sepharose. These lysates showed no response to any sugar
phosphate, and only a very weak response to DTT, but linear protein
synthesis at about the same rate as the parent lysate is obtained if
both DTT and either G6P, FDP or 2-deoxyglucose 6-phosphate are added
(Table I). No response was obtained with DTT and 6-phosphogluconate
even though the parent gel-filtered lysate was maximally reactivated
by 6PGA. Since 2-deoxyglucose 6-phosphate is probably metabolically
inert in these lysates, we conclude that the sugar phosphates must
be acting as cofactors in some way, and that 6PGA cannot fulfill this
function except if it can be converted to G6P via the pentose phosphate pathway.

     Most preparations of gel-filtered 2':5' ADP lysates don't respond
to the combination of sugar phosphates and a source of NADPH, but
they do show maximum response and sustained high rates of protein
synthesis if both the components of the *E. coli* thioredoxin system are
added as well. Evidently, thioredoxin reductase must be missing from
these lysates or is hypersensitive to inactivation.

     Since these results clearly point towards a role of NADPH in the
reduction of disulphide bonds, it becomes important to be able to
test the ability of various lysates to catalyse such reduction. One
approach we have used is to incubate lysates with insulin and then
with ^{14}C-N-ethylmaleimide. Gel electrophoresis is then used to detect
the labelling of insulin, which is, of course, dependent on the
reduction of the disulphide bonds. This test is a derivative of one
of the standard assays for the thioredoxin system (118). In the same
assay we can look for changes in the state of -SH groups in the
proteins of the lysate itself, since these proteins are also labelled
during the incubation with ^{14}C-NEM and are displayed by gel electrophoresis. Tim Hunt noticed that there is a protein of about 24,000
daltons which can be labelled in fresh gel-filtered lysates but can

no longer be labelled if the lysate is first incubated in the absence
of DTT or NADPH generating systems (Fig. 6). He has shown that the
exposed -SH groups of this protein are very sensitive to oxidation
to disulphide bonds, which prevents labelling of the protein by
^{14}C-NEM. By studying the labelling of this protein and of insulin
we can gain an impression of the capacity of the system to reduce
disulphide bonds. We observe that DTT and compounds which produce
NADPH promote reduction of disulphide bonds in gel-filtered lysates,
or in fresh 2':5' ADP lysates (Fig. 6). (It is interesting that NADPH
itself is rather inefficient in gel-filtered lysates, which correlates
with the rather feeble stimulatory effect on protein synthesis.) Gel-
filtered 2':5' ADP lysates which are not reactivated by the combination of isocitrate and G6P (in the absence of the *E. coli* thioredoxin
system) show little reduction of disulphide bonds; the presence of
the bacterial thioredoxin system is found to promote not only the reactivation of protein synthesis but also the reduction of disulphide
bonds. Thus, NADPH-generating systems prevent the shut-down in
initiation rates only in lysates which have the enzyme systems for
disulphide bond reduction.

In summary, there is a dual requirement for *both* sugar phosphates *and* reducing agents to prevent inhibition of initiation in
lysates. The sugar phosphates seem to act directly as cofactors
for some as yet undefined step. Glucose 6-phosphate, 2-deoxyglucose
6-phosphate and fructose 1:6 bisphosphate can fulfill this role. The
other requirement can be met by DTT or by NADPH generating systems,
and seems to be required to counteract the oxidation of certain
critical -SH groups to disulphide bonds. Whilst DTT probably reduces
disulphide bonds directly, the effect of NADPH seems to be mediated
by thioredoxin reductase and thioredoxin, or some system of equivalent
function.

Phosphorylation of eIF-2 is Controlled by Reducing Agents

The next question is the susceptible protein whose autoxidation
leads to the shut-off of protein synthesis ? The characteristics
of inhibition point to either eIF-2 itself or some protein involved
in the phosphorylation - dephosphorylation of eIF-2. The fact that
GTP and cAMP are fairly effective reactivators of gel-filtered
lysates and 2':5' ADP lysates would seem to favour the latter possibility. We therefore examined eIF-2 phosphorylation in gel-filtered
lysates and 2':5' ADP lysates which had been incubated under different
conditions and briefly incubated with γ-^{32}P-ATP before analysis by
SDS-polyacrylamide gel electrophoresis. The results show that sugar
phosphates *per se* do not affect the labelling of eIF-2α, but that the
lack of reducing power *does* lead to increased labelling (Fig. 7).
This distinction is most clearly seen in gel-filtered 2':5' ADP
lysates which require both DTT and sugar phosphates for full restoration of protein synthesis. Sugar phosphates alone do not affect
either protein synthesis or eIF-2 labelling, whilst DTT alone has a

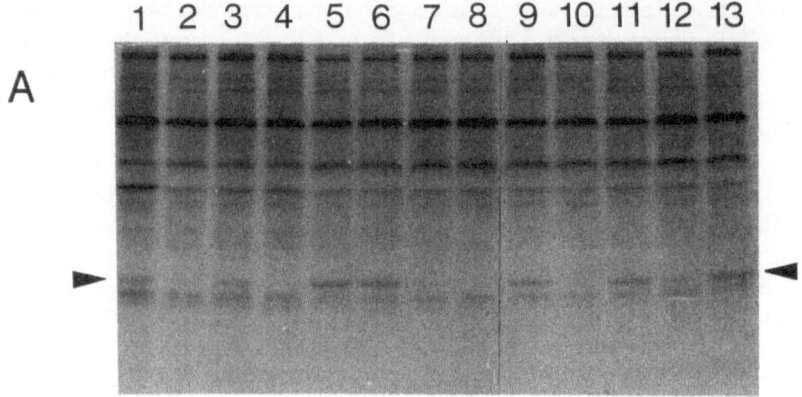

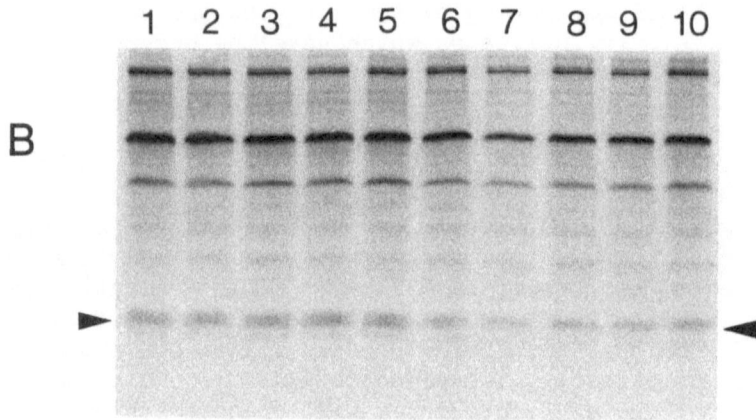

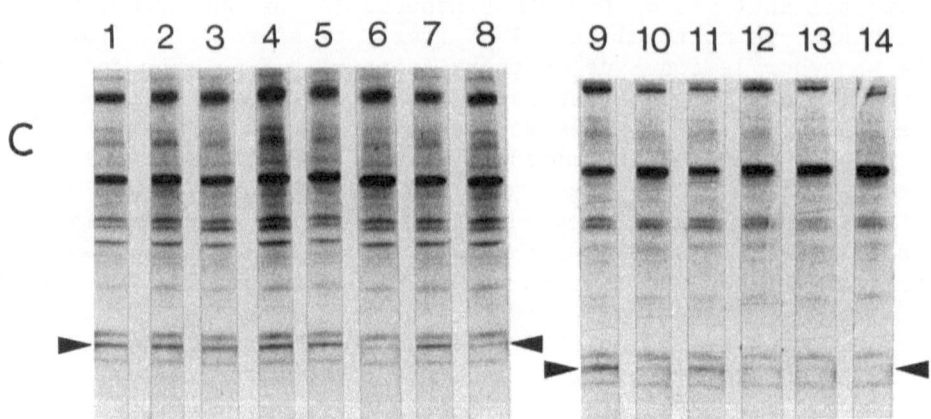

minimal reactivating effect on protein synthesis but gives maximal
suppression of the labelling of eIF-2α (Fig. 7). NADPH-generating
systems also give decreased labelling of eIF-2 provided the lysate
has an active thioredoxin system. In gel-filtered 2':5' ADP lysates
in which G6P and isocitrate did not reactivate protein synthesis
unless the *E. coli* thioredoxin system was added, the *E. coli* components were also required to suppress the labelling of eIF-2α.

The labelling observed in the absence of reducing agents is
comparable to that seen when normal lysates are incubated without
haemin or with dsRNA - two conditions that are known to lead to
activation of eIF-2 kinases. Thus the most obvious interpretation
is that the lack of a system for reducing disulphide bonds leads to
activation of an eIF-2 kinase. Since high levels of GTP prevent in-

---

Fig. 7.   Phosphorylation of eIF-2α in different lysates. Lysate
          samples (5 µl) were incubated under conditions for protein
          synthesis assay (see Fig. 4) for 35 min, then 3 µl of 0.2
          mM γ-^{32}P-ATP were added and incubation continued for 5 min.
          The material was then analysed by electrophoresis on SDS-
          polyacrylamide gels (15%) which were autoradiographed. In
          panel (A), tracks 1-4 are assays of gel-filtered (Sephadex
          G-25)lysate, tracks 5-8 of 2':5' ADP lysate, and tracks 9-13
          of untreated parent lysate. These lysate samples were
          incubated with 0.2 mM G6P (tracks 2, 6 and 7), 0.5 mM FDP
          (track 3), 0.2 mM isocitrate (tracks 4 and 8), reticulocyte
          glucose 6-phosphate dehydrogenase (track 7), 1.6 mM oxidised
          glutathione (track 12), 40 ng/ml dsRNA (track 11), purified
          HCR (track 13), and without haemin (track 9). Panel B shows
          assays of a gel-filtered (Sephadex G-25) lysate passed
          through 2':5' ADP-Sepharose before assay with the following
          additions: 0.5 mM DTT (tracks 6-10 inclusive), 0.2 mM G6P
          (tracks 2 and 7), 0.5 mM 2-deoxyglucose 6-phosphate (tracks
          3 and 8), 0.5 mM FDP (tracks 4 and 9), or 0.2 mM 6-phospho-
          gluconate (tracks 5 and 10). Panel C shows results of assays
          using the same type of lysate as in panel B but incubated
          with the following additions: 9 µM thioredoxin and 0.5 µM
          thioredoxin reductase from *E. coli* (tracks 2, 4, 6, 8, 10
          and 12), 0.2 mM G6P (tracks 3, 4, 9-12 inclusive and 14), 0.2
          mM isocitrate (tracks 5, 6, 9 and 10), 25 µM NADPH (tracks
          7, 8, 11 and 12) and 0.5 mM DTT (tracks 13 and 14). The
          experiment shown in panel C used two gels (tracks 1-8 and
          9-14). The arrow marks the position of eIF-2α; the mobility
          of this protein relative to other proteins is very much
          influenced by the batch of SDS used and by the exact con-
          ditions of electrophoresis, and so its position in relation
          to other proteins is somewhat variable between different
          experiments.

hibition of protein synthesis in gel-filtered lysates, 2':5' ADP
lysates and in normal lysates incubated without haemin (Section IV),
the haem-controlled kinase is an obvious candidate, especially as
previous studies have suggested that the status of -SH groups in the
pro-inhibitor may be an important factor in the activation mechanism
(Section IV). The main difficulty with this idea is that it is not
easy to demonstrate that gel-filtered lysates in which protein
synthesis has shut-off contain an inhibitor of initiation (see Section
VI), as is the case with normal lysates which have been incubated
without haemin. If it is the haem-controlled kinase that is activated
in the absence of a means of reducing disulphide bonds, then it must
be a novel form of this kinase - a form which is even more easily
reversible than the "classical" reversible inhibitor (see Section IV).

     In view of this problem it would be rash to overlook possible
alternative explanations. There are two which come to mind provided
one is allowed to presuppose that even in the presence of haemin there
is a low level of activated haem-controlled kinase. The first
proposes that the absence of reducing power renders the eIF-2 more
susceptible to phosphorylation by this low kinase activity, and hence
more readily inactivated. The other proposes that reducing power is
necessary to maintain the activity of the phosphatases which dephosphorylate eIF-2 and so reverse the effect of the hypothetical low
kinase level. Since GTP reverses the kinase activation step (Section
IV), whilst cAMP prevents the phosphorylation of eIF-2 by the haem-
controlled kinase (Section IV), both of these models can accommodate
the observation that GTP and cAMP rescue protein synthesis in gel-
filtered lysates, since these compounds would abolish the low kinase
activity. Our preliminary experiments point against the idea that
phosphatase activity is regulated by the availability of reducing
power, but further experiments are required to distinguish between
these hypotheses. One of the most important questions to be resolved
is whether it is -SH groups in the pro-inhibitor, or in eIF-2 or in
the phosphatase that tend to become oxidised in the absence of
reducing agents.

The effect of Oxidised Glutathione

     When oxidised glutathione (GSSG) is added in low concentrations
to normal (not gel-filtered lysates) it is quantitatively reduced via
GSSG reductase using NADPH generated mainly through the metabolism
of endogenous glucose and glucose 6-phosphate, and it has no effect
on protein synthesis. Addition of GSSG in excess over the capacity
of the system to reduce it all results in inhibition of initiation
after a lag of 5-15 minutes (63). Under these conditions the G6P and
glucose pools will be exhausted and the NADPH/NADP ratio will be low.
Is the inhibition solely due to the lack of G6P and NADPH, or does
GSSG itself act as an inhibitor ? This issue can be resolved by
using 2':5' ADP lysates in which linear rates of protein synthesis
can be obtained by the addition of isocitrate or G6PDH (see Table I)

and yet no reduction of GSSG can occur because GSSG reductase is
absent. Very low concentrations (less than 50 μM) GSSG strongly
inhibit protein synthesis under these conditions, and although the
addition of 0.2 mM additional G6P gives slight relief of inhibition
at 50 μM GSSG, no effect is seen at higher levels (100-200 μM). (The
antagonistic action of G6P at very low GSSG may be another manifesta-
tion of the weak reactivation activity seen in many inhibitory conditions, as discussed in Section VI). GSSG therefore seems to be an
inhibitor in its own right, and does not act merely through depletion
of G6P and NADPH pools.

When GSSG is added to G-25 lysates in the absence of any source
of reducing power, it causes an earlier shut-off of protein synthesis
than normal. On the other hand GSSG causes no such inhibition of G-50
lysates, which shut-off at about the same time irrespective of whether
GSSG is present or absent. One possible explanation for this difference
is that G-50 lysates differ from G-25 in their relative lack of glutaredoxin. Glutaredoxin acts as an intermediate in the GSH-driven
reduction if disulphide bonds (116). Is it possible that in the
presence of GSSG the flow of reducing equivalents via glutaredoxin is
reversed, so that the *reduction* of GSSG occurs coupled with the gene-
ration of disulphide bonds in proteins ? This could explain why GSSG
accelerates the onset of inhibition of initiation in G-25 lysates
(which contain glutaredoxin) but not in G-5? lysates.

When G-25 lysates are supplemented with G6P, judicious choice of
GSSG concentration can produce the interesting outcome of a transient
inhibition of protein synthesis from which the system recovers (Fig.
8). This again suggests that GSSG is an inhibitor in its own right,
quite independent of any effect on G6P or NADPH pools. However, the
consequences of this direct inhibitory action of GSSG seem to be
indistinguishable from those produced by a lack of reducing power:
the reduction of protein disulphide bonds is inhibited, and the
labelling of eIF-2α in the assays described in the previous section
is increased. Since London's group have presented evidence that GSSG
activates the haem-controlled inhibitory kinase (112), this can be
constructed as supporting the idea that the inhibition of initiation
in gel-filtered lysates and 2':5' ADP lysates is due to the activation
of some form of this kinase in the absence of reducing agents.

VII. ARE RETICULOCYTE CONTROL MECHANISMS RELEVANT TO OTHER CELLS ?

To what extent are the control mechanisms encountered in reticulocytes (other than the double stranded RNA effects which have been
discussed in Section V) also operative in nucleated cells ? A
reasonable assumption would be that the effect of haemin is probably
specific to reticulocytes. Reports that haemin stimulated protein
synthesis in cell-free extracts from nucleated cells therefore came
as quite a surprise (32, 119). In the absence of haemin, the initial
high rate of protein synthesis declined after a few minutes and

synthesis soon stopped. Haem prolonged the period of high rate and
delayed the shut-off, just as it does (though more dramatically) in
reticulocyte lysates (119). Moreover high GTP achieved the same
result in tissue culture cell extracts as in reticulocyte lysates.
In the absence of haemin, it was found that initiation ceased, and
that complexes between 40S ribosomal subunits and Met-tRNA$_f$ were no
longer detectable (119). The striking parallel with reticulocyte
lysates was extended by the finding that in the absence of haemin an
inhibitor accumulates in the post-ribosomal supernatant, although the
potency of this inhibitor seemed much lower than would be the case
if the same experiment had been carried out in reticulocyte lysates.

    Yet another point of similarity is that sugar phosphates stimu-
late protein synthesis in extracts from tissue culture cells, and
delay the onset of inhibition of initiation (114, 115). Although it
was once thought that the sugar phosphates might be acting as an
alternative and superior energy source via glycolysis(120), the fact
that 2-deoxyglucose 6-phosphate is effective eliminates this hypothesis and favours a more direct role of the sugar phosphate (114,
115).

    Perfused heart is another case where sugar phosphates may be
important in the control of initiation. After 1-2 hours perfusion
the rate of protein synthesis drops due to a decrease in the rate of
initiation. The presence of insulin in the perfusate prevents this

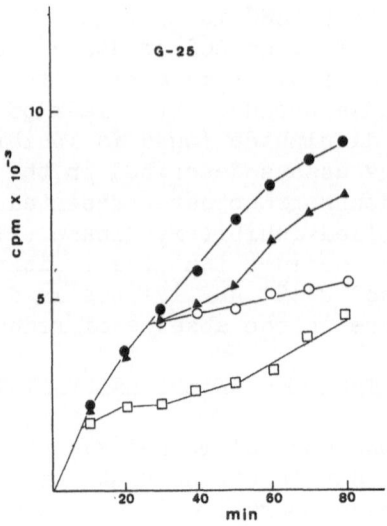

Fig. 8. Kinetics of protein synthesis in gel-filtered (Sephadex G-25)
         lysate assayed as described in Fig. 4 (O). Assays contained
         0.2 mM G6P (●) and 0.2 (▲) or 0.3 mM (□) oxidised
         glutathione.

shut-off irrespective of whether glucose is present or not (121).
The shut-off can also be prevented by perfusion with fatty acids but
in this case glucose must also be present, and in these conditions
the heart contains elevated glucose 6-phosphate pools (121). The
story is still a long way from proving that G6P has the same effect
as in reticulocyte lysates, but it holds promise that this might be
the case. How insulin acts to prevent the shut-off remains completely unknown and is a problem crying out for further investigation.

It has been reported that an inhibitor which appears superficially similar to HCR can be isolated from rat liver (122) and from
Ehrlich ascites tumour cells (123). In neither case have the properties of the inhibitor been examined in sufficient detail to be sure
that it is indeed identical to reticulocyte HCR. Even if such
identity were established, the inhibitor would seem to be present in
these cells at much lower levels than in reticulocytes and so it may
not play a particularly important role in overall control of protein
synthesis.

When nucleated cells are starved of essential amino acids the
response is a marked reduction in the rate of initiation with a
secondary, less significant reduction in elongation rates. In
Ehrlich ascites tumour cells starved of amino acids, the level of
$40S/Met-tRNA_f$ complexes was reduced to an extent comparable to the
degree of inhibition of initiation (124). The formation of these
complexes *in vitro* was also lower in extracts from starved cells as
opposed to fed cells, but addition of eIF-2 largely eliminated the
difference (125). Thus amino acid starvation seems to regulate the
same step in initiation as is the target of reticulocyte control
mechanisms, although there is no evidence yet for an increase in
eIF-2 phosphorylation following amino acid starvation. It is, moreover, worth bearing in mind that the rate of initiation in reticulocytes is *not* reduced on amino acid starvation, so that whatever
the regulatory system in nucleated cells may be, some essential part
of it must be missing from reticulocytes.

It is self-evident from this discussion that we really are very
much in the dark when it comes to understanding the mechanism of
control of initiation in nucleated cells. Most of the observations
have little significance on their own, and so the tendency is to seek
phenomenological parallels with the behaviour of the reticulocyte
lysate and to interpret the results in terms of reticulocyte control
mechanisms. That this might be a mistaken philosophy is indicated
by the fact that none of the superficial similarities between reticulocytes and nucleated cells (apart from the dsRNA-dependent regulation) has been successfully followed up to the point where it has
been shown that the general mechanism of control is the same. Many
situations in which there is a large change in the rate of initiation
in nucleated cells affecting the efficiency of translation of all
mRNA species are not really relevant to reticulocytes, e.g. mitosis,

hypertonic shock, amino acid starvation, serum starvation, and the
effects of hormones (reviewed in refs. 126, 127). There is no
compelling reason why regulation in these situations should show
similarities with reticulocyte control mechanisms. In fact recent
reports show that the effects of serum and of hypertonicity on the
rate of initiation correlate with the extent of phosphorylation of
ribosomal protein S6; high rates of initiation occur when S6 is
highly phosphorylated, and vice-versa (128, 129). Correlations,
however, do not help to distinguish cause from effect, and it remains
unproven whether an increase in the degree of phosphorylation of S6
really does cause an increase in the rate at which ribosomes can
initiate protein synthesis.

One of the main obstacles to unravelling the mechanisms of
translational control in nucleated cells is that cell-free protein
synthesis systems from these cells are not particularly good. The
rate of protein synthesis declines in the early stages of the
incubation, and their behaviour in general is not as close to that
of the intact cell as is the case with reticulocyte systems. Cell-
free systems from nucleated cells do, however, carry out some re-
initiation on endogenous and added mRNA, so all the initiation
factors must be active at least for a short period. Contrary to
widely held views, at least some such cell-free systems do *not*
degrade mRNA significantly. If a myeloma cell-free extract is
incubated until protein synthesis has stopped and the RNA is then
isolated from it by phenol extraction, translation of this RNA yields
the same full-length products in about the same yield as when the RNA
is extracted from a fresh, unincubated extract (130). Moreover there
is nothing pre-ordained or inevitable about the early shut-off in
protein synthesis activity, since the addition of rabbit reticulocyte
post-ribosomal supernatant to a crude myeloma extract prolongs the
period of high activity and delays the shut-off (130).

There are therefore good grounds for hoping that it only requires
some relatively small modifications to make these cell-free system
closely comparable to the reticulocyte lysate, and hence more useful
for the analysis of translational control mechanisms. It is worth
recalling that some 15 years ago the reticulocyte cell-free system had
the merit of a rapid initial rate of protein synthesis, and the
ability to reinitiate a few rounds of translation, but showed the
annoying and (at that time) inexplicable habit of stopping synthesis
after about 5 minutes (131). With the benefit of hindsight we now
know that the reason for this was that the haem-controlled inhibitor
was unwittingly being activated. It is far from impossible that the
early reduction in the rate of protein synthesis activity in extracts
from nucleated cells is due to the activation of some as yet unrecog-
nised control system, and this possibility makes it all the more
important that we try to understand the nature of the defect in these
systems and to improve their activity.

ACKNOWLEDGEMENT

I thank Tim Hunt for countless helpful discussions over many years of fruitful partnership, and Pam Herbert and Liz Cambell for invaluable technical assistance. The research carried out in our laboratory described in this review was supported by grants from the Medical Research Council and the Cancer Research Campaign.

REFERENCES

1. Revel, M. and Groner, Y. (1978), Ann. Rev. Biochem. 67, 1079-1126.

2. Safer, B. and Anderson, W.F. (1978), Crit. Rev. Biochem. 6, 261-289.

3. Ochoa, S. and De Haro, C. (1979), Ann. Rev. Biochem. 48, 549-580.

4. Austin, S.A. and Clemens, M.J. (1980), FEBS Lett. 110, 1-7.

5. Hunt, T. (1980), In *Protein Phosphorylation in Regulation* (P. Cohen ed.), Elsevier North-Holland Biomedical Press, Amsterdam, in press.

6. Jagus, R. and Safer, B. (1980), Progr. Nucleic Acid Research, in press.

7. Schreier, M.H., Erni, B. and Staehelin, T. (1977), J. Mol. Biol. 116, 727-753.

8. Trachsel, H., Erni, B., Schreier, M.H. and Stehelin, T. (1977), J. Mol. Biol. 116, 755-767.

9. Benne, R. and Hershey, J.W.B. (1978), J. Biol. Chem. 253, 3078-3087.

10. Ghosh-Dastidar, P., Yaghami, B., Das, H.K. and Gupta, N.K. (1980) J. Biol. Chem. 255, 365-368.

11. Trachsel, H., Sonenberg, N., Shatkin, A.J., Rose, J.K., Leong, K., Bergmann, J.E., Gordon, J. and Baltimore, D. (1980), Proc. Natl. Acad. Sci. U.S.A. 77, 770-774.

12. Sonenberg, N., Rupprecht, K.M., Hecht, S.M. and Shatkin, A.J. (1979), Proct. Natl. Acad. Sci. U.S.A. 76, 4345-4349.

13. Staehelin, T., Trachsel, H., Erni, B., Boschetti, A. and Schreier, M.H. (1975), *Proceedings 10th FEBS Meeting* 10, 309-323.

14. Golini, F., Thach, S.S., Birge, C.H., Safer, B., Merrick, W.C. and Thach, R.E. (1976), Proc. Natl. Acad. Sci. U.S.A. 73, 3040-3044.

15. Lodish, H.F. (1976), Ann. Rev. Biochem. 45, 39-72.

16. Walton, G.M. and Gill, G.N. (1975), Biochim. Biophys. Acta 390, 231-245.

17. Clemens, M.J., Pain, V.M., Tilleray, V.J. and Echetebu, C.O. (1980), Biochem. Soc. Trans. 8, 351-352.

18. Barrieux, A. and Rosenfeld, M.G. (1977), J. Biol. Chem. 252, 3843-3847.

19. Safer, B. and Jagus, R. (1979), Proc. Natl . Acad. Sci. U.S.A. 76, 1094-1098.

20. Kozak, M. (1978), Cell 15, 1109-1123.
21. Kozak, M. (1979), Nature, London, 280, 82-85.
22. Kozak, M. (1980), Cell 19, 79-90.
23. Ahlquist, P., Dasgupta, R., Shih, D.S., Zimmern, D. and Kaesberg,P. (1979), Nature, London, 281, 277-282.
24. Filipowicz, W. and Haenni, A.-L. (1979), Proc. Natl. Acad. Sci. U.S.A. 76, 3111-3115.
25. Fellner, P. (1979), In *The Molecular Biology of Picornaviruses* (R. Perez-Bercoff, ed.), pp. 25-47, Plenum Press, New York and London.
26. Brown, F. (1979), In *The Molecular Biology of Picornaviruses* (R. Perez-Bercoff ed.), pp. 49-72, Plenum Press, New York and London.
27. Erni, B. (1976), Ph.D. Thesis, ETH, Zurich.
28. Lucas-Lenard, J.M. (1979), In *The Molecular Biology of Picornaviruses* (R. Perez-Bercoff, ed.), pp. 73-99, Plenum Press, New York and London.
29. Moss, B. (1968), J. Virol. 2, 1028-1037.
30. Metz, D.H. and Esteban, M. (1972), Nature, London, 238, 385-388.
31. Rosemond-Hornbeak, H. and Moss, B. (1975), J. Virol. 16, 34-42.
32. Person, A. and Beaud, G. (1978), J. Virol. 25, 11-18.
33. Cooper, J.A. and Moss, B. (1979), Virology 96, 368-380.
34. Kaerlein, M. and Horak, I. (1976), Nature, London, 259, 150-152.
35. Person, A. and Beaud, G. (1980), Eur. J. Biochem. 103, 85-93.
36. Ashburner, M. and Bonner, J.J. (1979), Cell 17, 241-254.
37. McKenzie, S.L., Henikoff, S. and Meselson, M. (1975), Proc. Natl. Acad. Sci. U.S.A. 72, 1117-1121.
38. Mirault, M.-E., Goldschmidt-Clermont, M., Moran, L., Arrigo, A. P. and Tissieres, A. (1978), Cold Spring Harbor Symp. Quant. Biol. 42, 819-827.
39. Lindquist, S. (1980), J. Mol. Biol. 137, 151-158.
40. Brandhorst, B.P. (1976), Dev. Biol. 52, 310-317.
41. Ruderman, J.V., Woodland, H.R. and Sturgess, E.A. (1979), Dev. Biol. 71, 71-82.
42. Rosenthal, E.T., Hunt, T. and Ruderman, J.V. (1980), Cell 20, 487-494.
43. Lee, G.T.-Y. and Engelhardt, D.L. (1979), J. Mol. Biol. 129, 221-233.
44. Geoghegan, T., Cereghini, S. and Brawerman, G. (1979), Proc. Natl. Acad. Sci. U.S.A. 76, 5587-5591.
45. Vincent, A., Civelli, O., Buri, J.-F. and Scherrer, K. (1977), FEBS Lett. 77, 281-286.
46. Civelli, O., Vincent, A., Maundrell, K., Buri, J.-F. and Scherrer, K. (1980), Eur. J. Biochem. 107, 577-585.
47. Jackson, R.J. (1979), In *The Molecular Biology of Picornaviruses* (R. Perez-Bercoff, ed.), pp. 191-238, Plenum Press, New York and London.
48. Zahringer, J., Baliga, B.S. and Munro, H.N. (1976), Proc. Natl. Acad. Sci. U.S.A. 73, 857-861.

49. Chu, L.L.H. and Fineberg, R.A. (1969), J. Biol. Chem. 244, 3847-3854.

50. Kurtz, D.T., Chan, K.M. and Fiegelson, P. (1978), Cell 15, 743-750.

51. Kennedy, D.S., Siegel, E. and Heywood, S.M. (1978), FEBS Lett. 90, 209-214.

52. Jelinek, W. and Leinwand, L. (1978), Cell 15, 205-214.

53. Lerner, M.R., Boyle, J.A., Mount, S.M., Wolin, S.L. and Steitz, J.A. (1980), Nature, London, 283, 220-224.

54. Hunt, T. (1979), Miami Winter Symposia, Vol. 16, 321-346.

55. Hunt, T. (1976), Brit. Med. Bull. 32, 257-261.

56. Hunter, A.R. and Jackson, R.J. (1975), Eur. J. Biochem. 58, 421-430.

57. Gross, M. (1974), Biochim. Biophys. Acta 366, 319-332.

58. Mizuno, S. (1977), Arch. Biochem. Biophys. 179, 289-301.

59. Bonanou-Tzedaki, S.A., Smith, K.E., Sheeran, B.A. and Arnstein, H.R.V. (1978), Eur. J. Biochem. 84, 601-610.

60. Zehavi-Willner, T., Kosower, N.S., Hunt, T. and Kosower, E.M. (1970), Biochem. Biophys. Res. Commun. 40, 37-42.

61. Floyd, G.A. and Traugh, J.A. (1980), Eur. J. Biochem. 106, 269-320.

62. Adamson, S.D., Herbert, E. and Kemp, S.F. (1969), J. Mol. Biol. 42, 247-258.

63. Kosower, N.S., Vanderhoff, G.A. and Kosower, E.M. (1972), Biochim. Biophys. Acta 272, 623-637.

64. Giloh, H. and Mager, J. (1975), Biochim. Biophys. Acta 414, 293-308.

65. Giloh, H., Schochot, L. and Mager, J. (1975), Biochim. Biophys. Acta 414, 309-323.

66. Legon, S., Jackson, R.J. and Hunt, T. (1973), Nature New Biology 241, 150-152.

67. Darnbrough, C.H., Hunt, T. and Jackson, R.J. (1972), Biochem. Biophys. Res. Commun. 48, 1556-1564.

68. Gross, M. (1979), J. Biol. Chem. 254, 2370-2383.

69. Farrell, P.J. (1977), Ph. D. Thesis, University of Cambridge, England.

70. Ralston, R.O., Das, A., Dasgupta, A., Roy, R., Palmieri, S. and Gupta, N.K. (1978), Proc. Natl. Acad. Sci. U.S.A. 75, 4858-4862.

71. Legon, S., Brayley, A., Hunt, T. and Jackson, R.J. (1974), Biochem. Biophys. Res. Commun. 56, 745-752.

72. Cahn, F. and Lubin, M. (1975), Mol. Biol. Reports 2, 49-57.

73. Benne, R., Salimans, M., Goumans, H., Amesz, H. and Voorma, H.O. (1980), Eur. J. Biochem. 104, 501-509.

74. Gross, M. and Rabinovitz, M. (1972), Proc. Natl. Acad. Sci. U.S.A. 69, 1565-1568.

75. Trachsel, X., Ranu, R.S. and London, I.M. (1978), Proc. Natl. Acad. Sci. U.S.A. 75, 3654-3658.

76. Gross, M. and Rabinovitz, M. (1972), Biochim. Biophys. Acta 287, 340-352.

77. Gross, M. (1974), Biochim. Biophys. Acta 340, 484-497.

78. Farrell, P.J., Balkow, K., Hunt, T., Jackson, R.J. and Trachsel,
     H. (1977), Cell 11, 187-200.
79. Gross, M. and Mendelewski, J. (1978), Biochim. Biophys. Acta
     520, 650-663.
80. Ernst, V., Levin, D.H. and London, I.M. (1979), Proc. Natl. Acad.
     Sci. U.S.A. 76, 2118-2122.
81. Gross, M. (1974), Biochem. Biophys. Res. Commun. 57, 611-619.
82. Farrell, P.J., Hunt, T. and Jackson, R.J. (1978), Eur. J. Bio-
     chem. 89, 517-521.
83. Trachsel, H. and Staehelin, T. (1978), Proc. Natl. Acad. Sci.
     U.S.A. 75, 204-208.
84. Kramer, G., Henderson, A.B., Pinphanichakarn, P., Wallis, M.A.
     and Hardesty, B. (1977), Proc. Natl. Acad. Sci. U.S.A. 74, 1445-
     1449.
85. Cherbas, L. and London, I.M. (1976), Proc. Natl. Acad. Sci.
     U.S.A. 73, 3506-3510.
86. Safer, B., Peterson, D. and Merrick, W.C. (1977), In *Translation
     of Synthetic and Natural Polynucleotides* (A.B. Legocki, ed.)
     pp. 24-31, University of Agriculture, Poznan, Poland.
87. Jackson, R.J. and Hunt, T. (1978), FEBS Lett. 93, 235-238.
88. De Haro, C. and Ochoa, S. (1978), Proc. Natl. Acad. Sci. U.S.A.
     75, 2713-2716.
89. Das, A. Ralston, R.O., Grace, M., Roy, R., Ghosh-Dastidar, P.,
     Das, H.K., Yaghmai, B., Palmlieri, S. and Gupta, N.K. (1979),
     Proc. Natl. Acad. Sci. U.S.A. 76, 5076-5079.
90. Benne, R., Amesz, H., Hershey, J.W.B. and Voorma, H.O. (1979), J.
     Biol. Chem. 254, 3201-3205.
91. Grankowski, N., Lehmusvirta, D., Kramer, G. and Hardesty, B.
     (1980), J. Biol. Chem. 255, 310-317.
92. Gross, M. (1976), Biochim. Biophys. Acta 447, 445-459.
93. Ralston, R.O., Das, A., Grace, M., Das, H.K. and Gupta, N.K.
     (1979), Proc. Natl. Acad. Sci. U.S.A. 76, 5490-5494.
94. Amesz, H., Goumans, H., Haubrich-Morree, T., Voorma, H.O. and
     Benne, R. (1979), Eur. J. Biochem. 98, 513-520.
95. Gross, M. (1978), Biochim. Biophys. Acta 520, 642-649.
96. Datta, A., De Haro, C., Sierra, J.M. and Ochoa, S. (1977), Proc.
     Natl. Acad. Sci. U.S.A. 74, 3326-3329.
97. Datta, A., De Haro, C. and Ochoa, S. (1978), Proc. Natl. Acad.
     Sci. U.S.A. 75, 1148-1152.
98. Levin, D.H., Ernst, V. and London, I.M. (1979), J. Biol. Chem.
     254, 7935-7941.
99. Grankowski, N., Kramer, G. and Hardesty, B. (1979), J. Biol.
     Chem. 254, 3145-3147.
100. Gross, M. and Rabinovitz, M. (1972), Biochim. Biophys. Acta
     287, 340-352.
101. Balkow, K., Hunt, T. and Jackson, R.J. (1975), Biochem. Biophys.
     Res. Commun. 67, 366-375.
102. Ernst, V., Levin, D.H., Ranu, R.S. and London, I.M. (1976), Proc.
     Natl. Acad. Sci. U.S.A. 73, 1112-1116.

103. Henderson, B.A. and Hardesty, B. (1978), Biochem. Biophys. Res. Commun. 83, 715-723.

104. Henderson, B.A., Miller, A.H. and Hardesty, B. (1979), Proc. Natl. Acad. Sci. U.S.A. 76, 2605-2609.

105. Schmid, G., Ludemann, H.-D. and Jenicke, R. (1978), Eur. J. Biochem. 86, 219-224.

106. Hunter, T., Hunt, T., Jackson, R.J. and Robertson, H.D. (1975), J. Biol. Chem. 250, 409-417.

107. Petryshyn, R., Levin, D.H. and London, I.M. (1980), Biochem. Biophys. Res. Commun. 96, 1190-1198.

108. Petryshyn, R., Trachsel, H. and London, I.M. (1979), Proc. Natl. Acad. Sci. U.S.A. 76, 1575-1579.

109. Baglioni, C. (1979), Cell 17, 255-264.

110. Williams, B.R.G., Golgher, R.R., Brown, R.E., Gilbert, C.S. and Kerr, I.M. (1979), Nature, London, 282, 582-586.

111. Williams, B.R.G., Gilbert, C.S. and Kerr, I.M. (1979), Nucleic Acid Res. 6, 1335-1350.

112. Ernst, V., Levin, D.H. and London, I.M. (1978), Proc. Natl. Acad. Sci. U.S.A. 75, 4110-4114.

113. Ernst, V., Levin, D.H. and London, I.M. (1978), J. Biol. Chem. 253, 7163-7172.

114. Lenz, J.R., Chatterjee, G.E., Moroney, P.A. and Baglioni, C. (1978), Biochemistry 17, 80-87.

115. West, D.K., Lenz, J.R. and Baglioni, C. (1979), Biochemistry 18, 624-632.

116. Holmgren, A. (1979), J. Biol. Chem. 254, 3664-3671.

117. Holmgren, A. (1977), J. Biol. Chem. 252, 4600-4606.

118. Holmgren, A. (1979), J. Biol. Chem. 254, 9113-9119.

119. Weber, L., Feman, E. and Baglioni, C. (1975), Biochemistry 14, 5315-5321.

120. Baglioni, C. and Weber, L.A. (1978), FEBS Lett. 88, 37-40.

121. Jefferson, L.S., Boyd, T.A., Flaim, K.E. and Peavy, D.E. (1980), Biochem. Soc. Trans. 8, 282-283.

122. Delaunay, J., Ranu, R.S., Levin, D.H. Ernst, V. and London, I. M. (1977), Proc. Natl. Acad. Sci. U.S.A. 74, 2264-2268.

123. Clemens, M.J., Pain, V.M., Henshaw, E.C. and London, I.M. (1976), Biochem. Biophys. Res. Commun. 72, 768-775.

124. Pain, V.M. and Henshaw, E.C. (1975), Eur. J. Biochem. 57, 335-342.

125. Pain, V.M., Lewis, J.A., Huvos, P., Henshaw, E.C. and Clemens, M. J. (1980), J. Biol. Chem. 255, 1486-1491.

126. Jackson, R.J. (1975), MTP International Review of Science, Biochemistry, Series One, 7, 89-135.

127. Revel, M. (1977), In *Molecular Mechanisms of Protein Biosynthesis* (H. Weissbach and S, Pestka, eds.) pp. 246-321, Academic Press, New York.

128. Thomas, G., Siegmann, M. and Gordon, J. (1979), Proc. Natl. Acad. Sci. U.S.A. 76, 3952-3956.

129. Martini, O.H.W. and Kruppa, J. (1979), Eur. J. Biochem. 95, 349-358.

130. Oldfield, S. (1980), Ph. D. Thesis, University of Cambridge,
     England.
131. Lamfrom, H. and Knopf, P.M. (1964), J. Mol. Biol. $\underline{9}$, 558-575.

REGULATION OF eIF-2 ACTIVITY AND INITIATION OF PROTEIN SYNTHESIS IN MAMMALIAN CELLS

Naba K. Gupta

Department of Chemistry
The University of Nebraska
Lincoln, Nebraska 68588 , (U.S.A.)

INTRODUCTION

There is now convincing evidence that the eukaryotic peptide chain initiation factor 2 (eIF-2) plays a major role in regulation of protein syntehsis initiation in mammalian cells. Numerous reports indicate that eIF-2 activity changes under different physiological conditions with accompanying changes in protein synthesis activities in the cells. Austin and Clemens recently reviewed the literature concerning the role of eIF-2 in regulation of protein synthesis in different mammalian cells (1).

eIF-2 forms a ternary complex, Met-tRNA$_f$.eIF-2.GTP as the first step in peptide chain initiation (2-14). Several laboratories have purified eIF-2 to homogeneity (7, 8, 10-14). The homogeneous eIF-2 is composed of three subunits of approximate molecular weights: α, 32-38,000 daltons; β, 48-52,000 daltons and γ, 50-57,000 daltons. Recently Harbitz and Hauge (13) and Stringer *et al.* (14) have reported that active eIF-2 preparations from pig liver and rabbit reticulocytes contains only two subunits, presumably α and γ.

Operational Definitions

Recent work done in different laboratories indicates that several ancillary protein factors such as Co-eIF-2A (15-25), Co-eIF-2B (11, 26-28), Co-eIF-2C (29-33), and sRF (34, 35) are required for efficient ternary complex formation by eIF-2 and its proper functioning during peptide chain initiation.

a)   Co-eIF-2C (29-33) and sRF (34, 35) promote ternary complex formation by eIF-2 *in the presence of Mg^{2+}*;

b)   Co-eIF-2A binds to preformed ternary complex and forms a stable
quarternary complex, Met-tRNA$_f$.eIF-2.Co-eIF-2A.GTP;

c)   the precise function of Co-eIF-2B in peptide chain initiation is
not known.   In *partial* reactions, Co-eIF-2B promotes *dissociation* of
the ternary complexes *in the presence of high Mg^{2+}* (5 mM) and low
temperature (0°C).

An important mechanism of regulation of protein synthesis initiation in mammalian cells involves *phosphorylation* of the α-subunit
(38,000 daltons) of eIF-2 by one or more eIF-2 kinases such as HRI
(heme-regulated protein synthesis inhibitor, also called HCR, hemin-
controlled repressor (36-38)) (39-42) and dsI (double-stranded RNA
activated inhibitor (43-50)) (44-46) leading to loss of interaction
of the eIF-2α(P) thus formed with one or more ancillary protein
factors (Co-eIF-2B and Co-eIF-2C) (27-33) and concomitant inhibition
of overall protein syntehsis.

In this article, we will summarize the characteristics of the
above ancillary protein factors and also eIF-2 kinases and their
roles in the complex regulation of protein synthesis initiation in
mammalian cells.

Co-eIF-2A

The first indication that a protein factor with no detectable
eIF-2 activity can stimulate Met-tRNA$_f$ binding to eIF-2 was provided
by the discovery of Co-eIF-2A in reticulocyte ribosomal high salt
wash by Dasgupta *et al.* (15).   Reticulocyte Co-eIF-2A activity has
since been purified to homogeneity (17, 51).   The molecular weight of
homogeneous Co-eIF-2A preparation is approximately 25,000 daltons (51).

Table 1.   Properties of Co-eIF-2A

1. Mol. wt., 25,000 daltons; single polypeptide (51).

2. Stimulates (2-3-fold) Met-tRNA$_f$ binding to eIF-2 (15-25).

3. Binds to preformed Met-tRNA$_f$.eIF-2.GTP (24).

4. Prevents degradation of Met-tRNA$_f$.eIF-2.GTP complex
   by aurintricarboxylic acid, mRNAs and hemin (17, 25).

5. Required for Met-tRNA$_f$ binding to 40S ribosomes (17).

6. Required for overall protein synthesis in reticulocyte
   lysates (23).

Table 1 summarizes some characteristic properties of Co-eIF-2A.

Co-eIF-2A increases both the initial rate and total extent of Met-tRNA$_f$ binding to eIF-2 (15). In the presence of excess Co-eIF-2A, approximately 90 percent of input eIF-2 is bound to Met-tRNA$_f$ (15, 17). Also, Co-eIF-2A confers considerable *stability* to the Met-tRNA$_f$.eIF-2.GTP complexes (17). For example, the ternary complexes formed with homogeneous eIF-2 preparations dissociate extensively in the presence of aurintricarboxylic acid (3 x 10^{-5} M) whereas the same complexes formed in the presence of excess Co-eIF-2A are almost completely resistant to aurintricarboxylic acid under similar experimental conditions (17, 25).

Several laboratories have reported the presence of Co-eIF-2A-like activities in different eukaryotic cells such as rabbit reticulocytes (15, 17), mouse ascites tumor cells (16), wheat germ (19, 20) and *Artemia salina* (18, 21, 22). Osterhout *et al.*, have reported the presence of two factors in wheat germ that stimulate ternary complex formation by eIF-2 (20). Like Co-eIF-2A, both factors form aurintricarboxylic acid-resistant Met-tRNA$_f$.eIF-2 complexes. One of these factors, "Co-eIF-2α", like Co-eIF-2A is a low molecular weight protein (20,000 daltons) while the other factor, "Co-eIF-2β" is presumably a high molecular weight protein complex (20). Wahba and coworkers have reported the isolation of two protein factors, "Co-eIF-2A" (65,000 daltons) and "Co-eIF-2B" (two polypeptides, 105,000 and 120,000 daltons) from *Artemia salina* (22). Only one of these factors, Co-eIF-2B forms an aurintricarboxylic acid-resistant ternary complex (22). It should be noted, however, that in reticulocytes, two additional factors, Co-eIF-2C (29-33) and sRF (see below) (34, 35), besides Co-eIF-2A, stimulate ternary complex formation by eIF-2 under different physiological conditions. It is possible that the multiple eIF-2 stimulatory activities in different eukaryotic cells are due to Co-eIF-2C and sRF-like activities rather than Co-eIF-2A.

Requirement of Co-eIF-2A in Protein Synthesis

In partial reactions, Co-eIF-2A stimulates Met-tRNA$_f$ binding to eIF-2 approximately two-three fold. The question arose as to whether Co-eIF-2A is an *essential* component of protein synthesis or whether the Co-eIF-2A requirement can be overcome by addition of excess eIF-2. To answer these questions, Ghosh-Dastidar *et al.* prepared antibodies against homogeneous preparations of eIF-2 and Co-eIF-2A and studied the effects of these antibodies on protein synthesis in reticulocyte lysates (23). Addition of either anti-eIF-2 or anti-Co-eIF-2A to reticulocyte lysates strongly inhibited (>85%) protein synthesis and in each case, protein synthesis inhibition was specifically reversed by the addition of the corresponding homogeneous factor (23). More importantly, protein synthesis inhibition by anti-Co-eIF-2A was not reversed by the addition of eIF-2 at any of the concentrations tested. The specific requirement of Co-eIF-2A for restoration of protein synthesis activity in a lysate inhibited by anti-Co-eIF-2A cannot be explained by a stimulation by Co-eIF-2A of eIF-2 activity. If Co-eIF-2A were required simply to stimulate ternary complex formation by two-three fold, such a requirement would have been met by the addition of more eIF-2 in the anti-Co-eIF-2A treated lysates. The *specific* requirement of Co-eIF-2A for restoration of protein synthesis activity in anti-Co-eIF-2A treated lysates strongly indicates that Co-eIF-2A is an *integral* part of the protein synthesis machinery. It has been postulated (23) that the ternary complex formed in the absence of Co-eIF-2A is either extremely labile or lacks the proper conformation necessary for protein synthesis initiation.

Mechanism of Interaction of Co-eIF-2A with eIF-2

Ghosh-Dastidar *et al*. (24) studied the molecular mechanism of interaction of Co-eIF-2A with eIF-2 and other protein factors such as Co-eIF-2B and Co-eIF-2C using a fluorescence polarization technique. For these studies, Co-eIF-2A was fluorescently labelled by dansylation and the changes in fluorescence polarization were measured in the presence of different protein factors, Met-tRNA$_f$ and GTP (24). In these experiments, the fluorescence polarization value of dansyl-Co-eIF-2A remained essentially unchanged in the presence of eIF-2, GTP or Met-tRNA$_f$ and also remained unchanged in the presence of eIF-2 plus either GTP or Met-tRNA$_f$. However, when dansyl-Co-eIF-2A was incubated with eIF-2 in the presence of both Met-tRNA$_f$ and GTP, the polarization value increased significantly implying an increase in the effective molecular weight of Co-eIF-2A which would correlate with Co-eIF-2A binding to eIF-2. These results indicate that Co-eIF-2A binds specifically to the ternary complex but does not interact with free eIF-2, Met-tRNA$_f$ or GTP, or with eIF-2 plus either Met-tRNA$_f$ or GTP.

Based on these results, it was suggested that the first step in peptide chain initiation is the formation of the ternary complex, Met-tRNA$_f$.eIF-2.GTP, followed by the binding of Co-eIF-2A to form a stable quarternary complex, Met-tRNA$_f$.eIF-2.Co-eIF-2A.GTP. However, it should be noted that the above experiments were done *in the absence* of Mg^{2+}. As will be discussed in a later section, ternary complex formation in the presence of physiological Mg^{2+} concentrations require other high molecular weight protein complexes, Co-eIF-2C and/or sRF. The precise mechanism of ternary complex formation in the presence of these two high molecular weight protein factors is not known. Ghosh-Dastidar *et al*.(24) investigated the possibility that Co-eIF-2C binds to eIF-2 or the ternary complex during the complex formation. Such Co-eIF-2C binding should produce a further increase in fluorescence polarization of dansyl-Co-eIF-2A upon binding of dansyl-Co-eIF-2A to the ternary complex. In the presence of Mg^{2+}, but without Co-eIF-2C the ternary complex did not form and fluorescence polarization remained unchanged. Addition of Co-eIF-2C resulted in increased ternary complex formation and an increase in fluorescence polarization. However, there was no additional increase in fluorescence polarization which would be expected if a high molecular protein factor such as Co-eIF-2C were added to a Co-eIF-2A containing quarternary complex. This result indicates that Co-eIF-2C does not form a *stable* complex with Met-tRNA$_f$.eIF-2.Co-eIF-2A.GTP. However, it is possible that Co-eIF-2C may form a *transient* complex with eIF-2 and dissociate after ternary complex (or quarternary) formation or that there is a low molecular weight factor in Co-eIF-2C which interacts with eIF-2.

Based on the results of fluorescence polarization studies the following early steps of peptide chain initiation have been proposed:

(a) $\text{Met-tRNA}_f + \text{eIF-2} + \text{GTP} \xrightarrow{\text{Co-eIF-2C, Mg}^{2+}} \text{Met-tRNA}_f.\text{eIF-2.GTP}$

(b) $\text{Met-tRNA}_f.\text{eIF-2.GTP} + \text{Co-eIF-2A} \rightarrow \text{Met-tRNA}_f.\text{eIF-2.Co-eIF-2A.GTP}$

Stoichimetry of Co-eIF-2A Binding to eIF-2

Direct measurements of stoichimetry for Co-eIF-2A binding to eIF-2 have not been made. For maximum effect on the ternary complex, it is necessary to add a several-fold molar excess of Co-eIF-2A relative to eIF-2 (17, 24). More than one molecule of Co-eIF-2A may bind to the ternary complex or excess Co-eIF-2A may be required to drive the equilibrium toward the quarternary complex.

Oo-eIF-2A Confers Stability on the Ternary Complex

Co-eIF-2A binds to the ternary complex and forms a stable quarternary complex. As mentioned earlier, the ternary complex formed with homogeneous eIF-2 dissociates extensively in the presence of 3×10^{-5} M aurintricarboxylic acid whereas in the presence of excess Co-eIF-2A, the complex is almost completely *resistant* to aurintricarboxylic acid. Several laboratories have reported that mRNAs bind to eIF-2 and cause extensive degradation of ternary complexes (52-55). The significance of mRNA binding to eIF-2 is not clear. Malathi and Mazumdar have reported the presence of a heat-labile protein factor in *Artemia salina* which prevents degradation of the ternary complex by mRNAs (56).

Roy *et al.* (25) have recently observed that the ternary complex formed in the presence of excess Co-eIF-2A is almost completely resistant to mRNA degradation. These authors have also noted that another physiologically significant compound, hemin, also degrades ternary complexes and, addition of Co-eIF-2A prevents the degradation of ternary complexes by hemin.

It may be that under physiological conditions, the ternary complex does not exist in the free form because this complex is too easily degraded in the presence of such physiologically important

compounds as mRNAs and hemin. The ternary complex binding to
Co-eIF-2A may stabilize it under physiological conditions.

Co-eIF-2B (TDF, TERNARY COMPLEX DISSOCIATION FACTOR)

Co-eIF-2B is a high molecular weight protein complex (mol. wt.
450,000 daltons) composed of a number of polypeptides (11, 26).
Purified Co-eIF-2B preparations show very little eIF-2 activity.
These preparations efficiently promote dissociation of the $Met-tRNA_f$.
$eIF-2.GTP$ complex in the presence of the divalent cations; MG^{2+},
Mn^{2+}, Ca^{2+}, and Sr^{2+}. Some characteristic properties of Co-eIF-2B
are summarized in Table 2.

Co-eIF-2B activity is assayed by measuring its ability to dissociate ternary complexes at high Mg^{2+} (5 mM) and at low temperature
using a two-stage Millipore filtration method (11). In Stage I,
duplicate aliquots of ternary complex are prepared by incubating
mixtures of homogeneous preparations of eIF-2 and Co-eIF-2B with
$Met-tRNA_f$ and GTP. The aliquots are then mixed with either Mg^{2+}
(final concentration, 5 mM) or equal volume of water and the incubation continued at 0° for 15 min. The ternary complexes are then
analyzed by Millipore filtration and the extent of dissociation of
the ternary complexes in the presence of Co-eIF-B and Mg^{2+} is taken
as a measure of Co-eIF-2B activity (11).

The Co-eIF-2B-promoted dissociation of the ternary complex was
also observed when GTP was replaced by a non-hydrolyzable GTP-analog,
GMP-PCP, indicating that GTP hydrolysis is not necessary for the dissociation reaction. The significance of the ternary complex dissociation by Co-eIF-2B (TDF) in the overall peptide chain initiation
process is not clear. Presumably, under physiological conditions for
protein synthesis and at physiological Mg^{2+} concentrations, Co-eIF-2B
does not cause enough dissociation of the ternary complex to prevent

Table 2. Properties of Co-eIF-2B (TDF)

1. High molecular weight protein complex composed of a
   number of polypeptides. Mol. wt. 450,000 daltons.
   (11, 26)

2. Promotes dissociation of $Met-tRNA_f.eIF-2.GTP$ complex at
   high Mg^{2+} and low temperature (TDF, ternary complex
   dissociation factor activity) (11, 26).

3. Required for $Met-tRNA_f$ binding to 40S ribosomes (11, 26).

4. Co-eIF-2B does not promote dissociation of
   $Met-tRNA_f.eIF-2\alpha(P).GTP$ complex (27, 28, 32).

peptide chain initiation. Co-eIF-2B is not an inhibitor of peptide chain initiation. The importance of Co-eIF-2B (TDF) activity in peptide chain initiation is suggested by two observations. A partial-ly purified factor preparation containing Co-eIF-2B (TDF) activity was found necessary, along with eIF-2 for efficient Met-tRNA$_f$ binding to 40S ribosomes in the presence of AUG codon (11, 26). Further, eIF-2 kinases such as HRI (27, 28, 32) and dsI (57) which inhibit protein synthesis under physiological conditions, also inhibit Co-eIF-2B-promoted dissociation of the ternary complexes. The result of the studies of eIF-2 kinase inhibition of TDF activity will be discussed in a later section.

The precise role of Co-eIF-2B (TDF) in promoting Met-tRNA$_f$.40S complex formation is not apparent. It is possible that the ternary complex dissociation activity of Co-eIF-2B (TDF) is involved in the release and recycling of eIF-2 and Co-eIF-2A after the Met-tRNA$_f$.eIF-2.Co-eIF-2A.GTP complex is bound to 40S ribosomes.

There is an interesting similarity between the Mg^{2+}-lability of the peptide chain initiation complexes in pro- and eukaryotic cells. In prokaryotic cells, the initiation factor IF-2 forms a *binary* complex with the initiator tRNA, f.Met-tRNA$_f$.IF-2 (58, 60). This binary complex is formed *in the absence* of Mg^{2+} and is extremely *unstable in the presence* of Mg^{2+} (58, 60). In eukaryotic cells, the homogeneous eIF-2 preparation forms a *stable* ternary complex *in the absence* of Mg^{2+} and the preformed ternary complex is stable to further addition of Mg^{2+} (11), however, addition of Co-eIF-2B (TDF) makes the ternary complex labile to Mg^{2+}. This observation suggests that for-mation of an Mg^{2+}-labile initiation factor.initiator tRNA complex may be an integral part of peptide chain initiation in both pro- and eukaryotic cells. In prokaryotic cells, this requirement is served by one protein factor, namely, IF-2. In eukaryotic cells, two protein factors, namely eIF-2 and Co-eIF-2B (TDF), are necessary (presumably for regulation purposes) to serve the same function.

Millipore Filtration Assay for Met-tRNA$_f$ Binding to Ribosomes (40S and 40S + 60S)

Chatterjee *et al.* used the ternary complex dissociation activity of Co-eIF-2B to develop a rapid and convenient Millipore filtration assay for studies of Met-tRNA$_f$ binding to ribosomes (40S or 40S + 60S) (61). The assay is carried out in two stages. In Stage I, Met-tRNA$_f$ is bound to ribosomes (40S or 40S + 60S) in the presence of 2 mM Mg^{2+}, peptide chain initiation factors and AUG codon. In Stage II, more Mg^{2+} (5 mM, final concentration) is added to dissociate any Met-tRNA$_f$.eIF-2.GTP complexes not bound to ribosomes. Only ribosome-bound Met-tRNA$_f$ is left to be measured by Millipore filtration assay (61). Alternatively, Met-tRNA$_f$.ribosome complexes are first separated by sucrose density gradient centrifugation and the gradient fractions are analyzed by Millipore filtration (61).

The Millipore filtration assay method has been used to study the
requirements for Met-tRNA$_f$ binding to 40S ribosomes (11, 26). Some
significant observations are:

- Under the assay conditions, Met-tRNAf binding to 40S ribosomes,
  as in prokaryotes, is dependent upon addition of mRNA (AUG
  codon);

- Homogeneous eIF-2 preparations which efficiently form ternary
  complexes do not stimulate Met-tRNA$_f$ binding to 40S ribosomes.
  At least two factor preparations, eIF-2 and a partially purified
  factor preparation containing Co-eIF-2B (and also other activities) are required for efficient Met-tRNA$_f$ binding to 40S ribosomes.

Other laboratories have now used a similar Millipore filtration assay
method for studies of Met-tRNA$_f$ binding to 40S ribosomes and have
reached similar conclusions regarding the requirements of mRNAs and
factor preparations for such binding (28, 62).

Co-eIF-2C

Like Co-eIF-2B, Co-eIF-2C is also a high molecular weight protein
complex (mol. wt. 360,000 daltons) and is composed of a number of
polypeptides. deHaro, Datta and Ochoa first isolated a Co-eIF-2C-
like activity (termed ESP) from a 0.5 M KCl wash of reticulocyte ribosomes (29, 30). The Co-eIF-2C (ESP) activity is conveniently separated from eIF-2 activity using a CM-Sephadex column chromatographic
procedure; Co-eIF-2C (ESP) activity is *not* retained on the column in
the presence of 0.18 M KCl while eIF-2 activity, which is strongly
absorbed onto the column, is eluted by using a buffer containing 0.4
M KCl. Co-eIF-2C (ESP), stimulated ternary complex formation by eIF-2
in the presence of Mg^{2+} (29, 30). deHaro *et al.* reported that Co-eIF-
2C (ESP) stimulation of ternary complex formation was inhibited by HRI
and ATP (29, 30).

Das *et al.* (32) and Ranu and London (33) purified a factor with
Co-eIF-2C activity from reticulocyte ribosomal salt wash and studied
the characteristics of stimulation of ternary complex formation by
purified Co-eIF-2C. Both groups have reported that Co-eIF-2C stimulates ternary complex formation by eIF-2 *in the presence* of Mg^{2+},
and HRI and ATP inhibit such stimulation of eIF-2 activity by Co-eIF-
2C (32-33).

Two possible mechanisms for Co-eIF-2C stimulation of ternary
complex formation have been proposed. They differ chiefly in the
role of Mg^{2+} in ternary complex formation by eIF-2. deHaro and Ochoa
have suggested (31) that formation of the ternary complex is preceded
by formation of the binary complex, eIF-2.GTP and Co-eIF-2C acts at
the level of binary complex formation. These authors have suggested

that Co-eIF-2C (ESP) may act by binding to the binary complex, displacing the equilibrium toward increased binding of GTP by eIF-2 (reactions a and b, below). Co-eIF-2C (ESP) would then be released by Met-tRNA$_f$ to form the ternary complex proper (reaction c). The reaction sequences are as follows:

(a) $$eIF\text{-}2 + GTP \rightleftharpoons eIF\text{-}2.GTP$$

(b) $$eIF\text{-}2.GTP + Co\text{-}eIF\text{-}2C \text{ (ESP)} \rightleftharpoons eIF\text{-}2.GTP.Co\text{-}eIF\text{-}2C \text{ (ESP)}$$

(c) $$eIF\text{-}2.GTP.Co\text{-}eIF\text{-}2C \text{ (ESP)} + Met\text{-}tRNA_f \rightleftharpoons eIF\text{-}2.GTP.Met\text{-}tRNA_f + Co\text{-}eIF\text{-}2C \text{ (ESP)}$$

Attempts to detect any complex formation between eIF-2 and Co-eIF-2C (ESP) were not successful. The authors have suggested that such interaction between eIF-2 and Co-eIF-2C (ESP) may be very weak.

Das *et al.* (32) and Ranu and London (33) have reported that Co-eIF-2C stimulation of ternary complex formation is observed only *in the presence* of Mg^{2+}. It has been previously reported that addition of Mg^{2+} to eIF-2 *before* ternary complex formation, drastically inhibits subsequent binding of eIF-2 to Met-tRNA$_f$ (63). In the absence of Mg^{2+}, eIF-2 binds efficiently to Met-tRNA$_f$ and addition of Co-eIF-2C has very little effect on Met-tRNA$_f$ binding to eIF-2. Addition of 1 mM Mg^{2+}, however, strongly inhibits Met-tRNA$_f$ binding to eIF-2 and such inhibition is progressively relieved by increasing concentrations of Co-eIF-2C. In the presence of excess Co-eIF-2C, complete reversal of Mg^{2+} inhibition of ternary complex formation is observed. Das *et al.* (32) and Ranu and London (33) have suggested that Co-eIF-2C is required to maintain eIF-2 in an active conformation necessary for ternary complex formation in the presence of physiological Mg^{2+} concentrations.

Characteristic properties of Co-eIF-2C as have been reported by Das *et al.* (32), are summarized in Table 3.

Table 3.   Properties of Co-eIF-2C

1. High molecular weight protein complex, composed of a number of polypeptides. Mol. wt. 360,000 daltons (32).

2. Promotes ternary complex formation by eIF-2 in the presence of Mg^{2+} by relieving the Mg^{2+} inhibition of ternary complex formation (32).

3. Co-eIF-2C does not stimulate ternary complex formation by eIF-2α(P).

eIF-2 KINASES

Protein synthesis in animal cells is regulated by one or more
protein synthesis inhibitors which also act as eIF-2 kinases (1).
These inhibitors are present in inactive forms which are activated
under certain physiological conditions or by some external agents.
In reticulocyte lysates, these inhibitors are activated:

a)   during heme-deficiency (HRI, heme-regulated inhibitor (36-38));

b)   by double-stranded RNA (dsI, double-stranded RNA activated
      inhibitor (43-46));

c)   by pressure (64); and

d)   by oxidized glutathione (65).

In interferon-treated cells, reviewed by P. Lengyel in this
volume, the inhibitor is activated by double-stranded RNAs (dsI) (47-
50). All the inhibitors studied are protein kinases and phosphorylate
specifically the α-subunit (38,000 daltons) of eIF-2 (39-42, 44-46).

HRI has been extensively purified from rabbit reticulocytes (66).
The purified inhibitor is a single polypeptide of approximate molecular weight, 100, 000 daltons (66). The molecular weight of dsI purified from both rabbit reticulocytes (57) and interferon-treated cells
(67) is approximately 67,000 daltons. There are indications that both
HRI and dsI phosphorylate the same sites on eIF-2 (68, 69) and eIF-
2α(P) formed by either HRI and dsI behaves similarly during peptide
chain initiation (57).

The detailed mechanism of protein synthesis inhibition by HRI
has been studied in several laboratories. Legon, Jackson and Hunt
first observed that Met-tRNA$_f$ binding to 40S ribosomes was significantly inhibited during heme-deficiency in reticulocyte lysates (70).
Hardesty and coworkers reported inhibition of Met-tRNA$_f$ binding to
40S ribosomes by HRI in a reconstituted system from reticulocytes
using partially purified peptide chain initiation factors and isolated
40S ribosomal subunits (71-72).

Das and Gupta (27) and later Ranu et al. (28) first provided
clear evidence that HRI and ATP inhibit the activity of a specific
ancillary protein factor, namely Co-eIF-2B (TDF). As noted in Section
2 of this review, a partially purified peptide chain initiation factor
preparation containing Co-eIF-2B (TDF) activity is necessary along
with eIF-2 for efficient Met-tRNA$_f$ binding to 40S ribosomes (11, 26).
Co-eIF-2B promotes dissociation of the ternary complexes in the
presence of high Mg^{2+} (5 mM) and low temperature (0°C). Das and Gupta
(27) and Ranu et al. (28) reported that HRI inhibits Co-eIF-2B promoted dissociation of the ternary complexes and also inhibits eIF-2

and Co-eIF-2B (TDF) promoted Met-tRNA$_f$ binding to 40S ribosomes in the presence of AUG codon. In both cases, such inhibition required the presence of ATP.

deHaro, Datta and Ochoa reported that HRI also inhibits the activity of another eIF-2 ancillary protein factor, namely ESP (eIF-2 Stimulating Protein) (29, 30). These authors reported that at physiological eIF-2 concentrations, the formation of ternary complexes requires the presence of ESP and HRI inhibits ternary complex formation by blocking the interaction of eIF-2 and ESP (29, 30). An eIF-2 stimulatory protein factor similar to ESP has also been isolated by Das et al., (termed Co-eIF-2C) (32) and by Ranu and London (termed SF, Stablizing Factor) (33). In agreement with deHaro et al. (29, 30), both groups found significant inhibition of this factor activity by HRI and ATP (22, 33). Possible differences in the mechanism of action of this factor activity as have been reported by the above laboratories have been discussed in Section 3 of this chapter.

Based on the above results, it can be suggested that HRI phosphorylates eIF-2 and the eIF-2α(P) thus formed does not interact with two ancillary factors Co-eIF-2B and Co-eIF-2C (ESP, SF) and is defective at some step(s) in peptide chain initiation.

However, the above studies do not prove that HRI-promoted phosphorylation of eIF-2 is the sole cause of inhibition of Co-eIF-2B and Co-eIF-2C activities and is also the cause of inhibition of overall peptide chain initiation. The possibilities that phosphorylation of other components such as HRI, Co-eIF-2B and Co-eIF-2C is also involved in this complex inhibition are not ruled out. Further, the possibility that HRI can inhibit more than one step in peptide chain initiation has been indicated (73).

To more clearly define the mechanism of HRI inhibition of peptide chain initiation and the precise role of eIF-2α(P) in the inhibition process, Das et al. prepared eIF-2α(P) using purified HRI and studied its interaction with Co-eIF-2B and Co-eIF-2C (32). Partially purified eIF-2 was phosphorylated by using HRI and ATP, and the eIF-2α(P) thus formed was further purified by phosphocellulose chromatography. The extent of phosphorylation of eIF-2 was analyzed by isoelectric focusing. Almost complete phosphorylation of the 38,000 dalton subunit of eIF-2 was observed. The control experiment, without HRI, showed no phosphorylation of the 38,000 dalton subunit of eIF-2.

These authors compared the activities of eIF-2 and 2α(P) in different partial initiation reactions (32). The results of a typical experiment testing the effects of these cofactors, Co-eIF-2A, Co-eIF-2B and Co-eIF-2C on Met-tRNA$_f$ binding to eIF-2 and eIF-2α(P) are summarized in Table 4. Both eIF-2 and eIF-2α(P) formed ternary complexes with almost equal efficiencies. Also, both eIF-2 and eIF-2α(P) responded similarly to Co-eIF-2A and Mg^{2+} (Table 4).

Table 4. Effects of Phosphorylation of eIF-2 on its Response to
          Co-eIF-2A, Co-eIF-2B, and Co-eIF-2C (32)

Factors added	$[^{35}S]$Met-tRNA$_f$ bound, pmol	

Co-eIF-2A (Exp. 1)

	No Mg^{2+}	
eIF-2	0.50	
eIF-2 + Co-eIF-2A	1.25	
eIF-2α(P)	0.45	
eIF-2α(P) + Co-eIF-2A	1.20	

Co-eIF-2B (Exp. 2)

	No Mg^{2+}	+5 mM Mg^{2+} (Stage II)
eIF-2	1.35	1.40
eIF-2 + Co-eIF-2B (TDF)	1.70	0.52
eIF-2α(P)	1.66	1.50
eIF-2α(P) + Co-eIF-2B (TDF)	1.55	1.45

Co-eIF-2C (Exp. 3)

	No Mg^{2+}	+1 mM Mg^{2+}
eIF-2	1.70	0.60
eIF-2 + Co-eIF-2C	2.03	1.70
eIF-2α(P)	1.54	0.57
eIF-2α(P) + Co-eIF-2C	1.58	0.59

Co-eIF-2A stimulated ternary complex formation by both eIF-2 and
eIF-2α(P) (Exp. 1)   and prior addition of Mg^{2+} inhibited ternary
complex formation with both factors (Exp. 3). However, whereas the
ternary complexes formed with eIF-2 could be dissociated at 5 mM
Mg^{2+} and low temperature (Exp. 2) and also could be stimulated by
CO-eIF-2C at 1 mM Mg^{2+} (Exp. 3)) the ternary complexes formed with

eIF-2α(P) were not significantly affected by addition of Co-eIF-2B
or Co-eIF-2C under similar experimental conditions.

The above experiments were done with eIF-2 and eIF-2α(P) and no
HRI was present during incubation. The results thus clearly establish
that eIF-2α(P) does *not* interact with Co-eIF-2B or Co-eIF-2C and
neither further phosphorylation nor the continued presence of
HRI is necessary for such loss of interaction.

These authors also investigated the possibilities that HRI may
inhibit more than one step in peptide chain initiation (32). They
studied the effects of addition of HRI and ATP on the initial formation of Met-tRNA$_f$.40S.AUG complex and also on the subsequent joining
of Met-tRNA$_f$.40S.AUG with the 60S ribosomal subunit. As expected,
addition of HRI and ATP caused significant inhibition of Met-tRNA$_f$.
40S.AUG complex formation. However, when Met-tRNA$_f$.40S.AUG was preformed, addition of HRI and ATP had little effect on the subsequent
joining of the Met-tRNA$_f$.40S.AUG complex with 60S subunits to form
Met-tRNA$_f$80S.AUG complex.

The results so far obtained suggest that HRI inhibition is limited to phosphorylation of eIF-2 which blocks subsequent reaction of
eIF-2 with Co-eIF-2B and Co-eIF-2C and that HRI has no further inhibitory effect on other step in peptide chain initiation.

sRF

During heme-deficiency in reticulocyte lysates, the heme-regulated protein synthesis inhibitor (HRI) becomes activated and shuts
off the bulk of protein synthesis. Protein synthesis in the lysate,
however, continues at a very slow but definite rate. The mechanism
of this slow rate of protein synthesis in the presence of activated
HRI was not clearly understood.

Several laboratories have reported that protein synthesis inhibition in heme-deficient reticulocyte lysates can be reversed by
addition of a ribosomal salt wash factor preparation containing eIF-2
activity (39, 74-76). It has been proposed that HRI inactivates
eIF-2 by phosphorylation leading to loss of protein synthesis activity
of the system and addition of large excess of exogeneous eIF-2 reverses such inhibition. In other studies, Gross (77, 78) and Ranu
and London (79) reported the presence of a protein factor in reticulocyte lysates which also reverses protein synthesis inhibition in
heme-deficient reticulocyte lysates. The active component was further
purified and was freed from eIF-2 activity (78). However, the mechanism of reversal of protein synthesis inhibition by this supernatant
factor was not understood.

Ralston *et al.* investigated the characteristics of the factor(s)
that reverse protein synthesis inhibition in heme-deficient reticu-

locyte lysates. In the initial studies, ribosomal salt wash was used at the starting material for purification of the factor (termed rRF), that reverses protein synthesis inhibition in heme-deficient reticulocyte lysates (51). This rRF activity remained associated with the eIF-2 activity during several purification steps. However, the homogeneous eIF-2 preparation obtained after hydroxylapatite column chromatography (Fraction V (11)) was completely devoid of rRF activity (51). Also rRF activity was resolved from the bulk of eIF-2 activity using a glycerol density centrifugation procedure and phosphocellulose chromatography. The purified rRF preparation contained very little eIF-2 activity but reversed protein synthesis inhibition in heme-deficient reticulocyte lysates (51). From these results, it was concluded that rRF activity is *not* due to eIF-2 alone although it *may* require eIF-2 as a component of a complex factor.

More recently, Ralston *et al.* purified the post ribosomal supernatant factor, termed sRF, from reticulocytes and studied its mechanism of reversal protein synthesis inhibition in heme-deficient reticulocyte lysates (34). The sRF activity was purified using DEAE-cellulose and GTP-agarose column chromatography (34). Purified sRF contained very little eIF-2 activity but efficiently reversed protein synthesis inhibition in heme-deficient reticulocyte lysates.

The molecular mechanism of sRF action was studied using several partial reactions. As described in Section 3 of this review, partially purified peptide chain initiation factor preparation (eIF, Fraction III, (11)) contains Co-eIF-2C activity. Phosphorylation of eIF-2 by HRI and ATP leads to loss of interaction of eIF-2α(P) with Co-eIF-2C with resultant loss of Met-tRNA$_f$ binding to eIF-2. Ralston *et al.* (34) have noted that addition of purified sRF in HRI inhibited system using eIF, almost completely restores the ternary complex formation by eIF-2. This reversal of HRI inhibition of ternary complex formation by sRF was observed even when sRF was added after eIF-2 was phosphorylated by preincubation with HRI and ATP.

Ralston *et al.* provided evidence that sRF does *not* dephosphorylate eIF-2α(P) nor prevent phosphorylation of eIF-2 by HRI and ATP. The molecular mechanism of sRF action is indicated by the fact that sRF and not Co-eIF-2C can stimulate ternary complex formation by eIF-2α(P) in the presence of Mg^{2+}.

Some characteristic properties of sRF as have been reported by Ralston *et al.* (34) are summarized in Table 5.

Based on these observations, it has been suggested that protein synthesis inhibition in heme-deficient reticulocyte lysates by HRI is due to phosphorylation of eIF-2 and loss of interaction of eIF-2α(P) by a major ancillary factor namely Co-eIF-2C. The lysate also contains another factor, sRF, possibly in limiting quantities. sRF sustains protein synthesis in reticulocyte lysates at a slow rate

Table 5. Properties of sRF

  1. High molecular weight protein complex (mol. wt. 370,000
     daltons) composed of a number of polypeptides (34).

  2. sRF reverses protein synthesis inhibition in heme-
     deficient reticulocyte lysates (34, 35, 77-80).

  3. sRF promotes ternary complex formation by both eIF-2
     and also eIF-2α(P) in the presence of Mg^{2+} (34, 35).

  4. sRF reverses inhibition of ternary complex formation by
     HRI and ATP (34, 35).

  5. sRF does not dephosphorylate eIF-2α(P) nor blocks eIF-2
     phosphorylation by HRI and ATP (34, 35, 80).

and when added exogeneously in large excess can almost completely
reverse protein synthesis inhibition in heme-deficient lysates. In
partial reactions, sRF can stimulate ternary complex formation by
eIF-2α(P) preformed by the action of HRI and ATP and can thus presumably substitute Co-eIF-2C activity in peptide chain initiation reactions. In this respect, sRF provides and alternative route for
ternary complex formation under conditions where Co-eIF-2C becomes
ineffective due to phosphorylation of eIF-2 by eIF-2 kinases. These
dual pathways for regulation of protein synthesis initiation are
shown diagrammatically in Fig. 2.

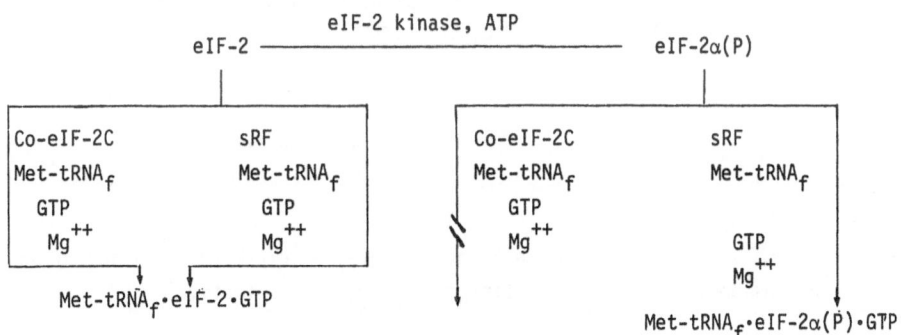

Fig. 2. A proposed mechanism for dual regulation of protein synthesis
         initiation in mammalian cells.

Independently Amsez *et al.* have also reported that a post ribosomal supernatant factor possibly similar to sRF, termed anti-inhibitor, that reverses protein synthesis inhibition in heme-deficient
reticulocyte lysates, does not dephosphorylate eIF-2α(P) nor prevent
HRI phosphorylation of eIF-2 (80). The anti-inhibitor reverses HRI-
inhibition of Met-tRNA$_f$.80S complex formation in reticulocyte lysates
but has no effect on any partial reactions using purified components.
These authors have postulated that HRI inhibition of protein synthesis in reticulocyte lysates involves phosphorylation of eIF-2 and
subsequent binding of eIF-2α(P) with an (as yet) unidentified component, X. The anti-inhibitor prevents the binding of such a component
to eIF-2 presumably by the formation of a complex between eIF-2 and
the anti-inhibitor. Their proposed mechanism is shown diagrammatically in Fig. 3.

Bennet *et al.* (81) have also reported that eIF-2α(P) is recycled
during protein synthesis in reticulocyte lysates and is, therefore, an
active component in protein synthesis.

Recently, Siekierka and Ochoa (35) have also isolated a supernatant factor from reticulocytes that reverses protein synthesis inhibition in reticulocyte lysates. This factor, also termed anti-
inhibitor, like sRF reverses HRI inhibition of ternary complex formation and also can stimulate Met-tRNA$_f$ binding to eIF-2α(P).

Thus, there is agreement among different laboratories that the
reticulocyte supernatant factor, sRF (anti-inhibitor (34, 35, 80)),
does *not* dephosphorylate eIF-2α(P) nor prevent phosphorylation of
eIF-2 by HRI. sRF can presumably promote peptide chain initiation
using eIF-2α(P) as an active component. Ralston *et al.* (34), and
Siekierka and Ochoa (35) agree that sRF can promote ternary complex
formation by eIF-2α(P) and thus provide an alternative route for
peptide chain initiation. Amsez *et al.* (80) observed that the anti-
inhibitor reverses inhibition of Met-tRNA$_f$.80S complex formation in
heme-deficient reticulocyte lysates but has no effect on any partial
reaction using purified factors. It is possible that such differences
may be due to use of different factor preparations or assay conditions. For example, the eIF-2 preparations used by Gupta and

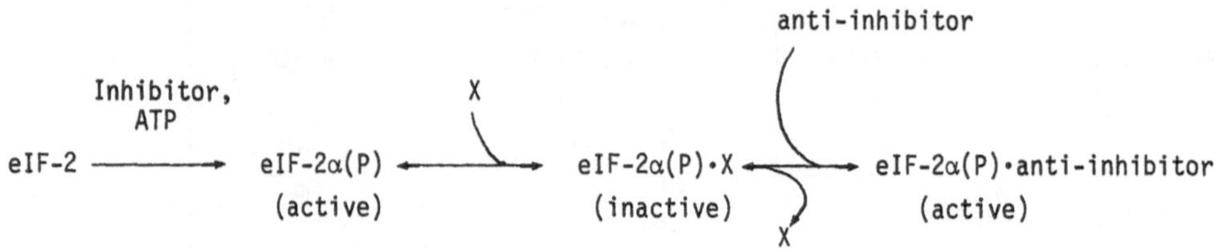

Fig. 3. Mechanism proposed by Amsez *et al.* for the action of the
         anti-inhibitor (80).

coworkers (32, 63) and also by Ranu and London (33) are extremely sensitive to Mg^{2+} and, according to these workers, Co-eIF-2C (32, 33) and also sRF (34) act by reversing Mg^{2+} inhibition of ternary complex formation by eIF-2. On the other hand, Benne *et al.* assays eIF-2 activity in the presence of 1 mM Mg^{2+} (82) and this eIF-2 preparation is presumably insensitive to Mg^{2+}. Such eIF-2 preparations would obviously not respond to cofactors such as Co-eIF-2C and sRF which are involved in reversal of Mg^{2+} inhibition of ternary complex formation. The reason for differences in Mg^{2+}-sensitivity of eIF-2 preparations as reported by different laboratories is not apparent, at present.

According to the proposed mechanism of dual regulation of protein synthesis inhibition (Fig. 2) protein synthesis in animal cells is regulated by eIF-2 kinases and also by the relative activities of Co-eIF-2C and sRF. In reticulocyte lysates, protein synthesis occurs at a very rapid rate in the presence of hemin and both Co-eIF-2C and sRF are active. During heme-deficiency, HRI phosphorylates eIF-2 and blocks protein synthesis by Co-eIF-2C pathway. Under these conditions, protein synthesis, presumably proceeds by the alternative sRF pathway. In reticulocyte lysates, Co-eIF-2C may be the major factor, since loss of interaction with Co-eIF-2C by eIF-2 phosphorylation leads to inhibition of the bulk of protein synthesis in this system.

Several *in vitro* protein synthesizing systems are known which incorporate amino acids into proteins at a very slow rate (83-85). These systems do not require hemin and promote several rounds of protein synthesis. An intriguing possibility is that protein synthesis in these systems proceeds by a HRI-*insensitive* pathway and is mediated by sRF using eIF-2α(P) as the active component.

CONCLUSIONS

In this article, we have discussed the possible mechanism of protein synthesis inhibition by an eIF-2 kinase, HRI and the roles of four ancillary protein factors, Co-eIF-2A, Co-eIF-2B, Co-eIF-2C and sRF in complex regulation of protein synthesis initiation in animal cells. There are many unanswered questions in these studies. HRI inhibits protein synthesis in reticulocyte lysates and also inhibits Met-tRNA$_f$ binding to 40S ribosomes. HRI phosphorylates eIF-2 and the eIF-2α(P) thus formed is not recognized by two ancillary factors Co-eIF-2B and Co-eIF-2C. The loss of interaction of eIF-2α(P) with these two ancillary factors has been suggested as the mechanism of protein synthesis inhibition by HRI. In partial reactions, Co-eIF-2C promotes ternary complex formation by eIF-2 in the presence of Mg^{2+}. However a requirement for Co-eIF-2C for Met-tRNA$_f$.40S complex formation or in overall protein synthesis initiation has not yet been demonstrated. Co-eIF-2B promotes dissociation of the ternary complexes in the presence of high Mg^{2+} (5 mM) and low temperature (TDF activity). The role of this TDF activity in protein synthesis initiation is not clear and also the requirement for Co-eIF-2B in overall protein synthesis has not been clearly demonstrated. A partially purified factor preparation containing Co-eIF-2B activity together with eIF-2 is required for Met-tRNA$_f$ binding to 40S ribosomes.

sRF reverses protein synthesis inhibition in heme-deficient reticulocyte lysates. The requirement for this factor in protein synthesis is clearly established. Because this factor is presumably present in limiting amounts in reticulocyte lysates, the lysates become dependent on exogeneously added sRF as an alternative factor when Co-eIF-2C becomes ineffective after HRI catalyzed phosphorylation of eIF-2. In partial reactions, sRF can promote ternary complex formation by eIF-2α(P) and thus can substitute for the Co-eIF-2C activity.

The role of Co-eIF-2B in sRF mediated protein synthesis is not clear. Co-eIF-2B (TDF) activity is inhibited by phsphorylation of eIF-2. It is not established whether sRF mediated protein synthesis occurs by a pathway independent of Co-eIF-2B activity or whether Co-eIF-2B inhibition is reversed by sRF.

Also, Co-eIF-2B, Co-eIF-2C and sRF preparations are high molecular weight protein complexes composed of a number of polypeptides. The precise roles of these polypeptide components in specific factor activity are not known. SDS-polyacrylamide gel electrophoresis data indicate that these complex proteins may contain several polypeptides of similar molecular weights. The roles of such common polypeptide components in different factor activities remain to be elucidated. It should be noted that these factors have been characterized mostly by their activities in partial reactions. Co-eIF-2B and Co-eIF-2C activities have been completely separated from each other and neither Co-eIF-2B nor Co-eIF-2C preparations contain sRF activity. Preparation of sRF, like Co-eIF-2C, stimulate ternary complex formation by eIF-2 in the presence of Mg^{2+}. sRF can recognize both eIF-2 and eIF-2α(P) whereas Co-eIF-2C can recognize only eIF-2.

Siekierka and Ochoa have suggested that Co-eIF-2C (ESP) and sRF are structurally related: in their view ESP may be a modified, less active form (perhaps a precursor) of the factor (35). To provide and unequivocal answer to any question regarding structure function relationship between Co-eIF-2C and sRF, it will be necessary to: a) extensively purify these factors, b) identify the polypeptide components responsible for these activities, and c) determine the structural characteristics of these factors.

A major problem in these studies has been the lack of availability of an efficient reconstituted protein synthesizing system in which to examine the requirement of specific factors in overall peptide chain initiation. Reticulocyte lysates very efficiently incorporate amino acids into proteins and presumably contain all the factors necessary for protein synthesis in nearly saturating amounts. Addition of exogeneous factors such as eIF-2, Co-eIF-2A, Co-eIF-2B and Co-eIF-2C had no effect on protein synthesis in reticulocyte lysates (23). This protein synthesizing system has, however, been successfully employed to demonstrate the requirement for eIF-2 and Co-eIF-2A using antibodies prepared against homogeneous factor preparations (23). The requirement for sRF for protein synthesis in reticulocytes would be demonstrated during heme-deficiency presumably because the system becomes dependent on sRF as the other factor, Co-eIF-2C, becomes ineffective (34).

In the absence of a suitable protein synthesizing system reconstituted from purified factors, the reticulocyte lysates can probably be used to study the requirements for high molecular protein complexes such as Co-eIF-2B, Co-eIF-2C and sRF, and also the requirements of individual polypeptide components in these complexes by using specific antibodies. The preparation of monoclonal antibodies (86, 87) against active polypeptide components of the complex protein factors can be used to induce requirements for their cognate factors for specific partial peptide chain initiation reactions and also in overall protein synthesis in reticulocyte lysates.

ACKNOWLEDGEMENTS

Work reported for the author's laboratory was supported by United States Public Health Service research grants GM-18790 and GM-22079. The author is grateful to Dr. George A. Vidaver for critical reading of this manuscript.

REFERENCES

1. Austin, S.A. and Clemens, M.J. (1980), FEBS Letters, 110, 1-6.
2. Chen, Y.C., Woodley, C.L., Bose, K.K. and Gupta, N.K. (1972), Biochem. Biophys. Res. Commun. 48, 1-9.
3. Gupta, N.K., Woodley, C.L., Chen, Y.C. and Bose, K.K. (1973), J. Biol. Chem. 248, 4500-4511.
4. Dettman, G.L. and Stanley, W.M., Jr. (1972), Biochim. Biophys. Acta 287, 124-133.
5. Levin, D.H., Kyner, D., and Acs. G. (1973), Proc. Natl. Acad. Sci. U.S.A. 70, 41-45.
6. Schreier, M.H. and Staehelin, T. (1973), Nature, London, New Biology, 242, 35-38.
7. Staehelin, T., Trachsel, H., Erni, B. Boschettie, A., and Schreier, M.H. (1975), FEBS Proc. Meet. 10, 309-323.
8. Safer, B., Anderson, W.F. and Merrick, W.C. (1975), J. Biol. Chem. 250, 9067-9075.
9. Ranu, R.S. and Wool, I.G. (1976), J. Biol. Chem. 251, 1926-1935.

10. Benne, R., Wong, C., Luedi, M. and Hershey, J.W.B. (1976), J. Biol. Chem. 251, 7675-7681.

11. Majumdar, A., Dasgupta, A., Chatterjee, B., Das, H.K. and Gupta, N.K. (1979), Methods Enzymol. 60, 35-52.

12. Lloyd, M.A., Osborne, J.C., Safer, B., Powell, G. and Merrick, W.C. (1980), J. Biol. Chem., 255, 1189-1194.

13. Harbitz, I., and Hauge, J.G. (1976), Arch. Biochem. Biophys. 176, 766-778.

14. Stringer, E.A., Chaudhuri, A., and Maitra, U. (1980), Proc. Natl. Acad. Sci. U.S.A. 77, 3356-3359.

15. Dasgupta, A., Majumdar, A., George, A.D. and Gupta, N.K. (1976), Biochem. Biophys. Res. Commun. 71, 1234-1241.

16. Reynolds, S., Dasgupta, A., Palmieri, S., Majumdar, A., and Gupta, N.K. (1977), Arch. Biochem. Biophys. 184, 325-335.

17. Dasgupta, A., Das, A., Roy, R., Ralston, R., Majumdar, A., and Gupta, N.K. (1978), J. Biol. Chem. 253, 6054-6059.

18. Malathia, V.G., and Majumdar, R. (1978), FEBS Letters, 86, 155-159.

19. Treadwell, B.V., Mauser, L., and Robinson, W.G. (1979), Methods Enzymol. 60, 181-193.

20. Osterhout, J.J., Phillips-Minton, J., and Ravel, J.M. (1979), Fed. Proc. 38, 327.

21. MacRae, T., Houston K.J., Woodley, C.L., and Wahba, A.J. (1979), Eur. J. Biochem. 100, 67-76.

22. Roth, W., Pardue, T.J., and Pao, A. (1980), Fed. Proc. 39, 2027.

23. Ghosh-Dastidar, P., Yaghmai, B., Das, A., Das, H.K., and Gupta, N.K. (1980), J. Biol. Chem. 255, 365-368.

24. Ghosh-Dastidar, P., Giblin, D., Yaghmai, B., Das, A., Das, H.K. Parkhurst, L.J., and Gupta, N.K. (1980), J. Biol. Chem. 255, 3826-3829.

25. Roy, R., Ghosh-Dastidar, P., Yaghmai, B., Das, A., and Gupta, N. K., Unpublished observation.

26. Majumdar, A., Roy, R., Das, A., Dasgupta, A., and Gupta, N.K. (1977), Biochem. Biophys. Res. Commun. 78, 161-169.

27. Das, A., and Gupta, N.K. (1977), Biochem. Biophys. Res. Commun. 78, 1433-1441.

28. Ranu, R.S., London, I.M., Das, A., Dasgupta, A., Majumdar, A., Ralston, R., Roy, R., and Gupta, N.K. (1978), Proc. Natl. Acad. Sci. U.S.A. 75, 745-749.

29. deHaro, C., Datta, A., and Ochoa, S. (1978), Proc. Natl. Acad. Sci. U.S.A. 75, 243-247.

30. deHaro, C., and Ochoa, S. (1978), Proc. Natl. Acad. Sci. U.S.A. 75, 2713-2716.

31. deHaro, C. and Ochoa, S. (1979), Proc. Natl. Acad. Sci. U.S.A. 76, 2163-2164.

32. Das, A., Ralston, R.O., Grace, M., Roy, R., Ghosh-Dastidar, P., Das, H.K., Yaghmai, B., Palmieri, S., and Gupta, N.K. (1979), Proc. Natl. Acad. Sci. U.S.A. 76, 5076-5079.

33. Ranu, R.S., and London, I.M. (1979), Proc. Natl. Acad. Sci. U S A. 76, 1079-1083.

34. Ralston, R.O., Das, A., Grace, M., Das, H.K., and Gupta, N.K. (1979), Proc. Natl. Acad. Sci. U.S.A. 76, 5490-5494.
35. Siekierka, J., and Ochoa, S. (1980), Fed. Proc. 39, 2028.
36. Rabinovitz, M., Freedman, M.L., Fisher, J.M., and Maxwell, C.R. (1969), Cold Spring Harbor Symp. Quant. Biol. 34, 567-578.
37. Howard, G.A., Adamson, S.D., and Herbert, E. (1970), Biochim. Biophys. Acta 213, 237-240.
38. Hunt, T., Vanderhoff, G., and London, I.M. (1972), J. Mol. Biol. 66, 471-481.
39. Farrell, P.J., Balkow, K., Hunt, T., Jackson, R.J., and Trachsel, H. (1977), Cell, 11, 187-200.
40. Levin, D.H., Ranu, R.S., Ernst, V., and London, I.M. (1976), Proc. Natl. Acad. Sci. U.S.A. 73, 3112-3116.
41. Kramer, G., Cimadevilla, J.M. and Hardesty, B. (1976), Proc. Natl. Acad. Sci. U.S.A. 73, 3078-3082.
42. Gross, M., and Mendelenski, J. (1977), Biochem. Biophys. Res. Comm. 74, 559-569.
43. Ehrenfeld, E., and Hunt, J. (1971), Proc. Natl. Acad. Sci. U.S.A. 68, 1075-1078.
44. Levin, D., and London, I.M. (1978), Proc. Natl. Acad. Sci. U.S.A. 75, 1121-1125.
45. Lenz, J.R., and Baglioni, C. (1978), J. Biol. Chem. 253, 4219-4223.
46. Levin, D.H., Petryshyn, R., and London, I.M. (1980), Proc. Natl. Acad. Sci. U.S.A. 77, 832-836.
47. Lebleu, B., Sen, G.C., Shaila, S., Carbrer, B., and Lengeyl, P. (1976), Proc. Natl. Acad. Sci. U.S.A. 73, 3107-3111.
48. Zilberstein, A., Fefermanm, P., Shulman, L., and Revel, M. (1976) FEBS Letters, 68, 119-124.
49. Roberts, W.K., Hovanessian, A.G., Brown, R.E., Clemens, M.J., and Kerr, I.M. (1976), Nature, London, 264, 477-480.
50. Sen, G.C., Taira, H., and Lengyel, P. (1978), J. Biol. Chem. 253, 5915-5921.
51. Ralston, R.O., Das, A., Dasgupta, A., Roy, R., Palmieri, S. and Gupta, N.K. (1978), Proc. Natl. Acad. Sci. U.S.A. 75, 4858-4862.
52. Kaempfer, R. (1974), Biochem. Biophys. Res. Commun. 61, 591-597.
53. Hellerman, J.G., and Shafritz, D.A. (1975), Proc. Natl. Acad. Sci. U.S.A. 72, 1021-1025.
54. Kaempfer, R., Rosen, H., and Israeli, R. (1978), Proc. Natl. Acad. Sci. U.S.A. 75, 650-654.
55. Barrieux, A., and Rosenfeld, M.G. (1978), J. Biol. Chem. 253, 6311-6314.
56. Malathi, V.G., and Mazumdar, R. (1979), Biochem. Biophys. Res. Commun. 89, 585-590.
57. Das, H.K., Das, A., Ghosh-Dastidar, P., Ralston, R.O., Yaghmai, B., Roy, R., and Gupta, N.K. Unpublished observation.
58. Majumdar, A., Bose, K.K., Gupta, N.K. and Wahba, A.J. (1976), J. Biol. Chem., 251, 137-140.
59. Sundari ,R.M., Stringer, E.A., Schulman, L.H. and Maitra, U. (1976), J. Biol. Chem. 251, 3338-3345.

60. Van der Hofstad, G.A.J.M., Foekens, J.A., Bosch, L., and Voorma, H.O. (1977), Eur. J. Biochem. 77, 69-75.

61. Chatterjee, B., Dasgupta, A., Palmieri, S., and Gupta, N.K. (1976), J. Biol. Chem. 251, 6379-6387.

62. Odom, O.W., Kramer, G., Henderson, B., Pinphanichakran, P., and Hardesty, B. (1978), J. Biol. Chem. 255, 1807-1816.

63. Dasgupta, A., Roy, R., Palmieri, S., Das, A., Ralston, R. and Gupta, N.K. (1978), Biochem. Biophys. Res. Commun. 82, 1019-1027.

64. Henderson, A.B. (1978), Fed. Proc. 37, 1623.

65. Kosower, N.S., Vanderhoff, G.A., Benerofe, B., Hunt, T., and Kosower, E. (1971), Biochem. Biophys. Res. Commun. 45, 816-821.

66. Ranu, R.S., and London, I.M. (1976), Proc. Natl. Acad. Sci. U.S.A. 73, 4349.4352.

67. Lengyel, P., this volume.

68. Ernst, V., and Leroux, A. (1980), Fed. Proc. 39, 2027.

69. Ranu, P.E., and Ranu, R.J. (1980), Fed. Proc. 39, 2027.

70. Legon, S., Jackson, R.J., and Hunt, T. (1973), Nature, London, New Biol. 241, 150-152.

71. Pinphanichakran, P., Kramer, G., and Hardesty, B. (1976), Biochem. Biophys. Res. Commun. 73, 625-631.

72. Kramer, G., Henderson, A.B., Pinphanichakran, P., Wallis, M.H., and Hardesty, B. (1977), Proc. Natl. Acad. Sci. U.S.A. 74, 1445-1449.

73. Gross, M. (1979), J. Biol. Chem. 254, 2370-2377.

74. Clemens, M.J., Henshaw, E.C., Rahamimoff, H., and London, I.M. (1974), Proc. Natl. Acad. Sci. U.S.A. 71, 2946-2950.

75. Kaempfer, R. (1976) Biochem. Biophys. Res. Commun. 61, 591-597.

76. Clemens, M.J. (1976), Eur. J. Biochem. 66, 413-422.

77. Gross, M. (1975), Biochem. Biophys. Res. Commun. 67, 1507-1515.

78. Gross, M. (1976), Biochim. Biophys. Acta 447, 445-449.

79. Ranu, R.S., and London, I.M. (1977), Fed. Proc. 36, 868.

80. Amesz, H., Gouman, S.H., Haudrich-Morre, T., Voorma, H.O. and Benne, R. (1979), Eur. J. Biochem. 98, 513-520.

81. Benne, R., Salimans, M.N., Gouman, S.H., Amesz, H., and Voorma, H.O. (1980), Eur. J. Biochem. 104, 501-509.

82. Benne, R., Amesz, H., Hershey, J.W.B., and Voorma, H.O. (1979), J. Biol. Chem. 254, 3201-3205.

83. Methews, M.B. and Korner, A. (1970), Eur. J. Biochem. 17, 328-338.

84. Schreier, M.H., and Staehelin, T. (1973), J. Mol. Biol. 73, 329-349.

85. Benne, R., Kasperaitis, M., Voorma, H.O., Ceglarz, E., and Legocki, A.B. (1980), Eur. J. Biochem. 104, 109-117.

86. Kohler, G., and Milstein, C. (1975), Nature, London, 255, 495-497.

87. Kohler, G., and Milstein, C. (1975), Eur. J. Immun. 6, 511-519.

MESSENGER RNA COMPETITION

Raymond Kaempfer

The Hebrew University
Hadassah Medical School
Jerusalem, Israel

MESSENGER RNA DISCRIMINATION VERSUS MESSENGER RNA COMPETITION

The molar translation yield of a given messenger RNA, that is, the frequency of translation of this mRNA, is not a constant property, but is subject to regulation. Experimentally, this is observed by a shift in the relative amounts of individual proteins synthesized as a function of (a) the rate of overall protein synthesis and (b) the amounts and types of mRNA present.

Selective translation of certain mRNA species over other ones is often involved in the regulation of eukaryotic gene expression during growth, differentiation or virus infection. The so-called *discrimination* of mRNA occurs mainly at the initiation step, which involves the recognition of mRNA and its binding to ribosomes. These observations have raised the question whether there exist mRNA-specific factors that promote the translation of certain mRNA species, but not of other ones. So far, no convincing evidence for such absolute specificity has been obtained, and most of the experiments on the subject can be explained in other ways, chiefly by different requirements, in terms of amount, for the same initiation factors by different mRNA species (reviewed in refs. 1-3). Restated more directly, this means that individual mRNA species possess different affinities for the standard initiation factors involved in the binding of mRNA. The subject of initiation factor/mRNA interactions has been reviewed elsewhere in this volume (4). Here, we shall examine the evidence that mRNA competition during protein synthesis does exist, and constitutes an essential aspect of translational control.

TRANSLATIONAL COMPETITION BETWEEN α- AND β-GLOBIN mRNA

Meaningful studies of translation frquency *in vitro* can be done only in cell-free systems that translate mRNA at high efficiency, such as the reticulocyte lysate and the micrococcal nuclease-treated reticulocyte (5) and cell lysates. In general, the rate-limiting step in translation is at initiation. Hence, differences in the rate of initiation can be detected only in systems where ribosomes can cycle repeatedly over mRNA.

A particularly clear case is the reticulocyte lysate, where, even though α- and β-globin are synthesized in equimolar amounts, translation on β-globin mRNA is initiated about 1.5 times more frequently than on α-globin mRNA (6). That the two mRNA species compete at the initiation step in binding to ribosomes, with β-globin mRNA the *stronger* species, follows from these observations.

α- and β-Globin mRNA Differ in Amount and in Rate of Initiation of Translation

The basic observation that β-globin is synthesized on polysomes containing about 1.5 times as many ribosomes as those engaged in α-globin synthesis (7) could be explained either by assuming that the rate of initiation on each mRNA is the same, while elongation is more rapid on α-globin mRNA, or, alternatively, by assuming that the elongation rates are identical, but initiation of β-globin synthesis is more frequent (6). In the latter case, there must be more α-globin mRNA than β-globin mRNA in the reticulocyte, in order to account for the observed equimolar synthesis of α- and β-globin. Lodish showed that the latter alternative is correct, by reducing the rate of elongation with drugs to the point that the rate of initiation no longer limited protein synthesis. Under such conditions, the ratio of α/β chains made increased from 1 to 1.5, and polysomes bearing nascent α- or β-globin contained the same number of ribosomes (6).

Any decrease in overall translation, such as caused by heme depletion, double-stranded RNA or salt, leads to a drop in the α/β globin synthetic ratio. Conversely, any increase in overall translation causes α-globin mRNA translation to rise more than translation of β-globin mRNA. This is the result expected if one assumes that each mRNA has its own rate constant for binding to ribosomes at initiation: any nonspecific reduction in the rate of initiation at or before binding of mRNA will then result in a preferential inhibition of the translation of the species with the lower rate constant, in this case α-globin mRNA (8). A quantitative analysis of such rate contants, and the effect of a limiting elongation rate on their determination, may be found in refs. 8 and 3.

Demonstration of mRNA Competition

Upon addition of increasing amounts of globin mRNA to mRNA-dependent translation systems, there is about equimolar synthesis of α- and β-globin as long as mRNA is not saturating. Beyond that point, the α/β synthetic ratio drops to a very low value even though overall translation remains constant (9, 10, 11).

There is sound experimental evidence showing that the ratio of α/β globin synthesized *in vitro* by a nuclease-treated lysate of reticulocytes in response to added globin mRNA, is constant as long as the amount of mRNA present does not exceed saturation level, but decreases sharply at higher levels of mRNA (11). This is the expected result if (a) α- and β-globin mRNAs compete each other in translation and (b) competition favors the translation of β-mRNA. The technique described there to demonstrate mRNA competition is both powerful and convenient and, in a larger sense, is generally applicable. Considering that the difference in initiation rate between α- and β-globin mRNA is only 1.5-fold in the intact reticulocyte lysate (6), it can be magnified tremendously by application of competition pressure, that is, by addition of greater than saturating amounts of mRNA. Applied in this manner, the technique provides a very sensitive tool for ordering mRNA species according to their "initiation strength".

eIF-2 is a Target of mRNA Competition

Eukaryotic initiation factor 2 (eIF-2) is known to reverse the block in initiation of protein synthesis observed in the absence of heme, in the presence of double-stranded RNA (13), or that induced by interferon (14). eIF-2 forms a ternary complex with Met-tRNA$_f$ and GTP (15-17 and chapters 3 and 16 of this volume) and directs the binding of Met-tRNA$_f$ to the 40S ribosomal subunit, a necessary pre-requisite to the subsequent binding of mRNA (18, 19). In addition to binding Met-tRNA$_f$, eIF-2 can also bind to mRNA (13, 20-24). Specificity in the binding of eIF-2 to mRNA was shown by the finding (22) that all mRNA species tested possess a high-affinity binding site for eIF-2, while RNA species not serving as mRNA, such as negative-strand RNA (22), tRNA, or rRNA (20) do not possess such a site. Indeed, eIF-2 recognizes and binds specifically to a 5'-terminal region of satellite tobacco necrosis virus (STNV) RNA that comprises the 40S ribosome binding site, and causes local unfolding of the RNA structure. These findings raised the possibility that the binding of a 40S ribosomal subunit to the initiation site in mRNA may be guided to an important extent by eIF-2. How eIF-2 may act during protein synthesis by binding first Met-tRNA$_f$ and then interacting on the 40S subunit with mRNA, is reviewed elsewhere in this volume (4). These properties point to an important role for eIF-2 in the recognition of mRNA and its binding to ribosomes, making it a protein uniquely suited for translational control.

The effect of added **initiation factor eIF-2 on the translational**
competition between α-mRNA and β-mRNA is shown in Fig. 1A. The addition of a constant amount of highly purified eIF-2 to translation
mixtures containing increasing amounts of globin mRNA causes a rise in
the α/β synthetic ratio that is more pronounced the lower the mRNA
concentration (Fig. 1B). At low levels of mRNA, eIF-2 acts to raise
the α/β synthetic ratio toward the value of about 1.5, the relative
content of α- and β-mRNA (6), as expected for a condition in which
α- and β-mRNA no longer compete. At very high levels of mRNA, the
effect of eIF-2 is no longer observed. Clearly, the effect of a given
amount of eIF-2 on mRNA competition can be abolished by saturation
with greater amounts of mRNA.

Even though addition of eIF-2 raises the α/β synthetic ratio as
much as 1.7-fold (Fig. 1B), it does not stimulate total protein synthesis at any mRNA concentration (11). Hence, eIF-2 does not limit
overall protein synthesis in this system. This result must mean that
in increasing the α/β synthetic ratio, eIF-2 does not act *only* by
stimulating the synthesis of α chains. Instead, the eIF-2-dependent
increase in α-chain synthesis is coupled with a *concomitant* decrease

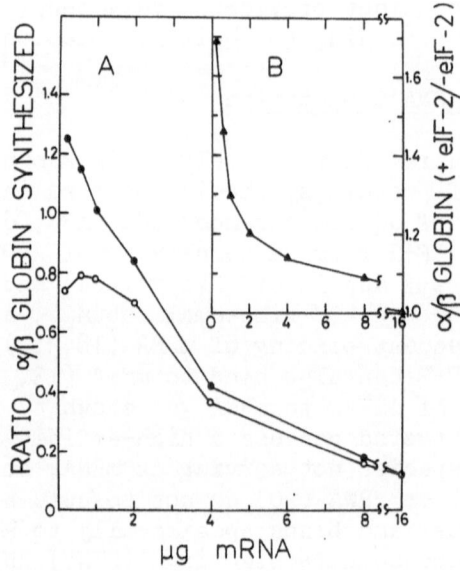

Fig. 1.   Effect of eIF-2 on the α/β synthetic ratio. (A) the α/β
          ratio is plotted as a function of the amount of mRNA present
          during translation (o). The ratio for the samples that re-
          ceived eIF-2 was determined likewise (●). (B) the effect of
          eIF-2 on the α/β ratio was determined from the data in A by
          dividing the ratio observed in the presence of added eIF-2 by
          that observed in its absence, and is plotted as a function of
          the amount of mRNA present. (Reprinted with permission from
          ref. 11)

in β-chain synthesis. In what follows, it will become clear that this
is precisely the result expected if eIF-2 acts to relieve mRNA competition.

The Effect of Salt on mRNA Competition

Synthesis of α chains is inhibited preferentially by increasing
amounts of globin mRNA. Another variable known to influence α- and
β-globin synthesis differentially is the salt concentration (9). This
is seen in Fig. 2A, which depicts total protein synthesis in a series
of lysates containing increasing concentrations of added KCl or KOAc.
KOAc supports protein synthesis at higher concentrations than KCl,
apparently because of an inhibitory effect of the Cl⁻ ion (26, 27).
The amounts of α- and β-globin synthesized in these samples are illustrated in Fig. 2C. It is seen that optimal translation of both
α-mRNA and β-mRNA occurs at the same salt concentration, but the
optimum is higher when KOAc is added. The sum of α- and β-globin syn-

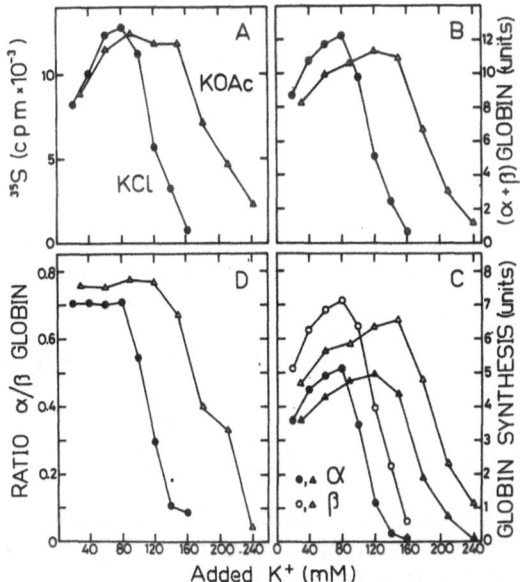

Fig. 2.   Effect of KCl and KOAc on the translation of α- and β-globin
          mRNA. Reaction mixtures for protein synthesis (12), con-
          taining 1.5 μg of globin mRNA, were incubated at the
          indicated concentrations of added KCl or KOAc. (A) Incorpo-
          ration of (^{35}S) methionine into protein ((●) KCl; (Δ) KOAc).
          (B) The sum of α- and β-globin synthesized (see (C))((●) KCl;
          (Δ) KOAc). (C) Synthesis of α- and β-globin was determined
          in the samples of (A), and is plotted in arbitrary units (α-
          globin, (●) KCl and (▲) KOAc; β-globin, (o) KCl and (Δ) KOAc)
          (D) α/β globin synthetic ratios, computed from the results of
          (C) ((●) KCl; (Δ) KOAc). (Reprinted with permission from
          ref. 11)

thesized (Fig. 2B) matches closely in its salt dependence with overall
protein synthesis (Fig. 2A).

In spite of the fact that α- and β-globin mRNA possessed the same
salt optimum for translation, they were shown to differ in their sensitivity to greater than optimal concentrations of salt: synthesis
of α chains decreased more drastically than that of β chains (cf. in
Fig. 2C the amounts of α- and β-globin synthesized at 120 mM added
KCl or 180 mM added KOAc). Indeed, the α/β synthetic ratio was
constant at salt concentrations up to the optimum but then decreased
steeply (Fig. 2D).

From a comparison of the behavior of globin synthesis in the
presence of KCl and KOAc, it was also clear that the decrease in the
α/β synthetic ratio between 80 and 140 mM of added KCl was caused by
the Cl⁻ ion and not by the K⁺ ion, for no such decrease was observed
when KOAc was used in the same range (Fig. 2C). Weber et al. had
shown that the Cl⁻ ion inhibits a step in initiation necessary for the
binding of mRNA to ribosomes, but does not significantly affect the
binding of Met-tRNA$_f$ to 40S ribosomal subunits (26). Since eIF-2 not
only binds to Met-tRNA$_f$ but can also bind to mRNA, it was logical to
ask if eIF-2 was able to overcome the inhibitory effect of Cl⁻ ions on
translation. This was indeed the case (see Fig. 3 and ref. 11).
Clearly eIF-2 was capable of relieving the translational competition
between α-mRNA and β-mRNA observed at high KCl concentrations. In
addition, it was shown that eIF-2 was able to relieve the inhibition
of overall protein synthesis at high KCl concentrations, in spite of
the fact that at optimal KCl concentration eIF-2 did not stimulate
overall synthesis. The conclusion to be drawn from the evidence so
far reviewed was that high salt concentrations inhibited the function
of eIF-2 in protein synthesis.

The Effect of Salt on Binding of mRNA to eIF-2

Ternary complex formation between eIF-2, Met-tRNA$_f$, and GTP is
much less sensitive to increasing KCl concentrations than is complex
formation between eIF-2 and mRNA (28). These findings, together with
the observation (26) that high KCl concentrations block the binding
of mRNA rather than Met-tRNA$_f$ to 40S ribosomal subunits, suggested
that KCl may inhibit the interaction of mRNA with eIF-2 during
translation (11). This concept is supported by the tight correlation
between the effects of KCl and KOAc on translation of globin mRNA on
one hand and on the binding of globin mRNA to eIF-2 on the other
(Fig. 4) (11).

Weaker Binding of α-Globin mRNA to eIF-2

It is clear that α-globin mRNA competes more weakly than β-globin
mRNA in translation. In particular, the differential inhibition of

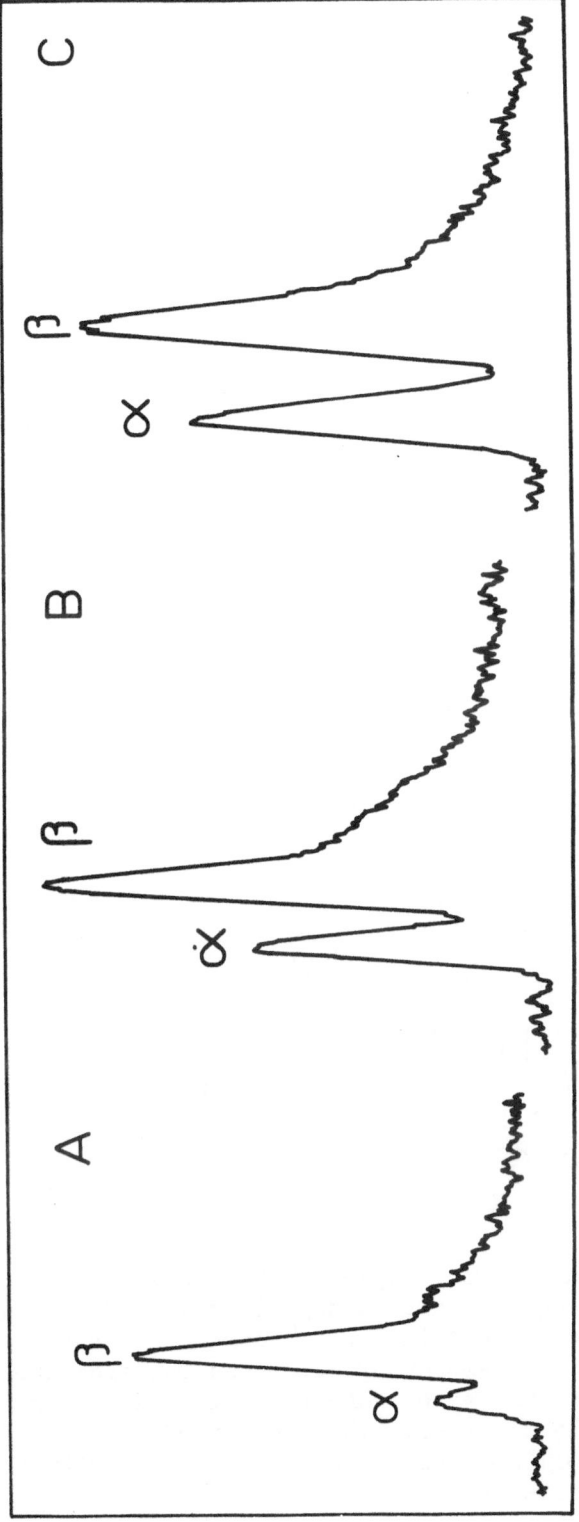

Fig. 3.   Effect of eIF-2 on the synthesis of α- and β-globin at high KCl concentration.
Densitometer scans of cellulose acetate electropherograms are depicted. Reaction
mixtures for protein synthesis of 26 µl contained 1.5 µg of globin mRNA, KCl to
an added concentration of 130 mM, and the following amounts of added eIF-2: none
(A), 1 µg (B), 1.5 µg (C). Incorporation of (^{35}S) methionine into protein per 3-µl
aliquot was 1,790 cpm in (A), 3,800 cpm in (B), and 4,130 cpm in (C). The α/β
synthetic ratio is 0.19 in (A), 0.42 in (B), and 0.57 in (C). (Reproduced with
permission from ref. 11).

the translation of α- and β-globin mRNA by elevated salt concentrations, and relief of this inhibition by eIF-2 suggest that the binding of α-mRNA to eIF-2 is more sensitive to salt than the binding of β-mRNA. Indeed, as seen in Fig. 5, the binding of purified α-mRNA (11) to eIF-2 is significantly more sensitive to salt than the binding of total globin mRNA (which contains 60% α-mRNA and 40% β-mRNA). This is precisely what would be expected if α-globin mRNA were to interact more weakly with eIF-2 than β-globin mRNA. These results strongly suggest that β-globin mRNA binds more tithly to eIF-2 than does α-globin mRNA.

Involvement of Other Initiation Factors

An initiation factor resembling eIF-4B was identified by its ability to stimulate α-globin mRNA translation selectively in a cell-free system from Krebs ascites cells (29), and was interpreted to be an α-globin mRNA-specific factor (29, 2). However, it acts to restore

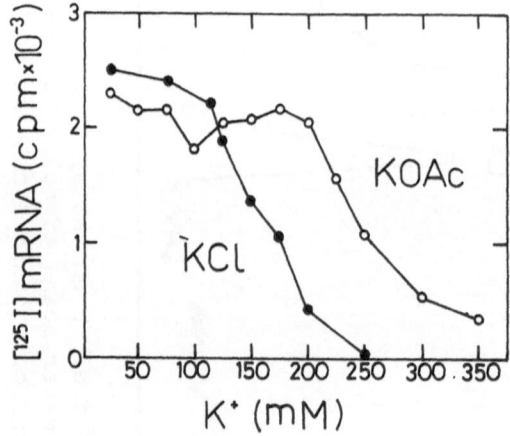

Fig. 4. Effect of KC1 and KOAc on complex formation between globin mRNA and eIF-2. mRNA-binding assay mixtures (12) contained ^{125}I-labelled globin mRNA (1.5 x 10^6 cpm/μg; input, 4,270 cpm) (12), a limiting amount of eIF-2 and the indicated concentrations of KC1 (•) or KOAc (o). In the presence of saturating amounts of eIF-2, 3,840 cpm of mRNA were bound in the presence of 50 mM KC1. Background binding (without eIF-2) was subtracted (11).

the α/β globin synthetic ratio to about one, and also can stimulate
β-globin translation. Thus, it apparently acts, like eIF-2, to
relieve mRNA competition. Support for this point came from another
study (10) showing that eIF-4B, and to a lesser extent eIF-4A, relieves translational competition between α- and β-globin mRNA without
stimulating overall translation. The initiation factor preparations
used in these studies were less well characterized than those now
available and may have contained cap-binding protein (20, 39) or
other factors. On the basis of translation data, it was suggested
that α- and β-globin mRNA compete for eIF-4B, but neither direct
binding nor the predicted greater affinity of the β-mRNA has been
demonstrated (10). Thus, it is not certain if eIF-4B and eIF-4A are
themselves the target of mRNA competition in this system, or if they
affect the binding of mRNA to another component, for instance eIF-2.

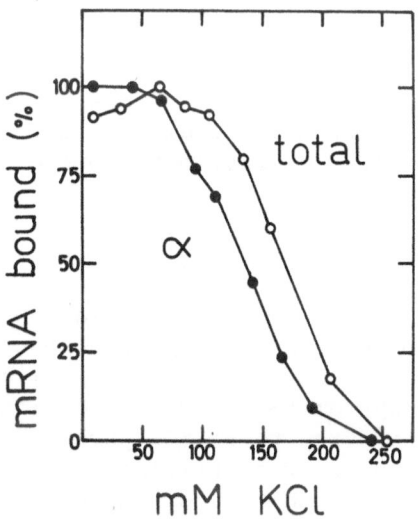

Fig. 5. Effect of KCl on complex formation between eIF-2 and α-globin
         mRNA or total globin mRNA. mRNA-binding assay (12) mixtures
         contained ^{125}I-labelled purified α-globin mRNA (1.5 x 10^5
         cpm/μg; input, 1,920 cpm) (●) or ^{125}I-labelled total globin
         mRNA (7.6 x 10^5 cpm/μg; input, 3,850 cpm) (o), a limiting
         amount of eIF-2, and the indicated concentrations of KCl.
         In the presence of saturating amounts of eIF-2, 1,380 and
         2,600 cpm, respectively, were bound at 50 mM KCl. Binding
         of 100% is 1,040 cpm for α-globin mRNA and 1,780 cpm for
         total globin mRNA. Background values were subtracted (11).

The High Affinity of β-Globin mRNA for eIF-2

It is not yet known what determines the favorable translation properties of β-globin mRNA, and its relatively high affinity for eIF-2, or other factors.

As discussed elsewhere in this Volume (4), eIF-2 binds specifically to STNV RNA at a 5'-terminal region that comprises the ribosome binding site (25). This region is located in the first 44 nucleotides of the RNA molecule, and contains the AUG initiation codon for translation. Possibly, a free AUG codon is one of the features in mRNA required for effective binding of eIF-2 (4). This concept is supported by the finding that the AUG initiation codon in rabbit or mouse β-globin mRNA also is readily accessible to ribonuclease T1, while by contrast in α-globin mRNA it is not (31). It is thus conceivable that the higher affinity of β-globin mRNA for eIF-2 is related to the accessibility of its AUG initiation codon.

TRANSLATIONAL COMPETITION BETWEEN HOST AND VIRAL mRNA

A second, well-studied example of translational competition concerns that between host and viral mRNA. Among the most effective mRNA species known are the RNA genomes of certain picornaviruses, such as Mengo and encephalomyocarditis (EMC) virus. Because of their very high efficiency of initiation, these RNA species provide sensitive probes for studies of translational competition. The available evidence indicates that both initiation factors eIF-4B and eIF-2 are involved.

eIF-4B is a Target of Competition

Simultaneous translation of cellular mRNA and RNA from EMC virus results in translation of EMC RNA at the expense of translation of cellular mRNA (37). Globin mRNA translation is likewise suppressed by EMC RNA in the same system (32). Upon addition of an initiation factor, at the time considered to be 50% pure eIF-4B, it was observed that overall protein synthesis was stimulated, and globin mRNA translation was stimulated more than that of EMC RNA (32).

As mentioned earlier, any component that increases the rate of overall translation is expected to cause greater stimulation of translation of the weaker mRNA species, in this case globin mRNA (2, 8). Thus, the results observed in this experiment did not show if globin mRNA and EMC RNA compete for eIF-4B. At greater than saturating levels of the initiation factor, overall protein synthesis was inhibited considerably; however, globin mRNA translation increased while EMC RNA translation decreased. On the basis of these results, it was suggested that EMC RNA out-competes globin mRNA for this factor (32).

More direct evidence for this idea came from equilibrium binding studies with purified eIF-4B (33). EMC RNA was found to bind about eight-fold more tightly than globin mRNA to eIF-4B. Whether there was a contribution of cap-binding protein, which tends to contaminate preparations of eIF-4B (4, 30, 39), in these experiments is at present not clear. Indeed, the cap-binding protein was shown to stimulate the translation of the capped mRNA species, globin mRNA or alfalfa mosaic virus (AMV)-4 RNA, when they were translated in the presence of (non-capped) STNV RNA (40). Translation of capped Sindbis virus RNA was also stimulated by cap-binding protein more than translation of EMC RNA, when they were translated together (40). In these experiments cap-binding protein led to an overall stimulation of protein synthesis, so that it is not clear if the mRNA species actually competed for cap-binding protein, or if the effects were due to non-specific stimulation (1,8).

eIF-2 is a Target of Competition

Recently, more quantitative results were obtained with the micrococcal nuclease-treated rabbit reticulocyte lysate, programmed with known amounts of globin and Mengo virus RNA (34). In this system, translation of Mengo virus RNA was shown to be as dependent on continued initiation as translation of globin mRNA. Addition of a molecule of Mengo RNA resulted in a 50% inhibition of translation of 35 molecules of globin mRNA, while overall translation remained constant. Hence, on a molar basis, the viral RNA apparently competes 35 times better than globin mRNA for a limiting component (34). The competition is relieved by eIF-2, in conditions where overall protein synthesis remains constant. That is, the rise in globin mRNA translation is coupled with a commensurate decrease in translation of Mengo virus RNA (34). The extent of relief is inversely proportional to the amount of competing Mengo virus RNA present, in a manner similar to that discussed for α- and β-globin mRNA. Direct mRNA-binding competition studies showed that Mengo RNA possesses a thirty-fold higher affinity for eIF-2 than does total globin mRNA. This high affinity was also reflected in the much greater salt resistance of complex formation between eIF-2 and Mengo RNA (34).

The direct correlation between the abilities of Mengo virus and globin mRNA to compete in translation, and their affinities for eIF-2, suggest strongly that there is direct competition for eIF-2 in this case.

These observations point to the existence, in Mengo virus RNA, of a specific site or sequence possessing an extremely high affinity

for eIF-2. Recent studies have revealed that eIF-2 binds to the same
specific sequences in Mengo virus RNA that are protected against
nuclease attack by 40S ribosomal subunits or 80S initiation complexes
(38).

The high affinity of Mengo virus RNA for eIF-2 may explain why
Mengo and other picornaviral RNA species, as well as STNV RNA, lack
the 5'-terminal cap structure. The existence of a sequence possessing
a very high affinity for eIF-2 would obviate the need for the additional stabilization imparted by binding at the cap.

These findings bear on the problem of host shut-off. Strikingly,
initiation on picornavirus mRNA depends more on eIF-4A than that on
globin or ovalbumin mRNA (36); this may reflect the unusual structure
of these viral RNAs.

REGULATION BY mRNA COMPETITION

The two cases discussed in this chapter are clear-cut examples
of mRNA competition for a *non-specific* component in translation that
results in the preferential translation of certain mRNA species over
other ones. Addition of the initiation factor leads to preferential
stimulation of translation of the weakly competing mRNA species.
Clearly, such preferential stimulation does *not* mean that the factor
is specific for the weak mRNA in question. The opposite is true
the stronger initiating mRNA has a higher affinity for the factor.

Indeed, there is no convincing evidence yet for the existence
of mRNA-specific factors (1). Rather, the relative translation rates
of mRNA species are determined (a) by their intrinsic affinity
properties for the initiation step, (b) by their molar ratios, and
(c) by the extent to which the target of competition (e.g., eIF-2)
is limiting. This means that both the introduction of new mRNA
species, as in differentiation or virus infection, and changes in
initiation factor activity, as produced by heme deprivation, dsRNA
or interferon, affect the relative expression of each mRNA in the
cell.

The Role of eIF-2

The ability of eIF-2 to relieve translational competition between
α- and β-globin mRNA (11), and likewise between globin mRNA and Mengo
virus RNA (34), in principle can be accounted for by two types of
explanations. One is that α- and β-globin mRNA possess different
affinities for eIF-2 and compete directly for this factor during
initiation of protein synthesis. The other is that eIF-2 acts nonspecifically to enlarge the pool of 40S/Met-tRNA$_f$ complexes; as
pointed out above, any increase in the rate of reactions at or before
the mRNA binding step that results in a nonspecific increase in the
rate of protein synthesis will lead to preferential stimulation of

the weakly competing mRNA species, in this case α-globin mRNA. Even though eIF-2 is responsible for the formation of 40S/Met-tRNA$_f$ complexes, several lines of evidence reviewed here suggest that direct mRNA competition for eIF-2 is actually involved in the regulation of α- and β-globin synthesis. First, addition of eIF-2 relieves mRNA competition without stimulating total protein synthesis at optimal salt concentration. This would not be expected if eIF-2 were to influence the rate-limiting step in protein synthesis nonspecifically. This finding is consistent with direct competition of mRNA for eIF-2, but does not prove it, for conceivably eIF-2 could affect a step preceding the binding of mRNA, while another step at or beyond the binding of mRNA could be rate-limiting. Second, the behavior of globin mRNA translation as a function of increasing concentrations of KCl or KOAc (Fig. 2) matches exactly with that observed for the direct binding of globin mRNA to eIF-2, both with respect to the final salt concentration giving 50% inhibition and the displacement between the response to KCl and to KOAc (Fig.4). This correlation between translation data and direct mRNA-binding experiments strongly suggest that the interaction of mRNA with eIF-2 is important in determining the overall rate of initiation of translation, as well as the relative rates of initiation on α- and β-globin mRNA. This is borne out by a third line of evidence demonstrating that the binding of α-globin mRNA to eIF-2 exhibits greater salt sensitivity than the binding of unfractionated globin mRNA. Thus, α-globin mRNA interacts more weakly with eIF-2 than a mixture of α- and β-globin mRNA (Fig. 5).

Fourth, eIF-2 relieves the inhibition of total protein synthesis, and the sharpened competition between α- and β-globin mRNA occurring at elevated KCl concentration (Fig. 3). Since high concentrations of Cl$^-$ ion inhibit initiation of translation by preventing primarily the binding of mRNA to 40S initiation complexes, while binding of Met-tRNA$_f$ to 40S subunits is only slightly inhibited, it seems likely that the ability of eIF-2 to relieve the inhibitory effect of KCl is based on its action at the mRNA-binding step rather than at the step involving 40S/Met-tRNA$_f$ complex formation. Indeed, KCl directly inhibits the binding of mRNA to eIF-2 at concentrations that hardly affect the formation of ternary complexes between eIF-2, Met-tRNA$_f$, and GTP (Figs. 4 and 5 and ref. 28).

The simplest interpretation, then, is that eIF-2 relieves translational competition by interacting with α- and β-globin mRNA, rather than by affecting the binding of Met-tRNA$_f$. While α- and β-globin mRNA could compete for free eIF-2 molecules, it is more likely that they compete for eIF-2 molecules located in 40S/Met-tRNA$_f$ complexes. Possible mechanisms for the interaction of eIF-2 with Met-tRNA$_f$ and subsequently, on the 40S ribosomal subunit, with mRNA, are discussed elsewhere in this volume (4). Any increase in the formation of 40S/eIF-2/Met-tRNA$_f$ complexes will lead to relief of competition between α- and β-globin mRNA, because it will increase the number of eIF-2 molecules available for mRNA competition.

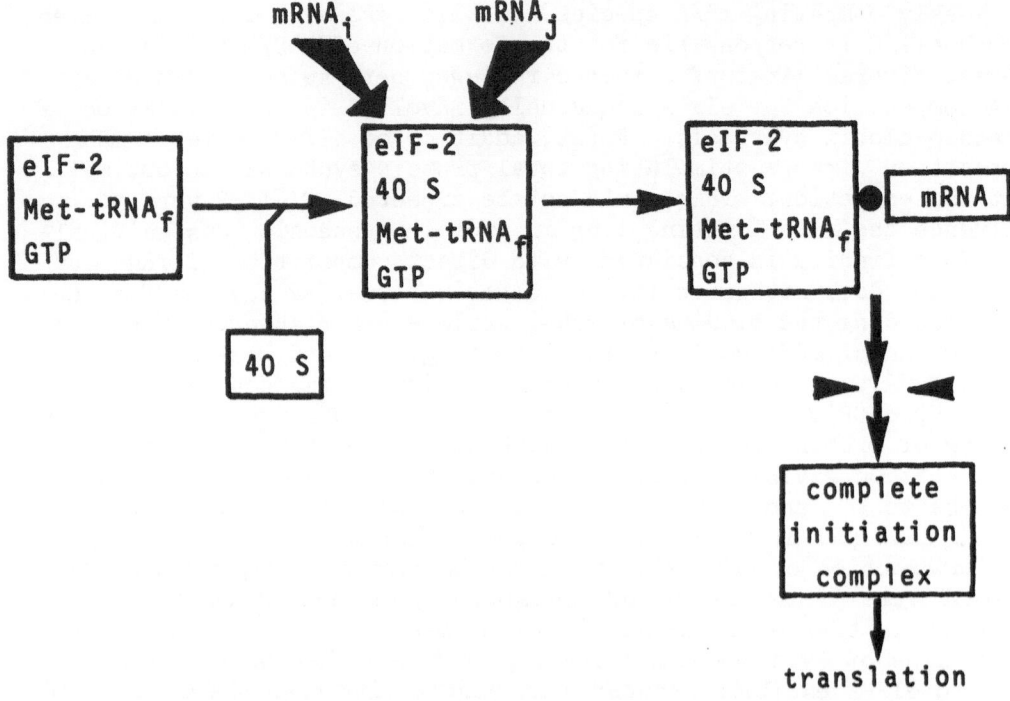

Fig. 6.   Diagram of mRNA competition for eIF-2 during protein
          synthesis (see text).

        This is illustrated more generally in the diagram of Fig. 6. For
any pair of competing mRNA species i and j, their relative frequency
of translation will not equal their molar ratio as long as the target
of competition, eIF-2, is limiting. Thus, if $mRNA_i$ is the species
with higher affinity and $mRNA_j$ the species with lower affinity, and
$mRNA_j$ is present at higher concentration (as is the case for the
examples discussed in this chapter), then any increase in the concentration of eIF-2 molecules located in $40S/eIF-2/Met-tRNA_f$ complexes will result in a relatively greater stimulation of binding of
$mRNA_j$, up to the point where binding of $mRNA_i$ and $mRNA_j$ is exactly
according to their molar ratio, and there no longer is competition.
Quite independently of this competition, the actual rate of protein
synthesis may be limited at a step beyond the interaction with eIF-2
(as indicated by the wedges in Fig. 6; these are placed arbitrarily
and in principle could occur at any later step). This is true for
the examples cited in this chapter, as addition of eIF-2 did not
lead to stimulation of translation at any mRNA concentration and hence
eIF-2 did not limit translation. Under these conditions, any increased binding of $mRNA_j$ must therefore be concomitant with decreased
binding of the stronger species, $mRNA_i$, precisely as is observed (11,
34).

Differentiation and mRNA Competition

mRNA competition may contribute to the regulation of gene expression during differentiation. The appearance of new mRNA species
must influence the expression of pre-existing mRNA species. The
actual effect will depend both on the relative amounts and the affinity properties of the new and pre-existing mRNA types present in
the cell at any given time. An appealing hypothesis is that differentiation may involve the sequential expression of genes encoding
mRNA species of increasingly greater initiation strength. Some
support for such a model is provided by the observation that synthesis
of immunoglobulin heavy and light chains, products of terminal differentiation, is directed by strong mRNA species, as judged by the
observation that in myeloma cells these mRNAs are still expressed
when synthesis of other proteins is inhibited by hypertonic salt concentrations that partially block the initiation of translation (41,
47).

Clearly, the appearance of new gene products will be affected
by the relative rates of transcription of individual genes, by
differential stability of pre-existing and new mRNA species, and by
positive or negative control mechanisms including those acting conceivably at the level of translation, but in addition, the contribution of mRNA competition could be very important.

During erythroid development, the synthesis of the constituents
of hemoglobin (heme and different globin chains) is exactly coordi-
nated. Initiation of protein synthesis in reticulocyte lysates is
dependent upon the continued supply of heme (42, 43). In the absence
of heme, the activity of eIF-2 is lost and protein synthesis can be
restored by addition of an excess of this initiation factor (13, 44-
46). Thus, eIF-2, through its property of heme dependence, may allow
the coordinate synthesis of heme and of globins during erythroid
development. The results discussed above indicate that, in addition,
eIF-2 acts to coordinate the synthesis of α- and β-globin by serving
as a target for competition between α- and β-globin mRNA. Thus,
eIF-2 may have a dual function in translational control of hemoglobin synthesis during late erythroid differentiation.

REFERENCES

1.   Lodish, H.F. (1976), Ann. Rev. Biochem. 45, 39-72.
2.   Revel, M. and Groner, Y. (1978), Ann. Rev. Biochem. 47, 1079-1126
3.   Birge, C.H., Golini, F. and Thach, R.E. (1979), Methods Enzymol.
     60, 375-380.
4.   Kaempfer, R. (1981), In *Protein Biosynthesis in Eukaryotes* (R.
     Perez-Bercoff, Ed.), Plenum Publishing Co., New York and London,
     Chapter 9.

5.   Pelham, H.R.B. and Jackson, R.J. (1976), Eur. J. Biochem. 67, 247-256.
6.   Lodish, H.F. (1971), J. Biol. Chem. 252, 7131-7138.
7.   Hunt, T., Hunter, T. and Munro, A. (1968), Nature, London, 220, 481-482.
8.   Lodish, H.F. (1974), Nature, London, 251, 385-388.
9.   McKeehan, W.L. (1974), J. Biol. Chem. 249, 6517-6526.
10. Kabat, D. and Chappell, M. (1977), J. Biol. Chem. 252, 2684-2690.
11. Di Segni, G., Rosen, H. and Kaempfer, R. (1979), Biochemistry 18, 2847-2854.
12. Kaempfer, R. (1979), Methods Enzymol. 60, 380-392.
13. Kaempfer, R. (1974), Biochem. Biophys. Res. Commun. 61, 591-597.
14. Kaempfer, R., Israeli, R., Rosen, H., Knoller, S., Zilberstein, A., Schmidt, A., Revel, M. (1979), Virology, 99, 170-173.
15. Dettman, G.L. and Stanley, W.M., Jr. (1972), Biochim. Biophys. Acta 287, 124-133.
16. Levin, D.M., Kyner, D. and Acs, G. (1973), Proc. Nat. Acad. Sci. U.S.A. 70, 41-45.
17. Schreier, M.H. and Staehelin, T. (1973), Nature New Biol. 242, 35-38.
18. Darnbrough, C.M., Legon, S., Hunt, T. and Jackson, R.J. (1973), J. Mol. Biol. 76, 379-403.
19. Trachsel, H., Erni, B., Schreier, M. and Staehelin, T. (1977), J. Mol. Biol. 116, 755-767.
20. Hellerman, J.G. and Shafritz, D.A. (1975), Proc. Nat. Acad. Sci. U.S.A. 72, 1021-1025.
21. Barrieux, A. and Rosenfeld, M.G. (1977), J. Biol. Chem. 252, 3843-3847.
22. Barrieux, A. and Rosenfeld, M.G. (1978), J. Biol. Chem. 253, 6311-6315.
23. Kaempfer, R., Hollender, R., Abrams, W.R. and Israeli, R. (1978), Proc. Nat. Acad. Sci. U.S.A. 75, 209-213.
24. Kaempfer, R., Hollender, R., Soreq, H. and Nudel, U. (1979), Eur. J. Biochem. 94, 591-600.
25. Kaempfer, R., Van Emmelo, J. and Fiers, W. (1981), Proc. Nat. Acad. Sci. U.S.A., in press.
26. Weber, L.A., Hickey, E.D., Maroney, P.A. and Baglioni, C. (1977), J. Biol. Chem. 252, 4007-4010.
27. Kemper, B. and Stolarsky, L. (1977), Biochemistry 16, 5676-5680.
28. Kaempfer, R., Rosen, H. and Israeli, R. (1978), Proc. Nat. Acad. Sci. U.S.A. 75, 650-654.
29. Nudel, U., Lebleu, B. and Revel, M. (1973), Proc. Nat. Acad. Sci. U.S.A. 70, 2139-2144.
30. Sonenberg, N., Rupprecht, K.M., Hecht, S.M. and Shatkin, A.J. (1979), Proc. Nat. Acad. Sci. U.S.A. 76, 4345-4349.
31. Pavlakis, G.N., Lockard, R.E., Vamvakopoulos, N., Rieser, L., Raj-Bhandary, U.L., Vournakis, J.M., (1980), Cell 19, 91-102.
32. Golini, F., Thach, S.S., Birge, C.H., Safer, B., Merrick, W.C. and Thach, R.E. (1976), Proc. Nat. Acad. Sci. U.S.A. 73, 3040-3044.

33. Baglioni, C., Simili, M. and Shafritz, D.A. (1978), Nature, London, 275, 240-243.

34. Rosen, H., Di Segni, G. and Kaempfer, R., submitted.

35. Rosen, H., Knoller, S., and Kaempfer, R. (1981), Biochemistry, in press.

36. Blair, G.E., Dahl, H.H.M., Truelsen, E., and Lelong, J.C. (1977), Nature, London, 265, 651-653.

37. Lawrence, C. and Thach, R. (1974), J. Virol. 14, 598-610.

38. Perez-Bercoff, R. and Kaempfer, R. (1981), submitted.

39. Bergmann, J.E., Trachsel, H., Sonenberg, N., Shatkin, A.J. and Lodish, H.F. (1979), J. Biol. Chem. 254, 1440-1443.

40. Sonenberg, N., Trachsel, H., Hecht, S. and Shatkin, A.J. (1980), Nature, London, 285, 331-333.

41. Nuss, D.L. and Koch, G. (1976), J. Mol. Biol. 102, 601-612.

42. Waxman, H. and Rabinovitz, M. (1966), Biochim. Biophys. Acta 129, 369-379.

43. Zucker, W. and Schulman, H. (1968), Proc. Nat. Acad. Sci. U.S.A. 59, 582-589.

44. Kaempfer, R. and Kaufman, J. (1972), Proc. Nat. Acad. Sci. U.S.A. 69, 3317-3321.

45. Raffel, C., Stein, S. and Kaempfer, R. (1974), Proc. Nat. Acad. Sci. U.S.A. 71, 4020-4024.

46. Clemens, M.J., Safer, B., Merrick, W.C., Anderson, W.F. and London, I.M. (1975), Proc. Nat. Acad. Sci. U.S.A. 72, 1286-1290.

47. Koch, F., Koch, G. and Kruppa, J. (1981), In: *Protein Biosynthesis in Eukaryotes* (R. Perez-Bercoff, Ed.), Plenum Publishing Co., New York and London.

INTERFERON ACTION: CONTROL OF

RNA PROCESSING, TRANSLATION AND DEGRADATION

Peter Lengyel

Department of Molecular Biophysics and Biochemistry
Yale University
New Haven, CT 06511 U.S.A.

INTRODUCTION

"During a study of the interference produced
by heat-inactivated influenza virus with the growth
of live virus in fragments of chick chorioallantoic
membrane it was found that following incubation of
heated virus with membrane a new factor was released.
This factor, recognized by its ability to induce
interference in fresh pieces of chorioallantoic membrane was called interferon. Following a lag phase
interferon was first detected in the membranes after
3 h incubation and thereafter it was released into
the surrounding fluid."

*Abstract from the 1957 paper of Isaacs and Lindenmann
in which they announced the discovery of interferons
(1).*

Studies in the last two decades revealed that interferons are
a set of polypeptides (2-5), several (and perhaps all) of which
share some sequence homology (6-13). They were found to be produced
in a large variety of vertebrates (2-5). In addition to their antiviral action, interferons affect numerous seemingly diverse biological phenomena. These include e.g., cell motion, cell proliferation,
and various immunological processes such as the antibody response,
delayed hypersensitivity, graft rejection, histocompatibility antigen
expression, natural killer cell recruitment, and macrophage activation (2-5, 14, 15).

459

The importance of interferons as antiviral and antitumor agents was clearly demonstrated. Mice treated with an antiserum to mouse interferons were killed by amounts of virus several hundred times less than needed to kill control mice (16, 17). Moreover, interferons injected were shown to protect animals from some viral diseases (2). The growth of various spontaneous, transplanted or virus-induced tumors in animals was found to be impaired by treatment with interferons (2, 15) and to be promoted by treatment with an antiserum to interferons (18).

In spite of these promising findings in animal experiments clinical tests with interferons are only at an early stage and are being performed only on a small scale (19). The lack of human interferon in sufficient amount and purity made it impossible to set up large scale tests in the past. During the last year, however, human genes specifying interferons were isolated, introduced into bacteria and proven to give rise in bacteria to biologically active human interferons (13, 20, 38). Thus, large amounts of human interferons synthesized by bacteria will become available for clinical trials in the near future.

After a short general survey of interferons this review will deal with the biochemistry of interferon action and especially with the control of mRNA processing, translation and turnover in cells treated with interferons. There are several recent books and reviews covering various aspects of interferon research (2-5, 15, 21-24).

SURVEY OF INTERFERONS

Assay

Most of the interferon assays are based on measuring antiviral activity in cell culture (2, 3). The simplest assay involves the determination of the protection of cells by interferons from the cytopathic effects caused by certain viruses. Other assays are based on the inhibition of infectious virus multiplication or the decrease caused by interferons in the accumulation of virus-specified macromolecules (DNA, RNA, and proteins). The assays usually consist of a comparison of one of these effects in cells treated with (various dilutions of) a sample whose interferon content is to be determined with the same effect in cells treated with a standard interferon preparation.

Induction

Interferons are formed upon induction in a large variety of vertebrates including fish, turtle, birds, and mammals. Within the organism and also in tissue culture, many though not all types of cells can produce interferons (2, 3, 22, 24). A mutant cell line forming small amounts of interferons constitutively, i.e., without an apparent need for inducer, was also described (25).

The inducers of interferons include members of most major virus groups, some species of protozoa, bacteria, mycoplasmas, tumor cells, as well as natural or synthetic double-stranded RNA·(e.g., poly(I).poly(C)), endotoxins, polysaccharides, antigens, (to which animals have been previously sensitized), mitogens, and some low molecular weight compounds. The same agent which induces interferon formation in certain species of animals or cells may not be an inducer in others (2, 3, 22, 24).

Interferons synthesized by different animal species are different. Moreover, one and the same animal or cell species can produce different interferons. Thus, at least three different classes of human interferons have been described with little or no immunological cross reactivity: α, β, and γ (26). These were designated earlier as Le or *leukocyte*, F or *fibroblast*, and type II, or *immune* interferons (2-4). The type of interferon produced may depend on the nature of the producing cell as well as that of the inducer. Thus, a human fibroblast cell line produces β type interferon and only little, if any, α type interferon if induced with poly(I).poly(C). However, upon infection with Newcastle disease virus (NDV), or Vesicular Stomatitis Virus (VSV), the same cell line produces about 25% as much α type interferon as β type (27). The human chromosomes 2, 5, and 9 each seem to harbor a gene specifying β type interferon (28).

Interferon mRNA

The first known intermediate in interferon induction is interferon mRNA (4, 29). This can be assayed by extracting poly(A) containing RNA (i.e., mRNA) from cells and adding the RNA to a cell free protein synthesizing system or to a culture of cells (from a species other than that from which the RNA was extracted) or injecting it into *Xenopus* oocytes. Interferon mRNA can be translated in each of these systems into biologically active interferon. Studies with various antisera specific for different types of interferons indicate that the interferons translated in each of the systems are characteristic of the species from which the interferon mRNA was extracted and not of the species in which the translation has taken place (4).

Control of Interferon Synthesis

The interferon synthesis following the treatment of animals or cells with interferon inducers is shut off after a period whose length depends on the nature of the inducer and of the animals or cells induced (4, 22, 24). The shutoff is accompanied (and most probably caused) by the disappearance of translatable interferon mRNA. At least in the case of the induction of cells of certain mammalian lines with poly(I).poly(C) the shutoff of interferon synthesis

and the disappearance of translatable interferon mRNA can be delayed
by treating the cells at the appropriate time *after* induction with
inhibitors of RNA and protein synthesis. This procedure results in
a greatly increased production of interferons and is called superinduction (3, 24, 30). The results of studies on superinduction
serve as the basis for a hypothesis according to which the shutoff
of interferon synthesis is a consequence of the inactivation of
interferon mRNA by a repressor. The hypothetical repressor is assumed
to be induced by poly(I).poly(C) and/or interferon and to have a
short half life (3).

After a round of interferon synthesis, or treatment with interferons at higher concentration, animals or cells in culture may become
temporarily refractive to repeated interferon induction (hyporespon-
siveness) (3, 4, 22). Treatment of certain cell types in culture
with interferons at low concentrations, however, enhances interferon
synthesis upon induction (priming) (3, 4, 22).

Mass Production of Human Interferons

A large portion of the human interferons used for the small-
scale clinical trials has been produced from leukocytes (19). These
are obtained from human blood by low-speed centrifugation. The
process results in a red cell-rich layer at the bottom of the centri-
fuge tube, a leukocyte-rich layer in the middle and plasma on top.
The leukocyte-rich fraction which had little clinical use in the past
is harvested and induced to produce interferons of the α type by
infection with Sendai virus (31, 32). Type β human interferon is
produced in human fibroblasts grown in monolayer cultures, induced
with poly(I).poly(C) and superinduced with inhibitors of RNA and
protein synthesis (33). The disadvantage of this system is the
limited life span of the fibroblasts: their ability to produce
interferon decreases after less than fifty passages (34). Human
interferons can also be produced on a large scale using established
cell lines. Lymphoblastoid cells give rise to the α type and fibro-
blastoid cells to β type interferons. Such cells may be passaged
and induced to form interferons indefinitely. However, such cell
lines may carry tumor viruses; thus the interferons produced require
through purification before taking into consideration clinical
application (35-37). The interferon yields of the human cell lines
used at present are low; high producers among mouse cell lines
synthesize twenty-forty times more interferon per cell.

A breakthrough in the technology of interferon production,
promising large amounts of pure interferons at low cost occurred at
the end of 1979, beginning of 1980: using recombinant DNA technology,
human interferon genes were isolated, inserted into *E. coli* and shown
to specify in the bacteria the synthesis of biologically active human
interferon (20, 38).

Isolation of Human Interferon Genes

The following approach was used to obtain bacterial plasmids
containing human interferon α genes (20): mRNA was isolated from
human leukocytes induced to synthesize interferons. The mRNA was
enriched in interferon mRNA by sucrose gradient fractionation. The
fraction with the highest interferon mRNA activity was identified
by the oocyte translation assay. The mRNA in this fraction was
converted to complementary double-stranded DNA using reverse transcriptase and DNA polymerase and inserted into bacterial plasmids.
These were propagated in *E. coli*. The plasmids (5,000 altogether)
were divided into pools of about 500. These were screened for interferon DNA segments in the following way: mRNA from induced leukocytes
was annealed to each DNA pool, the annealed mRNA was recovered and
assayed for interferon mRNA activity in oocytes. Out of the 12 pools
4 could anneal interferon mRNA. One of these four pools was sub-
divided into smaller pools, these were reassayed. After similar
further operations one hybrid plasmid was isolated which contained an
insert complementary to a segment of interferon mRNA. The insert
(320 base pairs) was shorter than the expected length of interferon
mRNA. Thus, it was used to search among the plasmids for those
containing longer inserts complementary to interferon mRNA. This
allowed the finding of a plasmid with a sufficiently long (910 base
pairs) insert complementary to interferon mRNA. Transformation of
E. coli with the plasmid containing this insert resulted in the
synthesis of a polypeptide with the biological activity of interferon
in human cells. Essentially similar approaches resulted in the
isolation of a human interferon β gene (11, 38).

Studies involving the hybridization to chromosomal DNA segments
of the above insert (complementary to interferon α mRNA) revealed
the existence of at least 12 chromosomal segments with interferon α
related sequences (13). Some of these are clearly linked. The DNA
sequence of one of the 12 segments was determined. It was found to
be identical to that of the cDNA insert complementary to interferon
α mRNA. This surprising finding indicates that *there are no introns
within the coding regions of this gene*. Further data obtained make
it probable that the entire transcription unit for this gene is
devoid of introns.

Studies with restriction enzymes revealed that the 12 interferon
α related sequences of the genome fall into not less than 8 distinct
classes. This indicates that there is a multiplicity of genes with
similar, but not identical sequences specifying different α interferons. Most, if not all, of these genes can direct the synthesis
of an interferon in *E. coli*. This makes it probably that many if not
all, of these genes might be devoid of introns. The apparent lack
of introns seems to distinguish interferon α genes from all other
gene coding for protein of higher eukaryotes described so far, with

the exception of sea urchin histone genes (13).

Isolation and Structure of Interferons

Two decades elapsed between the discovery of interferons (1) and the first isolation of pure interferon species (6-8, 39-45). This long delay was due to shortage of starting material for purification. The cells used for production, even if optimally induced synthesize only small amounts of interferons (3). One of the richest sources are mouse Ehrlich ascites tumor cells in monolayer culture infected with Newcastle disease virus. Even these synthesize only about 1 mg of interferons per liter of medium (45).

The isolation was facilitated by the stability of some interferons to acid pH and to sodium dodecylsulfate and also by an unusually sensitive biological assay allowing the detection of picogram quantities of interferons (3). Certain interferons were isolated using only conventional techniques of protein fractionation (6, 39, 41-45), whereas the purification of others involved affinity chromatography on antiinterferon immunoglobulins (8, 40, see also 46).

As determined by gel electrophoresis in the presence of sodium dodecylsulfate the apparent molecular weights of the different interferons vary between 18,000 and 40,000 daltons (3). Particular species, even though found to be pure by amino acid sequencing, may give rise to broad bands in gel electrophoresis in the presence of sodium dodecylsulfate (e.g. 6). Several interferons, including at least some of those giving rise to such broad bands are glycoproteins (3). The broadness of size distribution in the case of glycoproteins might perhaps be due to nonuniform substitution with carbohydrates (47, 48).

Interferons synthesized in animal cells in the presence of inhibitors of glycosylation (49, 50) (e.g., tunicamycin) and interferons synthesized in bacteria (and thus devoid of carbohydrate substitution) are biologically active (10, 51). This indicates that the sugar substitutions are not needed for (at least some of) the activities of (at least some) interferons.

Automated amino acid sequencing technique (52) requiring only microgram quantities of proteins was used to determine the N-terminal sequences of several human and mouse interferons (6-8). The prediction of the complete amino acid sequences of several human interferons became possible in consequence of the recent determinations of the entire nucleotide sequences of the corresponding interferon genes inserted into bacterial plasmids (11, 12, 53, 54).

A comparison of the predicted amino acid sequences of the human fibroblast interferon (β) (11, 54), two human leukocyte interferons (α 1 and α 2) (12, 54) and the N-terminal sequences of two mouse interferons (A and C) (6) revealed the following features (12):

1)   The chain length of human interferons α 1 and β is 166 amino acids, that of α 2 is 165 amino acids. This decrease in length is a result of a deletion of a codon specifying an amino acid residue in position 44 (ref 12)). The chain length of the two mouse interferons have not been determined;

2)   There is more homology between mouse interferon C and human interferon α than between mouse interferons A and C and more homology between mouse interferon A and a human interferon β than between mouse interferons A and C (ref 6). These findings allow the classification of mouse interferon A as β type and mouse interferon C as α type;

3)   The extent of sequence homology among the three human interferons in question is 10% among the first 28 to 30 amino acids residues 40% between residues 28 and 80, 15% between residues 80 and 122, and 40% between residues 122 and the C-terminal residue (12). The unequal extent of homology along the interferon polypeptides reveals the existence of conserved homology domains. It is likely that the conserved sequences may be essential for functions common to interferons. The sequence data also make it likely that the various interferon genes are derived from a common ancestral gene.

It remains to be established if there is a difference in function between the various interferons. A remarkable difference was already found between human interferons α 1 and α 2 in their relative virus protective activity in cells of different species: the ratio of activities on bovine and human cells was between 20 and 54 for interferon α 1 and 1 or less for interferon α 2 (ref 12). This difference in (an obviously nonphysiological) cross species reactivity could reflect a difference in specificity for cell receptors and raises the question as to whether different interferon species may be addressed to different target cells within the same host.

Nothing is known at this time about the structure of the γ type interferons (4, 55, 56). These interferons, previously designated as immune type or type 2 (26), are produced by T lymphocytes in response to mitogens or to antigens to which the cells were sensitized. They are serologically distinct from α and β interferons and more acid labile than these.

Interferon action. Establishment of the antiviral state

The interferon effect which has been examined the most extensively is the conversion of cells into an antiviral state. Much of our knowledge about the interaction of interferons with cells and the early phase of interferon action was gained from studies devoted to this effect.

Interferons are among the most potent biological agents. They impair virus replication in responsive cells at a concentration as low as 3×10^{-14} M (ref 3, 39). Interferons in general are species specific: in most cases they are more active on homologous than on heterologous cells (3).

Interferons are secreted from the producing cells and apparently have to interact with the cell surface to be active: cells in culture treated with the interferon inducer poly(I).poly(C) together with an antiserum to interferon are not converted into the antiviral state (57).

There is as yet no direct proof for the existence of interferon receptors on the cell surface. However, the following findings are consistent with the existence of such entities: a human-mouse somatic cell hybrid containing mouse chromosomes as well as a single human chromosome (number 21) is responsive to human interferons (58). Mice immunized by injection with this cell hybrid produce an antiserum which is presumably directed against cell surface antigens specified by the human chromosome 21. This antiserum impairs the action of human interferons on human cells (59). It is in line with the possibility that a human interferon receptor is specified by a gene (or genes) on chromosome 21 that cells trisomic for this chromosome require less interferon than normal (diploid) cells and cells monosomic for this chromosome require more interferon than normal cells to attain the same extent of virus growth inhibition (60). Cells of a mouse cell line (L1210R) not responsive to interferon were reported to bind much less (radiolabeled) interferon than cells of the parental interferon-responsive line (L1210S) (61).

The fact that particular gangliosides can bind interferons and that the responsiveness to interferons by ganglioside deficient cells is increased by treatment of the cells with gangliosides are consistent with the possibility that interferon receptors may share structural features with gangliosides (62-64).

It is not known if interferons have to be internalized to exert some or all of their effects.

An early effect of interferon treatment of mouse cells is a short lived increase in the cyclic GMP concentration. This takes place within five to ten minutes after exposing the cells to interferon (65). Neither the significance nor the mechanism of this increase has been established.

The conversion of cells into the antiviral state by interferons requires the presence of the nucleus, RNA synthesis and protein synthesis (66-69). This conversion is accompanied (and probably caused at least in part) by a change in the level of several enzymes (see later sections). The antiviral state is transient. It is dis-

sipated within a few days after the exposure of the cells to interferon (3, 70). Although in the presence of interferons the rate of
cell division may be decreased (71) after removal of interferons from
the medium the rate is restored to its normal value. Thus cells
exposed to interferons usually do not suffer lasting damage (72).

Cells in culture treated with interferons can be rescued from the
cytopathic effect of some viruses (e.g., Mengo) (73) but not of that
from other viruses (e.g., reovirus) (74). However, in both cases (as
long as the infecting virus is sensitive to interferons) virus replication is impaired by the treatment and consequently the infection is
localized.

THE INTERFERON-INDUCED TRANSLATIONAL REGULATION

The early attempts toward the understanding of the biochemistry
of interferon action aimed to discover the mechanism(s) by which the
replication of viruses is impaired by interferons. The question
usually asked was: which step in the replication of a particular
virus is blocked, or which is (or are) the viral component(s) or
virus-specific molecule(s) whose accumulation is diminished in cells
treated with interferon. Many of the results obtained pointed towards
viral RNA and protein accumulation as likely targets of interferon
action (75-77). This caused investigators to compare cell-free
extracts from interferon-treated cells with those from control cells
for their capacity to process, translate, and cleave viral and host
mRNAs. These comparisons resulted in the following findings:

1)   protein synthesis is more sensitive to inhibition by double-
stranded RNA in extracts from interferon-treated cells than in those
from control cells (78);

2)   the treatment of cells with interferons increases the level of
several enzymes (79, 85). Some (but not all) of these enzymes remain
latent unless activated by double-stranded RNA and ATP. The most
likely rationale for the involvement of double-stranded RNA in the
interferon response is that double-stranded RNA might be formed as
a side product or intermediate in the replication of various viruses
(86, 87) and thus might be a signal for the presence of replicating
viruses in cells.

The $(2'-5')(A)_n$ Synthetase-RNase L System

This system was discovered serendipitously in a comparison of
the rates of cleavage of reovirus mRNAs in extracts from interferon-
treated cells with those in extracts from control cells. The faster
cleavage in the extract from interferon-treated cells turned out to
depend on the presence of minute amounts of genomic *double-stranded*
RNA from reovirions contaminating the reovirus mRNA preparation.
After the removal of the double-stranded RNA the rates of cleavage

in the two types of extracts became identical; the readdition of
double-stranded RNA (from reovirions or of poly(I).poly(C)) accelerated the cleavage only in the extract from interferon-treated cells
and thereby restored the original difference (80). Subsequent studies
revealed that in addition to double-stranded RNA, ATP is also required
for accelerating RNA cleavage, and both of these compounds are needed
only for the activation of an endoribonuclease system (designated as
endoribonuclease$_{INT}$) not for its action (88). The endoribonuclease$_{INT}$
system consists of at least two enzymes. In the presence of double-
stranded RNA, the first of these catalyzes the production of a small
thermostable product from ATP. This in turn activates the second
enzyme, a latent endoribonuclease which was designated as RNase L
(ref 88).   The small thermostable product was identified as (2'-5')
(A)$_n$ (83, 89-92) a set of compounds discovered earlier as inhibitors
of protein synthesis that are formed from ATP in extracts from interferon-treated cells if the extracts are supplemented with double-
stranded RNA (81, 93, 94).

(2'-5')(A) Synthetase. The enzyme synthesizing this set of
compounds ((2'-5')(A)$_n$ synthetase) was found so far in human (90),
mouse (81), rabbit (95) and chicken cells (96). Treatment with
interferon boosts the level of the latent enzyme ten to a hundred
fold (97). Interferons α and β are equally effective as inducers
(98). (2'-5')(A)$_n$ synthetase was purified to homogeneity from interferon treated mouse Ehrlich ascites tumor (EAT) cells (99).

The molecular weight of the enzyme is 105,000 as determined by
gel electrophoresis in the presence of sodium dodecylsulfate and
85,000 as determined by sedimentation through a glycerol gradient.

In the presence of double-stranded RNA the purified enzyme can
convert the large majority (over 97%) of the ATP added to (2'-5')(A)$_n$
(n extends from 2 to about 15) and pyrophosphate, though it does not
cleave pyrophosphate (100). The stoichiometry of the reaction can be
formulated as:

$$(n + 1) \text{ ATP} \rightarrow (2'-5')\text{pppA(pA)}_n + n \text{ pyrophosphate}$$

(2'-5')pppA(pA)$_n$ stands for the same series of compounds that are
abbreviated otherwise as (2'-5')(A)$_n$. The different abbreviation is
used here to indicate that the 5' terminus of (2'-5')(A)$_n$ is a triphosphate.

Among the products of the enzyme usually di-, tri-, and tetra-
adenylates are the most abundant. The percent of longer oligomers
diminishes with increasing chain length (99, 101). The proportion
of different oligomers appears to reflect an increasing probability
of release from the enzyme as the oligomers are elongated. Dimers
added to the enzyme are substrates for adenylate addition but two
dimers do not become linked to each other (102).

$(2'-5')(A)_n$ synthetase was reported to be capable of adding adenylate residues to a variety of compounds with 3' terminal adenylate moities, e.g., NAD, ADPribose, and ApppppA (103). The affinity of the enzyme for ATP is low: the rate of the reaction increases when the ATP concentration is increased from 5 mM to 10 mM. The enzyme is maximally active when the concentration of double-stranded RNA (in w/v) is about half of that of the enzyme (100). Synthetic double-stranded RNA shorter than 30 base pairs does not activate the enzyme whereas double-stranded RNA longer than 65-80 base pairs causes maximal activation (104).

Phosphodiesterase Degrading $(2'-5')(A)_n$. An enzyme catalyzing the cleavage of $(2'-5')(A)_n$ into ATP and AMP was partially purified from L cells (85). The level of this enzyme increases four-five fold after the treatment of the cells with interferons. The enzyme can also cleave dinucleoside-monophosphates. Its activity is generally higher on (2'-5') than on (3'-5') phosphodiester bonds. The presence of a phosphate at the 5' terminus of an oligonucleotide does not affect the rate of enzymatic cleavage whereas the presence of a phosphate at the 3' end impairs the cleavage. These results suggest that the enzyme may attack oligonucleotides starting at their 3' end. The enzyme can also remove the CCA termini from tRNA (see a later section).

RNase L. This latent endoribonuclease which can be activated by $(2'-5')(A)_n$ was partially purified from EAT cells which had been treated with interferon (89, 105) and also from L cells (106). The interferon treatment raises the level of the enzyme only slightly (1.5 to 2 fold) (107). The enzyme purified from EAT cells is low in $(2'-5')(A)_n$ *independent* nuclease activity (105). Activation of the enzyme from EAT, L- or HeLa cells requires the presence of (2'-5') $(A)_n$ (the trimer of larger compounds), with a tri- or di-phosphate group at the 5' terminus (101). Interestingly, the RNase L from rabbit reticulocytes can be only activated by the *tetramer* of (2'-5') $(A)_n$ (108).

The activation of the enzyme by $(2'-5')(A)_n$ is *reversible*. Upon removal of $(2'-5')(A)_n$ from the activated enzyme (e.g., by gel filtration) the enzyme reverts to the latent state. Readdition of $(2'-5')(A)_n$ reactivates the enzyme (105).

The following results are consistent with the possibility that the activation of RNase L may involve the binding of $(2'-5')(A)_n$ to the enzyme:

1) a partially purified RNase L preparation retains $(2'-5')(A)_n$ on nitrocellulose filters (105);

2) the agent retaining $(2'-5')(A)_n$ copurifies with RNase L both during ion exchange chromatography and gel filtration (105);

3)   the treatment of the enzyme with N-ethylmaleimide abolishes both
its activatability by $(2'-5')(A)_n$ and its ability to retain $(2'-5')$
$(A)_n$ on nitrocellulose filters (109).

The molecular weight of the enzyme is 185,000 as estimated by
gel filtration. The activation does not seem to result in a large
size change of the enzyme. This and other data make it unlikely that
the activation should involve the binding or the release of a protein
(105).

The partially purified RNase L preparation if activated cleaves
a large variety of single-stranded RNAs (including poly(U)) but *not*
poly(C), poly(G), or poly(A) and *not* double-stranded RNA. The
products of cleavage are terminated by 3' phosphate and 5' OH residues
(G. Floyd-Smith, to be published).

The following observations are in line with a possible involvement of the endonuclease$_{INT}$ (i.e. the $(2'-5')$ synthetase-RNase L)
system in mediating at least some of the effects of interferon action:

1)   the replication of reovirus in mouse L929 cells is inhibited by
interferon; and reovirus mRNAs are degraded faster in interferon-
treated cells than in control cells (23);

2) $(2'-5')(A)_n$ occurs in intact cells and its level is higher in
virus infected, interferon-treated cells than in cells only treated
with interferon, or only infected with virus or in control cells
(108);

3)   the introduction into intact cells of $(2'-5')(A)_n$ or of $(2'-5')$
$(A)_n$ core (obtained by removing the 5' terminal triphosphate from
$(2'-5')(A)_n$ by treatment with alkaline phosphatase) results in a
transient impairment of protein synthesis and virus replication and
also in an accelerated cleavage of RNA (110).

In vitro the endonuclease$_{INT}$ system activated by double-stranded
RNA appears to cleave both viral and host RNAs without discrimination
(90,111). *In vivo*, however, at least in the case of some virus-cell
systems (e.g., reovirus-L cells), viral protein synthesis is inhibited
preferentially above host protein synthesis (74). An intriguing
series of experiments (112) revealed that, at least in principle,
the endonuclease$_{INT}$ system is capable of such discrimination. For
these experiments the poly(A) segment of a viral RNA (from vesicular
stomatitis virus) was annealed to poly(U) resulting in a mRNA in which
a single-stranded segment is covalently linked to a double-stranded
segment. In an extract from interferon-treated cells supplemented
with ATP this mRNA covalently bound to a double-stranded RNA segment
was degraded faster than an identical mRNA not bound to double-stranded RNA. This result is consistent with the possibility that in interferon-treated cells infected by an RNA virus the RNase L activity may

be localized in a region near to the partially double-stranded
replicative intermediate of the virus. This localization might be
a consequence of 1) the localized synthesis of $(2'-5')(A)_n$ by the
$(2'-5')(A)_n$ synthetase bound to the partially double-stranded viral
replicative intermediate and 2) the decrease in the concentration of
$(2'-5')(A)_n$ away from its site of formation that is caused by cleavage
by phosphodiesterase.

Protein Kinase

The need for double-stranded RNA and ATP for the activation of
the endonuclease$_{INT}$ system in extracts from interferon-treated cells,
the knowledge that several proteins are activated or inactivated by
phosphorylation (113) and other considerations resulted in tests on
the effect of double-stranded RNA on protein phosphorylation in
extracts from interferon-treated and control cells. The addition of
double-stranded RNA (from reovirus or of poly(I).poly(C)) to an
extract from interferon-treated EAT- or L-cells (but not or only to
a much lesser extent to an extract from control cells) was found to
cause the phosphorylation of at least two proteins: P_1 67,000 daltons
and P_2 37,000 daltons (82-84). Subsequently, a double-stranded RNA-
activatable protein kinase system was purified several thousand fold
from interferon-treated L- and EAT-cells (114, 115). The extent of
induction by interferon is three to ten fold. Interferons α, β, and
γ can all induce the enzyme system (98, 116).

The purified kinase preparation is essentially free of double-
stranded RNA independent kinase activity and is similar to a double-
stranded RNA dependent protein kinase structurally present in reticulocyte lysates (117). $(2'-5')(A)_n$ does not substitute for double-
stranded RNA in activating the enzyme and $(2'-5')(A)_n$ is not synthe-
tized by the activated enzyme (118).

The activation of a partially purified protein kinase by double-
stranded RNA and ATP was reported to be enhanced by an acidic protein
designated as factor A. The level of this factor in the cell is not
affected by treatment with interferon (115).

The P_2 protein that can be phosphorylated by the kinase is the
small subunit of the peptide chain initiation factor eIF-2 (ref 118,
119). The identity of P_1 has not been definitely established.
However, P_1 copurifies with the double-stranded RNA activatable
protein kinase system throughout a several thousand fold purification
and the most highly purified kinase preparation available at present
consists of P_1 as the major protein and two minor components (120).
It remains to be established whether or not P_1 is identical with the
protein kinase. The purified enzyme, if activated, can also phosphorylate some histones. Phosphoprotein phosphatase(s) which can remove
phosphate moieties from both phosphorylated eIF-2 and phosphorylated
P_1 is (are) present in cell extracts (115). This dephosphorylating

activity is impaired by double-stranded RNA (121).

The addition of the activated protein kinase preparation to a cell-free protein synthesizing system from EAT cells or reticulocyte lysates results in the inhibition of peptide chain initiation. The mechanism of this inhibition is under investigation. The inhibition can be overcome at least for a short while by the addition of further eIF-2 (ref 118) and might be a consequence at least in part of the phosphorylation of eIF-2 (ref 117).

The finding that double-stranded RNA added to intact, interferon-treated (but not to control) cells, results in the phosphorylation of P_1 protein indicates that the activation and action of the protein kinase is not an artifact of the *in vitro* system (122).

Double-Stranded RNA Does Not Have to be "Free" to Activate the Latent Enzymes

The genome of reovirus is *double-stranded* RNA. Apparently, however, it is always packaged in a protein coat (123). This fact prompted tests aiming to establish if double-stranded RNA has to be "free" to activate the endonuclease$_{INT}$ system and the protein kinase. This was not the case. Reovirions (which contain genomic double-stranded RNA), reovirus cores (i.e., reovirions partially uncoated by treatment with chymotrypsin), and reovirus subviral particles (formed from reovirions in infected cells by cleavage and removal of some outer coat proteins) purified by centrifugation through CsCl gradients were able to replace double-stranded RNA in promoting RNA degradation in an extract from interferon-treated cells.However, in enhancing RNA cleavage the reovirions are only 4% as efficient as free double-stranded RNA on a per mg of double-stranded RNA basis. Furthermore, reovirions treated with RNase III (an enzyme specific for cleaving double-stranded RNA) under conditions in which most free double-stranded RNA but little or no double-stranded RNA in virions is degraded, were no longer able to promote RNA cleavage. This suggested that double-stranded RNA on the surface of the reovirions may be responsible for the activation of the endonuclease$_{INT}$ system (23, 107). It remains to be established if the double-stranded RNA on the surface of the virions was of cellular origin or was an exposed segment of the viral genomic RNA.

In agreement with these findings, it was later shown that reovirus subviral particles (isolated by CsCl density gradient centrifugation from extracts of cells pre-treated with partially purified interferon), carried an endonuclease activity: If provided with the four ribonucleoside triphosphates, they produced *in vitro* mainly reovirus mRNAs *shorter* than full-size (124). The most likely explanation is that the interferon-induced $(2'-5')(A)_n$ synthetase binds to (and becomes activated by) the double-stranded RNA of such subviral particles. In a second stage, the RNase L would also bind

to the particles and would degrade the newly synthesized mRNAs. This
seems to be the case, since reovirus subviral particles similarly
isolated from extracts of untreated cells synthesized a set of full-
length reovirus mRNAs, and little (or no) endonuclease activity could
be detected.

Possible Rationale for the Multiple Roles of Double-Stranded RNA in Interferon Induction and Action

It remains to be understood why double-stranded RNA has become
an important modulator (i.e., both inducer and enzyme activator) of
the interferon system. It is conceivable that double-stranded RNA
serves as a signal revealing the presence of replicating viruses in
the cell. This might be the case for viruses in the replication of
which double-stranded RNA (even if partially coated) is an intermediate (e.g., viruses with a single-stranded RNA genome). It might
also be relevant that complementary virus RNA strands (which at least
in principle might anneal to a double-stranded form) were found in
cells infected with the DNA-containing vaccinia virus (86).

As described in previous sections, the treatment of cells with
interferon induces, among others, two latent enzyme systems. The
activation of these requires double-stranded RNA in some form. This
puzzling complexity might be rationalized in the following hypothesis
(108, 111): Localized virus infection results in the synthesis of
interferon in the infected cells and its secretion and spreading in
the body. In the cells exposed to interferon the two enzyme systems
are induced. Since they are latent they do not impair cell metabolism. When the cells previously exposed to interferon become infected
with a virus, however, this might result in the formation of some
double-stranded RNA and thereby, perhaps, in the activation of the
two enzyme systems. These in turn impair protein synthesis in the
virus-infected cells.

Whether or not the activator of these enzyme systems in intact
cells is double-stranded RNA remains to be established.

Two biochemical activities which are elicited by interferon
treatment of cells and are manifested in the cell extracts even
without double-stranded RNA are discussed in the following sections.

Impairment of Reovirus mRNA Methylation

The 5' termini of most eukaryotic mRNAs, including those of
reovirus mRNAs, are capped and methylated. The discovery that
methylated, capped reovirus mRNAs are usually more efficient as
messengers *in vitro* than unmethylated, capped reovirus mRNAs (125)
prompted a comparison of the process of mRNA methylation in extracts
from controls and interferon-treated cells. It was established that
the methylation of capped reovirus mRNAs by the cellular methylating

enzymes is impaired in extracts from interferon-treated cells (126).
The inhibitor is a macromolecule which is inactivated during the
incubation. The impairment is *not* due to cleavage of reovirus mRNA,
to depletion of the methyl donor (S-adenosylmethionine), or to the
irreversible inactivation of the methylating enzymes. It is possible
that the inhibitor acts by binding to the unmethylated, capped
reovirus mRNAs and making the cap region unavailable for the methylating enzymes. Apparently the impairment of methylation is not
affected by double-stranded RNA (127). It is conceivable that the
impairment of methylation is only one of the manifestations of the
"inhibitor(s) of methylation" and may not even be the most important
one. Thus it is imaginable that the same agent(s) might also interfere with the process of attachment of mRNA to ribosomes.

The finding of an impairment of reovirus mRNA methylation *in
vitro* prompted a comparison of the extent of cap methylation of viral
mRNAs in intact control and interferon-treated cells. It was
established that the proportion of reovirus mRNAs with cap 2 termini
(i.e., those carrying 3 methyl groups attached to the cap) is 35 to
50% lower in interferon-treated cells than in control cells (128).
The conversion of cap 1 into cap 2 termini (i.e., the attachment
to the cap of the third methyl group) is catalyzed *in vivo* by cellular
enzymes. The methylation of the cap of vaccinia virus mRNAs is also
impaired in cells treated with interferon (129). It remains to be
seen if the decrease in viral mRNA cap methylation in interferon-
treated cells *in vivo* is mediated by the same inhibitor(s) which
impair(s) methylation of reovirus mRNA caps *in vitro*.

Impairment of Exogenous mRNA Translation: The tRNA Effect

This impairment is most pronounced in an extract from interferon-
treated cells which had been "preincubated" to decrease protein
synthesis directed by endogenous mRNA and passed through Sephadex
G-25 to remove small molecules. It is mediated by an inhibitor
loosely attached to ribosomes. The impairment is manifested in the
cessation of translation of exogenous mRNA in an extract from interferon-treated cells at a time when the translation in an extract from
control cells still continues. It is apparently a consequence of a
block in peptide chain elongation but is not due to the cleavage of
the mRNA.

After the translation in the extract from interferon-treated
cells has ceased it can still be restored by adding tRNA (130-132).
It was established later that in extracts from interferon-treated
EAT cells which had been preincubated and passed through Sephadex
G-25 some species of aminoacyl tRNAs (e.g., leucyl tRNAs) were
inactivated faster than in extracts from control cells. This result
suggested that the requirement for added tRNA to restore translation
in extracts from interferon-treated cells was due to a faster inactivation of tRNA (79, 133). As noted in an earlier section of this

review, the treatment of cells with interferon increases the level of an enzyme which cleaves $(2'-5')(A)_n$; it was also reported that the same enzyme inactivates tRNAs by removing their 3' terminal CCA residues (85). Thus it is conceivable that the increase in the level of this enzyme may account for the faster tRNA inactivation in extracts from interferon-treated cells (85). It should be noted, however, that in order to restore the translation ability of pre-incubated lysate of interferon-treated cells, no significant difference has been detected so far in the amount of tRNA needed, whether it had been extracted from control or interferon-treated cells (79, 134). Thus the *in vivo* relevance of these findings remains to be seen.

MESSENGER RNAs AND PROTEINS INDUCED BY INTERFERON

The increase in the level of various cell enzymes after treatment with interferon together with the fact that the establishment of the interferon-induced antiviral state requires RNA (67) and protein synthesis (68, 69) prompted investigators to test whether interferons induced the synthesis of new mRNAs and new proteins. The treatment of various human, mouse, and chicken cells with homologous interferons was found to induce several mRNAs and the corresponding proteins (96, 135-140). The correspondence was established by translating the mRNAs into the proteins *in vitro* (136, 139). The induction of both, the mRNAs and proteins, could be inhibited by actionomycin D indicating *de novo* synthesis of RNA. Short term labeling at different times after treatment with interferons indicated that the mRNAs and proteins were induced within a few hours after the beginning of the treatment with interferon. The syntheses of the different proteinswere maximal at different times and thereafter tended to decline (140). These results made it likely that the interferon-induced mRNAs had short half lives. The functions and identities of the interferon-induced proteins, and their relationship to the enzymes whose level is increased by interferon treatment, remain to be established. It is of interest that some of the proteins induced can be retained on poly(I).poly(C) agarose columns indicating their high affinity to double-stranded RNA (138).

CONCLUSIONS

As described in the earlier sections most of our knowledge about the enzymology of interferon action is based on the comparison of enzyme activities in extracts from interferon-treated cells and extracts from control cells. This approach led to the following major conclusions: Interferon treatment of cells enhances the level of two distinct enzymatic pathways which remain latent unless activated by double-stranded RNA.

One of the pathways, if activated, results in the accelerated cleavage of mRNA. This pathway involves two latent enzymes: $(2'-5')$

(A)$_n$ synthetase and RNase L. If activated by double-stranded RNA
$(2'-5')(A)_n$ synthetase produces $(2'-5')(A)_n$. $(2'-5')(A)_n$ in turn
activates RNase L. The activation of RNase L ceases upon removal
of $(2'-5')(A)_n$. The level of phosphodiesterase that can degrade
$(2'-5')(A)_n$ is also raised by interferon treatment.

The second pathway if activated leads to the impairment of
peptide chain initiation. This is apparently in consequence of the
phosphorylation of peptidyl chain initiation factor eIF-2. It remains
to be established if the protein kinase system catalyzing this pathway
consists of one or several enzymes. The significance of the fact
that the same enzyme system can phosphorylate some histones is unclear
and so is the relationship of the kinase to P_1 protein. This protein
is present even in the most purified kinase preparations and becomes
phosphorylated in the presence of double-stranded RNA. The impairment of peptide chain initiation resulting from the action of the
kinase system can be reversed by a protein phosphatase which can
remove phosphate from phosphorylated eIF-2.

The treatment of cells with interferons also boosts pathways
whose functioning does not appear to depend on double-stranded RNA.
Thus the methylation of the cap structure of mRNA is impaired in an
extract from interferon-treated cells. The lability of the agents
responsible for this impairment makes their characterization difficult.

Finally, in extracts from interferon-treated cells, especially
if these have undergone gel filtration, the inactivation of some
tRNA species is accelerated. This happens presumably in consequence
of the increase in the level of the phosphodiesterase which in
addition to cleaving $(2'-5')(A)_n$ can also remove the CCA termini from
tRNA.

Some of the data which are in accord with a role of the two
double-stranded RNA dependent pathways and of the inhibition of mRNA
cap methylation in mediating some of the antiviral effects of interferons were outlined in the earlier sections. A definitive test for
the function of each of the above pathways would require mutants
defective (at least conditionally) in the pathway.

The following features of interferon inducible enzymes and of
their actions need further comments: agents other than interferon
can also affect the level of the enzymes which can be induced by
interferons. Thus withdrawal of estrogen from chick oviducts results
in a several fold increase in $(2'-5')(A)_n$ synthetase (97). This
observation together with the fact that the level of RNase L is high
even in cells not treated with interferon makes it probable that the
$(2'-5')(A)_n$ synthetase-RNase L system has functions besides that of
mediating interferon action.

The levels of both protein kinase and $(2'-5')(A)_n$ synthetase are remarkably high in reticulocytes from rabbits not treated with interferon (95, 117). It remains to be established whether or not this is a consequence of an exposure of these cells to interferon in the bone marrow.

At present the only characterized activity of $(2'-5')(A)_n$ in cell-free systems is the activation of RNase L. However, it is not known whether or not RNase L activation can account for the impairment of DNA synthesis in mitogen stimulated lymphocytes by $(2'-5')(A)_n$ and by $(2'-5')(A)_n$ from which the 5' terminal triphosphate moiety had been removed (141, see also 110).

As note earlier the increase in the activity of the various enzymes upon interferon treatment is most likely a consequence of the induction of the synthesis of mRNAs and proteins. However the mechanism of this induction has not been elucidated.

Footnotes: Nomenclature of interferons and enzymes

The interferons of type α, β, and γ used to be designated earlier as leukocyte or Le type, fibroblast or F type, and immune or type II, respectively (26). $(2'-5')(A)_n$ is also designated as oligoisoadenylate and is sometimes abbreviated as 2-5A. The enzyme synthesizing $(2'-5')(A)_n$ is designated by different authors as $(2'-5;)(A)_n$ synthetase or $(2'-5')$ oligo(A)polymerase or oligo-isoadenylate synthetase E. The endoribonuclease which can be activated by $(2'-5')(A)_n$ is designated as RNase L or RNase F. The protein kinase which can be activated by double-stranded RNA is also designated as protein kinase PK-i or eIF-2 kinase. The phosphodiesterase degrading $(2'-5')(A)_n$ is also designated as phosphodiesterase 2'-PDi.

ACKNOWLEDGEMENTS

Studies in the author's laboratory described in this review have been supported by USPHS NIH research grants AI 12320 and CA 16038. The help of R. Broeze in preparing this manuscript is appreciated.

REFERENCES

1. Isaacs, A. and Lindenmann, J. (1957), Proc. R. Soc. London Ser. B. 147, 258-267.

2. S. Baron and F. Dinazani eds, *The Interferon System* (1977), Tex. Rep. Biol. Med. 35, 1-573.

3. W.E. Stewart II, ed. *The Interferon System* (1979), Springer Verlag, New York.

4. De Maeyer, E. and De Maeyer-Guignard, J. (1979), In *Comprehensive Virology* (H. Fraenkel-Conrat and R.R. Wagner, eds.) 15, 205-284.

5. J. Vilcek, ed. (1980), *Regulatory Functions of Interferons*, New York Academy of Sciences, New York.

6. Taira, H., Broeze, R.J., Jayaram, B.M., Lengyel, P., Hunkapiller, M. and Hood L.E. (1980), Science 207, 528-530.

7. Knight, E., Hunkapiller, M.W., Korant, B.D., Hardy, R.W.F. and Hood, L.E. (1980), Science 207, 525-526.

8. Zoon, K.C., Smith, M.E., Bridgen, P.J., Aufinsen, C.B., Hunkapiller, M.W. and Hood, L.E. (1980), Science 207, 527-528.

9. Okamura, H., Berthol, W., Hood, L., Hunkapiller, M., Inoue, M., Smith-Johansen, H. and Tan, Y.H. (1980), Biochemistry 19, 3831-3835.

10. Taniguchi, T., Mantei, N., Schwarzstein, M., Nagata, S., Muramatsu, M. and Weissmann, C. (1980), Nature, London, 285, 547-549.

11. Drynck, R., Content, J., Declerq, E., Volkaert, G., Tavernier, J., Devos, R. and Fiers, W. (1980), Nature, London, 285, 542-547.

12. Streuli, M., Nagata, S. and Weissmann, C. (1980), Science 209, 1343-1347.

13. Nagata, S., Mantei, N. and Weissmann, C. (1980), Nature, London, 287, 401-408.

14. Gresser, I. (1977), Cell. Immunol. 34, 406-415.

15. Gresser, I. and Tovey, M.G. (1980), Biochim. Biophys. Acta 516, 231-247.

16. Gresser, I., Tovey, M.G., Bandu, M.T., Maury, C. and Brouty-Boye, D. (1976), J. Exp. Med. 144, 1305-1315.

17. Gresser, I., Tovey, M.G., Maury, C. and Bandu, M.T. (1976), J. Exp. Med. 144, 1316-1323.

18. Gresser, I., Maury, C., Bandu, M.T., Tovey, M. and Maunoury, M.T. (1978), Int. J. Cancer 21, 72-77.

19. Cantell, K. (1979), In *Interferon I* (I. Gresser, ed.), Academic Press, New York.

20. Nagata, S., Taira, H., Hall, A., Johnsrud, L., Streuli, M., Ecsodi, J., Boll, W., Cantell, K., and Weissmann, C. (1980), Nature, London, 284, 316-320.

21. I. Gresser, ed. (1978) *Interferon I*, pp. 1-163, Academic Press New York.

22. Torrence, P.F. and DeClercq, E. (1977), Pharmac. Ther, A. 2, 1-88.

23. Lengyel, P., Desrosiers, R., Broeze, R., Slattery, E., Taira, H., Dougherty, J., Samanta, H., Pichon, J., Farrell, P., Ratner, L. and Sen, G. (1980), In *Microbiology 1980* (D. Schlessinger, ed.), pp 219-226, Am. Soc. Microbiol. Washington D.C..

24. Sehgal, P.B., Pfeffer, L.M. and Tamm, I. (1980), In *Chemotherapy of Viral Infections* (P.E. Came and L.A. Caliguiri, eds.), in press.

25. Jarvis, A.P. and Colby, C. (1978), Cell 14, 355-363.

26. Stewart, W.E. II (1980), Nature, London, 285, 111.

27. Havell, E.A., Hayes, T.G., and Vilcek, J. (1978), Virology 89, 330-334.
28. Slate, D. and Ruddle, F.H. (1979), Pharmac. Ther, 4, 221-230.
29. De Maeyer-Guignard, J., De Maeyer, E. and Montagnier, L. (1972), Proc. Natl. Acad. Sci. U.S.A. 69, 1203-1207.
30. Sehgal, P.B., Lyles, D.S. and Tamm, I. (1978), Virology 89, 186-198.
31. Mogensen, K.E. and Cantell, K. (1977), Pharmac. Ther I C, 369-381.
32. Cantell, K. (1978), Endeavour, New Series 2, 27-30.
33. Havell, E.A. and Vilcek, J. (1972), Antimicrob. Ag. Chemother. 2, 476-484.
34. Horoszewicz, J.S., Leong, S.S., Ito, M., DiBerardino, L. and Carter, W.A. (1978), Infect. Immunity 19, 720-726.
35. Strander, H., Mogensen, K.E. and Cantell, K. (1974), J. Clin. Microbiol 1, 116-117.
36. Zoon, K.C., Buckler, C.E., Bridgen, P.J., Gurari-Rotman, D. (1978), J. Clin. Microbiol. 7, 44-51.
37. Berthold, W., Tan, C., Tan, Y.H. (1978), J. Biol. Chem. 253, 5206-5212.
38. Taniguchi, T., Sakai, M., Fujii-Kuriyama, Y., Muramatsu, M., Kobayashi, S. and Sudo, T. (1979), Proc. Japan Acad. Ser. B. 55, 464-469.
39. Kawakita, M., Cabrer, B., Taira, H., Rebello, M., Slattery, E., Weideli, H. and Lengyel, P. (1978), J. Biol. Chem. 253, 598-602.
40. DeMaeyer-Guignard, J., Tovey, M.C., Gresser, I. and DeMaeyer, E. (1978), Nature, London, 271, 622-625.
41. Knight, E. (1976), Proc. Natl. Acad. Sci. U.S.A. 73, 520-523.
42. Iwakura, Y., Yonehara, S. and Kawade, Y. (1978), J. Biol. Chem. 253, 5074-5079.
43. Cabrer, B., Taira, H., Broeze, R.J., Kempe, T.D., Williams, K., Slattery, E., Konigsberg, W.H. and Lengyel, P. (1979), J. Biol. Chem. 254, 3681-3684.
44. Rubinstein, M., Rubinstein, S., Familletti, P.C., Miller, R.S., Waldman, A.A. and Pestka, S. (1979), Proc. Natl. Acad. Sci. U.S.A. 76, 640-644.
45. Taira, H., Broeze, R.J., Slattery, E. and Lengyel, P. (1980), J. Gen. Virol. 49, 231-234.
46. Secher, D.S. and Burke, D.C. (1980), Nature, London, 285, 446-450.
47. Bose, S., Gurari-Rotman, D., Ruegg, U.J., Corley, L. and Anfinsen, C.B. (1976), J. Biol. Chem. 251, 1659-1662.
48. Stewart II, W.E., Chudzio, T., Lin, L.S. and Wiranowska-Stewart, M. (1978), Proc. Natl; Acad. Sci. U.S.A. 75, 4814-4818.
49. Havell, E., Yamazaki, S. and Vilcek, J. (1977), J. Biol. Chem. 252, 4425-4427.
50. Fujisawa, J., Iwakura, Y., and Kawade, Y. (1978), J. Biol. Chem. 253, 8677-8679.
51. Masucci, M.G., Szigeti, R., Klein, E., Klein, G., Gruest, J., Montagnier, L., Taira, H., Hall, A., Nagata, S., Weissmann, C. (1980), Science 209, 1431-1435.

52. Hunkapiller, M.W. and Hood, L.E. (1980), Science 207, 523-525.
53. Mantei, N., Schwarzstein, M., Streuli, M., Panem, S., Nagata, S. and Weissmann, C. (1980), Gene 10, 1-10.
54. Taniguchi, T., Ohno, S., Fujii-Kuriyama, Y. and Muramatsu, M. (1980), Gene 10, 11-15.
55. Whelock, E.F. (1965), Science 149, 310-311.
56. Green, J.A., Cooperband, S.R. and Kibrick, S. (1969), Science 164, 1415-1417.
57. Vengris, V.E., Stollar, B.D. and Pitha, P.M. (1975), Virology 65, 410-417.
58. Tan, Y.H., Tischfield, J. and Ruddle, F.H. (1973), J; Exp. Ped. 137, 317-330.
59. Revel, M., Bash, D. and Ruddle, F.H. (1976), Nature, London, 260, 139-141.
60. Tan, Y.H., Schneider, E.L., Tischfield, J., Epstein, C.J. and Ruddle, F.H. (1974), Science 186, 61-63.
61. Aguet, M. (1980), Nature, London, 284, 459-461.
62. Besançon, F. and Ankel, H. (1974), Nature, London, 252, 478-480.
63. Bensançon, F., Ankel, H., and Basu, S. (1976), Nature, London, 259, 576-578.
64. Vengris, V.E., Reynolds, F.H.Jr., Hollenberg, M.D. and Pitha, P.M. (1976), Virology 72, 486-496.
65. Tovey, M.G., Rochette-Egly, C. and Castagna, M. (1979), Proc. Natl. Aâad. Sci. U.S.A. 76, 3890-3893.
66. Radke, K.L., Colby, C., Kates, J.R., Krider, H.M. and Prescott, D.M. (1974), J. Virol. 13, 623-630.
67. Taylor, J. (1964), Biochem. Biophys. Res. Commun. 14, 447-451.
68. Lockart, R.Z.Jr. (1964), Biochem. Biophys. Res. Commun. 15, 513-518.
69. Friedman, R.M. and Sonnabend, J.A. (1965), J. Immunol. 95, 696-703.
70. Baron, S., Buckler, C.E. and Dianzani, F. (1968), In: *Interferon* (G.E.W. Wolstenholme and M. O'Connor, eds.) p. 186, J. and A. Churchill, London.
71. Paucker, K., Cantell, K. and Henle, W. (1962), Virology 17, 324-334.
72. Ratner, L., Nordlund, J.J. and Lengyel, P. (1980), Proc. Soc. Exp. Biol. Med. 163, 267-272.
73. Falcoff, R. and Sanceau, J. (1979), Virology 98, 433-438.
74. Gupta, S.L., Graziadei, W.D., Weideli, H., Sopori, M.L. and Lengyel, P. (1974), Virology 57, 49-63.
75. Friedman, R.M. (1970), J. Gen. Physiol. 56, 149s-171s.
76. Sonnabend, J.A. and Friedman, R.M. (1973), In: *Interferons and Interferon Inducers* (N.B. Finter, ed.) pp. 201-239.
77. Metz, D.H. (1975), Cell 6, 429-439.
78. Kerr, I.M., Brown, R. and Ball, L.A. (1974), Nature, London, 250, 57-59.
79. Sen, G.C., Gupta, S.L., Brown, G.E., Lebleu, B., Rebello, M.A. and Lengyel, P. (1976), J. ViroL. 17, 191-203.

80. Brown, G.E., Lebleu, B., Kawakita, M., Shaila, S., Seng, G.C. and Lengyel, P. (1976), Biochem. Biophys. Res. Commun. 69, 114-122.
81. Hovanessian, A.G., Brown, R.E. and Kerr, I.M. (1977), Nature, London, 268, 537-539.
82. Lebleu, B., Sen, G.C., Shaila, S., Cabrer, B. and Lengyel, P. (1976), Proc. Natld. Acad. Sci. U.S.A. 73, 3107-3111.
83. Zilberstein, A., Kimchi, A., Schmidt, A. and Revel, M. (1978). Proc. Natl. Acad. Sci. U.S.A. 75, 4734-4738.
84. Roberts, W.K., Hovanessian, A., Brown, R.E., Clemens, M.J. and Kerr, I.M. (1976), Nature, London, 264, 477-480.
85. Schmidt, A., Chernajovsky, Y., Shulman, L., Federman, P., Berissi, H. and Revel, M. (1979), Proc. Natl. Acad. Sci. U.S.A. 76, 4788-4792.
86. Colby, C. and Duesberg, P.H. (1969), Nature, London, 222, 940-944.
87. Boone, R.F., Parr, R.P. and Moss, B. (1979), J. Virol. 30, 365-374.
88. Sen, G.C., Lebleu, B., Brown, G.E., Kawakita, M., Slattery, E. and Lengyel, P. (1976), Nature, London, 264, 370-373.
89. Ratner, L., Wiegand, R., Farrell, P., Sen, G.C., Cabrer, B. and Lengyel, P. (1978), Biochem. Biophys. Res. Commun. 81, 947-954.
90. Baglioni, C., Minks, M.A. and Maroney, P.A. (1978), Nature, London, 273, 684-687.
91. Clemens, M.J. and Williams, B.R.G. (1978), Cell 13, 565-572.
92. Epstein, D.A. and Samuel, C.E. (1978), Virology 89, 240-251.
93. Kerr, I.M., Brown, R.E. and Hovanessian, A.G. (1977), Nature, London, 268, 540-542.
94. Kerr, I.M. and Brown, R.E. (1978), Proc. Natl. Acad. Sci. U.S.A. 75, 256-260.
95. Hovanessian, A.G. and Kerr, I.M. (1978), Eur. J. Biochem. 81, 149-159.
96. Ball, L.A. (1978), Proc. Natl. Acad. Sci. U.S.A. 75, 1167-1171.
97. Stark, G.R., Dower, W.J., Schimke, R.T., Brown, R.E. and Kerr, I.M. (1979), Nature, London, 278, 471-473.
98. Broeze, R.J., Dougherty, J.P. and Lengyel, P. (1980), Fed. Proc. 39, 2205.
99. Dougherty, J.P., Samanta, H., Farrell, P.J. and Lengyel, P. (1980), J. Biol. Chem. 255, 3813-3816.
100. Samanta, H., Dougherty, J.P. and Lengyel, P. (1981), J. Biol. Chem. in press.
101. Martin, E.M., Birdsall, N.J.M., Brown, R.E. and Kerr, I.M. (1979) Eur. J. Biochem. 95, 295-307.
102. Minks, M.A., Benvin, S. and Baglioni, C. (1980), J. Biol. Chem. 255, 5031-5035.
103. Ball, L.A. and White, C.N. (1979), In: Regulation of Macromolecular Synthesis by Low Molecular Weight Mediators (G. Koch and D. Richter, eds.) pp. 303-317, Academic Press, New York and London.

104. Minks, M.A., West, D.K., Benvin, S. and Baglioni, C. (1979), J. Biol. Chem. $\underline{254}$, 10180-10183.
105. Slattery, E., Ghosh, N., Samanta, H. and Lengyel, P. (1979), Proc. Natl. Acad. Sci. U.S.A. $\underline{76}$, 4778-4782.
106. Revel, M. (1979), In: *Interferon I* (I. Gresser, ed.) pp. 101-163, Academic Press, New York.
107. Ratner, L. (1979), Ph. D. dissertation, Yale Univ.
108. Williams, B.R.G., Golgher, R.R., Brown, R.E., Gilbert, C.S. and Kerr, I.M. (1979), Nature, London, $\underline{282}$, 582-586.
109. Slattery, E., Ghosh, N., Samanta, H. and Lengyel, P. (1979), In *Interferon: Properties and Clinical Uses* (A. Khan and G.L. Dorn, eds.), pp. 521-528, Wadley Institutes of Molecular Medicine.
110. Hovanessian, A.G. and Wood, J.N. (1980), Virology $\underline{101}$, 81-90.
111. Ratner, L., Sen, G.C., Brown, G.E., Lebleu, B., Kawakita, M., Cabrer, B., Slattery, E. and Lengyle, P. (1977), Eur. J. Biochem. $\underline{79}$, 565-577.
112. Nilsen, T.W. and Baglioni, C. (1979), Proc. Natl. Acad. Sci. U.S.A. $\underline{76}$, 2600-2604.
113. Rubin, C.S. and Rosen, O.M. (1975), Ann. Rev. Biochem. $\underline{44}$, 831-887.
114. Sen, G.C., Taira, H. and P. Lengyel, (1978), J. Biol. Chem. $\underline{253}$, 5915-5921.
115. Kimchi, A., Zilberstein, A., Schmidt, A., Shulman, L. and Revel, M. (1979), J. Biol. Chem. $\underline{254}$, 9846-9853.
116. Hovanessian, A.G., Meurs, E., Aujean, O., Vaquero, C., Stefanos, S. and Falcoff, E. (1980), Virology $\underline{104}$, 195-204.
117. Farrell, P.J., Balkow, K., Hunt, T. and Jackson, R.J. (1977), Cell 11, 187-200.
118. Farrell, P.J., Sen, G.C., Dubois, M.-F., Ratner, L., Slattery, E. and Lengyel, P. (1978), Proc. Natl. Acad. Sci. U.S.A. $\underline{75}$, 5893-5897.
119. Samuel, C.E. (1979), Proc. Natl. Acad. Sci. U.S.A. $\underline{76}$, 600-604.
120. Lengyel, P., Samanta, H., Pichon, J., Dougherty, J., Slattery, E. and Farrel, P. (1980), In *Regulatory Functions of Interferons*, (J. Vilcek, I. Gresse and T.C. Merigan eds.) pp. 441-447, New York Acad. Sci., New York.
121. Epstein, D.E., Torrence, P.F. and Friedman, R.M. (1980), Proc. Natl. Acad. Sci. U.S.A. $\underline{77}$, 107-111.
122. Gupta, S.L. (1979), J. Virol. $\underline{29}$, 301-311.
123. Joklik, W.K. (1974), In *Comprehensive Virology* (H. Fraenkel-Conrat and R. Wagner, eds.), $\underline{1}$, pp. 231-234, Plenum Press, New York.
124. Glaster, R.J. and Lengyel, P. (1976), Nuc. Acids Res. $\underline{3}$, 581-598.
125. Shatkin, A.J. and Both, G.W. (1976), Cell $\underline{7}$, 305-313.
126. Sen, G.C., Lebleu, B., Brown, G.E., Rebello, M.A., Furuichi, Y. Morgan, M., Shatkin, A.J. and Lengyel, P. (1975), Biochem. Biophys. Res. Commun. $\underline{65}$, 427-434.
127. Sen, G.C., Shaila, S., Lebleu, B., Brown, C.E., Desrosiers, R.C. and Lengyel, P. (1977), J. Virol. $\underline{21}$, 68-83.

128. Desrosiers, R.C. and Lengyel, P. (1979), Biochim. Biophys. Acta 562, 471-480.
129. Kroath, H., Gross, H.J., Jungwirth, C. and Bodo, G. (1978), Nuc. Acids Res. 5, 2441-2454.
130. Gupta, S.L., Sopori, M.L. and Lengyel, P. (1974), Biochem. Biophys. Res. Commun. 57, 763-770.
131. Content, J., Lebleu, B., Zilbertstein, A., Berissi, H. and Revel, M. (1974), FEBS Lett. 41, 125-130.
132. Zilberstein, A., Dudock, B., Berissi, H. and Revel, M. (1976), J. Mol. Biol. 108, 43-54.
133. Falcoff, R., Falcoff, E., Sanceau, J. and Lewis, J.A. (1978), Virology 86, 507-515.
134. Colby, C., Penhoet, E.E. and Samuel, C.E. (1976), Virology 74, 262-264.
135. Knight., E. Jr., and Korant, B.D. (1979), Proc. Natl. Acad. Sci. U.S.A. 76, 1824-1827.
136. Farrell, P.J., Broeze, R.J. and Lengyel, P. (1979), Nature, London, 279, 523-525.
137. DeLey, M., Billiau, A. and DeSomer, P. (1979), Biochem. Biophys. Res. Commun. 89, 701-705.
138. Gupta, S.L., Rubin, B.Y. and Holmes, S.L. (1979), Proc. Natl. Acad. Sci. U.S.A. 76, 4817-4821.
139. Farrell, P.J., Broeze, R.J. and Lengyel, P. (1980), In *Regulatory Functions of Interferons* (J. Vilcek, I. Gresser and T.C. Merigan, eds.) pp. 615-616, New York Acad. Sci. New York.
140. Rubin, B.Y. and Gupta, S.L., (1980), J. Virology 34, 446-454.
141. Kimchi, A., Shure, H. and Revel, M. (1979), Nature, London, 282, 849-851.

PARTICIPANTS

ADLER, C.   Dept.Microbiology.-S.U.N.Y. at Stony Brook. STONY BROOK,
NY, 11794, USA

ALMIS, G.   Dept.Biophysics.-Faculty of Medicine. University of
Istanbul.- CAPA-ISTANBUL, Turkey

ALONI, Y.   Dept.Genetics.-The Weizmann Institute of Sciences
REHOVOT,   Israel

AZOU, Y.   Centre Biochimie & Biol.Moléculaire CNRS.-31, Chemin
Joseph Aiguier 13274-MARSEILLE CEDEX 2, France

BALDUCCI, L.   Centro Virus Respiratori CNR.-Clinica Pediatrica Univ.
Roma.-V.le R.Elena 324, 00161-ROMA, Italy

BOSSART, W.   The Institute of Microbiology.-University of Basel.
Peterplatz 10, 4000-BASEL, Switzerland

BROT, N.   Roche Institute of Molecular Biology.-NUTLEY,NJ,07110,
USA

BUHL, W-J. Institut für Biologie III.-Albert-Ludwig Universität.
Schlänzerst.1.-D-7800 FREIBURG, Germany

CARRI', M.T.   Centro degli Acidi Nucleici CNR.-Università di Roma.
Città Universitaria.-00100-ROMA, Italy

CAZILLIS, M.   Insitut Curie.-Section Biologie.-Centre Universitaire
Bât.110.-91405-ORSAY CEDEX, France

COLBURN, Th. Institut Biologie Physico-Chimique.-13,rue Pierre &
          Marie Curie.-75005-PARIS, France

CONTENT, J. Institut Pasteur du Brabant.-28,rue du Remorqueur.
          B-1040-BRUSSELS, Beigium

COWGILL, C.A. Dpt.Biochemistry & Biophysics.-Oregon State University.
          CORVALLIS,OR,97331, USA

DE FERRA, F. Dept.Biology.-S.U.N.Y. at Albany.-1400 Washington Ave.
          ALBANY,NY, 12222, USA

DEGENER, A.M. The Institute for Virology.-University of Rome.-V.le
          Porta Tiburtina 28.-00185-ROMA, Italy

DELPINO, A. Laboratorio Biofisica.-Ist.Regina Elena.-V.le Regina
          Elena 291.-00161-ROMA, Italy

DOLLAK, H. Institut für Biologie III Universität Tübingen.-Auf der
          Morgenstelle 28.-D-7400 TUEBINGEN, Germany

DWORKIN, M.B. Dept.Biological Chemistry University of California.
          DAVIS,CA, 95616, USA

EFRAT, S. Dept.Molecular Biology.-Hadassah Medical School.-P.O.Box
          1172.-JERUSALEM, Israel

ERNST, H.M. Institut für Biologie III Albert-Ludwig Universität.
          Schlanzerst.1.-D-7800 FREIBURG, Germany

ETTINGER,L. Dept.Biological Chemistry.-Institut of Life Sciences.
          Hebrew University of Jerusalem.-JERUSALEM, Israel

FERBUS, D. Institut Biologie Physico-Chimique.-13,rue Pierre &
          Marie Curie.-F-75005-PARIS, France

FLANDROY, L. IRIBHN.-School of Medicine.-Free Univ.Brussels.-B-1000
          BRUSSELS, Belgium

GEORGE, S. Dept.Microbiology & Immunology.-Washington University.
          School Medicine.-ST-LOUIS,MO, 63110, USA

GEVERT, D.R. Dept.Biochemistry.-St George Hosp.Med.School.-Cranmer
          Terrace.-LONDON SW17 ORE, England

GOEMAN, W. Institut Physiologisches Chemie Universität Hamburg.
          Grindelallee 117.-D-2000 HAMBURG 20, Germany

DE GROOT, N. Dept.Biological Sciences.-Institut Life Sciences.
          Hebrew University Jerusalem.-JERUSALEM, Israel

GUPTA, N.K. Dept.Chemistry.-University of Nebraska.-LINCOLN,NE,
         68588, USA

HERSHEY, J.W.B. Dept.Biological Chemistry.-Univ.California
         DAVIS,CA, 95616, USA

HUEZ, G. Laboratory Biological Chemistry.-Univ.Brussels.-67, rue
         des Chevaux.-B-1640 RHODE-ST-GENESE, Belgium

IRVIN, J.D. Dept.Chemistry.-Southwest Texas State University.
         SAN MARCOS,TX, 78666, USA

ISSINGER, O-G. Biologisches Institut Universität Stuttgart.
         Ulmerstr.227.-D-7000 STUTTGART 60, Germany

IVELL, R. Institut für Physiologisches Chemie.-Univ. Hamburg.
         Martinistr.52.-D-2000 HAMBURG 20, Germany

JACKSON, J.R. Dept.Biochemistry.-Univ.Cambridge.-CAMBRIDGE CB2
         1QW, England

JAUREGUIBERRY, G. Insitut Pasteur.-28, rue du Dr. Roux.-F-75724
         PARIS CEDEX 15, France

JEWELL, J.E. Biophysics Laboratory.-Univ.Wisconsin.-1525 Linden
         Drive.-MADISON,WI,53706, USA

KAEMPFER, R. Laboratory Molecular Biology.-Hadassah Medical School.
         P.O.Box 1172.-JERUSALEM, Israel

KALKINEN, N. Dept.Biochemistry.-Univ.Helsenki.-Unionikatu 35.
         SF-00170-HELSENKI 17, Finland

KOCH, G. Physiologisch-Chemisch Institut.-Universität Hamburg.
         Grindelallee 117.-D-2000-HAMBURG 13, Germany

KOZAK, M. Dept.Biological Sciences.-Univ. Pittsburgh-PITTSBURGH,PA,
         15260, USA

KURU, A. Dept.General Botany.-Univ. Istambul.-SULEYMANIYE-ISTANBUL,
         Turkey.

LENGYEL, P. Dept Mol. Biophysics & Biochemistry-Yale Univ.,
         260 Whitney Ave., P.O. Box 6666-NEW HAVEN,CT, 06511, USA

MÄENPÄÄ, P. Dept.Biochemistry.-Univ. Kuopio.-P.O.Box 138.-SF-70101
         KUOPIO 10, Finland

MALY, P. Max-Planck-Institut Molekulargenetik.-Ihnestr.63/73
         D-1 BERLIN DAHLEM, Germany

MARBAIX, G.   Dept.Biologie Moléculaire Université Bruxelles.-67, rue
          des Chevaux.-B-1640-RHODE-ST-GENESE, Belgium

MARTI, J.   Lab.Biochimie Protéines USTL.-Place Eugène Bataillon..
          F-34060 MONTPELLIER CEDEX, France

MARTIN PEREZ, J.   Friedrich-Miescher Institut.-Postfach 273.-4002-
          BASEL, Switzerland

MIDULLA, M.   Centro Virus Respiratori CNR.-Via R.Pereira 21.
          00136-ROMA, Italy

MONK, R.   Institute for Cancer Research.-7701 Burholm Ave.-Fox Chase
          PHILADELPHIA, PA, 19111, USA

MORADAS FERREIRA, P.   Instituto Ciencias Biomédicas.-Dept.Biochemistry.
          Largo da Scola Medica 2.-P-4000 PORTO, Potugal

MORROW, J.   Dept.Biochemistry Texas Tech.Univ.Health Center.
          LUBBOCK, Texas, 79403, USA

NIESSEN, P.J.   Friedrich-Miescher Institut.-Postfach 273.-CH-4002
          BASEL, Switzerland

NURTEN, R.   Dept.Biophysics.-Faculty of Medicine.-Univ. Istanbul
          CAPA-ISTANBUL, Turkey

OFENGAND, J.   Roche Institute Molecular Biology.-NUTLEY,NJ, 07110
          USA

PALME, K.   Dept.Biochemistry.-Univ. Ulm.-D-79 ULM/DONAU, Germany

PEREZ-BERCOFF, R.   The Institute for Virology.-University of Rome
          V.le di Porta Tiburtina 28.-00185-ROMA, Italy

PERRY, R.   The Institute for Cancer Research.-7701 Bruholm Ave.-Fox
          Chase.-PHILADELPHIA,PA, 19111, USA

PERSON, A.   Institut Biologie Moléculaire CNRS.-Univ.Paris VII.-2,
          Place Jussieu.-F-75221-PARIS CEDEX 05, France

PINCK, L.   Institut Biologie Moléculaire & Cellulaire.-15 rue Descartes
          F-67000 STRASBOURG, France

POLLARD, J.   Dept.Biochem. Quenn Elisabeth College Univ.London.
          Campden Hill.-LONDON W8 7AH, England

POWER, F.   Dept.Anatomy McGill University.-3640 University Streeet
          MONTREAL PQ, H3A 2B2, Canada

RHOADS, R.R.   Dept.Biochemistry Univ. Kentucky.- Medical Center
            LEXINGTON,KE, 40536, USA

ROMBAUTS, W.   Dept.Biochemistry.-Medical Faculty.-49 Heerestraat
            B-3000 LEUVEN, Belgium

ROSENTHAL, A.   Dept.Molecular Virology.-Hadassah Med.School.-P.O.
            Box 1172.-JERUSALEM, Israel

ROWLANDS, D.J.   Animal Virus Res. Institute.-PIRTBRIGH, Surrey GU24
            ONF, England

RYDNINGEN, B.   Narvik Gynnasium.-Snorre gt 89.-N-8500 NARVIK, Norway

SANGAR, D.   Animal Virus Research Institute.-PIRTBRIGH, Surrey GU24
            ONF, England

SHATKIN, A.J.   Roche Institute Molecular Biology.-NUTLEY,NJ, 07110,
            USA

SIGLER, P.   Dept.Biophysics & Theoretical Biochemistry.-Univ.Chicago.
            920 East 58th.St.-CHICAGO,IL, 60637, USA

SLEGERS, H.   Dept. Cell Biology Univ. Antwerp.-Universiteitsplein 1.
            B-2610 WILRIJK, Belgium

SPAHR, P.F.   Dept.Biologie Moléculaire Univ.Genève.-30 Quai E.Ansermet
            CH-1211-GENEVE 4, Switzerland

TEIXEIRA, F.P.   Istituto de Farmacologia Faculdade Medicina Coimbra.
            P-3000-COIMBRA, Portugal

ULBRICH, N.   Dept.Biochemistry.-Univ.Chicago.-920 East 58th.Street.
            CHICAGO,IL, 60637, USA

VAINSTEIN, A. Dept.Biological Chemistry.-Inst.Life Sciences.-Hebrew
            Univ. Jerusalem.-JERUSALEM, Israel

VANDEN BERGHE, D.A.   Dept.Geneeskunde Univ.Antwerp.-Universiteitsplein
            1.-B-2610-WILRIJK, Belgium

VAQUERO, C.   Pavillon Pasteur Institut du Radium.-26, reu d'Ulm.
            F-75231-PARIS CEDEX 05, France

VASCONCELOS, M.R.   Laboratorio Bioquimica.-Av.Almirante Joao Azevzdo
            Coutinho.-MURTAL SAN PEDRO DO ESTORIL, Portugal

VAZQUEZ, D.   Instituto Bioquimica Macromoléculas.-Univ.Autonoma
            Madrid.-Canto Blanco.-MADRID 34, Spain

VINCENT, A.   Service Biochimie Différentiation.-Inst.Biol.Mol.CNRS.
              2, Place Jussieu.-F-75221-PARIS CEDEX 05, France

VRIJSEN, R.   Dept.Microbiol.& Hygiene.-Free Univ. Brussels.-
              Laarbeklaan 103.-B-1090-BRUSSELS, Belgium

WEBER, L.D. Dept.Biology.-Room 16-744.-M.I.T.-CAMBRIDGE,MA, 02139,
              USA

WEINMANN, R.   The Wistar Institute.-36th. & Spruce.-PHILADELPHIA,PA,
              19104, USA

WIMMER, E.   Dept. Microbiology.-S.U.N.Y. at Stony Brook.-STONY BROOK,
              NY,11794, USA

WOOL, I.   Dept.Biochemistry.-Univ.Chicago.-920 East 58th.street.
              CHICAGO,IL, 60637, USA

ZARBL, H.   Dept.Biochemistry McGill University.-3655 Drummond Street
              MONTREAL, PQ, H3G 1Y6, Canada

MARATEA, the small fishing harbor   MARATEA

N. BROT & G. KOCH   R. IVELL & J. POLLARD

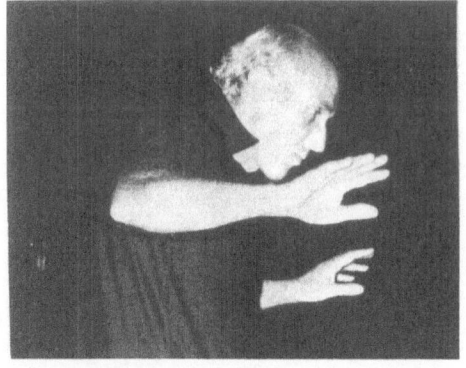

M.T. CARRI, R. PEREZ-BERCOFF, R.   P. SIGLER
KAEMPFER, and J. POLLARD

N. BROT and D. VAZQUEZ

R.J. JACKSON and A. PERSON

R. PERRY, M. BALDUCCI and A.M.
DEGENER

D. VAZQUEZ, J. MORROW, two guides,
P. LENGYEL, and Mrs. ULBRICH

Coffee break during night session
N.K. GUPTA and R.P. PERRY

F. POWER and G. JAUREGUIBERRY

INDEX